Exogenous Dermatoses: Environmental Dermatitis

Editors

Torkil Menné, M.D.
Associate Professor
Department of Dermatology
Gentofte Hospital
Hellerup, Denmark

Howard I. Maibach, M.D.
Professor and Vice-Chair
Department of Dermatology
School of Medicine
University of California
San Francisco, California

CRC Press
Boca Raton Ann Arbor Boston

Library of Congress Cataloging in Publication Data

Exogenous dermatoses: environmental dermatitis/editors, Torkil
Menné, Howard I. Maibach.
p. cm.
Includes bibliographical references.
Includes index.
ISBN 0-8493-5969-4
1. Contact dermatitis. I. Menné, Torkil. II. Maibach, Howard I.
[DNLM: 1. Dermatitis, Contact. 2. Occupational Dermatitis. WR
175 E96]
RC593.C6E96 1990
616.5'1--dc20
DNLM/DLC
for Library of Congress 90-2483
CIP

This book represents information obtained from authentic and highly regarded sources. Reprinted material is quoted with permission, and sources are indicated. A wide variety of references are listed. Every reasonable effort has been made to give reliable data and information, but the author and the publisher cannot assume responsibility for the validity of all materials or for the consequences of their use.

Direct all inquiries to CRC Press, Inc., 2000 Corporate Blvd., N.W., Boca Raton, Florida 33431.

International Standard Book Number 0-8493-5969-4

Library of Congress Card Number 90-2483
Printed in the United States

PREFACE

Patients presenting with exogenous dermatoses are an everyday challenge for the dermatologist. Many scientists working within the field of exogenous dermatoses have produced an increasing understanding of the interface between the environment and the human skin. This increasing knowledge is published in a wide range of medical journals. The concept of *Exogenous Dermatoses: Environmental Dermatitis* is to provide the reader with comprehensive essays written by experts who are scientifically active within the specific research field. It is hoped that *Exogenous Dermatoses: Environmental Dermatitis* will fill in the gap which exists between the time the original research is published in the journals and the time it appears in the textbook.

Torkil Menné
Howard Maibach

THE EDITORS

Torkil Menné, M.D., Ph.D. is Associate Professor, Department of Dermatology, Gentofte Hospital, Hellerup, Denmark.

Dr. Menné obtained his M.D. degree from the University of Copenhagen in 1971 and received a Ph.D. in 1983 at the same university for investigations on genetic and epidemiologic aspects of nickel dermatitis.

Dr. Menné is a member of the Danish Dermatological Society and the American Academy of Dermatology. He is Chairman of the European Environmental Contact Dermatitis Research Group.

Dr. Menné has presented over 15 invited lectures at international meetings. He is the author of 135 articles and the author and/or editor of 3 books.

Howard I. Maibach, M.D. is Professor of Dermatology at the School of Medicine, University of California, San Francisco.

Dr. Maibach graduated from Tulane University, New Orleans, Louisiana (A.B. and M.D.) and received his research and clinical training at the University of Pennsylvania, Philadelphia. He received an honorary doctorate from the University of Paris Sud in 1988.

Dr. Maibach is a member of the International Contact Dermatitis Research Group, the North American Contact Dermatitis Group, and the European Environmental Contact Dermatitis Group. He has published more than 800 papers and 35 textbooks.

CONTRIBUTORS

Christian Avnstorp, M.D., Ph.D.
Senior Registrar
Department of Dermatology
Gentofte Hospital
Hellerup, Denmark

Ole Baadsgaard, M.D.
Assistant Professor
Department of Dermatology
University of Copenhagen
Gentofte Hospital
Hellerup, Denmark

Thomas R. Bender, M.D., M.P.H.
Director
Division of Safety Research
National Institute for Occupational Safety and Health
Centers for Disease Control
Morgantown, West Virginia

Enzo Berardesca, M.D.
Assistant Professor
Department of Dermatology
University of Pavia
Pavia, Italy

Stephen P. Berardinelli, Ph.D.
National Institute of Occupational Safety and Health
Centers for Disease Control
Morgantown, West Virginia

Bret Berner, Ph.D.
Director
Basic Pharmaceutics Research
CIBA-GEIGY Corporation
Ardsley, New York

Bert Björkner, M.D., Ph.D.
Professor
Department of Occupational Dermatology
General Hospital
Malmö, Sweden

Ernest Bloom, M.D.
Cellular Pharmacology Laboratory
University of California
San Francisco, California

P. W. Bowers, F.R.A.C.P., M.R.C.P.
Consultant Dermatologist
Department of Dermatology
Royal Cornwall Hospital
Truro, Cornwall
England

Derk P. Bruynzeel, M.D.
Department of Occupational Dermatology
Free University
Amsterdam, The Netherlands

Magnus Bruze, M.D., Ph.D.
Associate Professor
Department of Occupational Dermatology
University Hospital
Lund, Sweden

Desmond Burrows, M.D., F.R.C.P.
Consultant Dermatologist
Department of Dermatology
Royal Victoria Hospital
Belfast, Northern Ireland

Frank H. Clarke, Ph.D.
Distinguished Research Fellow
Pharmaceutical Research
CIBA-GEIGY Corporation
Summit, New Jersey

Richard B. Crook, Ph.D.
Assistant Research Biochemist
Cellular Pharmacology Laboratory
University of California
San Francisco, California

An Dooms-Goossens, Ph.D.
Lector
Faculty of Medicine
Department of Dermatology
Katholieke Universiteit Leuven
Leuven, Belgium

Peter Elsner, M.D.
University Lecturer
Department of Dermatology
University of Wuerzburg
Wuerzburg, West Germany

Tuula Eslander, M.D.
Specialist in Dermatology and
Occupational Dermatology
Section of Dermatology
Institute of Occupational Health
Helsinki, Finland

Donald J. Fauss, M.D.
Cellular Pharmacology Laboratory
University of California
San Francisco, California

Torkel Fischer, M.D.
Professor
National Institute of Occupational Health
Solna, Sweden

Alexander A. Fisher, M.D.
Clinical Professor
Department of Dermatology
New York University Postgraduate
Medical School
New York, New York

Sigfrid Fregert, M.D.
Professor Emeritus
Department of Occupational Dermatology
University of Lund
Lund, Sweden

Chee Leok Goh, M.Med., M.R.C.P.
Medical Director/Consultant
Dermatologist
Contact and Occupational Dermatoses
Clinic
National Skin Centre
Singapore

Dennis F. Gross, M.D.
Senior Resident
Skin and Cancer Unit
Department of Dermatology
New York University
New York, New York

Birgitta Gruvberger, M.S.
Industrial Hygienist
Department of Occupational Dermatology
University Hospital
Lund, Sweden

Richard H. Guy, Ph.D.
Associate Professor
School of Pharmacy
University of California
San Francisco, California

Lars Halkier-Sørensen, M.D.
Department of Dermatology
Marselisborg Hospital
Aarhus, Denmark

Matti Hannuksela, M.D.
Chairman and Professor
Department of Dermatology
University of Oulu
Oulu, Finland

Jens Hansen, Ph.D.
Area Leader, Dermatology
Department of Pharmaceutical Research
and Development
Pharmacia AS
Hillerød, Denmark

Niels Hjorth, M.D.
Professor Emeritus
Department of Dermatology
Gentofte Hospital
Hellerup, Denmark

Daniel J. Hogan, M.D.
Associate Professor
Department of Dermatology and
Cutaneous Surgery
University of Miami
Miami, Florida

D. Linn Holness, M.D., M.H.Sc.
Assistant Professor
Occupational and Environmental
Health Unit
University of Toronto
Toronto, Ontario
Canada

WeiDong Huang, M.D.
Cellular Pharmacology Laboratory
University of California
San Francisco, California

Riitta Jolanki, M.Sc.
Specialist Researcher
Section of Dermatology
Institute of Occupational Health
Helsinki, Finland

Lasse Kanerva, M.D., Ph.D.
Chief
Section of Dermatology
Institute of Occupational Health
Helsinki, Finland

Bradford R. Katchen, M.D.
Senior Resident
Department of Dermatology
Mt. Sinai Hospital
New York, New York

Tuula Kinnunen, M.D.
Department of Dermatology
University of Oulu
Oulu, Finland

Arto Lahti, M.D., Ph.D.
Assistant Professor
Department of Dermatology
University of Oulu
Oulu, Finland

Kaija Lammintausta, M.D.
Assistant Professor
Department of Dermatology
Turku University Central Hospital
Turku, Finland

Gerard C. Mazzenga, M.S.
Research Scientist
Basic Pharmaceutics Research
CIBA-GEIGY Corporation
Ardsley, New York

Howard I. Maibach, M.D.
Professor and Vice-Chair
Department of Dermatology
University of California School of Medicine
San Francisco, California

Torkil Menné, M.D.
Associate Professor
Department of Dermatology
Gentofte Hospital
Hellerup, Denmark

Kee Chan, Moon, M.D.
Associate Professor
Department of Dermatology
Korea University College of Medicine
Seoul, Korea

James R. Nethercott, M.D.
Professor
Department of Environmental Health Sciences
School of Hygiene and Public Health
The Johns Hopkins University
Baltimore, Maryland

Howard A. Oriba, M.D.
Postdoctoral Fellow
Department of Dermatology
University of California
San Francisco, California

Jon R. Polansky, M.D.
Associate Professor
Cellular Pharmacology Laboratory
University of California
San Francisco, California

Timo Reunala, M.D.
Associate Professor
Department of Dermatology
University of Tampere
Tampere, Finland

Robert L. Rietschel, M.D.
Chairman
Department of Dermatology
Ochsner Clinic
New Orleans, Louisiana

Darrell S. Rigel, M.D.
Clinical Assistant Professor
Department of Dermatology
New York University
New York, New York

Berndt Stenberg, M.D.
Department of Dermatology
University Hospital
Umeå, Sweden

Raija Tammi, M.D.
Department of Anatomy
University of Kuopio
Kuopio, Finland

Kristian Thestrup-Pedersen, M.D.
Professor
Department of Dermatology
University of Aarhus
Aarhus, Denmark

Kristiina Turjanmaa, M.D., Ph.D.
Department of Dermatology
University Central Hospital
Tampere, Finland

Niels K. Veien, M.D., Ph.D.
Dermatology Clinic
Aalborg, Denmark

Jan E. Wahlberg, M.D., Ph.D.
Professor
Department of Occupational Dermatology
Karolinska Hospital and
National Institute of Occupational Health
Stockholm, Sweden

Klaus-P. Wilhelm, M.D.
Research Fellow
Department of Dermatology
University of California School of
Medicine
San Francisco, California

Donald R. Wilson, M.S.
Group Leader, Toxicology
Research and Development
Cygnus Research Corporation
Redwood City, California

TABLE OF CONTENTS

Section I. Mechanisms

Chapter 1

IMMUNOLOGIC MECHANISMS OF CONTACT DERMATITIS

Ole Baadsgaard

TABLE OF CONTENTS

I. INTRODUCTION

Many factors and cell types, both resident and nonresident, are involved in the immune regulation of the skin. T-Lymphocytes are central to immunoregulation and are continuously trafficking through the skin surveying for exogenous and endogenous antigens including bacterial, virus, and tumor antigens. In normal skin, negative and positive signals from the resident epidermal cells are balanced such that significant T-cell activation does not occur. However, antigens or irritants can perturb this balance, resulting in T-cell activation lymphokine release and, thus, causing an inflammatory immune reaction.

Activation of T-lymphocytes can occur via both antigen dependent and independent pathways. The Langerhans' cells represent an antigen-dependent mode of T-cell activation. By presenting antigens to T-cells and thereby eliciting T-cell activation, the epidermal Langerhans' cells can provide positive signals for up-regulation of the immune response. However, other epidermal antigen-presenting cell subtypes, which either are constitutively present or appear in the epidermis in response to hypersensitivity reactions, may activate suppressor pathways and result in down-regulation of the immune reaction. Keratinocytes are not passive bystanders in the immune response; they can transmit both positive and negative signals and, thus, function as a modulator of the immune system, up-regulating or down-regulating T-cell responses. In addition to this regulatory role, injury- or irritant-activated keratinocytes can elaborate factors and cytokines that may be responsible for antigen-independent T-cell activation and lymphokine release resulting in an inflammatory skin reaction. Such antigen-independent ways of T-cell activation may be the mechanisms responsible for irritant-induced contact dermatitis. The cell types that are involved and the mechanism that governs their interplay in the development of both allergic and irritant contact dermatitis form the focus of this chapter.

II. EPIDERMAL IMMUNOCOMPETENT CELLS

To understand the pathogenetic mechanism of contact dermatitis, it is important to appreciate the different immunocompetent cells present in the skin. The epidermis represents the outermost part of the immune system and is mainly comprised of keratinocytes, Langerhans' cells, and melanocytes. Of these, keratinocytes and Langerhans' cells are known to be immunocompetent. Unlike the Langerhans' cells and the keratinocytes, the immunological function of the melanocytes, if any, is not yet clear.

Keratinocytes are the major cell population of the epidermis (90% of total epidermal cells) and have only recently been considered a part of the immune system. The first indication of their immunocompetent nature was the unexpected discovery of the keratinocyte-derived cytokine, epidermal cell-derived thymocyte activating factor (ETAF).[1,2] Later it became clear that activated keratinocytes contain the ability to produce a variety of distinct T-cell activation and growth-stimulatory cytokines, including interleukin-1α (IL-1α), interleukin-1β (IL-1β),[3,4] interleukin-3 (IL-3),[5,6] keratinocyte-derived T-cell growth factor (KTGF),[7] granulocyte macrophage colony stimulating factor (GM-CSF),[8] epidermal cell-derived natural killer cell activating factor (ENKAF),[9] and interleukin-6 (IL-6).[10] In addition to these growth-promoting and activating cytokines, keratinocytes can also synthesize T-cell-inhibitory cytokines and factors such as epidermal cell-derived lymphocyte differentiation inhibiting factor (ELDIF),[11] keratinocyte lymphocyte inhibitory factor (KLIF),[12,13] eicosanoids,[14-16] IL-1 inhibitor,[17] and transforming growth factor β (TGF-β).[18] The functions of some of these epidermal factors and cytokines as they may relate to contact dermatitis are briefly introduced below. Epidermal IL-1 release can be induced by specific stimuli, i.e., antigens,[19] or by traumatic epidermal injury, including irritants and ultraviolet (UV) irradiation.[19-21] IL-1 can induce keratinocyte proliferation,[22] directly enhance T-cell proliferation, and induce the secretion of GM-CSF.[23] However, high concentration of IL-1 also may stimulate the generation of prostaglandins and thereby indirectly inhibit the effector arm of the contact hypersensitivity response, as shown by Daynes et al.[25] IL-6 is synergistic with IL-1 in T-cell activation and thereby may enhance IL-1-induced T-cell proliferation and lymphokine release.[10] KTGF has been found to be identical to GM-CSF and possesses the capacity to induce bone marrow cell proliferation; it also functions as a T-cell growth factor.[26] In contrast to the growth-stimulatory cytokines, both ELDIF and KLIF inhibit T-cell proliferation and interleukin-2 (IL-2) production.[11-13] The effect of KLIF may be due to TGF-β since the inhibition by KLIF is reversible by an antibody against TGF-β.[18,27] Eicosanoids also released by the keratinocytes interact with antigen-presenting cells and $CD4^+$ and $CD8^+$ lymphocytes such that $CD8^+$ suppressor effector T-cells are preferentially activated.[14]

Another immunocompetent epidermal cell type is the dendritic bone marrow-derived Langerhans cell which comprises 2 to 5% of the total epidermal cell population.[28] It has a suprabasal localization in the epidermis and was the first described resident epidermal cell type with immune function. Ultrastructurally, the Langerhans' cell is distinguished from other cell types by its intracytoplasmic granules known as Birbecks granules. Cells that have the ultrastructural features of Langerhans' cells, but lack the Birbecks granules, are termed indeterminant cells. Langerhans' cells and indeterminant cells are the only cell types in normal human epidermis that demonstrate the capacity to present both self and foreign antigens to T-cells, eliciting T-cell activation.[29,30] The Langerhans' cell expresses the CD1 determinant and class II major histocompatibility complex (MHC) molecules such as HLA-DR.[31-33] In keeping with its functions as an antigen-presenting cell, the Langerhans' cell synthesizes IL-1.

III. DERMAL IMMUNOCOMPETENT CELLS

The immunocompetent cell types found in the dermis are fibroblasts, endothelial cells, dermal antigen-presenting cells, mast cells, and lymphocytes. The functions of these cell types will be discussed briefly at this point.

Structurally, the dermis consists of a framework of collagen, elastic, and reticular fibers embedded in a substance of glycosaminoglycans. All these elements are produced by the fibroblast. IL-1 can stimulate fibroblasts to produce prostaglandins,[34] which may in turn be involved in the activation of suppressor T-cell mechanisms. Thus, fibroblasts, in addition to the production of a dermal framework, also may have an immune function.

The dermal vessels are lined by endothelial cells that constitutively express the inducible intercellular adhesion molecule (ICAM-1).[35] This molecule is a natural ligand for LFA-1,[36] a glycoprotein expressed on the surface of most leukocytes. These adhesion molecules are recognized to be important for cell-to-cell interactions and may be involved in the migration of leukocytes from the vessels into the dermis and the epidermis.

Dermal HLA-DR^+ dendritic antigen-presenting cells (termed dendrocytes) localized around the papillary vascular plexus may be involved in antigen presentation. This is supported by the findings in the murine system which demonstrate that dermal Ia^+ (HLA-DR^+) cells contain the capacity to present soluble antigens to T-lymphocytes.[37]

The mast cell, another dermal cell type, can be stimulated by the immune system to release histamine and platelet activating factor.[38,39] Histamine generates prostaglandins, stimulates suppressor lymphocytes, and is a potent vasoconstrictor.[40,41] Platelet activating factor is a potent stimulus for cell migration and production of lipoxygenase products of arachidonic acid metabolism.[42] These products are potent mediators of inflammation.

The few lymphocytes present in the dermis are mainly T-cells and include virtually no B-cells. The T-cells are localized around the postcapillary venules of the papillary vascular plexus. Most of these cells express HLA-DR and IL-2 receptors and are, therefore, in a functionally activated state. They are distributed evenly between $CD4^+$ helper/inducer T-cells and $CD8^+$ suppressor/cytotoxic T-cells.[43] The majority of the $CD4^+$ T-cells are $4B4^+$

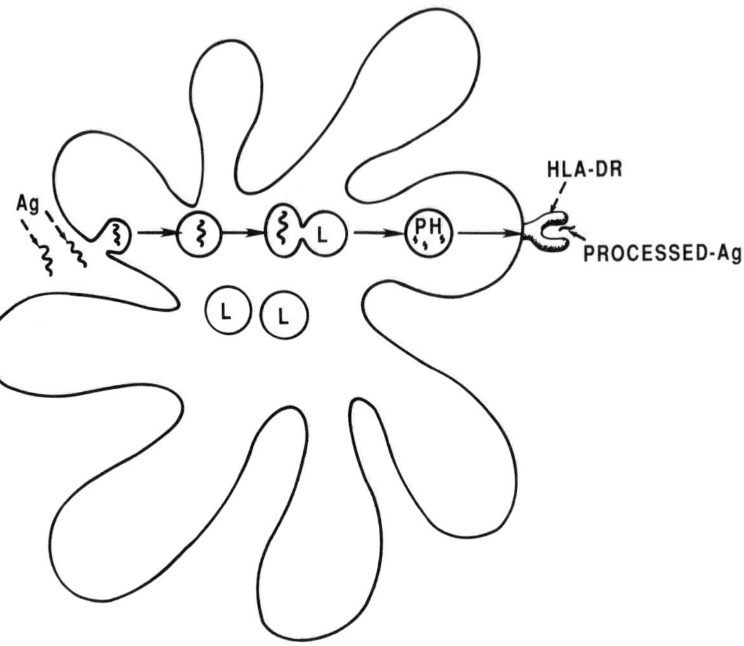

FIGURE 1. Antigen processing and presentation. Soluble antigens are taken up by the Langerhans' cell by pinocytosis. The internalized antigen contained within the cell membrane vesicle is fused with a lysosome (L), creating a phagolysosome (PH). Enzymes within the phagolysosome partially digest the antigen, which is subsequently expressed on the surface of the cell membrane in the context of class II major histocompatibility complex molecules such as HLA-DR. The antigen/HLA-DR complex will now be recognized by the T-cell receptor eliciting T-cell activation.

(helper/inducer T-lymphocytes), whereas only a small minority are $2H4^+$ (suppressor/inducer T-lymphocytes).[43] This distribution is in contrast to that found in the peripheral blood, wherein the $CD4^+2H4^+$ and the $CD4^+4B4^+$ lymphocyte subpopulations are the same size.

As stated previously, T-lymphocytes can be activated both through antigen dependent and independent pathways. Activated T-lymphocytes elaborate a variety of lymphokines potentially important for contact dermatitis reactions, including IL-2 (T-cell growth factor),[44] γ-interferon,[45] IL-3,[46,47] IL-4,[48] and GM-CSF.[49,50] Antigen-dependent T-cell activation is a critical event in the initiation of contact hypersensitivity reactions and therefore will be described below.

IV. ANTIGEN-DEPENDENT T-LYMPHOCYTE ACTIVATION

Antigen-specific activation of T-cells is critically dependent on antigen-presenting cells which express class II MHC molecules such as HLA-DR or HLA-DQ. The Langerhans' cells, which express CD1 and HLA-DR (i.e., $CD1^+DR^+$), are the only cells in normal human epidermis capable of presenting antigens to T-cells, eliciting T-cell activation.

It is thought that the Langerhans' cells internalize the antigen by pinocytosis. The pinocytotic antigen-containing vesicle then fuses with a lysosome to create a phagolysosome, and enzymes contained within the phagolysosome partly digest the antigen. Immunogenic antigen fragments (i.e., processed antigen) — also termed nominal antigens — are combined with cytoplasmic HLA-DR molecules and transported to the cell surface membrane, where they are expressed together (Figure 1). The $CD4^+$ T-cells are unable to recognize nominal antigens alone. The antigen is only recognized when presented in the context of class II MHC molecules. Antigen presentation, and the resulting T-cell activation, is enhanced by the presence of IL-1 which is secreted by the antigen-presenting Langerhans' cell and surrounding keratinocytes.

The structure on the T-cell responsible for antigen recognition is the T-cell receptor, an α/β peptide heterodimer which is rearranged during thymic development and which generates a unique antigen specific receptor for each T-cell clone. The signal produced in response to antigen recognition by the T-cell receptor is transmitted to the interior of the cell via the transduction peptide CD3 which is complexed with the T-cell receptor. The resulting activation of the T-cell causes both expression of IL-2 receptors on the cell membrane and the release of IL-2 as well as other lymphokines. Subsequent binding of IL-2 to the IL-2 receptor induces T-cell proliferation.

V. PATHOGENETIC MECHANISM OF ALLERGIC CONTACT DERMATITIS

Most allergens are low-molecular-weight, highly reactive, lipid-soluble haptens that must be conjugated to an epidermal carrier protein before they can function as complete antigens. The $CD1^+DR^+$ Langerhans' cells are able to process and present hapten/carrier conjugates in an immunogenic form to $CD4^+$ inducer T-lymphocytes.[29,30,51] This results in the generation of effector T-cells capable of mediating delayed type hypersensitivity reactions,[52-54] as well as the generation of cytotoxic T-cells directed against haptenized epidermal cells.[55,56] However, immunoregulatory suppressor T-cells that down-modulate the effector responses are also induced.[57,58] Thus, the net effect of the immune response is dependent on the balance between the effector and regulatory cells.

Except for very potent allergens, the primary sensitization does not normally result in clinical skin lesions, probably due to the low numbers of responder T-lymphocytes present. However, subsequent rechallenges resulting in clonal T-cell expansion and re-presentation of the antigen to already primed (memory) T-cells may result in cytokine release and cytotoxicity, generating an acute contact dermatitis lesion characterized by dermal perivascular mononuclear cell infiltration and epidermal mononuclear cell infiltration, spongiosis, and hyperplasia. The sensitization and elicitation phase of the immune response in allergic contact dermatitis will now be described in more detail.

A. SENSITIZATION IN ALLERGIC CONTACT DERMATITIS

As already mentioned, the Langerhans cell is critical for induction of contact hypersensitivity. It has also been shown that sensitization of skin depleted of Langerhans' cells, such as murine tail skin, UV-irradiated skin, or UV-irradiated epidermal cells, results in the development of antigen-specific tolerance.[59-61] Likewise, peroral or intravenous administration of the allergen may also result in antigen-specific tolerance. Thus, when the antigen is presented by non-Langerhans' cell antigen-presenting cells, induction of T-suppressor lymphocytes and development of tolerance, rather than contact hypersensitivity, is the result.

In animals, Langerhans' cells are found in the dermal lymphatics and in the draining regional lymph nodes following sensitization.[62] The Langerhans' cells therefore seem to take up the antigen and migrate to the paracortical area of the regional lymph node, where the antigen is presented to unprimed T-cells which possess specificity for that particular antigen. This results in the generation and expansion of antigen-specific memory T-cells (i.e., primed T-cells) which will then distribute throughout the entire circulation and home to the site of antigen challenge (Figure 2).

B. ELICITATION OF ALLERGIC CONTACT DERMATITIS

Rechallenge of the skin will result in re-presentation of the antigen by the Langerhans' cells, and possibly non-Langerhans' cell antigen-presenting cells, to primed T-cells. This

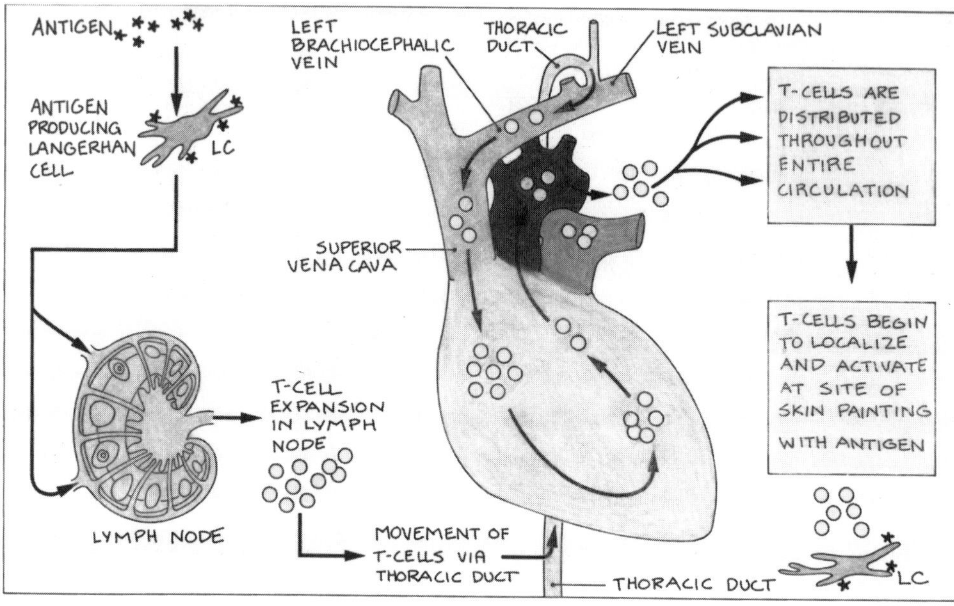

FIGURE 2. Sensitization, the efferent phase of the immune response in allergic contact dermatitis. Lipid-soluble reactive haptens applied to the skin cross the stratum corneum and bind to protein carriers in the skin, creating complete antigens. The antigen (hapten-carrier conjugate) is internalized by the Langerhans' cells, then processed and expressed on the surface membrane in the context of class II major histocompatibility complex molecules. The Langerhans' cell moves out of the epidermis through the afferent draining lymphatic to the regional lymph nodes. Following Langerhans' cell antigen presentation to T-cells in the paracortical area, the T-cells undergo expansion and then exit the node. The T-cells travel via the thoracic duct to the superior vena cava. From the superior vena cava they are distributed throughout the entire circulatory system and home to the skin challenged with hapten. Re-presentation of antigens to primed (memory) T-cells can now occur and will result in amplification of the immune response. (From Cooper, K., Roitt, I. M., Du Vivier, A., Brostoff, J., and Male, D. K., *Immunology Slide Atlas,* Gower Medical Publishing, New York, 1989, Figure 3.8. With permission.)

may take place in the epidermis, the dermis, or the regional draining lymph nodes and results in T-cell activation. These activated T-cells release a variety of lymphokines, such as IL-2 (T-cell growth factor), γ-interferon, GM-CSF, IL-3, and IL-4. These lymphokines have a variety of immunological effects.

IL-2 can induce proliferation and activation of both antigen-specific and antigen-non-specific immunoregulatory T-cells[44] and thereby up- or down-modulate the immune response.

γ-Interferon released from lesional activated T-cells induces the expression of intercellular adhesion molecules, such as ICAM-1, on both keratinocytes and endothelial cells.[27,63,64] This is thought to be important for the recruitment of T-cells and other leukocytes which express LFA-1, the ligand (i.e., receptor) for ICAM-1.[36] These recruited cells will then participate in the ongoing immune response.

IL-3 is present in the early phases of a delayed-type hypersensitivity reaction[65] and is released from both activated T-cells and keratinocytes. It induces the proliferation of mast cells and is involved in the activation of monocytes, two important cell types which are present in increased numbers in contact hypersensitivity reactions. Furthermore, lymphokines including IL-3 and GM-CSF enhance keratinocyte proliferation.[66] Thus, these cytokines may be partly responsible for the epidermal hyperplasia seen in contact dermatitis (Figure 3).

In addition to lymphokines released by activated T-cells, other cytokines are important in the elicitation of contact dermatitis. IL-1, a cytokine released from Langerhans' cells, macrophages, and keratinocytes, has been found in increased amounts in the epidermis during an allergic patch test reaction.[19] The binding of IL-1 to keratinocyte IL-1 receptors

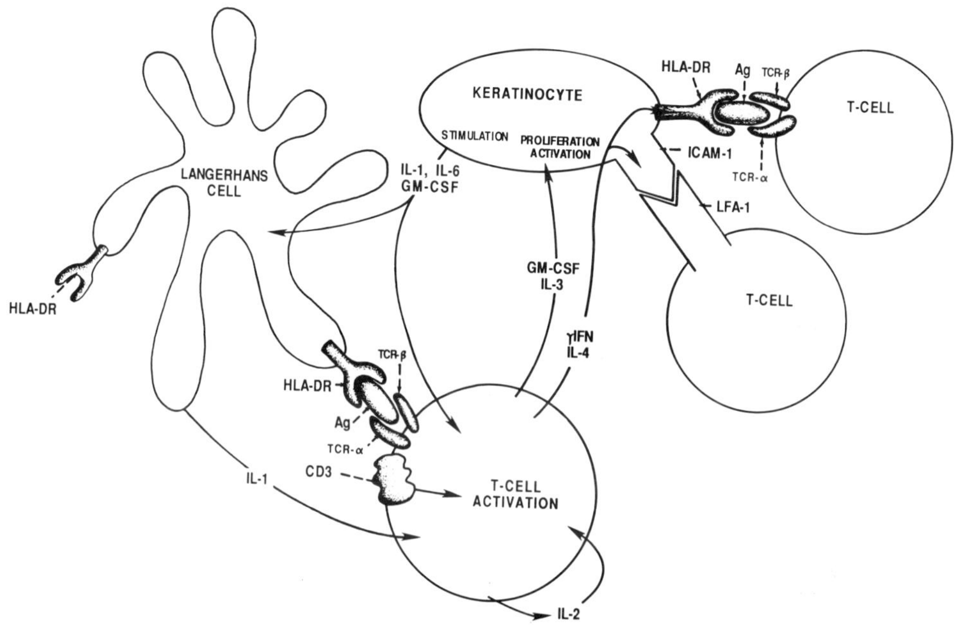

FIGURE 3. Elicitation, the afferent phase of the immune response in allergic contact dermatitis. Re-presentation of the antigen by Langerhans' cells and other antigen-presenting cells in the skin to primed (memory) T-cells results in T-cell activation and expansion and lymphokine release. The antigen (Ag) is presented in the context of HLA-DR class II major histocompatibility complex molecules to the T-cell receptor, a heterodimer composed on an α (TCR-α) and β (TCR-β) peptide chain. The signal is transmitted from the T-cell receptor via a transduction peptide termed CD3 to the interior of the cell. Activated T-cells release lymphokines including IL-2, IL-4, γ-interferon (γIFN), and GM-CSF. IL-2 induces T-cell proliferation. IL-3 and GM-CSF induce keratinocyte proliferation and may be responsible for the epidermal hyperplasia seen in contact dermatitis. γ-IFN and possibly IL-4 induce keratinocyte expression of HLA-DR and intercellular adhesion molecules (ICAM-1). ICAM-1 is important for cell-to-cell interaction and may be responsible for compartmentalization of leukocytes in the skin during a contact dermatitis reaction. Keratinocytes that express HLA-DR may, under certain circumstances, present protease-digested antigen fragments to already primed T-cells, thereby enhancing the ongoing immune response. HLA-DR$^+$ keratinocytes also may be a target for certain cytotoxic effector T-cells. Cytokines such as IL-1, IL-6, and GM-CSF released from activated keratinocytes can stimulate activation and proliferation of both Langerhans' cells and T-cells. Monocytes and mast cells also may be activated and release proinflammatory mediators. The net effect of all these interactions is spongiosis, cell injury, and cell death, resulting in the clinical picture of allergic contact dermatitis.

stimulates keratinocyte growth and thereby may contribute to the epidermal hyperplasia seen in contact hypersensitivity.[22,23] Furthermore, IL-1 may stimulate keratinocyte release of GM-CSF, a Langerhans' cell growth factor which may result in the expansion of the Langerhans' cell population.[23] Indeed, the number of Langerhans' cells are increased in allergic contact dermatitis.[67-70]

In normal human epidermis, the only antigen-presenting cell is the CD1$^+$DR$^+$ Langerhans' cell. However, in allergic skin reactions, CD1$^-$DR$^+$ non-Langerhans' cell antigen-presenting cells appear in the epidermis.[67,70] These non-Langerhans' cell antigen-presenting cells are able to activate autologous T-cells and are responsible for up to 50% of the total epidermal T-cell activating capacity during the allergic reaction.[70] These antigen-presenting cells may be analogous to the UV-induced non-Langerhans' cell antigen-presenting cell population which functions to down-regulate immune responses. Thus, non-Langerhans' antigen-presenting cells may function as a negative feedback mechanism down-regulating the hypersensitivity reaction. This hypothesis is currently under investigation.

Keratinocytes in allergic and irritant skin reactions express both ICAM-1 and HLA-DR; these immunocompetent surface molecules are probably induced by lymphokines such as

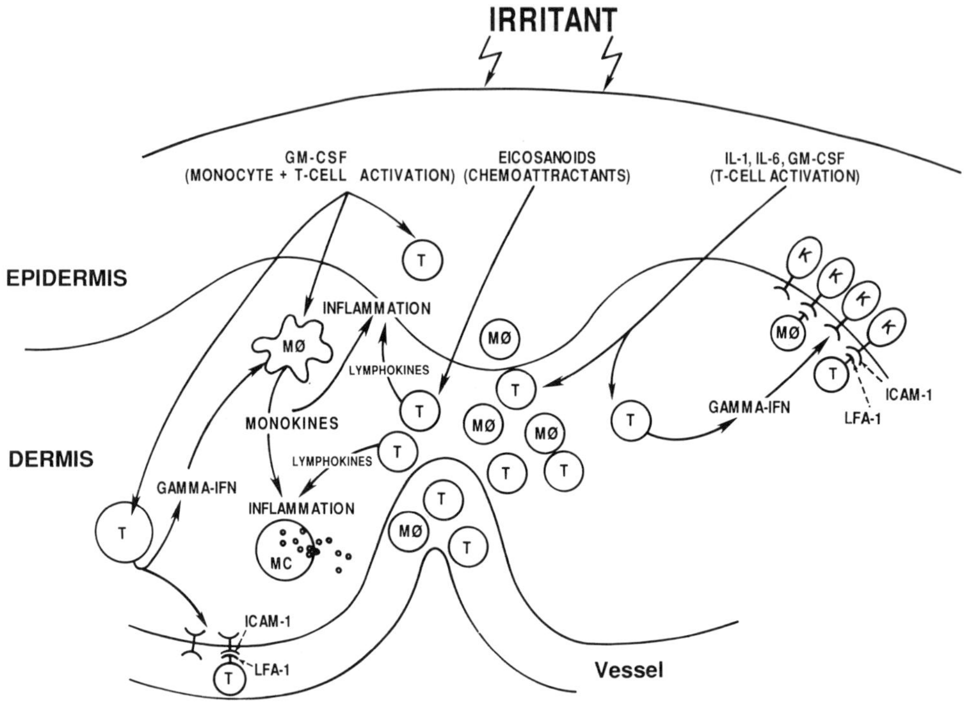

FIGURE 4. Irritants and injury trigger release of cytokines and factors from keratinocytes. Epidermal injury triggers the release of eicosanoids from keratinocytes. Eicosanoids function as potent chemoattractants and may be responsible for the migration of inflammatory leukocytes into the challenged area of irritant-induced contact dermatitis. Irrespective of the specific mediator involved, irritant-challenged epidermis contains a chemotactic factor for T-lymphocytes (T). IL-1 is also released by epidermal injury and may function to enhance activation of local T-cells, possibly in combination with IL-6 and GM-CSF, two other T-cell stimulatory cytokines elaborated by keratinocytes. Lymphokines released from activated T-cells, such as γ-interferon, activate macrophages (MØ) capable of monokine release, and lymphokines induce the expression of the intercellular adhesion molecule (ICAM-1) on keratinocytes (K) and endothelial cells, thus enhancing trapping of immunocompetent cells in the area. IL-3 released from activated T-cells can induce mast cell (MC) proliferation. Cytokines and factors released from the immunocompetent cells cause inflammation and result in irritant contact dermatitis.

γ-interferon released from local T-cells.[27,63,64,71-73] In contrast to HLA-DR$^+$ antigen-presenting cells, HLA-DR$^+$ keratinocytes are unable to stimulate allogeneic (nonself) T-cells. However, under certain circumstances they are able to present antigenic peptide fragments to memory T-cells.[74] Additionally, HLA-DR$^+$ keratinocytes can also act as targets for class II restricted cytotoxic T-cell clones.[74] Thus, HLA-DR expression may allow the keratinocytes to participate in certain immunological responses such as antigen presentation and cytotoxicity resulting in enhancement of the immune reaction.

VI. PATHOGENETIC MECHANISM OF IRRITANT CONTACT DERMATITIS

Allergic and irritant dermatitis share many clinical and histopathological features and are most often indistinguishable; however, clear differences exist. For example, although both responses are characterized by a similar distribution of T-cells, the number of CD1$^+$ Langerhans' cells is decreased in irritant reactions, whereas it is increased in the allergic response.[67,68,70,75] These findings are highlighted by our observation of decreased epidermal antigen-presenting cell capacity to activate T-cells following irritant challenge and our further findings of increased antigen-presenting cell function after allergic challenge.[67,70] Injured

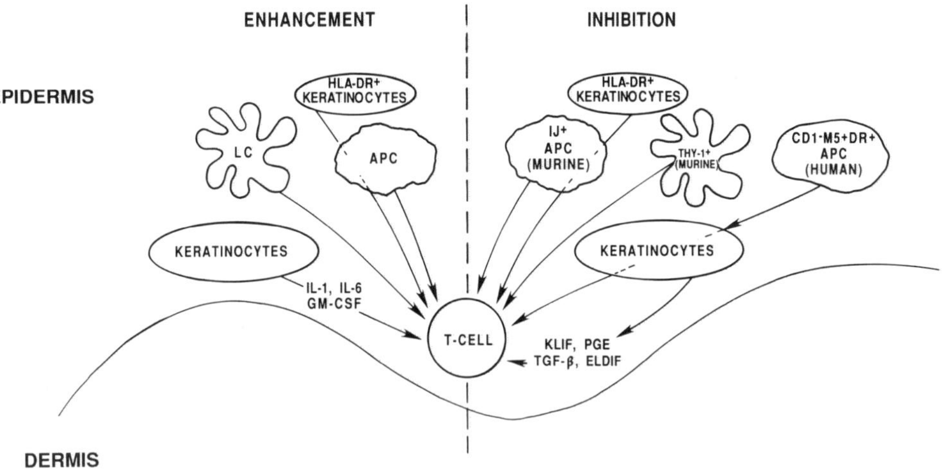

FIGURE 5. Mechanisms involved in skin immune regulation. Multiple factors and mechanisms are responsible for immune regulation in the skin. Langerhans' cells (LC) provide positive signals for T-cell activation; other nonresident antigen-presenting cells (APC) present in inflamed skin may enhance this signal. Keratinocytes releasing T-cell stimulatory cytokines, such as IL-1, IL-6, and GM-CSF, may also function to enhance the immune response. Although the alloantigen-presenting capacity of HLA-DR$^+$ keratinocytes is weak, the keratinocytes can, under certain circumstances, present antigens leading to T-cell activation and possibly also lymphokine release and, thus, potentiate an ongoing immune response. On the other hand, the epidermis also contains the capacity to down-regulate the immune response. UV-induced non-Langerhans' cell CD1$^-$OKM5$^+$DR$^+$ APC seem to activate suppressor pathways preferentially. Furthermore, HLA-DR$^+$ keratinocytes can present antigen to unprimed T-cells, resulting in antigen-specific tolerance to subsequent antigen presentation by Langerhans cells. A variety of T-cell inhibitor cytokines are released from the keratinocytes during *in vitro* immune responses, including KLIF, PGE, TGF-β, and ELDIF. The mechanism responsible for regulation of the release of stimulatory and inhibitory cytokines from the keratinocytes is unknown. In the murine system, IJ$^+$ APC and Thy1$^+$ T-cells also have the capability of down-regulating immune responses. However, clear human analogues to the two murine cell types await identification.

keratinocytes release eicosanoids which are potent chemoattractants for leukocytes (Figure 4). This may be the initiating event in irritant-induced contact dermatitis. The mechanism through which the attracted T-cells obtain and maintain their activated state is unknown.

As mentioned earlier, T-lymphocytes can be activated via both antigen dependent and independent pathways. Allergic contact dermatitis is a classical example of antigen-dependent T-cell activation. However, the decreased capacity of irritant-challenged epidermis to activate T-cells indicates that antigen-presenting cells and, thus, antigen-dependent pathways may not be critical for the immunological response in this disease. This author proposes that irritant contact dermatitis represents an antigen-independent way of T-cell activation, that cytokines released from mechanically or chemically injured keratinocytes induce T-cell activation via antigen independent pathways (Figure 4). This hypothesis is supported by the observation that epidermal-derived chemoattractants and growth-enhancing factors are present in irritant skin reactions.[19,20] Following T-cell activation and lymphokine release, the cellular immunological event and inflammatory response in allergic and irritant contact dermatitis seem to be similar.

VII. REGULATION OF THE INFLAMMATORY RESPONSE IN ALLERGIC CONTACT DERMATITIS

Multiple cell types, factors, and cytokines are involved in the regulation of the inflammatory response in allergic contact dermatitis. Both inhibitory and stimulatory factors interact to determine whether the immune response is up- or down-regulated (Figure 5). Normally, the stimulatory signals provided by the Langerhans' cells are counteracted by inhibitory signals provided by the keratinocytes such that the immune response is not activated. Introduction of antigens or irritants can perturbate this balance, resulting in the development of contact dermatitis.

As mentioned earlier, antigen administration that bypasses the Langerhans' cells (such as intravenous administration) results in activation of antigen-specific suppressor T-cells capable of down-regulating the immune response.[61,76,77] This effect has been utilized by Kligman and others[78] to treat patients with poison ivy dermatitis with oral urishiol extract. However, urishiol extract desensitization is short lived, and the side effects are troublesome.

Sensitization results in antigen-specific activation of both effector and suppressor pathways. The net effect is dependent on both the physical and chemical properties of the antigens as well as the route of administration. High concentrations of antigen induce antigen-specific unresponsiveness which can be transferred with donor T-lymphocytes to syngeneic recipients.[79] The mechanism responsible for activation of these antigen-specific suppressor T-cells in this system is unknown, but may be due to an overload of the skin with antigen and subsequent systemic antigen absorption through the skin, bypassing the Langerhans' cells. Supraoptimal doses of antigens resulting in T-suppressor cell activation via bypassing of the epidermal Langerhans' cells also may be responsible for the rare incidence of contact dermatitis seen in workers heavily exposed to allergens, e.g., nickel workers.

The type of antigen is also important for the effect of sensitization. For example, application of 2,4-dinitrothiocyanatebenzene (DNTB), an antigen that cross-reacts with 2,4-dinitrofluorobenzene (DNFB), to the skin results in activation of suppressor T-cells which down-regulate subsequent reactions to DNFB.[80,81]

As mentioned earlier, during the course of a contact dermatitis reaction, cytokines released from activated T-cells induce HLA-DR expression on keratinocytes. Since presentation of antigen by HLA-DR^+ keratinocytes to unprimed T-cells results in the development of antigen-specific unresponsiveness to later antigen presentation by Langerhans' cells, the HLA-DR^+ keratinocytes may represent a negative feedback mechanism regulating the inflammatory response (Figure 5).[82]

Cytokines derived from keratinocytes can exert both inhibitory and stimulatory effects on T-cells and may be a critical factor determining T-cell activation state. Keratinocyte factors that down-regulate T-cell responses include prostaglandins,[25] eicosanoids,[14] ELDIF,[11] KLIF,[13] and TGF-β (Figure 5).[18]

Bone marrow-derived cellular sources for down-regulating signals also exist within the epidermis. In the murine system, Granstein[83,84] has demonstrated an IJ^+ UV-resistant antigen-presenting cell type that functions to activate suppressor T-cell pathways instead of effector T-cell pathways activated by the Langerhans' cell (Figure 5). Also, in murine epidermis a dendritic bone marrow-derived cell type distinct from the Langerhans' cell is constitutively present.[85-88] In contrast to Langerhans' cells, these cells lack class II MHC expression and express Thy-1 and T-cell receptors composed of γ- and δ-chains.[89-91] In contrast to Langerhans' cells, which provide stimulatory signals for the immune system, dendritic intraepidermal γ/δ T-cell receptor-positive cells exert a negative effect on the immune system, resulting in induction of antigen-specific tolerance (Figure 5).[92]

Despite intensive research, clear human analogues to these cell types have not yet been found. However, bone marrow-derived non-Langerhans' cell antigen-presenting cells do appear in the course of a contact dermatitis reaction and may be involved in suppressor T-cell activation.[67,70] This hypothesis is supported by the observation that these non-Langerhans' cells demonstrate the $OKM1^-OKM5^+$ phenotype.[70] This is a minor subset of blood monocytes that contains cells responsible for activation of autoreactive immunoregulatory T-cells capable of immune down-regulation.[93-95]

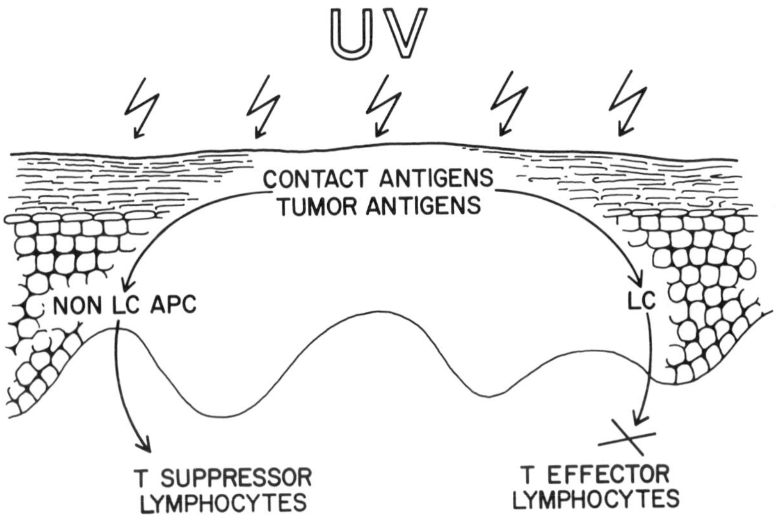

FIGURE 6. Ultraviolet irradiation of the skin induces activation of suppressor T-cell pathways. Ultraviolet (UV) irradiation of the skin abrogates the function of Langerhans cells (LC) and induces the appearance of an epidermal non Langerhans' cell antigen-presenting cell (non-LC APC) type. This UV-induced non-Langerhans' cell antigen-presenting cell is distinct from Langerhans' cells in that it lacks Bierbecks granules and the CD1 determinant. When Langerhans' cells present antigens to T-lymphocytes, activation of effector T-lymphocytes and up-regulation of the immune response is the result. In contrast, when non-Langerhans' cell antigen-presenting cells present the antigen, T-suppressor lymphocytes are activated, resulting in down-regulation of the immune response.

VIII. MECHANISM OF ACTION OF THERAPEUTIC AGENTS IN THE TREATMENT OF CONTACT DERMATITIS

A. CORTICOSTEROIDS

Corticosteroids have a profound effect on the immune system. Topical corticosteroids inhibit both the afferent and the efferent limbs of the allergic contact dermatitis reaction.[96,97] Steroids inhibit antigen-specific lymphocyte activation and proliferation.[98] This effect is probably due to direct inhibition of both antigen-presenting cells and T-cells. Topical application of steroids to the skin depletes Langerhans' cells of CD1 and HLA-DR molecules and abrogates their antigen-presenting function.[99-101] Corticosteroids also block T-cell release of IL-2, thereby inhibiting T-cell proliferation.[98] These immunomodulatory effects turn off the ongoing immune response in contact dermatitis.

B. ULTRAVIOLET RADIATION

UV irradiation of the skin has a profound effect on the immune system. Sensitization of rodents through skin previously exposed to low doses of UV radiation results in antigen-specific unresponsiveness. This effect is due to the appearance of antigen-specific T-suppressor cells that operate at the afferent limb of the immune response to prevent the activation of effector cells for delayed-type hypersensitivity.[102] High doses of UV radiation, in contrast to low doses, result in systemic suppression of the immune response; thus, sensitization through either exposed or unexposed skin results in induction of antigen-specific tolerance.[60,103] Although the mechanism responsible for induction of these suppressor T-cells is not clear, it seems to be due to an altered epidermal antigen presentation caused by a changed distribution in the antigen-presenting cells between the skin and lymph nodes.

UV exposure of the skin results in disappearance of Langerhans' cell surface molecules (CD1 and HLA-DR) and abrogates their antigen-presenting cell function (Figure 6).[104,105]

However, HLA-DR$^+$ bone marrow-derived antigen-presenting melanophages appear in the epidermis in response to the irradiation.[106-109] These non-Langerhans' antigen-presenting cells (non-LC APC), also termed melanophages, are ultrastructurally distinct from Langerhans' cells in that they lack Birbeck granules and express melanosomes within phagolysosomes. Phenotypically, they lack the CD1 determinant expressed on epidermal Langerhans' cells and bear the OKM1$^-$OKM5$^+$ phenotype demonstrating analogy to a minor ($<5\%$ of blood monocyte) subpopulation of antigen-presenting cells in the blood[93] which are uniquely capable of activating regulatory T-cells in the autologous mixed lymphocyte reaction.[95]

The T-cell subsets activated by UV-exposed epidermal cells (non-LC APC) include both antigen-specific CD4$^+$2H4$^-$ T-cells, which provide help for B-cell immunoglobulin production, as well as autoreactive CD4$^+$2H4$^+$ T-cells, which induce the maturation of suppressor T-cells.[110,111] Although both T-cell subsets are activated by UV-exposed epidermal cells, CD4$^+$2H4$^+$ suppressor-inducer T-cells dominate, and the net functional effect is, therefore, the induction of suppression (Figure 6).[112,113] This may be one explanation for decreased skin reaction to allergens applied following UV irradiation.[112,113]

In the murine system, UV-resistant γ/δ T-cell receptor-positive T-cells and IJ$^+$ epidermal antigen-presenting cells have been shown to be involved in activation of T-suppressor cell pathways.[83,84,92] The UV-induced epidermal CD1$^-$OKM5$^+$DR$^+$ non-Langerhans' cell antigen-presenting cell in the human system may be the functional analogue of the epidermal cells in the murine system that provide antigen-specific downregulating signals.

C. CYCLOSPORINE A

Cyclosporine A is a cyclic undecapeptide that causes suppression of immune stimulation *in vivo* and *in vitro*. Systemic cyclosporine A is very efficient in the treatment of psoriasis and will probably demonstrate efficacy in all skin diseases dependent on activated T-cells.[114] Cyclosporine A has a differential effect on T-lymphocyte subsets. The T-helper-inducer cell and its production of lymphokines such as IL-2, IL-4, γ-interferon, and tumor necrosis factor are sensitive to very low doses of cyclosporine A.[115-117] Maturation and generation of cytotoxic T-cells are also affected by cyclosporine A.[118] In contrast, T-suppressor cells are not sensitive to low doses of cyclosporine A.[115,119]

These immunological effects of cyclosporine A indicate that it may be useful in the treatment of contact dermatitis. Accordingly, animal studies have demonstrated that systemic cyclosporine A blocks the effector function of delayed-type hypersensitivity T-cells, although the priming of these cells is cyclosporine A insensitive.[120] Topical cyclosporine A inhibits the elicitation of contact hypersensitivity in guinea pigs.[121] However, application of topical cyclosporine A to human skin demonstrates only a minimal effect which is probably due to lack of absorption or epidermal and dermal inactivation of the drug.[122] Systemic cyclosporine A may play a role in the treatment of severe recalcitrant contact dermatitis: however, due to its nephrotoxicity, widespread use of this agent in contact dermatitis awaits an efficient topical formulation.

ACKNOWLEDGMENT

This paper was supported in part by the Babcock Foundation.

REFERENCES

1. **Sauder, D. N., Carter, C. S., Katz, S. I., and Oppenheim, J. J.,** Epidermal cell production of thymocyte activating factor (ETAF), *J. Immunol.*, 127, 1493, 1982.
2. **Luger, T. A., Stadler, B. M., Katz, S. I., and Oppenheim, J. J.,** Epidermal cell derived thymocyte activating factor (ETAF), *J. Immunol.*, 127, 1493, 1981.
3. **Kupper, T. S., Ballard, D. W., Chua, A. O., et al.,** Human keratinocytes contain mRNA indistinguishable from monocyte interleukin I alpha and beta mRNA, *J. Exp. Med.*, 164, 2095, 1986.
4. **Cooper, K. D., Baadsgaard, O., Elder, J. T., Sauder, D. N., Voorhees, J. J., and Fisher, G.,** Increased levels of an interleukin-1 inhibitor block interleukin-1 activity but not immunoreactivity or mRNA expression of interleukin-1 beta in psoriasis skin, *J. Invest. Dermatol.*, 90, 552, 1988.
5. **Luger, T. A., Wirth, V., and Kock, A.,** Epidermal cells synthesize a cytokine with interleukin 3-like properties, *J. Immunol.*, 134, 915, 1985.
6. **Danner, M. and Luger, T. A.,** Human keratinocytes and epidermoid carcinoma cell lines produce a cytokine with interleukin 3-like activity, *J. Invest. Dermatol.*, 88, 353, 1987.
7. **Kupper, T. S., Coleman, D. L., McGuire, J., Goldminz, D., and Horowitz, D.,** Keratinocyte derived T cell growth factor: a T cell growth factor functionally distinct from interleukin-2, *Proc. Natl. Acad. Sci. U.S.A.*, 83, 4451, 1986.
8. **Chodakewitz, J. A., Kupper, T. S., and Coleman, D. L.,** Keratinocyte-derived granulocyte/macrophage colony-stimulating factor induces DNA synthesis by peritoneal macrophages, *J. Immunol.*, 140, 832, 1988.
9. **Luger, T. A., Uchida, A., Kock, A., Golot, M., and Micksche, M.,** Human epidermal cells and squamous carcinoma cells synthesize a cytokine that augments natural killer cell activity, *J. Immunol.*, 134, 2477, 1985.
10. **Kupper, T. S., May, L., Birchall, N., and Sehgal, P.,** Keratinocytes produce interleukin 6, a cytokine which can provide a 2nd signal in the activation of T-cells, *Clin. Res.*, 3, 665A, 1987.
11. **Nicolas, J. F., Kaiserlian, D., Dardenne, M., et al.,** Epidermal cell derived lymphocyte differentiating factor (ELDIF) inhibits *in vitro* lymphoproliferative responses and Interleukin-2 production, *J. Invest. Dermatol.*, 88, 161, 1987.
12. **Nickoloff, B. J., Basham, T. Y., Merigan, T. C., and Morhenn, V. B.,** Cultured keratinocytes secrete a substance which inhibits allogeneic lymphocyte proliferation, *J. Invest. Dermatol.*, 84, 302, 1985.
13. **Nickoloff, B. J., Basham, T. Y., Torseth, J., Merigan, T. C., and Morhenn, V. B.,** Human keratinocyte-lymphocyte reactions *in vitro, J. Invest. Dermatol.*, 87, 11, 1986.
14. **Rola-Pleszczynski, M.,** Differential effects of leukotriene B4 on T4 and T8 lymphocyte phenotype and immunoregulatory functions, *J. Immunol.*, 135, 1357, 1985.
15. **Hammerstrom, S., Lindgren, J. A., Marcelo, C., Duell, E. A., Anderson, T. F. and Voorhees, J. J.,** Arachidonic acid transformations in normal psoriatic skin, *J. Invest. Dermatol.*, 73, 180, 1979.
16. **Grabbe, J., Czarnetzki, B. M., and Mardin, M.,** Release of lipoxygenase products of arachidonic acid from freshly isolated human keratinocyte, *Arch. Dermatol. Res.*, 276, 128, 1984.
17. **Walsh, L. J., Lander, P. E., Seymour, G. J., and Powell, R. N.,** Isolation and purification of ILS, an interleukin 1 inhibitor produced by human gingival epithelial cells, *Clin. Exp. Immunol.*, 68, 366, 1987.
18. **Nickoloff, B. J.,** Keratinocytes produce a lymphocyte inhibitory factor which is partically reversible by an antibody to transforming growth factor-beta, *Ann. N. Y. Acad. Sci.*, 548, 312, 1988.
19. **Larsen, C. G., Ternowitz, T., Larsen, F. G., and Pedersen, K. T.,** Epidermis and lymphocyte interactions during an allergic patch test reaction. Increased activity of ETAF/IL-1, epidermal derived lymphocyte chemotatic factor and mixed skin lymphocyte reactivity in persons with type IV allergy, *J. Invest. Dermatol.*, 90, 230, 1988.
20. **Larsen, C. G., Ternowitz, T., Larsen, F. G., and Thestrup-Pedersen, K.,** Epidermal derived lymphokines and their presence in allergic and irritant skin reactions, *J. Invest. Dermatol.*, 91, 405, 1988.
21. **Kupper, T. S., Chua, A. O., Flood, P., McGuire, J., and Gubler, U.,** Interleukin-1 gene expression in cultured human keratinocytes is augmented by ultraviolet irradiation, *J. Clin. Invest.*, 80, 430, 1987.
22. **Ristow, H. J.,** A major factor contributing to epidermal proliferation in inflammatory skin diseases appears to be interleukin 1 or a related protein, *Proc. Natl. Acad. Sci. U.S.A.*, 84, 1940, 1987.
23. **Kupper, T. S., Lee, F., Birchall, N., Clark, S., and Dower, S.,** Interleukin 1 binds to specific receptors on human keratinocytes and induces granulocyte macrophage colony-stimulating factor mRNA and protein: a potential autocrine role of interleukin 1 in epidermis, *J. Clin. Invest.*, 82, 1787, 1988.
24. **Mizel, S. B., Dayer, J. M., and Krane, S. M.,** Stimulation of rheumatoid synovial cell collagenase and prostaglandin production by partially purified lymphocyte activating factor (IL-1), *Proc. Natl. Acad. Sci. U.S.A.*, 78, 2474, 1981.
25. **Robertson, B., Gahring, L., Newton, R., and Daynes, R. A.,** *In vivo* administration of interleukin 1 to normal mice depresses their capacity to elicit contact hypersensitivity response: prostaglandins are involved in this modification of immune function, *J. Invest. Dermatol.*, 88, 380, 1988.

26. **Kupper, T. S., Lee, F., Coleman, D. L., Chodakewitz, J. A., Flood, P., and Horowitz, M. C.,** Keratinocyte derived T-cell growth factor (KTGF) is identifical to granulocyte macrophage colony stimulating factor (GM-CSF), *J. Invest. Dermatol.,* 91, 185, 1988.
27. **Nickoloff, B. J.,** Keratinocytes produce the lymphocyte inhibitory factor which is partially inhibited by an antibody to TGF-beta, *J. Am. Acad. Sci.,* 548, 312, 1988.
28. **Katz, S. I., Tamaki, K., and Sachs, D. H.,** Epidermal Langerhans cells are derived from cells originating in bone marrow, *Nature,* 282, 324, 1979.
29. **Stingl, G., Katz, S. I., Clement, L., Green, I., and Shevach, E. M.,** Immunologic functions of Ia-bearing epidermal Langerhans cells, *J. Immunol.,* 121, 2005, 1978.
30. **Braathen, L. R.,** Studies on human epidermal Langerhans cells. III. Induction of T lymphocyte response to nickel sulphate in sensitized individuals, *Br. J. Dermatol.,* 103, 517, 1980.
31. **Rowden, G., Lewis, M. G., and Sullivan, A. K.,** Ia antigen expression on human epidermal Langerhans cells, *Nature,* 268, 245, 1977.
32. **Klareskog, L., Tjernlund, U. M., Forsum, U., and Peterson, P. A.,** Ia antigen expression on human epidermal Langerhans cells, *Nature,* 268, 247, 1977.
33. **Fithian, E., Kung, P., Goldstein, G., Rubenfeld, M., Fenoglio, C., and Edelson, R.,** Reactivity of Langerhans cells with hybridoma antibody, *Proc. Natl. Acad. Sci. U.S.A.,* 78, 2541, 1981.
34. **Raz, A., Wyche, A., Siegel, N., and Needleman, P.,** Regulation of fibroblast cyclooxygenase synthesis by interleukin-1, *J. Biol. Chem.,* 263, 3022, 1988.
35. **Dustin, M. L., Rothlein, R., and Bahn, A. K.,** Induction by IL-1 and interferon: tissue distribution, biochemistry, and function of a natural adherence molecule (ICAM-1), *J. Immunol.,* 137, 245, 1986.
36. **Marlin, S. D. and Springer, T. A.,** Purified intercellular adhesion molecule-1 (ICAM-1) is a ligand for lymphocyte function-associated antigen-1 (LFA-1), *Cell,* 51, 813, 1987.
37. **Kang, S., Duraiswamy, N., Crespo, J., Lee, W., and Cooper, K. D.,** Phenotype and antigen presenting cell function of murine dermal cells, *Clin. Res.,* 35, 814A, 1987.
38. **Camussi, G., Aglietta, M., Coda, R., Bussolino, F., Piacibello, W., and Tetta, C.,** Release of platelet activating factor and histamine. II. The cellular origin of human PAF: monocytes, polymorphonuclear neutrophils, and basophils, *Immunology,* 42, 191, 1981.
39. **Paterson, N. A. M., Wasserman, S. I., Said, J. W., and Austen, K. F.,** Release of chemical mediators from partially purified human lung cells, *J. Immunol.,* 117, 1356, 1976.
40. **Alhenc-Gelas, F., Tsai, S. J., Callahan, K. S., Campbell, W. B., and Johnson, A. R.,** Stimulation of prostaglandin formation by vasoactive mediators in cultured human endothelial cells, *Prostaglandins,* 24, 723, 1982.
41. **Rocklin, R. E. and Haberek-Davidson, E.,** Histamine activates suppressor cells *in vitro* using a co-culture technique, *J. Immunol.,* 117, 1356, 1976.
42. **Shaw, J. O., Pinckard, R. N., Ferrigni, K. S., McManus, L. M., and Hanahan, D. J.,** Activation of human neutrophils with 1-*O*-hexadecyl/octadecyl-2-acetyl-sn-glycerol-3-phosphorylcholine; the active moiety of platelet activating factor, *J. Immunol.,* 127, 1250, 1981.
43. **Bos, J. D., Zonneveld, I. N., Das, P. K., Krieg, S. R., van der Loos, C. M., and Kapsenberg, M. L.,** The skin immune system (SIS): Distribution and immunophenotype of lymphocyte subpopulations in normal human skin, *J. Invest. Dermatol.,* 88, 569, 1987.
44. **Morgan, D. A., Ruscetti, F. W., and Gallo, R.,** Selective *in vitro* growth of T lymphocytes from normal human bone marrows, *Science,* 193, 1007, 1976.
45. **Herberman, R. B., Ortaldo, J. R., Timonen, T., Reynolds, C. W., Djeu, J. Y., Pestka, S., and Stanton, G. J.,** Interferon and natural killer (NK) cells, *Tex. Rep. Biol. Med.,* 41, 590, 1981.
46. **Suzuki, R., Suzuki, S., Takahashi, T., and Kumagai, K.,** Production of a cytokine with interleukin 3-like properties and cytokine-dependent proliferation in human autologous mixed lymphocyte reaction, *J. Exp. Med.,* 164, 1682, 1986.
47. **Suzuki, R., Suzuki, S., Igarashi, M., and Kumagai, K.,** Induction of interleukin 3 but not interleukin 2 or interferon production in the syngeneic mixed lymphocyte reaction, *J. Immunol.,* 137, 1564, 1986.
48. **Paliard, X., de Waal, M. R., Yssel, H., Blanchard, D., Chretien, I., Abrams, J., de Vries, J., and Spits, H.,** Simultaneous production of IL-2, IL-4, and IFN-gamma by activated human CD4+ and CD8+ T cell clones, *J. Immunol.,* 141, 849, 1988.
49. **Lee, K., Yokota, T., Otsuka, T., Gemmell, L., Larson, N., Luh, J., Arai, K. I., and Rennick, D.,** Isolation of cDNA for a human granulocyte-macrophage colony-stimulating factor by functional expression in mammalian cells, *Proc. Natl. Acad. Sci. U.S.A.,* 82, 4360, 1985.
50. **Lee, K., Yokota, T., Otsuka, T., Meyerson, P., Villaret, D., Coffman, R. L., Mosmann, T., Rennick, D., Roehm, N., Smith, C., Zlotnik, A., and Arai, K. I.,** Isolation and characterization of a mouse interleukin cDNA clone that expresses B-cell stimulatory factor 1 activities and T-cell and mast cell stimulating activities, *Proc. Natl. Acad. Sci. U.S.A.,* 83, 2061, 1986.
51. **Ptak, W., Rozychka, D., Askenase, P. W., and Gershon, R.,** Role of antigen presenting cells in the development and persistence of contact hypersensitivity, *J. Immunol.,* 132, 2273, 1984.

52. **Asherson, G. L. and Ptak, W.,** Contact and delayed hypersensitivity in the mouse. I. Active sensitization and passive transfer, *Immunology,* 15, 405, 1968.
53. **Marchal, G., Seman, M., Milon, G., Truffa-Bachi, P., and Zilberfarb, V.,** Local adoptive transfer of skin delayed-type hypersensitivity initiated by a single T lymphocyte, *J. Immunol.,* 129, 954, 1982.
54. **Scheper, R. J., Ans, C. H. M., Dinther-Janssen, V., and Polak, L.,** Specific accumulation of hapten-reactive T cells in contact sensitivity reaction sites, *J. Immunol.,* 134, 1333, 1985.
55. **Pehamberger, H., Stingl, L. A., Pogantsch, S., Steiner, G., Wolff, K., and Stingl, G.,** Epidermal cell induced generation of cytotoxic T lymphocyte responses against alloantigens or TNP-modified syngeneic cells: requirement for Ia-positive Langerhans cells, *J. Invest. Dermatol.,* 81, 208, 1983.
56. **Tamaki, K., Fujiwara, H., and Levy, R. B.,** Hapten-specific TNP-reactive cytotoxic effector cells using epidermal cells as targets, *J. Invest. Dermatol.,* 77, 225, 1981.
57. **Kalish, R. S. and Morimoto, C.,** Urushiol (poison ivy)-triggered suppressor T cell clone generated from peripheral blood, *J. Clin. Invest.,* 82, 825, 1988.
58. **Claman, H. N., Miller, S. D., Man-Sun, S., and Moorhead, J. W.,** Suppressive mechanisms involving sensitization and tolerance in contact allergy, *Immunol. Rev.,* 50, 105, 1980.
59. **Sauder, D., Tamaki, K., Moshell, A., Fujiwara, H., and Katz, S. I.,** Induction of tolerance to topically applied TNCB using TNCB-conjugated ultraviolet light-irradiated epidermal cells, *J. Immunol.,* 127, 261, 1981.
60. **Noonan, F. P., Kripke, M. L., Pedersen, G. M., and Greene, M. I.,** Suppression of contact hypersensitivity in mice by ultraviolet irradiation is associated with defective antigen presentation, *Immunology,* 43, 527, 1981.
61. **Toews, G., Bergstresser, P., Streilein, J., and Sullivan, S.,** Epidermal Langerhans cell density determines whether contact hypersensitivity or unresponsiveness follows skin painting with DNFB, *J. Immunol.,* 124, 445, 1980.
62. **Silberberg-Sinakin, I., Thorbecke, G. J., Baer, R. L., Rosenthal, S. A., and Berezowsky, V.,** Antigen-bearing Langerhans cells in skin, dermal lymphatics and in lymph nodes, *Cell. Immunol.,* 25, 137, 1976.
63. **Griffiths, C. E. M., Voorhees, J. J., and Nickoloff, B. J.,** Characterization of intercellular adhesion molecule-1 and HLA-DR expression in normal and inflamed skin, *Am. Acad. Dermatol.,* 20, 617, 1989.
64. **Baadsgaard, O., Hansen, E., Elder, J. T., Ho, V., Voorhees, J. J., Wantzin, G. L., Fox, D. A., Fisher, G., and Cooper, K. D.,** UM4D4 the surface molecule of an antigen independent T cell activation pathway in psoriasis, Alan R. Liss, New York, 9, 469, 1989.
65. **van Loveren, H., Meade, R., and Askenase, P. W.,** An early component of delayed-type hypersensitivity mediated by T cells and mast cells, *J. Exp. Med.,* 157, 1604, 1983.
66. **Hancock, G. E., Kaplan, G., and Cohn, Z. A.,** Keratinocyte growth regulation by the products of immune cells, *J. Exp. Med.,* 168, 1395, 1988.
67. **Lisby, S., Baadsgaard, O., Avnstorp, C., Clemmensen, O., Cooper, K. D., and Wantzin, G. L.,** Irritant skin reactions demonstrates, in contrast to allergic reactions, decreased number and function of epidermal T6+DR+ Langerhans cells, *J. Invest. Dermatol.,* 91, 405, 1988.
68. **Gawkrodger, D. J., McVittie, E., Carr, M. N., Ross, J. A., and Hunter, J. A. A.,** Phenotypic characterization of the early cellular responses in allergic and irritant contact dermatitis, *Clin. Exp. Immunol.,* 66, 590, 1986.
69. **Christensen, O. B., Daniels, T. E., and Maibach, H. I.,** Expression of OKT6 antigen by Langerhans cells in patch test reactions, *Contact Dermatitis,* 14, 26, 1986.
70. **Baadsgaard, O., Lisby, S., Clemmensen, O., Avnstorp, C., and Kejlsgaard, G.,** Antigen presenting activity of non-Langerhans epidermal cells in contact hypersensitivity reactions, *Scand. J. Immunol.,* in press.
71. **Basham, T. Y., Nickoloff, B. J., Merigan, T. C., and Vorhenn, V. B.,** Recombinant gamma interferon induces HLA-DR expression on cultured human keratinocytes, *J. Invest. Dermatol.,* 83, 88, 1984.
72. **Volc-Platzer, B., Majdic, O., Knapp, W., Wolff, K., Hinterberger, W., Lechner, K., and Stingl, G.,** Evidence of HLA-DR antigen biosynthesis by human keratinocytes in disease, *J. Exp. Med.,* 159, 1784, 1984.
73. **Lampert, I. A.,** Expression of HLA-DR (Ia like) antigen on epidermal keratinocytes in human dermatoses, *Clin. Exp. Immunol.,* 57, 93, 1984.
74. **Gaspari, A. A. and Katz, S. I.,** Induction and functional characterization of Class II MHC (Ia) antigens on murine keratinocytes, *J. Immunol.,* 140, 2956, 1988.
75. **Lisby, S., Baadsgaard, O., Cooper, K. D., and Wantzin, G. L.,** Decreased number and function of epidermal antigen presenting cells in the skin following application of irritant agents: relevance to cutaneous tumors?, *J. Invest. Dermatol.,* 92, 842, 1989.
76. **Tamaki, K., Fujiwara, H., and Katz, S. I.,** The role of epidermal cells in the induction and suppression of contact sensitivity, *J. Invest. Dermatol.,* 76, 275, 1981.
77. **Miller, S. D. and Claman, H. N.,** The induction of hapten-specific T cell tolerance by using hapten-modified lymphoid cells. I. Characteristics of tolerance induction, *J. Immunol.,* 117, 1519, 1978.

78. **Kligman, A. M.,** Hyposensitization against rhus dermatitis, *Arch. Dermatol.,* 78, 47, 1958.
79. **Sy, M. S., Miller, S. D., and Claman, H. N.,** Immune suppression with supraoptimal doses of antigen in contact sensitivity. I. Demonstration of suppressor cells and their sensitivity to cyclophosphamide, *J. Immunol.,* 119, 240, 1977.
80. **Sommer, G., Parker, D., and Turk, J. L.,** Epicutaneous induction of hyporeactivity in contact sensitization: demonstration of suppressor cells induced by contact with 2-4-dinitrothiocyanatebenzene, *Immunology,* 29, 517, 1975.
81. **Iijima, M. and Katz, S. I.,** Specific immunologic tolerance to dinitrofluorobenzene following topical application of dinitrothiocyanobenzene: modulation by suppressor T cells, *J. Invest. Dermatol.,* 81, 325, 1983.
82. **Gaspari, A., Jenkins, M., and Katz, S. I.,** Class II MHC-bearing keratinocytes induce antigen-specific unresponsiveness in hapten-specific TH1 clones, *J. Immunol.,* 141, 2216, 1988.
83. **Granstein, R. D., Lowy, A., and Greene, M. I.,** Epidermal antigen-presenting cells in activation of suppression: identification of a new functional type of ultraviolet radiation-resistant epidermal cell, *J. Immunol.,* 132, 563, 1985.
84. **Granstein, R. D.,** Epidermal I-J bearing cells are responsible for transferable suppressor cell generation after immunization of mice with ultraviolet radiation-treated epidermal cells, *J. Invest. Dermatol.,* 84, 206, 1985.
85. **Aberer, W. and Katz, S. I.,** Thy-1+ dendritic cells in murine epidermis are bone-marrow derived, *J. Invest. Dermatol.,* 83, 74, 1984.
86. **Bergstresser, P., Tigelaar, R. E., Dees, J. H., and Streilein, J. W.,** Thy-1 antigen-bearing dendritic cells populate murine epidermis, *J. Invest. Dermatol.,* 81, 286, 1983.
87. **Bergstresser, P., Tigelaar, R. E., and Streilein, J. W.,** Thy-1 antigen bearing dendritic cells in murine epidermis are derived from bone marrow precursors, *J. Invest. Dermatol.,* 83, 83, 1984.
88. **Tschachler, E., Schuler, G., Hutterer, J., Leibl, H., Wolff, K., and Stingl, G.,** Expression of Thy-1 antigen by murine epidermal cells, *J. Invest. Dermatol.,* 81, 282, 1983.
89. **Konig, F., Stingl, G., Yokoyama, W. M., Yamada, H., Maloy, W. L., Tschachler, E., Shevach, E. M., and Coligan, J. E.,** Identification of a T3-associated gamma-delta T-cell receptor on Thy-1+ dendritic epidermal cell lines, *Science,* 235, 834, 1987.
90. **Stingl, G., Konig, F., Yamada, H., Yokoyama, W. M., Tschachler, E., Bluestone, J. A., Steiner, G., Samelson, L. E., Lew, A. M., Coligan, J. E., and Shevach, E. M.,** Thy-1+ dendritic epidermal cells express T3 and the T-cell receptor gamma chain, *Proc. Natl. Acad. Sci. U.S.A.,* 84, 4586, 1987.
91. **Juziel, W. A., Takashima, A., Bonyhadi, M., Bergstresser, P., Allison, J. P., Tigelaar, R. E., and Tucker, P. W.,** Regulation of T-cell receptor gamma-chain RNA expression in murine Thy-1+ dendritic epidermal cells, *Nature,* 328, 263, 1987.
92. **Sullivan, S., Bergstresser, P., Tigelaar, R. E., and Streilein, J. W.,** Induction and regulation of contact hypersensitivity by resident, bone marrow-derived, dendritic epidermal cells: Langerhans cells and Thy-1+ epidermal cells, *J. Immunol.,* 137, 2460, 1986.
93. **Shen, H. H., Talle, M. A., Goldstein, G., and Chess, L.,** Functional subsets of human monocytes defined by monoclonal antibodies; a distinct subset of monocytes contain the cells capable of inducing the autologous mixed lymphocyte culture, *J. Immunol.,* 130, 698, 1983.
94. **Smith, J. B. and Knowlton, R.,** Activation of suppressor T cells in human autologous mixed lymphocyte culture, *J. Immunol.,* 123, 419, 1979.
95. **James, S. P., Yenokida, G. G., Graeff, A. S., Elson, C. O., and Strober, W.,** Immunoregulatory function of T cells activated in the autologous mixed lymphocyte reaction, *J. Immunol.,* 127, 2605, 1981.
96. **Burrows, W. M. and Stoughton, R. B.,** Inhibition of induction of human contact sensitization by topical glucocorticosteroids, *Arch. Dermatol.,* 112, 175, 1976.
97. **Lynch, D. H., Gurish, M. F., and Daynes, R. A.,** Relationship between epidermal Langerhans cell density ATPase activity and the induction of contact hypersensitivity, *J. Immunol.,* 126, 1892, 1981.
98. **Larsson, E. L.,** Cyclosporine A and dexamethasone suppress T cell responses by selectively acting at distinct sites of the triggering process, *J. Immunol.,* 124, 2828, 1980.
99. **Aberer, W., Stingl, L., Pogantsch, S., and Stingl, G.,** Effect of glucocorticosteroids on epidermal cell-induced immune responses, *J. Immunol.,* 133, 792, 1984.
100. **Ashworth, J., Booker, J., and Breathnach, S. M.,** Effects of topical corticosteroid therapy on Langerhans cell antigen presenting function in human skin, *Br. J. Dermatol.,* 118, 457, 1988.
101. **Belsito, D. V., Flotte, T. J., Lim, H. W., Baer, R. L., Thorbecke, G. J., and Gigli, I.,** Effect of glucocorticosteroids on epidermal Langerhans cells, *J. Exp. Med.,* 155, 291, 1982.
102. **Elmets, A., Bergstresser, P., Tigelaar, R. E., and Wood, P.,** Analysis of the mechanism of unresponsiveness produced by haptens painted on skin exposed to low-dose ultraviolet irradiation, *J. Exp. Med.,* 158, 781, 1983.
103. **Noonan, F. P., Bucana, C., Sauder, D. N., and DeFabo, E. C.,** Mechanism of systemic immune suppression by UV irradiation *in vivo, J. Immunol.,* 132, 2408, 1984.

104. **Aberer, G., Schuler, G., Stingl, G., Honigsmann, H., and Wolff, K.,** Ultraviolet light depletes surface markers of Langerhans cells, *J. Invest. Dermatol.,* 76, 202, 1981.

105. **Stingl, G., Gazze-Stingl, L. A., Aberer, W., and Wolff, K.,** Antigen presentation by murine epidermal Langerhans cells and its alteration by ultraviolet B light, *J. Immunol.,* 127, 1707, 1981.

106. **Cooper, K. D., Neises, G. R., and Katz, S. I.,** Antigen-presenting OKM5+ melanophages appear in human epidermis after ultraviolet radiation, *J. Invest. Dermatol.,* 86, 363, 1986.

107. **Cooper, K. D., Fox, P. F., and Katz, S. I.,** Effects of ultraviolet radiation on human epidermal cell alloantigen presentation: initial depression of Langerhan cell function is followed by the appearance of T6-DR+ cells which enhance epidermal alloantigen presentation, *J. Immunol.,* 134, 129, 1984.

108. **Baadsgaard, O., Cooper, K. D., Lisby, S., Wulf, H. C., and Wantzin, G. L.,** Dose response and time course for induction of T6-DR+ human epidermal antigen presenting cells by *in vivo* UVA, UVB and UVC irradiation, *J. Am. Acad. Dermatol.,* 17, 792, 1987.

109. **Baadsgaard, O., Wulf, H. C., Wantzin, G. L., and Cooper, K. D.,** UVB and UVC, but not UVA, potently induce the appearance of T6-DR+ antigen-presenting cells in human epidermis, *J. Invest. Dermatol.,* 89, 113, 1987.

110. **Morimoto, C., Letvin, N., Boyd, A., Hagan, M., Brown, H., Kornacki, M., and Schlossman, S.,** The isolation and characterization of the human helper inducer subset, *J. Immunol.,* 134, 3762, 1985.

111. **Morimoto, C., Letvin, N., Distaso, J., Aldrich, W., and Schlossman, S.,** The isolation and characterization of the human suppressor inducer T cell subset, *J. Immunol.,* 134, 1508, 1985.

112. **Baadsgaard, O., Fox, D. A., and Cooper, K. D.,** Human epidermal cells from ultraviolet light-exposed skin potently activate autoreactive CD4+2H4+ suppressor-inducer lymphocytes and CD8+ suppressor/cytotoxic lymphocytes, *J. Immunol.,* 140, 1738, 1988.

113. **Baadsgaard, O., Mannie, A., Dass, B., Fox, D. A., and Cooper, K. D.,** Increased T suppressor cell activity in humans after stimulation by UV-exposed epidermal cells: involvement of the 2H4 molecule, *Clin. Res.,* 36, 631A, 1988.

114. **Ellis, C. N., Gorsulowsky, D. C., Hamilton, T. A., Billings, J. K., Brown, M. D., Headington, J. T., Cooper, K. D., Baadsgaard, O., Duell, E. A., Annesley, T. M., Turcotte, J. G., and Voorhees, J. J.,** Cyclosporine improves psoriasis in a double-blind study, *J. Am. Med. Assoc.,* 256, 3110, 1986.

115. **Hess, A. D., Tutschka, P. J., and Santos, G. W.,** Effect of cyclosporin A on human lymphocyte responses *in vitro, J. Immunol.,* 126, 961, 1981.

116. **Reem, G. H. and Cook, L. A.,** Gamma interferon synthesis by human thymocytes and T lymphocytes inhibited by cyclosporin A, *Science,* 1, 63, 1983.

117. **Shevach, E. M.,** The effects of cyclosporin A on the immune system, *Ann. Rev. Immunol.,* 3, 397, 1985.

118. **Orosz, C. G., Fidelus, R. K., Roopenian, D. C., Widmer, M. B., Ferguson, R. M., and Bach, F. H.,** Analysis of cloned T cell function. I. Dissection of cloned T cell proliferative responses using cyclosporin A, *J. Immunol.,* 129, 1865, 1982.

119. **Leapman, S. B., Filo, R. S., Smith, E. J., and Smith, P. G.,** *In vitro* effects of cyclosporin A on lymphocyte subpopulations, *Transplantation,* 30, 404, 1980.

120. **Braida, M. and Knop, J.,** Effect of cyclosporin A on the T-effector and T-suppressor cell response in contact sensitivity, *Immunology,* 59, 503, 1986.

121. **Nakagawa, S., Oka, D., Jinno, Y., Takei, Y., Bang, D., and Ueki, H.,** Topical application of cyclosporine on guinea pig allergic contact dermatitis, *Arch. Dermatol.,* 124, 907, 1988.

122. **Aldridge, R. D., Sewell, H. F., King, G., and Thomson, A. W.,** Topical cyclosporin A in nickel contact hypersensitivity: results of a preliminary clinical and immunohistochemical investigation, *Clin. Exp. Immunol.,* 66, 582, 1986.

Chapter 2

IMMEDIATE CONTACT REACTIONS

Arto Lahti and Howard I. Maibach

TABLE OF CONTENTS

I. INTRODUCTION

Among immediate contact reactions of the skin, various types of contact urticaria have attracted increasing interest in clinical work and research during the last 10 years. The accumulation of new information on the etiologies and the clinical and pathophysiological features of these reactions in the skin has expanded the concept of immediate contact reactions to include not only wheals and flares, but also, for example, transient erythematous and eczematous reactions.

The epidemiology of immediate contact reactions is not well known at present. The first studies on this issue have been performed in Hawaii, Sweden, and Denmark.[1-5] These studies suggest that immediate contact reactions are common in a dermatology practice.

II. SYMPTOMS

Immediate contact reactions appear on normal or eczematous skin within some minutes to 1 h after agents capable of producing this type of contact reaction have been in touch with the skin; they disappear within 24 h (usually within a few hours). The symptoms can be classified according to morphology and severity; itching, tingling, or burning accompanied by erythema is the weakest type of immediate contact reaction and is often seen from cosmetics[6] and from fruits and vegetables.[30] Local wheal and flare is the prototype reaction of contact urticaria. Generalized urticaria after a local contact is a rare phenomenon, but can occur with strong allergens. Tiny vesicles frequently appear on the fingers in protein contact dermatitis. Apart from the skin, symptoms may appear in other organs in cases of very strong hypersensitivity (contact urticaria syndrome). In some cases, immediate contact reactions can be demonstrated only on slightly or previously affected skin, and it can be part of the mechanism responsible for maintenance of chronic eczemas.[5,7,8]

The difference between an immediate irritant (toxic) reaction and contact urticaria is far from clear; therefore, there is much confusion in using terms such as contact urticaria, immediate contact reaction, atopic contact dermatitis, and protein contact dermatitis (Table 1). Immediate contact reaction includes both urticarial and other reactions, whereas protein contact dermatitis means allergic or nonallergic eczematous dermatitis caused by proteins or proteinaceous materials. Atopic contact dermatitis is an immediate type of allergic contact reaction in atopic people.[7] It may be regarded as a special form of immunologic or allergic contact urticaria.

According to the mechanisms underlying contact reactions, they are divided into two main types — namely, immunologic and nonimmunologic immediate contact reactions. However, there are substances causing immediate contact reactions, but it is not known whether the mechanism is immunologic or not.

III. IMMUNOLOGIC IMMEDIATE CONTACT REACTIONS

Immunologic contact urticaria and other immunologic immediate contact reactions are immediate allergic reactions in people who have previously become sensitized to the causative agent. In many cases of immunologic contact reactions, the respiratory and gastrointestinal tracts have obviously been the routes of sensitization. However, natural latex and some foods may sensitize people through the skin. In skin challenge, the molecules of a contact reactant penetrate through the epidermis and react with specific immunoglobulin E molecules (IgE) attached to mast-cell membranes. The cutaneous symptoms — erythema and edema — are elicited by vasoactive substances, mainly histamine released from mast cells. The role of histamine is central, but other mediators of inflammation (e.g., prostaglandins, leukotrienes, and kinins) also may influence the degree of response. However, little is known regarding the dynamics of their interplay in clinical situations.

TABLE 1
Terminology of Immediate Contact Reactions

Term	Remarks
Immediate contact reaction	Includes urticarial, eczematous, and other immediate reactions
Contact urticaria	Allergic and nonallergic contact urticaria reactions
Protein contact dermatitis	Allergic or nonallergic eczematous immediate reactions caused by proteins or proteinaceous material
Atopic contact dermatitis	Immediate urticarial or eczematous IgE-mediated immediate contact reaction
Contact urticaria syndrome	Includes both local and systemic immediate reactions precipitated by contact urticaria agents

TABLE 2
Substances which Have Caused Immediate Local Reactions and Anaphylactic Symptoms in Skin Tests

Aminophenazone
Ampicillin
Balsam of Peru
Bacitracin
Chloramphenicol
Diethyltoluamide
Egg
Epoxy resin
Mechlorethamine
Neomycin
Penicillin
Streptomycin

Sometimes an immunologic contact reaction extends beyond the contact site, and generalized urticaria or symptoms in other target organs, such as rhinitis, conjunctivitis, asthmatic attack, or even anaphylactic shock, may appear. The term "contact urticaria syndrome" was therefore suggested by Maibach and Johnson.[9] The appearance of systemic symptoms is rare, but they may occur in cases of very strong hypersensitivity or in widespread exposure and abundant percutaneous absorption of an allergen. Table 2 gives a list of substances that have produced a local reaction and anaphylactic symptoms in association with skin testing.

The substances causing immunologic immediate contact reactions are usually proteins or protein complexes. As a proof of immediate hypersensitivity, specific IgE antibodies against the causative agent can typically be found in the patient's serum using the RAST (radioallergosorbent test) technique and with skin tests for immediate allergy. Also, the passive transfer test often gives a positive result.

IV. PROTEIN CONTACT DERMATITIS

The term "protein contact dermatitis" was introduced by Hjorth and Roed-Petersen[10] for people with hand eczema demonstrating immediate symptoms when the skin was exposed to certain food proteins. Most of these individuals handled food products as a main part of their job for a protracted period before the symptoms appeared. This type of dermatitis is also seen in veterinarians and slaughterers. Itching, erythema, urticarial swelling, or small vesicles appear on the fingers or dorsa of hands within 30 min of contact with fish, shellfish, and, for example, with cow amnion fluid. "Baker's dermatitis" is another type of immediate contact dermatitis which is usually caused by wheat flour. Protein contact dermatitis may appear without previous urticarial rashes, but it also may be a result of repeatedly occurring contact urticaria.[11]

Atopic contact dermatitis is an immediate type of allergic contact reaction in atopic people,[7] and it may be regarded as a special form of immunologic or allergic contact reaction. There is some evidence that, in atopic dermatitis, allergens which cause immediate type allergy (such as pollens) may cause eczematous dermatitis when applied directly to the skin.[12] This dermatitis might well be only a sign of a continuous, immediate-type reaction in the skin, but it is also possible that Langerhans' cells could participate in this type of inflammation.[13,14] Eczema-like reactions are indistinguishable from primary irritant or allergic dermatitis, and careful study of the patient's history and performing skin tests are the only ways to ensure a correct diagnosis.

V. NONIMMUNOLOGIC IMMEDIATE CONTACT REACTIONS

Nonimmunologic immediate contact reactions occur without previous sensitization in most exposed individuals, and they are the most common type of immediate contact reaction. The reaction remains localized and does not spread to become generalized urticaria, nor does it cause systemic symptoms. Typically, the strength of the reaction varies from erythema to an urticarial response, depending on the concentration, the skin area exposed, the mode of exposure, and the substance itself.[15]

The face, the back skin, and the extensor sides of the upper extremities react more readily than other parts of the body; the soles and palms are the least sensitive areas.[15,16] Scratching does not enhance the reactivity; neither does the use of occlusion, at least where reactions to benzoic acid are concerned.

The mechanism of nonimmunologic immediate contact reactions has not been established, but possible mechanisms are a direct influence upon dermal vessel walls or a non-antibody-mediated release of histamine, prostaglandins, leukotrienes, substance P, or other inflammatory mediators.[17] No specific antibodies against the causative agent can be found in the serum.

Nonimmunologic contact reactions to benzoic acid, cinnamic acid, cinnamic aldehyde, methyl nicotinate, and diethyl fumarate can be inhibited by peroral nonsteroidal anti-inflammatory drugs such as acetylsalicylic acid and indomethacin[18,19] or by topical application of diclofenac or naproxene gels[20] and anesthetic cream,[21] but not by H_1-antagonists, hydroxyzine, or terfenadine[15,22] or by topical capsaicin treatment.[21] These studies support the concept that nonspecific histamine release from mast cells is not the main mechanism of nonimmunologic contact urticaria from those substances. The mechanism by which acetylsalicylic acid and indomethacin inhibit contact reactions in human skin has not been defined yet, but it may be ascribed to their common pharmacological action, i.e., inhibition of prostaglandin bioformation. However, cinnamic aldehyde can result in histamine release from leukocytes *in vitro*.[23]

Immediate contact reactions to benzoic acid can also be inhibited by ultraviolet B and A light exposure; this effect lasts at least 2 weeks.[24] Nonimmunologic contact reactions from different agents are probably linked to different combinations of mediators.

Distinguishing between irritant reactions and nonimmunologic contact urticaria may be difficult. Strong irritants, such as hydrochloric acid, formaldehyde, and phenol, can cause immediate whealing, but the reaction does not disappear within a few hours. Instead, it is followed by erythema, scaling, or crusting which may still be visible 24 h later. Some substances are only contact urticants (e.g., nicotinic acid esters), some are pure irritants (e.g., sodium lauryl sulfate), and some have both these features (e.g., phenol and dimethyl sulfoxide).

Little is known about the histology of immediate contact reactions. In recent studies, the accumulation of mononuclear cell perivascular infiltrate was seen from 15 min and that of neutrophils from 2 h onward, persisting up to 48 h in normal subjects. Leucocytoclasis was also observed. The cell infiltrate was seen to a lesser degree in one of six atopic eczema patients, but not in normal subjects treated with 600 mg of acetylsalicylic acid before the nicotinate application.[25,26]

VI. ETIOLOGY

Tables 2, 3, and 4 present lists of agents which have been reported to cause immediate contact reactions, and they include numerous chemicals in medications, industrial contactants, cosmetics, foods, and many chemically undefined agents. These tables are mainly based on references cited in recent review articles on immediate contact reactions.[27-29] The pathogenetic classification (nonimmunologic vs. immunologic) is also given, but in many instances it is arbitrary because the mechanisms of various types of contact reactions are unclear or because a pathogenetic evaluation was not performed.

Increasing awareness of immediate contact reactions will expand the list of etiologic agents, and a more thorough understanding of pathophysiologic mechanisms will lead to a better and more rational classification of these reactions than is possible at present.

Foodstuffs are the most common causes of immediate allergic contact reactions (Table 3). The orolaryngeal area is a site where immediate reactions often are provoked by food allergens, most frequently among atopic individuals. Of 230 patients allergic to birch pollen, 152 (66%) gave a history of itching, tingling, and/or edema of the lips and tongue and hoarseness or irritation of the throat when eating such raw fruits and vegetables as apple, potato, carrot, and tomato.[30] Plum, peach, cherry, kiwi, celery, and parsnip can also elicit immediate contact reactions in birch pollen allergic people. Positive results in the scratch chamber test with suspected raw fruits and vegetables were noted in 36% of 230 patients. Apple, carrot, parsnip, and potato elicited reactions more often than swede (rutabaga), tomato, onion, celery, and parsley. The clinical relevance of the skin test results with apple, potato, and carrot was 80 to 90%. Only 7 of 158 (4%) atopic patients who were not allergic to birch pollen had positive skin test reactions to any of the fruits and vegetables.

RAST and RAST-inhibition studies have confirmed the real cross-allergy between birch pollen and fruits and vegetables. All immunologic determinants in apple, carrot, and celery tuber appeared to be present also in birch pollen, but not vice versa.[31,32]

Veterinarians can get contact urticaria on their hands after contact with cow amnion fluid, but do not have reactions to cow dander in clinical provocation tests or in skin prick tests with cow epithelium extracts. In RAST investigations it has been shown that antibodies to cow amnionic fluid and serum, but not to epithelia, can be found in veterinarians' sera. An interesting finding was that the allergen causing contact urticaria in these cases was a compound present in the amniotic fluid and serum but not in the epithelium of cows.[33]

Carbonless copy paper is suspected of causing several types of symptoms, but it has been difficult to prove the causal relationships. Contact urticaria to carbonless copy paper with eye, throat, and respiratory symptoms has been found to be related to prostaglandin release.[34]

Contact urticaria to formaldehyde appears on healthy skin following repeated open applications; it can occur after a single application on slightly diseased skin. The mechanism of this kind of delayed onset contact urticaria is not known.[35]

Contact urticaria to latex occurs more commonly than what was thought earlier.[36-37] There have been cases of anaphylactic symptoms and generalized urticaria after contact with surgical[38] and household rubber gloves.[39] These reactions have been shown to be immediate-type allergic and IgE-mediated.[39,40] The allergens are among the proteins which natural latex

TABLE 3
Agents Producing Immunologic Immediate Contact Reactions

Agent	Ref.
Animal products	
Amnion fluid	
Blood	
Cercariae	
Cheyletus malaccensis	52
Chironomidae, *Chironomus thummi thummi*	53
Cockroaches	
Dander	54
Dermestes maculatus Degeer	55
Gut	
Hair	
Listrophorus gibbus	56
Liver	
Mealworm	
Placenta	
Saliva	
Serum	
Silk	
Spider mite	
Wool	
Food	
Dairy	
Cheese	
Egg	
Milk	57,58
Fruits	
Apple	31
Apricot	
Banana	
Kiwi	
Mango	
Orange	
Peach	
Plum	
Grains	
Flour	
Maize	
Malt	
Wheat bran	
Honey	
Nuts	
Peanut butter	
Sesame seed	
Sunflower seed	
Meats	
Beef	
Chicken	
Lamb	
Liver	
Turkey	
Seafood	
Fish	59,60
Prawns	
Shrimp	
Vegetables	
Beans	
Cabbage	
Carrot	61
Celery	
Chives	
Cucumber	
Endive	
Lettuce	
Onion	
Parsley	
Parsnip	
Potato	
Rutabaga (swede)	
Tomato	
Soybean	
Fragrances and flavorings	
Balsam of Peru	
Menthol	
Vanillin	
Medicaments	
Acetylsalicylic acid	
Antibiotics	
Ampicillin	
Bacitracin	
Cephalosporins	
Chloramphenicol	62
Gentamicin	
Iodochlorhydroxyquin	
Neomycin	
Nifuroxime	63
Penicillin	64
Rifamycin	65
Streptomycin	
Virginiamycin	66
Benzocaine	67
Benzoylperoxide	
Clobetasol-17-propionate	68
Dinitrochlorobenzene	69,70
Mechlorethamine	
Phenothiazines	
Chlorpromazine	71
Promethazine	
Pyrazolones	
Aminophenazone	
Methamizole	
Propylphenazone	
Tocopherol	72
Metals	
Copper	73
Nickel	74
Platinum	
Rhodium	
Plant products	75
Algae	

TABLE 3 (continued)
Agents Producing Immunologic Immediate Contact Reactions

Agent	Ref.	Agent	Ref.
Plant products		Preservatives and disinfectants	
Birch		Phenylmercuric propionate	
Camomile		Polysorbates	
Castor bean		Sodium hypochlorite	
Chrysanthemum	76	Sorbitan monolaurate	
Cinchona	77	Tropicamide	106
Colophony	78	Miscellaneous	
Corn starch	79,80	Acetyl acetone	107
Cotoneaster		Acrylic monomer	
Emetin		Alcohols (amyl-, butyl-, ethyl-, isopropyl-)	108
Fennel	81	Aliphatic polyamide	
Garlic		Ammonia	
Grevillea juniperina	82	Ammonium persulphate	
Hakea suaveolens	83	Aminothiazole	
Hawthorn	84	Benzophenone	
Henna		Butylated hydroxytoluene	
Latex rubber	37,40,85,86	Carbonless copy paper	
Lichens		Cu(II)-acetyl acetonate	107
Lily	87	Denatonium benzoate	
Lime	88	Diethyltoluamide	
Mahogany		Epoxy resin	109
Mustard	59	Formaldehyde resin	
Papain	89	Lanolin alcohols	
Perfumes		Lindane	
Pickles	90	Methyl ethyl ketone	110
Rose	28	Monoamylamine	
Rouge		Naphtha	111
Spices	91	Naphthylacetic acid	112
Strawberry	92	Nylon	113
Teak		Oleylamide	
Tobacco	93	Paraphenylenediamine	114,115
Tulip	87	Patent blue dye	
Winged bean	94	Perlon	
Preservatives and disinfectants		Phosphorous sesquisulfide	116
Benzoic acid	95	Plastic	
Benzyl alcohol		Polypropylene	117
Chlorhexidine	96,97	Polyethylene glycol	
Chloramine		Potassium ferricyanide	
Chlorocresol	98	Seminal fluid	118
1,3-Diiodo-2-hydroxypropane	99	Sodium silicate	
Formaldehyde	35,100	Sodium sulfide	
Gentian violet	101	Sulfur dioxide	
Hexantriol	102	Terpinyl acetate	
p-Hydroxybenzoic acid	103	Vinyl pyridine	
ortho-Phenylphenate	104	Zinc diethyldithiocarbamate	
Parabens	105		

contains in concentrations of 1 to 2%. Allergy to latex can be proved by skin tests and by latex RAST, which is also commercially available (Pharmacia, Malmö, Sweden).

Ammonium persulfate, an oxidizing hair bleach, is a contradictory substance. The fact that some individuals react on the first exposure and that the passive transfer test is negative suggests a nonimmunologic reaction, whereas negative skin tests in most control subjects

TABLE 4
Agents Producing Nonimmunologic Immediate Contact Reactions

Agent	Ref.	Agent	Ref.
Animals		Medicaments	
Arthropods		Methyl salicylate	
Caterpillars	119,120	Methylene green	
Corals		Myrrh	
Jellyfish		Nicotinic acid esters	
Moths		Resorcinol	
Sea anemones		Tar extracts	
Foods		Tincture of benzoin	
Cayenne pepper		Witch hazel	
Fish		Metals	
Mustard		Cobalt	
Thyme		Plants	
Fragrances and flavorings		Nettles	
Balsam of Peru		Sea weed	
Benzaldhyde		Preservatives and disinfectants	
Cassia (cinnamon oil)		Benzoic acid	
Cinnamic acid		Chlorocresol	122
Cinnamic aldehyde	6,42	Formaldehyde	
Medicaments		Sodium benzoate	
Alcohols	121	Sorbic acid	41
Benzocaine		Miscellaneous	
Camphor		Butyric acid	
Cantharides		Diethyl fumarate	48,123
Capsaicin		Histamine	
Chloroform		Pine oil	
Dimethyl sulfoxide		Pyridine carboxaldehyde	124,125
Friar's balsam		Sulfur	103
Iodine		Turpentine	

and systemic symptoms (generalized urticaria, rhinitis, asthma, and even vascular collapse) suggest an immunologic contact reaction.

The most potent and best studied substances producing immediate nonimmunologic contact reactions (Table 4) are benzoic acid, sorbic acid, cinnamic acid, cinnamic aldehyde, and nicotinic acid esters. Under optimal conditions more than half of the individuals react with local erythema and edema to these substances within 45 min of application if the concentration is high enough. Benzoic acid, sorbic acid, and sodium benzoate, which are used as preservatives for cosmetics and other topical preparations, are capable of producing immediate contact reactions at concentrations from 0.1 to 0.2%.[15,41]

Cinnamic aldehyde at a concentration of 0.01% may elicit erythema with a burning or stinging feeling in the skin. Some mouthwashes and chewing gums contain cinnamic aldehyde at concentrations high enough to produce a pleasant tingling or ''lively'' sensation in the mouth and enhance the sale of the product. Higher concentrations produce lip swelling or contact urticaria in normal skin.

Eugenol in the mixture inhibits delayed contact sensitization by cinnamic aldehyde, and it also inhibits immediate contact reactions from this same substance.[42] The mechanism of the quenching effect is not certain, but a competitive inhibition at the receptor level may be an explanation.

VII. DIAGNOSTIC TESTS

The diagnosis of immediate contact reactions is based on a full medical history and on skin tests with suspected substances.

A. TESTS FOR BOTH IMMUNOLOGIC AND NONIMMUNOLOGIC CONTACT REACTIONS

1. Rub Test

The simplest test is the rub test. For this test, the suspected substance (i.e., fish, apple, or carrot) is applied and gently rubbed on either normal-looking or slightly affected skin, usually the hand. The test site is observed for 60 min.[11] A positive result is seen as an edema and erythema reaction or as tiny intraepidermal spongiotic vesicles typical for acute eczema.

2. Use Test

The use test is a test where the patient tries to handle the suspected agent precisely as he or she handled it when symptoms appeared. Wearing surgical gloves on wet hands to provoke contact urticaria to latex is a typical use test.

3. Open Application Test

In the open application test or open patch test, 0.1 ml of the test substance is spread on a 3 × 3 cm area of the skin of the upper back or on the extensor side of the upper arm, about 10 $\mu l/cm^2$. The test should first be performed on nondiseased skin and then, if negative, on previously or currently affected skin.[17] Even in immunologic contact urticaria there may be a marked difference between skin sites in the ability to elicit contact urticaria,[43] a feature which is typical of nonimmunologic contact urticaria. The face has been considered the most sensitive skin area.[16] Often it is necessary to apply contact urticants to skin sites suggested by the patients's history. The immunologic contact reactions usually appear within 15 to 20 min and nonimmunologic ones within 45 to 60 min after application. A positive reaction comprises a wheal-and-flare reaction and sometimes a vesicular eruption indistinguishable from that seen in eczema.[10]

4. Chamber Test

Occlusive patch and chamber tests on nondiseased or slightly affected skin have been said to be somewhat more sensitive than the open application test in immunologic immediate contact reactions.[44] The occlusion time is 15 to 20 min, and a positive reaction is usually seen within 45 min after application.

B. TESTS FOR IMMUNOLOGIC CONTACT REACTIONS

1. Prick Test

Prick testing is often the method of choice for testing patients with suspected allergic contact reactions.

2. Scratch Test

The scratch test is a less standardized method than the prick test, but it is still useful when nonstandardized allergens must be used. The allergen solutions used in scratch testing are the same as those used in prick testing. Also, freeze-dried and other powdered allergens moistened with 0.1 *N* aqueous sodium hydroxide solution and fresh foodstuffs (e.g., potato, apple, carrot) can be used. When testing with poorly standardized or nonstandardized substances, control tests should be made on at least 20 people to avoid false interpretation of the test results.

3. Scratch Chamber Test

The scratch chamber test was introduced for testing foodstuffs when commercial allergens with proven efficacy are not available.[30] Potato, apple, and carrot are examples of fruits and vegetables which lose their allergenicity easily when cooked, deep frozen, or made into juice, and it is therefore best to use them fresh for skin testing.

In the scratch chamber test the procedure is that of the ordinary scratch test, but the scratch and the foodstuff are covered with a small aluminum chamber (Finn Chamber®, Epitest Ltd. Oy, Hyrylä, Finland) for 15 min. The result is read 5 min after the removal of the test chamber according to the criteria of the scratch test. Reactions at least the size of a similarly produced histamine reaction are usually clinically significant. Histamine dihydrochloride (10 mg/ml) is the positive reference and aqueous 0.1 *N* sodium hydroxide the negative reference.

4. Passive Transfer Test

The Prausnitz-Küstner test or passive transfer test is only used for special purposes. It has been used in occupational dermatology for detecting immunologic contact urticaria to potato[45] and to rubber.[46]

5. RAST

RAST is seldom needed for contact urticaria diagnosis, but RAST inhibition tests are used in investigating cross-allergenicity.[31] For this purpose crossed radioimmunoelectrophoresis and its inhibition are also used.

Nonsteroidal anti-inflammatory drugs and antihistamines should not be used by patients during tests for immediate contact reactions because these drugs may inhibit the reactions. Using the same test site repeatedly may result in the tachyphylaxis phenomenon and cause false negative results.

When examining patients with a suspected allergic contact reaction, the prick, scratch, or scratch chamber tests are done first because the test procedures are fast and reliable. The final diagnosis should be based on the result of the rub, open application, or use test.

VIII. IMMEDIATE CONTACT REACTIONS IN ANIMAL SKIN

Animal skin test methods for immediate contact reactions to screen putative agents and to study the pathophysiology are needed, but have not been available. Guinea pig body skin reacts with rapidly appearing erythema to cinnamic aldehyde, methyl nicotinate, and dimethyl sulfoxide, but not to benzoic acid, sorbic acid, or cinnamic acid. If these substances are applied to the guinea pig earlobe, erythema and edema appear with all of them. Quantification of edema by measuring changes in the ear thickness with a micrometer is an accurate, reproducible, and quick method.[47]

Analogous reactions can be elicited in the earlobes of other laboratory animals. Cinnamic aldehyde and dimethyl sulfoxide produce ear swelling in rat and mouse, but benzoic acid, sorbic acid, cinnamic acid, diethyl fumarate, and methyl nicotinate produce no response. This suggests that either there are several mechanisms involved in immediate contact reactions from different substances or there are differences in the activation of mediators of inflammation between guinea pig, rat, and mouse.[48,49]

The swelling response in the guinea pig earlobe is dependent on the concentration of the eliciting substance. The maximal response is about a 100% increase in the ear thickness which appears within 50 min after application.

In biopsies taken from the guinea pig earlobe 40 min after application of test substances, marked dermal edema and intra- and perivascular infiltrates of heterophilic granulocytes are seen, and they appear to be characteristic of nonimmunologic contact urticaria in the guinea pig ear.[47,50]

A decrease in reactivity to contact urticants is noticed after reapplication to the guinea pig ear on the following day.[51] The tachyphylaxis is not specific to the substance which produces it, and the reactivity to other agents decreases as well. The length of the refractory period varies with the compound used (4 d for methyl nicotinate, 8 d for diethyl fumarate and cinnamic aldehyde, and up to 16 d for benzoic acid, cinnamic acid, and dimethyl sulfoxide).

The guinea pig earlobe seems to react to nonimmunologic contact urticants in a way similar to human skin. The similarities include the morphology, the time course of maximal response, the concentrations of the eliciting substances, the tachyphylaxis phenomenon,[15] and the lack of the inhibitory effect of antihistamines on contact reactions.[22,50]

IX. SUMMARY

In clinical practice, patients often complain about immediate contact reactions after applying skin lotions, other cosmetics or therapeutic agents, and after handling food products, for example. Not only have dermatologists and allergists been uncertain about the character of these reactions, but manufacturers, their toxicologists, and other safety personnel have had great difficulty in understanding this type of reaction and in developing less irritating products.

Studies on the mechanisms of immediate contact reactions from different substances and the standardization of human and animal tests for these reactions are a challenge for future research. Dermatologists, allergists, and toxicologists will have to ask the medical authorities and industry to investigate the ability of various environmental agents in contact with the skin to produce immediate contact reactions, as has been the case for many years in the search for delayed-type contact sensitizers.

REFERENCES

1. **Elpern, D. J.,** The syndrome of immediate reactivities (contact urticaria syndrome). An historical study from a dermatology practice. I. Age, sex, race and putative substances, *Hawaii Med. J.,* 44, 426, 1985.
2. **Elpern, D. J.,** The syndrome of immediate reactivities (contact urticaria syndrome). An historical study from a dermatology practice. II. The atopic diathesis and drug reactions, *Hawaii Med. J.,* 44, 466, 1985.
3. **Elpern, D. J.,** The syndrome of immediate reactivities (contact urticaria syndrome). An historical study from a dermatology practice. III. General discussion and conclusions, *Hawaii Med. J.,* 45, 10, 1986.
4. **Nilsson, E.,** Contact sensitivity and urticaria in "wet" work, *Contact Dermatitis,* 13, 321, 1985.
5. **Veien, N. K., Hattel, T., Justesen, O., and Norholm, A.,** Dietary restrictions in the treatment of adult patients with eczema, *Contact Dermatitis,* 17, 223, 1987.
6. **Emmons, W. W. and Marks, J. G.,** Immediate and delayed reactions to cosmetic ingredients, *Contact Dermatitis,* 13, 258, 1985.
7. **Hannuksela, M.,** Atopic contact dermatitis, *Contact Dermatitis,* 6, 30, 1980.
8. **Maibach, H. I.,** Immediate hypersensitivity in hand dermatitis: role of food contact dermatitis, *Arch. Dermatol.,* 112, 1289, 1976.
9. **Maibach, H. I. and Johnson, H. L.,** Contact urticaria syndrome: contact urticaria to diethyltoluamide (immediate type hypersensitivity), *Arch. Dermatol.,* 111, 726, 1975.
10. **Hjorth, N. and Roed-Petersen, J.,** Occupational protein contact dermatitis in foodhandlers, *Contact Dermatitis,* 2, 28, 1976.
11. **Hannuksela, M.,** Contact urticaria from foods, in *Nutrition and the Skin,* Roe, D., Ed., Contemporary Issues in Clinical Nutrition Series, Rivlin, R. S., Ed., Alan R. Liss, New York, 1986, 153.
12. **Reitamo, S., Visa, K., Kähönen, K., Käyhkö, K., Stubb, S., and Salo, O. P.,** Eczematous reactions in atopic patients caused by epicutaneous testing with inhalant allergens, *Br. J. Dermatol.,* 114, 303, 1986.
13. **Barker, J. N. W. N., Alegre, V. A., and MacDonald, D. M.,** Surface-bound immunoglobulin E on antigen presenting cells in cutaneous tissue of atopic dermatitis, *J. Invest. Dermatol.,* 90, 117, 1988.

14. **Bruynzeel-Koomen, C.**, IgE on Langerhans cells: new insights into the pathogenesis of atopic dermatitis, *Dermatologica,* 172, 181, 1986.
15. **Lahti, A.**, Non-immunologic contact urticaria, *Acta Derm. Venereol.*, 60 (Suppl. 91), 1, 1980.
16. **Gollhausen, R. and Kligman, A. M.**, Human assay for identifying substances which induce non-allergic contact urticaria: the NICU-test, *Contact Dermatitis,* 13, 98, 1985.
17. **Lathi, A., von Krogh, G., and Maibach, H. I.**, Contact urticaria syndrome. An expanding phenomenon, in *Dermatologic Immunology and Allergy,* Stone, J., Ed., C. V. Mosby, St. Louis, 1985, 379.
18. **Lahti, A., Oikarinen, A., Ylikorkala, O., and Viinikka, L.**, Prostaglandins in contact urticaria induced by benzoic acid, *Acta Derm. Venereol.*, 63, 425, 1983.
19. **Lahti, A., Väänänen, A., Kokkonen, E.-L., and Hannuksela, M.**, Acetylsalicylic acid inhibits non-immunologic contact urticaria, *Contact Dermatitis,* 16, 133, 1987.
20. **Johansson, J. and Lahti, A.**, Topical non-steroidal anti-inflammatory drugs inhibit non-immunologic immediate contact reactions, *Contact Dermatitis,* 19, 161, 1988.
21. **Larmi, E., Lahti, A., and Hannuksela, M.**, Ultraviolet light inhibits nonimmunologic immediate contact reactions to benzoic acid, *Arch. Dermatol. Res.*, 280, 420, 1988.
22. **Lahti, A.**, Terfenadine does not inhibit non-immunologic contact urticaria, *Contact Dermatitis,* 16, 220, 1987.
23. **Nater, J. P., de Jong, M. C. J. M., Baar, A. J. M., and Bleumink, E.**, Contact urticarial skin responses to cinnamaldehyde, *Contact Dermatitis,* 3, 151, 1977.
24. **Larmi, E., Lahti, A., and Hannuksela, M.**, Ultraviolet light inhibits nonimmunologic immediate contact reactions to benzoic acid, *Arch. Dermatol. Res.*, 280, 420, 1988.
25. **English, J. S. C., Winkelmann, R. K., Louback, J. B., Greaves, M. W., and MacDonald, D. M.**, The cellular inflammatory response in nicotinate skin reactions, *Br. J. Dermatol.*, 116, 341, 1987.
26. **Daroczy, J. and Temesvari, E.**, Light microscopic and electron microscopic (EM) examination of contact urticaria, *Contact Dermatitis,* 19, 156, 1988.
27. **Burdick, A. E. and Mathias, C. G. T.**, The contact urticaria syndrome, *Dermatol. Clin.*, 3(1), 71, 1985.
28. **Kleinhans, D.**, Kontakt-Urtikaria, *Dermatosen,* 33, 198, 1985.
29. **Lahti, A. and Maibach, H. I.**, Immediate contact reactions: contact urticaria syndrome, *Semin. Dermatol.*, 6, 313, 1987.
30. **Hannuksela, M. and Lahti, A.**, Immediate reactions to fruits and vegetables, *Contact Dermatitis,* 3, 79, 1977.
31. **Halmepuro, L. and Løvenstein, H.**, Immunological investigation of possible structural similarities between pollen antigens and antigens in apple, carrot and celery tuber, *Allergy,* 40, 264, 1985.
32. **Halmepuro, L., Vuontela, K., Kalimo, K., and Björksten, F.**, Cross-reactivity of IgE antibodies with allergens in birch pollen, fruits and vegetables, *Int. Arch. Allergy Appl. Immunol.*, 74, 235, 1984.
33. **Kalveram, K.-J., Kästner, H., and Frock, G.**, Detection of specific IgE antibodies in veterinarians suffering from contact urticaria, *Z. Hautkr.*, 61, 75, 1986.
34. **Marks, J. G., Trautlein, J. J., Zwillich, C. W., and Demers, L. M.**, Contact urticaria and airway obstruction from carbonless copy paper, *J. Am. Med. Assoc.*, 252, 1038, 1984.
35. **Andersen, K. E. and Maibach, H. I.**, Multiple application delayed onset contact urticaria: possible relation to certain unusual formalin and textile reactions, *Contact Dermatitis,* 10, 227, 1984.
36. **Wrangsjö, K., Mellström, G., and Axelsson, G.**, Discomfort from rubber gloves indicating contact urticaria, *Contact Dermatitis,* 15, 79, 1986.
37. **Turjanmaa, K.**, Incidence of immediate allergy to latex gloves in hospital personnel, *Contact Dermatitis,* 17, 270, 1987.
38. **Carrillo, T., Cuevas, M., Muñoz, T., Hinojosa, M., and Moneo, I.**, Contact urticaria and rhinitis from latex surgical gloves, *Contact Dermatitis,* 15, 69, 1986.
39. **Seifert, H. U., Wahl, R., Vocks, E., Borelli, S., and Maasch, H. J.**, Immunoglobulin E-vermittelte Kontakturtikaria bzw. Asthma bronchiale durch Latex-enthaltende Haushaltsgummihandschuhe, *Dermatosen,* 35, 137, 1987.
40. **Frosch, P. J., Wahl, R., Bahmer, F. A. and Maasch, H. J.**, Contact urticaria to rubber gloves is IgE-mediated, *Contact Dermatitis,* 14, 241, 1986.
41. **Soschin, D. and Leyden, J. J.**, Sorbic acid-induced erythema and edema, *J. Am. Acad. Dermatol.*, 14, 234, 1986.
42. **Guin, J. D., Meyer, B. N., Drake, R. D., and Haffley, P.**, The effect of quenching agents on contact urticaria caused by cinnamic aldehyde, *J. Am. Acad. Dermatol.*, 10, 45, 1984.
43. **Maibach, H. I.**, Regional variation in elicitation of contact urticaria syndrome (immediate hypersensitivity syndrome): shrimp, *Contact Dermatitis,* 15, 100, 1986.
44. **von Krogh, G. and Maibach, H. I.**, The contact urticaria syndrome — 1982, *Semin. Dermatol.*, 1, 59, 1982.
45. **Tuft, L. and Blumstein, G. I.**, Studies in food allergy. II. Sensitization to fresh fruits: clinical and experimental observations, *J. Allergy,* 13, 574, 1942.

46. **Köpman, A. and Hannuksela, M.,** Contact urticaria to rubber, *Duodecim,* 99, 221, 1983.
47. **Lahti, A. and Maibach, H. I.,** An animal model for nonimmunologic contact urticaria, *Toxicol. Appl. Pharmacol.,* 76, 219, 1984.
48. **Lahti, A. and Maibach, H. I.,** Contact urticaria from diethyl fumarate, *Contact Dermatitis,* 12, 139, 1985.
49. **Lahti, A. and Maibach, H. I.,** Species specificity of nonimmunologic contact urticaria: guinea pig, rat and mouse, *J. Am. Acad. Dermatol.,* 13, 66, 1985.
50. **Lahti, A., McDonald, D. M., Tammi, R., and Maibach, H. I.,** Pharmacological studies on nonimmunologic contact urticaria in guinea pig, *Arch. Dermatol. Res.,* 279, 44, 1986.
51. **Lahti, A. and Maibach, H. I.,** Long refractory period after one application of nonimmunologic contact urticaria agents to the guinea pig ear, *J. Am. Acad. Dermatol.,* 13, 585, 1985.
52. **Yoshikawa, M.,** Skin lesions of papular urticaria induced experimentally by *Chyletus malaccensis* and *Chelacaropsis* sp. (Acari, Cheyletidae), *J. Med. Entomol.,* 22, 115, 1985.
53. **Mittelbach, F.,** Urticaria and Quincke's edema caused by Chironomidae *(Chironomus thummi thummi)* as fishfood, *Z. Hautkr.,* 58, 1548, 1983.
54. **Agrup, G. and Sjöstedt, L.,** Contact urticaria in laboratory technicians working with animals, *Acta Derm. Venereol.,* 65, 111, 1985.
55. **Lewis-Jones, M. S.,** Papular urticaria caused by Dermestes maculatus Degeer, *Clin. Exp. Dermatol.,* 10, 181, 1985.
56. **Burns, D. A.,** Papular urticaria produced by the mite *Listrophorus gibbus, Clin. Exp. Dermatol.,* 12, 200, 1987.
57. **Boso, E. B. and Brestel, E. P.,** Contact urticaria to cow milk, *Allergy,* 42, 151, 1987.
58. **Salo, O. P., Mäkinen-Kiljunen, S., and Juntunen, K.,** Milk causes a rapid urticarial reaction on the skin of children with atopic dermatitis and milk allergy, *Acta Derm. Venereol.,* 66, 438, 1986.
59. **Kavli, G. and Moseng, D.,** Contact urticaria from mustard in fish stick production, *Contact Dermatitis,* 17, 153, 1987.
60. **Melino, M., Toni, F., and Riguzzi, G.,** Immunologic contact urticaria to fish, *Contact Dermatitis,* 17, 182, 1987.
61. **Muñoz, D., Leanizbarrutia, I., Lobera, T. and de Corres, F.,** Anaphylaxis from contact with carrot, *Contact Dermatitis,* 13, 345, 1985.
62. **Schewach-Millet, M. and Shpiro, D.,** Urticaria and angioedema due to topically applied chloramphenicol ointment, *Arch. Dermatol.,* 121, 587, 1985.
63. **Aaronson, C. M.,** Generalized urticaria from sensitivity to nifuroxime, *J. Am. Med. Assoc.,* 210, 557, 1969.
64. **Rudzki, E. and Rebandel, P.,** Occupational contact urticaria from penicillin, *Contact Dermatitis,* 13, 192, 1985.
65. **Grob, J. J., Pommier, G., Robaglia, A., Collet-Villette, A. M., and Bonerandi, J. J.,** Contact urticaria from rifamycin, *Contact Dermatitis,* 16, 284, 1987.
66. **Baes, H.,** Allergic contact dermatitis to virginiamycin, *Dermatologica (Basel),* 149, 231, 1974.
67. **Kleinhans, D. and Zwissler, H.,** Anaphylaktischer Schock nach Anwendung einer Benzocainhaltigen Salbe, *Z. Hautkr.,* 55, 945, 1980.
68. **Gottmann-Lückerath, I.,** Kontakturticaria nach Dermoxin®. Society Proceedings, *Dermatosen,* 30, 124, 1982.
69. **Valsecchi, R., Foiadelli, L., Reseghetti, A., and Cainelli, T.,** Generalized urticaria from DNCB, *Contact Dermatitis,* 14, 254, 1986.
70. **Van Hecke, E. and Santosa, S.,** Contact urticaria to DNCB, *Contact Dermatitis,* 12, 282, 1985.
71. **Lovell, C. R., Cronin, E., and Rhodes, E. L.,** Photocontact urticaria from chlorpromazine, *Contact Dermatitis,* 14, 290, 1986.
72. **Kassen, B. and Mitchell, J. C.,** Contact urticaria from a vitamin E preparation in two siblings, *Contact Dermatitis Newsl.,* 16, 482, 1974.
73. **Shelley, W. B., Shelley, E. D., and Ho, A. K. S.,** Cholinergic urticaria: acetylcholine-receptor-dependent immediate-type hypersensitivity reaction to copper, *Lancet,* 1, 843, 1983.
74. **Valsecchi, R. and Cainelli, T.,** Contact urticaria from nickel, *Contact Dermatitis,* 17, 187, 1987.
75. **Lahti, A.,** Contact urticaria to plants, *Clin. Dermatol.,* 4(2), 127, 1986.
76. **Tanaka, T., Moriwaki, S., and Horio, T.,** Occupational dermatitis with simultaneous immediate and delayed allergy to chrysanthemum, *Contact Dermatitis,* 16, 152, 1987.
77. **Dooms-Goossens, A., Deveylder, H., Duron, C., Dooms, M., and Degreef, H.,** Airborne contact urticaria due to cinchona, *Contact Dermatitis,* 15, 258, 1986.
78. **Rivers, J. K. and Rycroft, R. J. G.,** Occupational allergic contact urticaria from colophony, *Contact Dermatitis,* 17, 181, 1987.
79. **Fisher, A. A.,** Contact urticaria and anaphylactoid reaction due to corn starch surgical glove powder, *Contact Dermatitis,* 16, 224, 1987.

80. **Assalve, D., Cicioni, P., Perno, P., and Lisi, P.,** Contact urticaria and anaphylactoid reaction from cornstarch surgical glove powder, *Contact Dermatitis,* 19, 61, 1988.
81. **La Rosa, M., Crea, G. F., Di Francesco, S., Di Paola, M., and Castiglione, N.,** Fennel allergy: case report, in Abstr. 3rd Int. Symp. Immunological and Clinical Problems of Food Allergy, Taormina, Giardini Naxos, Italy October 1 to 4, 1986.
82. **Apted, J.,** Acute contact urticaria from *Grevillea juniperina, Contact Dermatitis,* 18, 126, 1988.
83. **Apted, J.,** Acute contact urticaria from *Hakea suaveolens, Contact Dermatitis,* 18, 126, 1988.
84. **Steinman, H. K., Lovell, C. R., and Cronin, E.,** Immediate-type hypersensitivity to *Crataegus monogyna* (hawthorn), *Contact Dermatitis,* 11, 321, 1984.
85. **Van der Meeren, H. L. M. and van Erp, P. E. J.,** Life-threatening contact urticaria from glove powder, *Contact Dermatitis,* 14, 190, 1986.
86. **Axelsson, J. G. K., Johansson, S. G. O., and Wrangsjö, K.,** IgE-mediated anaphylactoid reactions to rubber, *Allergy,* 42, 46, 1987.
87. **Lahti, A.,** Contact urticaria and respiratory symptoms from tulips and lilies, *Contact Dermatitis,* 14, 317, 1986.
88. **Picardo, M., Rovina, R., Cristaudo, A., Cannistraci, C., and Santucci, B.,** Contact urticaria from Tilia (lime), *Contact Dermatitis,* 19, 72, 1988.
89. **Santucci, B., Cristaudo, A., and Picardo, M.,** Contact urticaria from papain in a soft lens solution, *Contact Dermatitis,* 12, 233, 1985.
90. **Edwards, E. K. and Edwards, E. K.,** Contact urticaria provoked by pickels, *Cutis,* 33, 230, 1984.
91. **Niinimäki, A.,** Scratch-chamber tests in food handler dermatitis, *Contact Dermatitis,* 16, 11, 1987.
92. **Grattan, C. E. H. and Harman, R. R. M.,** Contact urticaria to strawberry, *Contact Dermatitis,* 13, 191, 1985.
93. **Tosti, A., Melino, M., and Veronesi, S.,** Contact urticaria to tobacco, *Contact Dermatitis,* 16, 225, 1987.
94. **Lovell, C. R. and Rycroft, R. J. G.,** Contact urticaria from winged bean *(Psophocarpus tetragonolobus), Contact Dermatitis,* 10, 310, 1984.
95. **Nethercott, J. R., Lawrence, M. J., Roy, A.-M., and Gibson, B. L.,** Airborne contact urticaria due to sodium benzoate in a pharmaceutical manufacturing plant, *J. Occup. Med.,* 26, 734, 1984.
96. **Bergqvist-Karlsson, A.,** Delayed and immediate-type hypersensitivity to chlorhexidine, *Contact Dermatitis,* 18, 84, 1988.
97. **Nishioka, K., Doi, T., and Katayama, I.,** Histamine release in contact urticaria, *Contact Dermatitis,* 11, 191, 1984.
98. **Goncalo, M., Gongalo, S., and Moreno, A.,** Immediate and delayed sensitivity to chlorocresol, *Contact Dermatitis,* 17, 46, 1987.
99. **Löwenfeld, W.,** Überempfindlichkeit gegen Iodthion mit gleichzeitiger urtikarieller reaction, *Derm. Wochenschr.,* 78, 502, 1928.
100. **Lindskov, R.,** Contact urticaria to formaldehyde, *Contact Dermatitis,* 8, 333, 1982.
101. **Francois, A., Henin, P., Carli Basset, C., and Ginies, G.,** Anaphylactic shock following applications of Milian's solution, *Bull. Soc. Fr. Dermatol. Syphiligr.,* 77, 834, 1970.
102. **Tachibana, S., Horio, T., and Hayakawa, M.,** Contact urticaria and dermatitis due to fluocinonide cream, *Acta Dermatol. Kyoto Engl. Ed.,* 72, 141, 1977.
103. **Böttger, E. M., Mücke, C., and Tronnier, H.,** Kontaktdermatitis auf neuere Antikykotika und Kontakturtikaria, *Akta Dermatol.,* 7, 70, 1981.
104. **Tuer, W. F., James, W. D., and Summers, R. J.,** Contact urticaria to *O*-phenyl-phenate, *Ann. Allergy,* 56, 19, 1986.
105. **Henry, J. C., Tschen, E. H., and Becker, L. E.,** Contact urticaria to parabens, *Arch. Dermatol.,* 115, 1231, 1979.
106. **Guill, A., Goette, K., Knight, C. G., Peck, C. C., and Lupton, G. P.,** Erythema multiforme and urticaria, *Arch. Dermatol.,* 115, 742, 1979.
107. **Sterry, W. and Schmoll, M.,** Contact urticaria and dermatitis from self-adhesive pads, *Contact Dermatitis,* 13, 284, 1985.
108. **Rilliet, A., Hunziker, N., and Brun, R.,** Alcohol contact urticaria syndrome (immediate type hypersensitivity), *Dermatologica (Basel),* 161, 361, 1980.
109. **Jolanki, R., Estlander, T., and Kanerva, L.,** Occupational contact dermatitis and contact urticaria caused by epoxy resins, *Acta Derm. Venereol.,* 134, 90, 1987.
110. **Varigos, G. A. and Nurse, D. S.,** Contact urticaria from methyl ethyl ketone, *Contact Dermatitis,* 15, 259, 1986.
111. **Goodfield, M. J. D. and Saihan, E. M.,** Contact urticaria to naphtha present in a solvent, *Contact Dermatitis,* 18, 187, 1988.
112. **Camarasa, J. G.,** Contact urticaria to naphthylacetic acid, *Contact Dermatitis,* 14, 113, 1986.
113. **Dooms-Goossens, A., Duron, C., Loncke, J., and Degreef, H.,** Contact urticaria due to nylon, *Contact Dermatitis,* 14, 63, 1986.

114. **Edwards, E. K. and Edwards, E. K.,** Contact urticaria and allergic contact dermatitis caused by paraphenylenediamine, *Cutis,* 34, 87, 1984.
115. **Temesvari, E.,** Contact urticaria from paraphenylenediamine, *Contact Dermatitis,* 11, 125, 1984.
116. **Payero, M. L. P., Correcher, B. L., and Garcia-Perez, A.,** Contact urticaria and dermatitis from phosphorous sesquisulphide, *Contact Dermatitis,* 13, 126, 1985.
117. **Tosti, A., Bettoli, V., Iannini, G., and Forlani, L.,** Contact urticaria from polypropylene, *Contact Dermatitis,* 15, 51, 1986.
118. **Blair, H. and Parish, W. E.,** Asthma and urticaria induced by seminal plasma in a woman with IgE antibody and T-lymphocyte responsiveness to a seminal plasma antigen, *Clin. Allergy,* 15, 117, 1985.
119. **Ducombs, G., Lamy, M., Michel, M., Pradinaud, R., Jamet, P., Vincendeau, P., Maleville, J., and Texier, L.,** La papillonite de Guyane Francaise. Étude clinique et épidémiologique, *Ann. Dermatol. Venereol.,* 110, 809, 1983.
120. **Edwards, E. K., Edwards, E. K., and Kowalczyk, A. P.,** Contact urticaria and allergic contact dermatitis to saddleback caterpillar with histologic correlation, *Int. J. Dermatol.,* 25, 467, 1986.
121. **Wilkin, J. K. and Fortner, G.,** Ethnic contact urticaria to alcohol, *Contact Dermatitis,* 12, 118, 1985.
122. **Freitas, J. P. and Brandão, F. M.,** Contact urticaria to chlorocresol, *Contact Dermatitis,* 15, 252, 1986.
123. **White, I. R. and Cronin, E.,** Irritant contact urticaria to diethyl fumarate, *Contact Dermatitis,* 10, 315, 1984.
124. **Archer, C. B. and Cronin, E.,** Contact urticaria induced by pyridine carboxaldehyde, *Contact Dermatitis,* 15, 308, 1986.
125. **Hannuksela, A., Lahti, A., and Hannuksela, M.,** Non-immunologic immediate contact reactions to three isomers of pyridine carboxaldehyde, in *Current Topics in Contact Dermatitis,* Frosch, P., Ed., Springer-Verlag, Heidelberg, 1989, 448.

Chapter 3

THE RELATIONSHIP OF PRIMARY SKIN IRRITATION AND pK_a IN MAN

Bret Berner, Donald R. Wilson, Richard H. Guy, Gerard C. Mazzenga, Frank H. Clarke, and Howard I. Maibach

TABLE OF CONTENTS

I. INTRODUCTION

A. BACKGROUND

An understanding of the complex relationship between chemical structure and skin irritation would aid in the selection of drugs and excipients for use in transdermal or topical systems and in the occupational setting. As an initial step toward this understanding, a study in man of the role of pK_a in skin irritation is reported.

Skin irritation is a complex phenomenon consisting of solution properties of the vehicle, transport phenomena, biological response, and clearance. Any attempt to elucidate the variables important to the biological response must carefully control or measure the solution and transport properties and the clearance. The biological response of skin irritation as observed by erythema and edema[1] seems to have multiple mechanisms, both nonspecific and specific in nature.[2] To isolate nonphysiological pH as a cause of skin irritation is a gross oversimplification, but it is a usefule variable to investigate first. Furthermore, the local pH changes caused by a drug permeating the skin may again be related to the skin flux, the partitioning, the pK_a, and the clearance of the compound.

Studies of the effects of pH on skin irritation are discussed in References 3 to 6. A pH variation from 4 to 8 in a hydrocortisone cream had no effect on skin irritation in man.[3] On the other hand, intracutaneous injections in mice at varying pH values and buffer capacities produced some irritation at pH 4 and 10 and clear-cut irritation at pH 3 and 11.[4-6] Aliphatic amines produced greater skin irritation in rabbits than aromatic amines, suggesting that the pK_a of the compound influences skin irritation.[7]

B. STUDY DESIGN

To isolate pK_a as a variable, a homologous series of four compounds with a wide range of pK_a values and comparable melting points, molecular weights, and partition coefficients was selected to ascertain the irritation in man. The criteria of comparable *in vitro* melting points and partition coefficients are predictive of the oil and water solubilities of the compound,[8] and the melting point, molecular weight, and partition coefficient are predictive of the skin permeation of the compound.[9] Thus, for this series the skin permeation and solubilities are fixed, and pK_a should be the dominant independent variable influencing skin irritation. Although the nitrophenols have the widest pK_a range,[10] this series has not been tested in man. Certain derivatives of benzoic acids form a homologous series found to approximate the following criteria: literature pK_a 2.9 to 8.1, partition coefficients 20 to 170, predicted skin permeation 30 to 80 mcg/cm^2/h, melting points ranging from 103 to 158°C, well-established LD_{50},[11] and a history of use in man.[12] Furthermore, the pK_a range selected varies on the acid end from pHs which should be nonirritating to irritating.[4-6] The greatest concentration of pK_a values for this series occurs where irritation should become striking. The four test compounds were salicylic acid, salicylamide, *m*-nitrobenzoic acid, and *m*-anisic (3-methoxybenzoic) acid.

In the skin irritation study in man, monolithic reservoir devices of 1 cm^2 area containing the test compound as an infinite reservoir at unit activity were attached onto the scapular region of the upper back for 24 h. Erythema, edema, laser Doppler velocimetric (LDV) recordings, primary irritation index (PII), and color measurements were studied.

II. MATERIALS AND METHODS

A. MATERIALS

1. Test Compounds

Salicylic acid (Fisher, Springfield, NJ), salicylamide (Eastman, Rochester, NY), *m*-nitrobenozic acid (Eastman) and *m*-anisic acid (Aldrich, Milwaukee, WI) were used as obtained. No extraneous peaks were observed in the high-pressure liquid chromatography (HPLC) methods described below.

2. Hydroxyethyl Methacrylate Copolymer Hydrogels

A stock 65/35 mixture (see Section II.B.3) was obtained (CIBA-GEIGY). Hydroxyethyl methacrylate (Sipomer, HEMA) and *t*-butylperoctoate initiator were obtained courtesy of Dr. K. F. Mueller, CIBA-GEIGY, Ardsley, NY.

B. METHODS FOR PHYSICAL CHARACTERIZATION OF TEST COMPOUNDS AND DEVICES

1. Partitioning and pK_a

The octanol-water partition coefficient, K, and pK_a of each of the compounds were measured using differential titration and nonlinear curve-fitting. This method is detailed elsewhere.[13]

2. Analytical

a. HPLC Assays

Instrument:	HPLC, Waters (Milford, MA) ''WISP'' Model 710 with Waters Model 530 programmable solvent pump.
Detector:	Kratos (Paramus, NJ) ''Spectroflow'' Model 783.
HPLC Method	
Column:	Waters Nova-pak C-18, (3.9 mm × 15 cm, 4 μm).
Mobile phase:	35% methanol, 65% sodium acetate (Fisher, Springfield, N.J.) buffer, 0.005 *M* adjusted to appropriate pH with acetic acid.
UV detector:	230 nm
Flow rate:	1.0 ml/min
Injection volume:	20 μl
Range:	1 to 200 μg/ml

Compound	pH of Mobile phase	Retention time (min)
Salicylic Acid	4	1.9
Salicylamide	5	3.6
m-Nitrobenzoic acid	4	2.7
m-Anisic acid	5	2.4

3. Hydrogel Polymerization

The hydrogels were prepared by dilution of the 65/35 monomer/polymer mixture with HEMA monomer to give an 80:20 mixture. After degassing and addition of 0.20% *t*-butylperoctoate, films were polymerized by heating at 100°C for 1 h in polyester-lined glass molds. The films were soaked in water for 24 h, punched into 1 cm^2 discs, and vacuum dried. These discs were Soxhlet extracted in ethanol for 48 h and again vacuum dried.

Three methods were used to assess the quality control of the hydrogel discs: gas chromatography, swelling, and physical inspection. (1) Dried hydrogel discs (5 g) were Soxhlet extracted in methylene chloride for 4 h. The methylene chloride extract was reduced to 1 ml volume and analyzed for residual HEMA monomer, butylated hydroxytoluene (BHT), and monoethylhydroquinone (MEHQ) by gas chromatography. All three components were not detected at 1 ppm sensitivity. (2) The degree of swelling of the hydrogel discs was checked in water and ethanol-water (7:3 v/v) and found to be 42 and 144%, respectively. The degree of swelling is expressed as precent solvent weight gain per dry weight of polymer. (3) Only discs without visual defects between 33 and 37 mg dry weight were used.

TABLE 1
Physical Properties of the Test Compounds

	Salicyclic acid	Salicylamide	*m*-Nitrobenzoic acid	*m*-Anisic acid
Melting point (°C)	158	140	142	103
Molecular weight	138.1	137.1	167.1	152.1
Log K (octanol-water)				
Experimental	2.22	1.31	1.79	1.90
Literature	2.21—2.26	1.25—1.28	1.83	2.02
pK_a				
Experimental	2.95	8.10	3.56	3.99
Literature	2.97	8.66 (35°C)	3.45	4.09
Water solubility (mg/ml)	2.5	0.23	3.9	0.18
pH of saturated solution	2.41	5.91	2.57	3.06
Ethanol-water solubility (7:3 v/v) at 38°C (mg/ml)	273	133	555	263

Note: Literature values for K and pK_a are from References 12, 16, and 17.

4. Hydrogel Loading with Test Compound

Hydrogel discs were soaked for 24 h with ethanol-water (7:3 v/v) saturated at 38°C with the test compound. The solubilities of the four test compounds are in Table 1. Upon removal from the loading solution, the disc was quickly rinsed (approximately 2 s) in ethanol-water (7:3 v/v) at 38°C to remove surface test compound. The discs were blotted and air dried for 24 h. Since some of the test compounds sublime, vacuum drying was not appropriate. Immediately prior to use, the loaded discs were soaked for 2 h in a saturated aqueous solution of the test compound at room temperature. The loading was determined by differences in the weight of the membrane disc before and after loading. These observations were checked by extracting the hydrogels into a known volume of ethanol-water (7:3 v/v).

5. Solubility

Solubility measurements for solvents were performed by rotating capped centrifuge tubes of the saturated solutions at the appropriate temperature for 24 h. Duplicate vials were prepared, and duplicate samples from each vial were analyzed by HPLC.

6. *In Vitro* Release Studies

In vitro release from the hydrogel discs was determined using a U.S.P. basket method dissolution apparatus rotating at 100 rpm. Each basket contained 400 ml HPLC-grade water pre-equilibrated to 37°C.

7. Skin Permeation

Male human skin was obtained frozen from skin banks. The samples had been cascade frozen and stored at liquid nitrogen temperatures at the skin bank. These samples were stored at 90°C until the epidermis was separated.

The epidermis was peeled away from the dermis after exposure to 60°C water for 80.[14] The isolated epidermis was rinsed rapidly with hexane to remove surface lipid, rinsed with water, and placed on aluminum foil.

Modified static skin diffusion cells maintained at 32°C were used.[15] The epidermis was mounted on the lower receptor compartment by floating it on the receiver solution (0.01% aqueous gentamicin). Either a 1-ml reservoir of saturated solutions or the loaded hydrogel discs were the donor phase. Data were analyzed by a linear least-squares fit to the steady-state region of the cumulative amount penetrated vs. time curve.

TABLE 2
Experimental and Theoretical Skin Permeation

Compound	Flux ($\mu g/cm^2/h$)		Time lag (h)		Flux from hydrogel discs ($\mu g/cm^2/h$)
	Experiment	Theory	Experiment	Theory	
Salciylic acid	88 ± 10	30	2.6 ± 0.3	1.3	130 ± 50
Salicylamide	15 ± 4	45	3.5 ± 1.4	1.3	14 ± 6
m-Nitrobenzoic acid	60 ± 20	30	4.1 ± 1.6	2.0	80 ± 27
m-Anisic acid	75 ± 2	80	3.1 ± 0.7	1.6	49 ± 30

C. METHODS TO STUDY HUMAN SKIN IRRITATION

1. Subjects

A total of 16 female subjects ranging from 35 to 45 years of age consented to the study; 10 were of hispanic origin possessing various degrees of olive-complected skin, 5 were Caucasian with type II skin, and 1 was a Caucasian/Polynesian mixture with a slightly olive skin tone. All subjects had reasonably clear backs with no or few comedones, moles, or freckles.

Preliminary skin irritation studies on rabbits were performed, and none of the compounds were found to be severe irritants.

2. Procedure

Day 1: the subjects acclimated for 0.5 h, during which time the sites were marked on the back with a felt tip marker. The scapular regions of the upper back unoccluded by undergarments were tested. A column of five test sites was marked on each side, avoiding the midline. This accomodated two replicates per subject for each of the four compounds and a distilled water control. The position of each treatment within a side of the back was randomized. Baseline scoring and measurements were taken. The relative humidity and temperature were noted. The 12 loaded test discs were applied to the skin occluded with 3% ethylene vinyl acetate (EVA) membrane and secured with tape (Scanpore Norgeplaster 6, Oslo, Norway).

Day 2: the patches were removed and the sites marked again; 30 min later, the measurements were repeated. Erythema and edema were scored by standard 0 to 4 visual scales,[1] with an additional scale at 0.5 to signify borderline reaction. LDV (LDV MedPacific LD 5000, Seattle, WA)-assessed blood flow and color (reflected light color analyzer, Minolta® Chroma Meter CR-100) were measured.

Day 3: the procedure of day 2 was repeated.

PII was calculated as the average of the sum of the erythema and edema scores for the two time points.

III. RESULTS

A. PHYSICAL CHARACTERIZATION AND SKIN PERMEATION

The selected relevant physical properties for skin permeation and skin irritation are summarized in Table 1. Experimentally determined pK_a and K values are in general agreement with literature values.[12,16,17]

Skin permeation rates from saturated solutions fall in the range 15 to 88 $\mu g/cm^2/h$; lag times for the establishment of steady state are on the order of 3 h for all compounds (Table 2). The observed steady-state fluxes and time lags agree with the theoretical predictions for these chemicals within a three- to fourfold limitation of the structure-skin permeation model based on melting point, molecular weight, and partition coefficient.[9] Thus, this model[9] served its purpose — to obtain acids with comparable skin permeation — but the predicted

TABLE 3
Dissolution of Test Compounds from Hydrogel Systems at 37°C

Compound	Slope[a] (mg/ min$^{1/2}$)	Loading dose (mg/ml)	Percent uptake per dry weight
Salicylic acid	4.9 ± 1.1	403	67.5 ± 5.9
Salicylamide	1.9 ± 0.2	268	36.6 ± 5.9
m-Nitrobenzoic acid	2.39 ± 0.04	505	101.9 ± 6.6
m-Anisic acid	2.8 ± 0.1	336	50.6 ± 8.1

[a] The slope is the slope of the initial portion of an amount released vs. square root of time plot.

fluxes were no better than an estimate. *In vitro* skin permeation from the loaded hydrogel systems was comparable to skin permeation from saturated solutions (Table 2).

The loading and *in vitro* release of these test compounds from the hydrogel discs were assessed (Table 3). The loading from ethanol-water (7:3) was extremely large and varied from 36 to 102% of the polymer dry weight. The loading was sufficiently large that the diameter of the dried, loaded systems greatly exceeded the diameter of the dried, unloaded systems. The uniformity of the loading was generally excellent ($\leq \pm 5\%$). Dissolution of the test compounds (Table 3) from the discs was rapid (50% release in 30 to 60 min) compared to skin transport (Table 2).

B. THE SKIN IRRITATION STUDY

There were no statistically significant differences between the left and right sides; consequently, the results were pooled.

The means for each treatment for the various measures of skin irritation and times are summarized in Table 4. pK_a was correlated ($p < 0.05$ for linear correlation model and possibly better for a nonlinear model) with both erythema and edema at 24 h and with the primary irritation index. The correlations of pK_a with the 48-h measurements and the instrumental measurements were not statistically significant. In Figure 1, the average measurements for each treatment are plotted against pK_a, and a trend of a steep increase in irritation for $pK_a < 4$ is evident. Treatment with salicylamide, which has a pK_a of 8.1, was not significantly different from a water control in any of the observables. In Table 4, the statistically significant different pairwise groups ($p < 0.05$ by pairwise *t*-tests) for each measurement are presented. Erythema at 24 h appeared to be the most sensitive method of detecting differences between pK_a treatments, with edema at 24 h a close second. None of the instrumental parameters were particularly sensitive to differences in pK_a, and the 48-h measurements were somewhat less sensitive than the 24-h measurements.

All of the measurements of irritation were strongly correlated with each other ($p < 0.001$ for a linear regression except for 48-h color with edema and LDV, for which $p < 0.02$). In Figure 2, the averages for the treatments at a given time are plotted against each other for the different time points. Erythema and edema appeared strongly correlated. Color measurements were relatively insensitive, but strongly correlated with the other parameters. LDV was insensitive to small changes in erythema. While the instrumental parameters are strongly correlated with the subjective grading, the ability to predict a given subjective score is not accurate and varies greatly among subjects.

TABLE 4
Mean Irritation Scores for Different Chemicals*

	24-h scores			
Applied chemical	**Erythema**	**Edema**	**Color**	**LDV**
Salicylic acid	1.0 ± 0.9 (a)	0.8 ± 0.9 (a)	7.3 ± 2.8 (a)	150 ± 140 (a)
m-Nitrobenzoic acid	0.5 ± 0.8 (b)	0.5 ± 0.8 (a)	5.8 ± 2.3 (b)	80 ± 110 (b)
m-Anisic acid	0.3 ± 0.5 (b)	0.1 ± 0.3 (b)	5.8 ± 2.1 (b)	60 ± 50 (b)
Salicylamide	0.05 ± 0.15 (c)	0.03 ± 0.12 (c)	5.3 ± 2.2 (b)	44 ± 15 (b)
Water control	0.03 ± 0.18 (c)	0.03 ± 0.18 (c)	5.1 ± 1.5 (b)	44 ± 13 (b)

	48-h scores				
Applied chemical	**Erythema**	**Edema**	**Color**	**LDV**	**PII**
Salicylic acid	0.7 ± 0.8 (a)	0.6 ± 0.8 (a)	7.2 ± 3.2 (a)	150 ± 170 (a)	1.5 ± 1.4 (a)
m-Nitrobenzoic acid	0.6 ± 0.7 (a)	0.4 ± 0.6 (a)	6.4 ± 2.6 (a)	130 ± 150 (a)	1.0 ± 1.3 (a)
m-Anisic acid	0.016 ± 0.09 (b)	0 ± 0 (b)	4.9 ± 1.6 (b)	45 ± 13 (b)	0.2 ± 0.4 (b)
Salicylamide	0.08 ± 0.22 (b)	0 ± 0 (b)	4.8 ± 1.6 (b)	43 ± 11 (b)	0.08 ± 0.18 (b)
Water control	0.06 ± 0.21 (b)	0 ± 0 (b)	4.7 ± 1.4 (b)	42 ± 12 (b)	0.06 ± 0.24 (b)

Note: LDV = laser doppler velocimetry. PII (primary irritation index) is the average of the 24- and 48-h scores for the sums of erythema and edema.

* The letters in each column represent the statistically significant pairwise groups ($p < 0.05$ in *t*-tests) for each column. That is, within a column, treatments marked "a" are statistically different than those marked "b", which are different from "c", etc.

IV. DISCUSSION

While skin irritation is a complex biological response related to solution and transport properties, partitioning, and clearance, pK_a appears predictive of the degree of skin irritation in this one homologous series. For the range of oil and water solubilities and skin permeation of these compounds, structures with $pK_a < 4.1$ appear to induce skin irritation. The degree of irritation increases rapidly for smaller pK_a values. These results agree with subcutaneous injection studies in mice[4-6] and the study of vehicle pH in hydrocortisone cream.[3]

The authors have recently observed that for a nonhomologous series of basic compounds, where the concentrations have been adjusted to achieve comparable skin permeation (15 to 70 μg/cm²/h), pK_a and irritation are strongly correlated, and for conjugate acids with $pK_a \geq 8$, skin irritation becomes significant (Figure 3). To predict skin irritation adequately from a chemical structure, flux across skin vs. pK_a vs. skin irritation curves would need to be constructed from approximately 35 to 45 acids and bases. This might be the minimal variable approach to prediction. Melting point, molecular weight, partition coefficient, and pK_a also may be an appropriate set of variables.[9]

The effect of pK_a on skin irritation is presumably due to pH effects at the epidermal level and is consistent with a cellular damage mechanism of skin irritation. Such pH-related

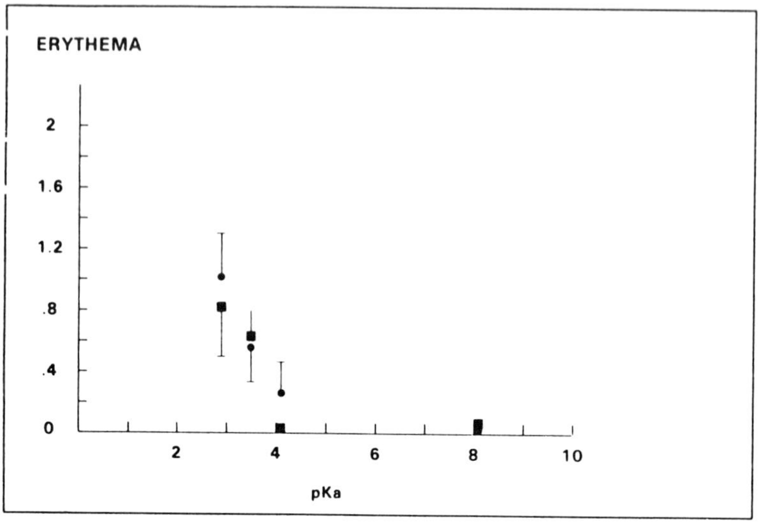

A

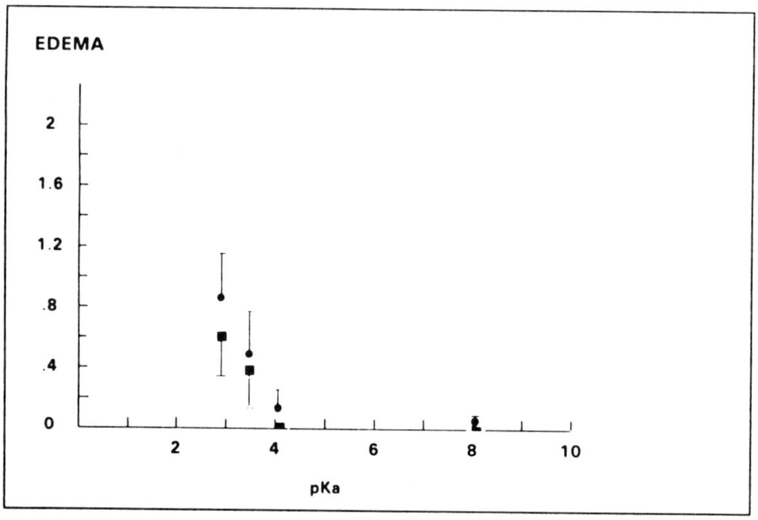

B

FIGURE 1. Measurements of skin irritation vs. pK_a of the compounds as observed at 24 (●) and 48 h (■). Error bars represent 95% confidence limits. The measurements of skin irritation used are (A) erythema, (B) edema, (C) primary irritation index, (D) LDV, and (E) color meter.

effects at the epidermal level could result from two potential sources of H^+: (1) different activities of H^+ in the solutions applied to the skin, and these dissociated ions must cross the stratum corneum; or (2) intact drug which crosses the stratum corneum and then dissociates into ions in the epidermis. In the experiment reported here, the hydrogen ion would have to traverse the stratum corneum with the only available source of counterion, the ionized chemical. Ionized species transport across stratum corneum is sufficiently slow that this mechanism does not apply. Therefore, the local dissociation of the chemical in the epidermis must predominate.

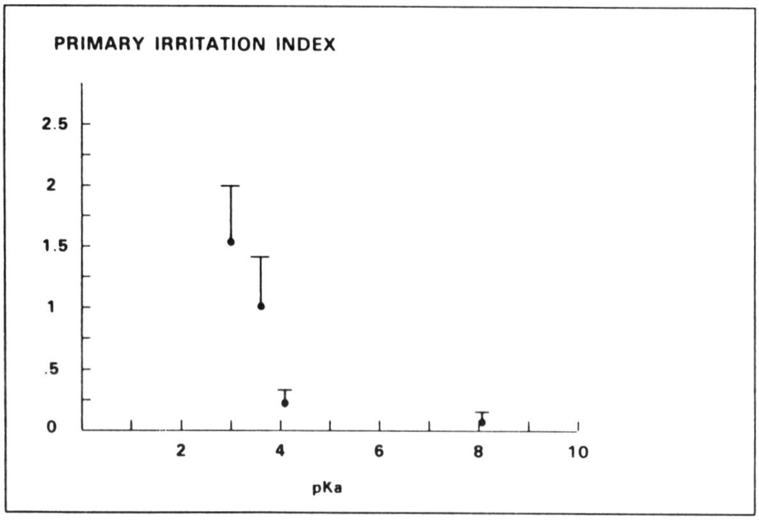

FIGURE 1C.

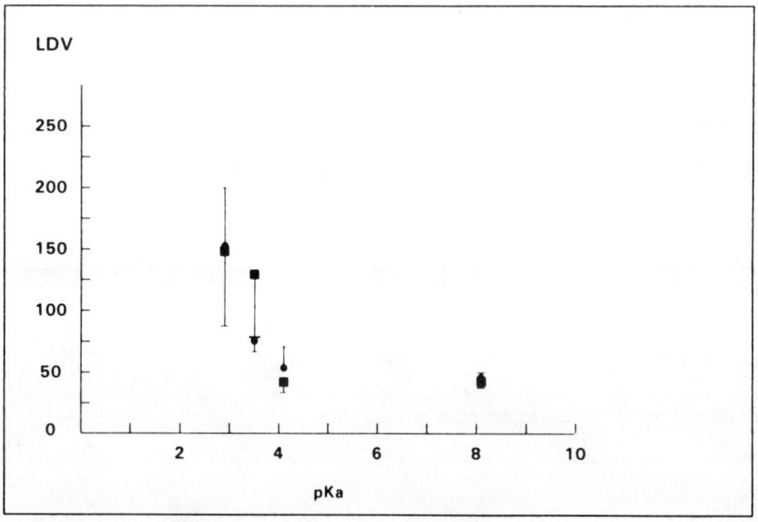

FIGURE 1D.

In a more recent experiment, the skin irritation from a pH 11.8 sodium hydroxide solution, a pH 11.8 imipramine solution, and water were compared. The observed mean erythema scores at 24 h were 0.5 ± 0.9, 0.5 ± 0.6, and 2.1 ± 1.0, respectively ($n = 16$). While the sodium hydroxide treatment is not statistically different from the water control, the imipramine solution is statistically different ($p < 0.001$) from the other two treatments. It appears that the surface pH effect is minor in comparison.

V. CONCLUSION

All of the parameters used to assess skin irritation are highly linearly correlated, and any one might be used. Erythema at 24 h appears to be the most sensitive of these parameters.

pK_a is highly correlated with skin irritation for this homologous series of acids. In more

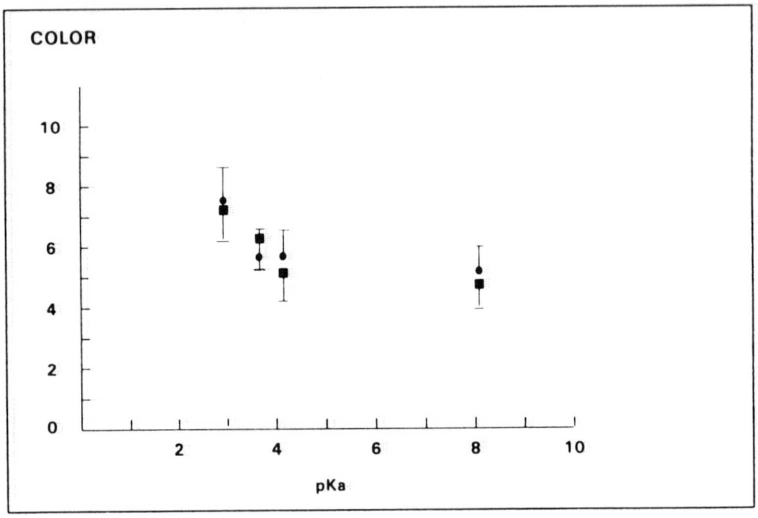

FIGURE 1E.

recent work by the authors, the skin irritation of a series of bases with conjugate acids of varying pK_a but comparable skin permeation was investigated, and pK_a was also highly correlated ($p < 0.001$) with irritation for this second series, which was not homologous (unpublished data). Characterizing the relationship among flux, pK_a, and skin irritation is certainly justified.

ACKNOWLEDGMENT

R.H.G. is the recipient of a special Emphasis Research Career Award (K 01-OH-00017) from the National Institute of Occupational Safety and Health. He is supported in part by NIH Grant GM-33395.

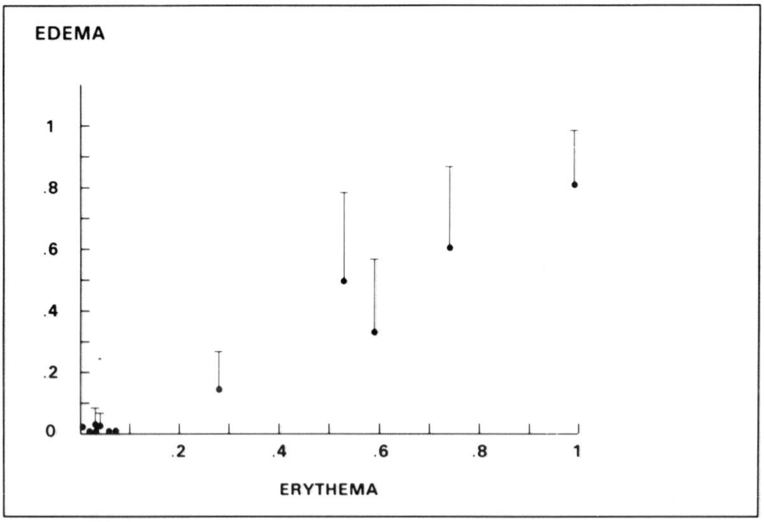

A

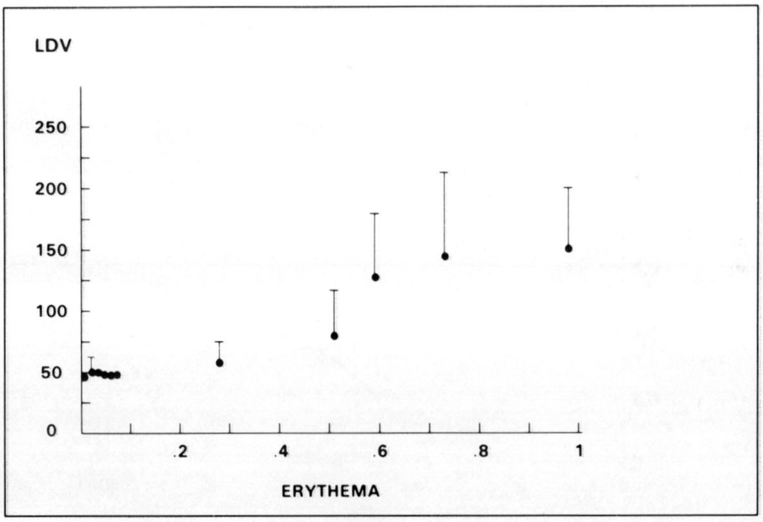

B

FIGURE 2. The correlation of (A) edema, (B) LDV, and (C) color meter with erythema for the compounds at both time points.

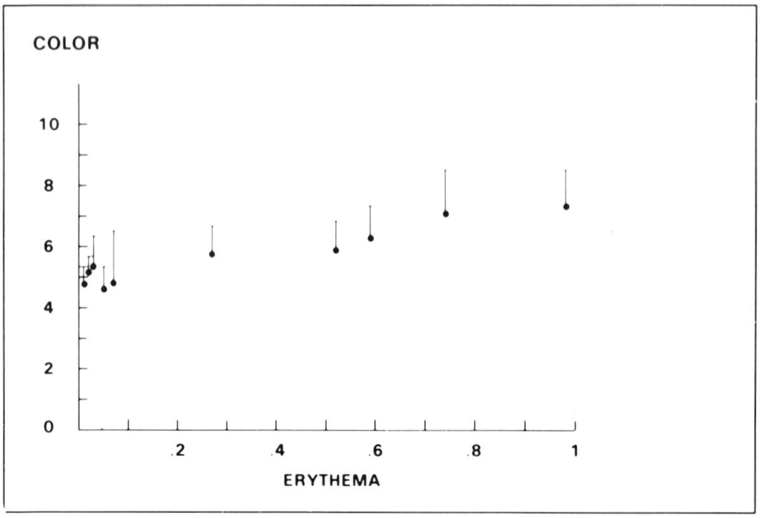

FIGURE 2C.

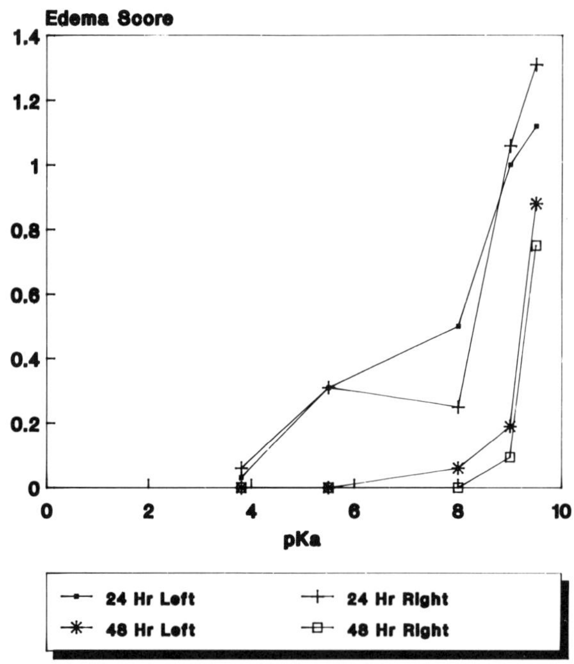

FIGURE 3. Skin edema vs. pKa of the conjugate acid for the series of bases (8-aminoquinoline, benzamidizole, nicotine, norephedrine, and imipramine) as observed at 24 and 48 h on the left and right back of 16 female subjects.

REFERENCES

1. **Draize, J. H., Woodard, G., and Calvary, H. O.,** Methods for the study of irritation and toxicity of substances applied topically to the skin and mucous membranes, *J. Pharmacol. Exp. Ther.*, 82, 377, 1944.
2. **Patrick, E.,** Investigations of the Mechanisms of Chemically Induced Skin Irritation, Ph.D. dissertation #8509113, University Microfilms Int., Ann Arbor, MI, 1984.
3. **Koch, V. E., Loebich, F., and Frosch, P. J.,** Die Hautverträngigkeit von corticoid-externa in Abhängigkeit rom pH-wart der Emulsionen, *Derm. Beruf Umwelt,* 31, 122, 1983.
4. **Bucher, K., Bucher, K. E., and Walz, D.,** Irritant actions of unphysiological pH values. A controlled procedure to test for topical irritancy, *Agents Actions,* 9, 124, 1979.
5. **Bucher, K., Bucher, E. E., and Walz, D.,** The topically irritant substance: essentials — bio — tests — predictions, *Agents Actions,* 11, 515, 1981.
6. **Walz, D. and Bucher, K. E.,** The quantitative assessment of topical irritancy and its application to unphysiological pH values, *Agents Actions,* 12, 552, 1982.
7. **Lunter T.,** Draize Rabbit Skin Irritation Data Base, Release 1.0, Health Designs, Inc., Rochester, NY, 1985.
8. **Yalkowsky, S. H.,** Solubility and solubilization of nonelectrolytes, in *Techniques of Solubilization of Drugs,* Yalkowsky, S. H., Eds., Marcel Dekker, New York, 1981, 1.
9. **Berner, B. and Cooper, E. R.,** Penetration enhancers, in *The Transdermal Delivery of Drugs,* Kydonieus, A. F. and Berner, B., Eds., CRC Press, Boca Raton, FL, 1987, 41.
10. **Martin, Y. C.,** The quantitative relationships between pK_a, ionization, and drug potency: utility of model-based equations, in *Physical Chemical Properties of Drugs,* Yalkowsky, S. H., Sinkula, A. A., and Valvani, S. C., Eds., Marcel Dekker, New York, 1980, 49.
11. *Registry of Toxic Effects of Chemical Substances,* April 1985.
12. *The Merck Index,* 10th ed. Merck and Company, Rahway NJ, 1983, 97, 596, 945, 1199, 1200.
13. **Clarke, F. H.,** Ionization constants by curve-fitting: application to the determination of partition coefficients, *J. Pharm. Sci.*, 73, 226, 1984.
14. **Kligman, A. M. and Christopher, E.,** Preparation of isolatged sheets of human stratum corneum, *Arch. Dermatol.*, 88, 702, 1963.
15. **Merritt, E. W. and Cooper, E. R.,** Diffusion apparatus for skin penetration, *J. Contr. Rel.*, 1, 161, 1984.
16. **Fasman, G. D., Ed.,** *Handbook of Biochemistry and Molecular Biology,* Vol. 1, 3rd ed., CRC Press, Boca Raton, FL, 1976, 176.
17. **Hansch, C. and Albert, L.,** *Substituent Constants for Correlation Analysis in Chemistry and Biology,* John Wiley & Sons, New York, 1979, 212.

Chapter 4

IMMEDIATE-TYPE CONTACT REACTION: IMMUNOLOGIC CONTACT URTICARIA

Bradford R. Katchen and Howard I. Maibach

TABLE OF CONTENTS

I. DEFINITION

Contact urticaria is classically defined as a wheal and flare reaction which appears within minutes to 1 hour after agents contact the skin and disappears within hours. In 1975, Maibach and Johnson[1] coined the term "contact urticaria syndrome" to define a response with a broad spectrum of local and systemic clinical manifestations. It may extend beyond the contact site to include rhinoconjunctivitis, gastrointestinal dysfunction, asthma, dyspnea, and anaphylactic shock.

This syndrome can be subdivided into nonimmunologic contact urticaria and immunologic contact urticaria according to its pathogenesis. Nonimmunologic contact urticaria is a more frequent phenomena. It can occur in individuals not previously sensitized to the contactant. The mechanism of action is the result of a direct release of vasoactive substances which causes a localized response (urticaria) and has a benign outcome. Immunologic contact urticaria has been documented less frequently. It occurs in those previously exposed to a specific antigen. When it makes contacts with the skin, an IgE-mediated release of vasoactive substances results in a localized or generalized response which potentially can be fatal. A staging of it effects has been proposed by Maibach and Johnson[1] and van Krogh and Maibach.[2] Stage 1 includes localized urticaria without distant lesions or systemic symptoms. A wheal and flare, or simply erythema (itching, tingling, or burning), is experienced. Stage 2 involves generalized urticaria which may include angioedema. Stage 3 includes urticaria with extracutaneous reactions of asthma, rhinoconjunctivitis, and gastrointestinal or orolaryngeal dysfunction. Stage 4 is urticaria with anaphylaxis (Table 1).

II. INCIDENCE

There has been little epidemiologic evaluation of the incidence of immunologic or nonimmunologic contact urticaria. Elpern[3-5] interviewed 1020 patients in his Hawaii dermatology practice; 26% noted that topical substances elicited immediate symptoms including pruritis, dermatitis, local and generalized urticaria, asthma, orolaryngeal edema, rhinoconjunctivitis, gastrointestinal symptoms, headache, and anaphylactic reactions. There was no statistical difference in race and age between the second through the eighth decade. Of 261 substances, those that most frequently caused symptoms were 40 plants, 42 fruits, vegetables, and nuts, 20 of animal origin, and 14 animal products. He coined the term "syndrome of immediate reactivities".

van Krogh and Maibach[2] surveyed 86 contact urticaria patients and found that 12% suggested an immunologic mechanism; 15% demonstrated extracutaneous, immediate-type reactions, including rhinoconjunctivitis, asthma, orolaryngeal symptoms, dysphagia, and swelling of the oral mucosa.

III. PATHOGENESIS

As illustrated in Figure 1, when an antigen is applied to the skin, two IgE antibodies adhere via the F_{ab} region of its variable region. These antibodies are attached to the mast cells via the F_c receptor. This causes the mast cells to degranulate and release their granules, which contain histamine, neutrophil chemotactic factor, eosinophil chemotactic factor, platelet activating factor, leukotrienes, and prostaglandins.[6,7] These in turn produce the wheal and flare response as well as extracutaneous manifestations. It is a type 1 hypersensitivity reaction of the Coombs and Gel classification.[8]

Furthermore, many (but not all) antigens that cause immunologic contact urticaria tend to be haptens (low-molecular-weight proteins). They must first be bound to a large protein

TABLE 1
The Contact Urticaria Syndrome: Staging by Symptomatology[78]

Cutaneous reactions only
- Stage 1
  - Localized urticaria
  - Dermatitis
  - Nonspecific symptoms (itching, tingling, burning, etc.)
- Stage 2
  - Generalized urticaria

Extracutaneous reactions
- Stage 3
  - Bronchial asthma
  - Rhinoconjunctivitis
  - Orolaryngeal symptoms
  - Gastrointestinal symptoms
- Stage 4
  - Anaphylactoid reactions

From von Krogh, G. and Maibach, H. I., in Dermatology, Vol. 1, 2nd ed., Moschella, S. and Hurley, H., Eds., W. B. Saunders, Philadelphia, 1985, 323. With permission.

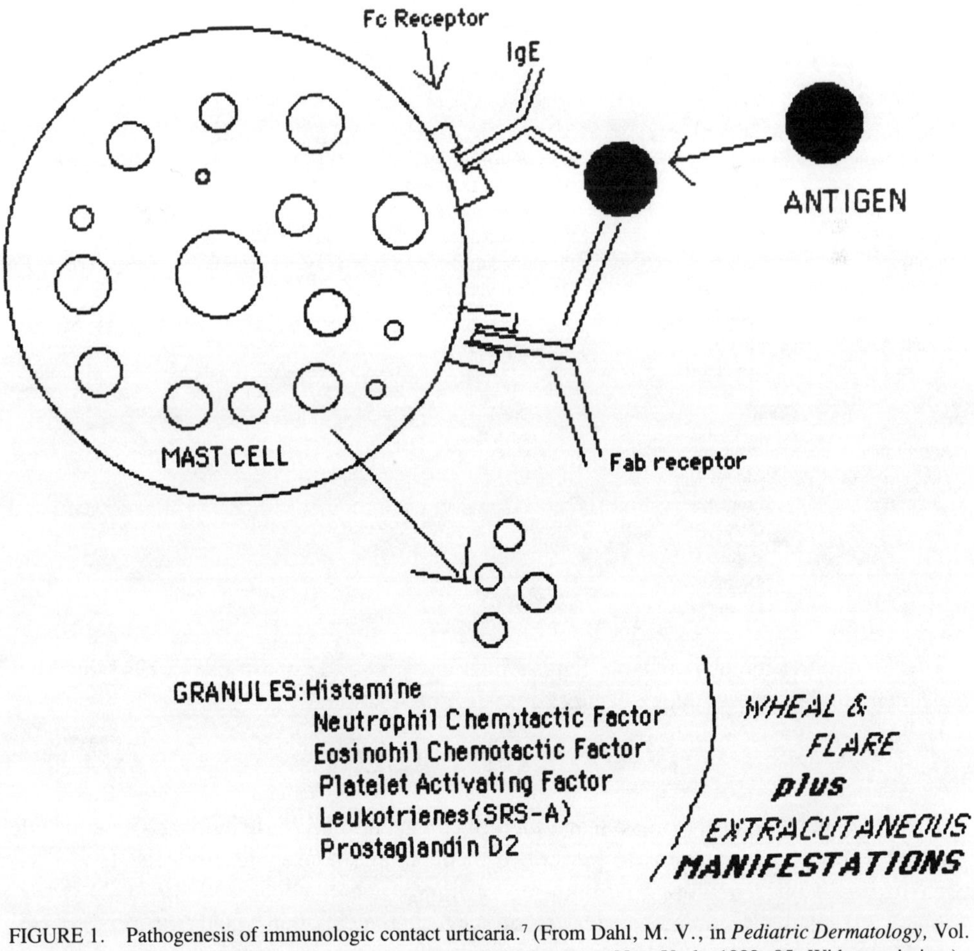

FIGURE 1. Pathogenesis of immunologic contact urticaria.[7] (From Dahl, M. V., in *Pediatric Dermatology,* Vol. 1, Schachner, L. A. and Hanson, R. C., Eds., Churchill Livingstone, New York, 1988, 85. With permission.)

TABLE 2
Immunologic Contact Urticaria

Occupational	Food	Medicaments	Biologic	Plants
Henna	Handler or consumer flour	Neomycin	Seminal fluid	Tulips
p-Phenylenediamine	Vegetables carrots	Bacitracin	Animal byproducts	Lillies
Cow placenta	Meats	Chlorhexidine	Human sweat	Chrysanthemums
Cattle dander	Lamb	Mechlorethamine hydrochloride		
Animal byproducts	Turkey	Lindane		
Rat	Wheat	Aminophenazone		
Mouse	Garlic	Cephalosporin		
Cat	Onion	Streptomycin		
Dog	Potato	Diethyl toluemide		
Guinea pig	Milk	Dinitrochlorobenzene		
Mouse urine	Fish	*o*-Phenylphenate		
Safety matches	Egg	Penicillin		
Metals	Fresh fruits	Chlorpromazine		
Platinum	Peach	Papain		
Rubber	Apple	Aminophenazine		
Latex	Tomato	Formaldehyde		
Wood	Endive	Merthiolate		
Carbonless carbon paper	Celery	Aminopyrine		
Glove powder	Parsnip	Pentachlorophenate		
Tobacco	Strawberry	Chloramphenicol		
Cinchona	Caraway seed	Gentamycin		
Castor beans	Wine	Streptomycin		
Colophony	Alcohol	Phenylmercuric propionate		
Inhalants	Spices	Sulfur dioxide		
Platinum	Fish	Sodium sulfide		
Anhydrides	Cod	Monoamylamine		
Diisocyanates	Anchovy			
Cobalt dust	Mullet			
Western red cedar *(Thuja plicata)*	Pilchard			
Plicatic acid	Watermelon			
Chromium	Honeydew melon			
Silk	Cantaloupe			
	Pineapple			
	Grape			
	Bananas			
	Oranges			
	Green peas			

(i.e., albumin) to become antigenic. Landsteimer and Chase[9] demonstrated in 1941 that low-molecular-weight haptens alone cannot cause IgE production.[9]

IV. CAUSATIVE AGENTS

The literature contains increasing numbers of contact urticaria case reports. These include patients with adverse reactions to prescribed medications, household products, foods, antigens exposed in their work environment, or simply agents encountered in daily living. Table 2 is a relatively comprehensive list of suggested agents that putatively cause immunologic contact urticaria.

A large percentage of cases are occupationally acquired. Hair dyes such as henna and *p*-phenylenediamine have caused hairdressers as well as their clients to become anaphylactic.[10-13] Veterinarians, lab technicians, and animal care takers, if previously sensitized, can develop contact urticaria to specific animal by-products including hair, urine, saliva, placenta, and milk.[14-17] This phenomenon has been provoked by simply touching a rat's tail. Numerous reports concern people allergic to rubber, latex, and glove powder.[18-32] Those affected include physicians, nurses, dentists, housecleaners, and others who use gloves to protect their hands during work. Also, those who come into contact with others who wear the gloves (i.e., patients examined by the gynecologist and dentist) have been affected. Other rubber products such as cofferdams (used by the dentist), condoms, balloons, baby pacifiers, and elastic waistbands have elicited allergic reactions. Furthermore, food handlers have been especially noted to develop hand dermatitis and subsequent itching, burning, and tingling after contact with certain meat, fruit, or vegetable products.[33-43]

There exists an arsenal of topical medications prescribed daily to patients. Infrequent but serious manifestations have occurred after certain individuals have used preparations to which they have developed a sensitivity. The most common include neomycin, bacitracin, and chlorhexidine,[44-58] but there remain others which have been reported less often. Furthermore, there are certain plants[59-63] as well as biological products (such as seminal fluid[64-69]) which have caused anaphylaxis when contact with the skin is made.

The contact urticaria syndrome also can be applied to those substances that contact the orolaryngeal and gastrointestinal mucosa, as well as the respiratory tract, that subsequently elicit cutaneous and extracutaneous manifestations. For example, consumed food products may cause itching, tingling, edema, and irritation of the lip, mouth, and tongue, as well as hoarseness, abdominal disturbances, rhinitis, and angioedema.[70-77] A variety of inhalants, especially those related to industry (including platinum, isocyanates, and anhydrides) have elicited similar symptoms, plus asthma.[79-101]

Of all the agents proposed to cause immunologic contact urticaria, the pathogenesis of only a few has been proven (Table 3). The deciding factor is whether an IgE mechanism can be demonstrated. This can be performed in a few manners, either by a specific radioallergosorbent test (RAST) or by the passive transfer test (explained in Section VI). Each of the specific antigens is categorized according to the clinical symptoms and the stage of the contact urticaria syndrome.

V. DIAGNOSIS

A difficult dilemma for a dermatologist or emergency physician is to determine the etiology of an urticarial or anaphylactoid reaction, since the causative agents can be limitless. A complete history must be obtained from the patient. This should include any complaints about immediate reactions after applying lotions, creams, cosmetics, and medicines. It is essential to elicit information regarding familial or personal history of atopy, bronchial asthma, rhinoconjunctivitis, orolaryngeal or gastrointestinal dysfunction, development of wheal and flare, and angioedema with wheezing, hypotension, or syncope. Subsequently, the physical exam is usually normal, but localized wheals, angioedema, or extracutaneous manifestations may be observed.

When there is a suggestion of a causative agent, a variety of tests may be instituted[1,2,78,116,117] (Table 4). In the open test, approximately 0.1 ml of the suspected chemical is applied with a glass rod to the ventral aspect of a forearm. This area is observed for 60 min for erythema, wheal, and flare. A closed patch test can be performed by applying the agent to the patch and then adhering it to various sites on the body. The patch should be removed after 15 to 20 min and the site observed. Unlike the standard patch test, when evaluation time is at 48 and 72 h, an evaluation for immunologic contact urticaria should be made within 60 min,

TABLE 3
Proven Immunologic Contact Urticaria

Agent	Reaction	Test	Ref.
Diethyltoluamide	LU	PT, 48/80	1
Platinum	GU,RC,BA,AR	RAST, intradermal, PT	83,87,99
Nickel	BA	RAST, IgG specific	94
Latex	RC,GU,BA	indirect ELISA, prick, patch, epicut, hist rel, RAST	18,21
Rubber	AR,GU,BA	PT, RAST, patch, scratch	20,23,26
Phenylmercuric propionate	BA,GU,RC	Scratch, inhalation, PT, hist rel	108
Alcohol	LU	Patch, PT with scratch	109
Seminal plasma protein	GU,AR,OL	Intradermal, hist rel, PT, RAST inhibition	64,65,67,68
Rat tail, mouse urine, cat saliva	RC,BA,LU	Scratch, RAST	14
Cow placenta	GU,LU,	RAST, prick	17
Streptomycin	GU,BA,AR	PT	50
p-Phenylenediamine	RC,GU,BA,AR	Patch, PT, prick	12
o-Phenylphenate	LU	Patch, PT	102
Chlorpromazine	OL,RC,BA	PT, patch	55,107
Formaldehyde	RC,GU,BA,OL,AR	Scratch, RAST, patch	110
Chlorhexidine	GU,AR	PT, RAST, open, scratch hist rel	44,54,56-58
Bacitracin	GU,OL,BA,AR	PT, patch, prick	46,49,51,53
Carrot, celery	GU,RC	RAST, scratch	41
Fish	GU,OL	RAST, prick, patch	36,39
Potato	RC,OL,BA	Scratch, intradermal PT, RAST, hist rel	33,74
Peach, apple, pear, pineapple, plum, cantaloupe, watermelon, grape	OL,GU	Scratch, PT	77

Note: LU (local urticaria), GU (generalized urticaria), OL (orolaryngeal), RC (rhinoconjunctivitis), BA (bronchial asthma), AR (anaphylactic reaction), PT (passive transfer); PT (prick test), RAST (radioallergosorbent test), ELISA (enzyme-linked immunosorbent assay), hist rel (histamine release)

TABLE 4
Tests for Causation and Pathogenesis of Immediate Contact Urticaria

Prove causation	Prove pathogenesis
Open test	RAST
Closed patch	Passive transfer (Prausnitz-Küstner)
Scratch	Passive Cutaneous Anaphylaxis
Intradermal	Immunodiffusion
	Compound 48/80
	Chlorpheniramine
	Atropine

the typical time frame for the reaction. A chamber test is similarly used. Here the vehicle for the agent is a clear plastic chamber with cotton or filter paper soaked with the agent, taped to the skin, and left in place for 30 to 45 min. Scratch and prick tests and intradermal tests have traditionally been used by allergists for determination of a patient's immune status to a variety of allergens. Here the skin is scratched or pricked with a needle, and antigen is either applied topically or a small amount is injected intradermally. The utilization of prick, scratch, and intradermal testing requires highly trained personnel and adequate controls to rule out any false-positive (irritant) responses. Such disturbance of the skin may lead to false positives as well as potentially fatal outcomes. This is a more systemic administration of a substance that may produce anaphylaxis. In increasing order of morbidity, open $<$ closed and patch $<$ prick $<$ scratch $<$ intradermal. It is important to test normal, uninvolved skin, then slightly affected skin if the previous test is negative (Figure 2). Many of the eliciting agents, being large protein molecules, will pass through damaged skin more easily. However, it is often difficult to assess a wheal and flare on the damaged skin. Furthermore, sites of the body differ in regard to reactivity. The upper back, extensor sides of arms, antecubital fossa, and forearm are most sensitive.

It is often helpful if the patient is advised to bring from home the agent believed to cause urticaria and demonstrate its use to observe a reaction within 1 h. The patient should be requested to bring in all agents that induce burning, stinging, and itching. Providing a convenient container such as an ice cube tray simplifies the task for the patient. Commercial antigens either lack active ingredients or lose antigenicity in processing or storage.

A. PRECAUTIONS

If the patient has extracutaneous symptoms of asthma, laryngeal edema, or anaphylaxis, then the patient should be tested in a facility with full resuscitative facilities readily available. When extracutaneous symptoms are present, do serial dilutions, one at a time, starting with as little as 1×10^7 to avoid the induction of anaphylaxis, asthma, or laryngeal edema.

VI. PATHOGENETIC DETERMINATION

Once the causative agent has been determined, the mechanism of its action can be researched to identify whether it is immunologic contact urticaria. An excellent test for this is the RAST.[120] The patient's serum is applied to the specific antigen-coated disc, and labeled anti-human IgE is added. The result is the identification of antigen-specific IgE. To verify this, the serum can be heated to 56°C and retested. Heating deactivates this reaginic antibody, whether it is IgE or IgG4. Therefore, a negative result with heat confirms a previous positive. A problem in the performance of this test is that often the causative agents are haptens; hence, it is difficult to form the coated disc needed for the RAST. An alternate test is the Prausnitz-Küstner test (passive transfer).[6] The patient's serum is injected intradermally into a volunteer or animal who is subsequently challenged with a patch or intradermal injection 24 h later at the initial injection site. A wheal and flare response is proof of a reaginic antibody, IgE, or IgG4 in the patient. This test is not practical in man in our present climate of easily transmissible diseases for which questionable blood screening still exists. A similar test, the passive cutaneous anaphylaxis test, consists of injecting serum intradermally into an animal, followed 48 h later with an intravenous injection of the antigen in question plus a blue dye. A blue spot developing at the site of the initial injection verifies the presence of a reaginic antibody.

Other tests include immunodiffusion, whereby the agglutination of a patient's serum and antigen in separate wells of an Ouchterlony plate verifies an antigen-specific antibody.

To test the presence of a histaminic reaction, compound 48/80, which depletes mast cells of histamine, can be injected intradermally into the ventral forearm at 8-h intervals for 1 day and then patched at this site. A wheal and flare from a patch site not previously treated with compound 48/80 and a negative response from the treated site are indications of histamine. Similar injections may be made with blocking agents, including chlorpheniramine or atropine, to identify histamine- or atropine-dependent mechanisms. Finally, when patch testing, the blood of the limb treated can be assayed for histamine, complement, prostaglandins, leukotrienes, and chemotactic factors.

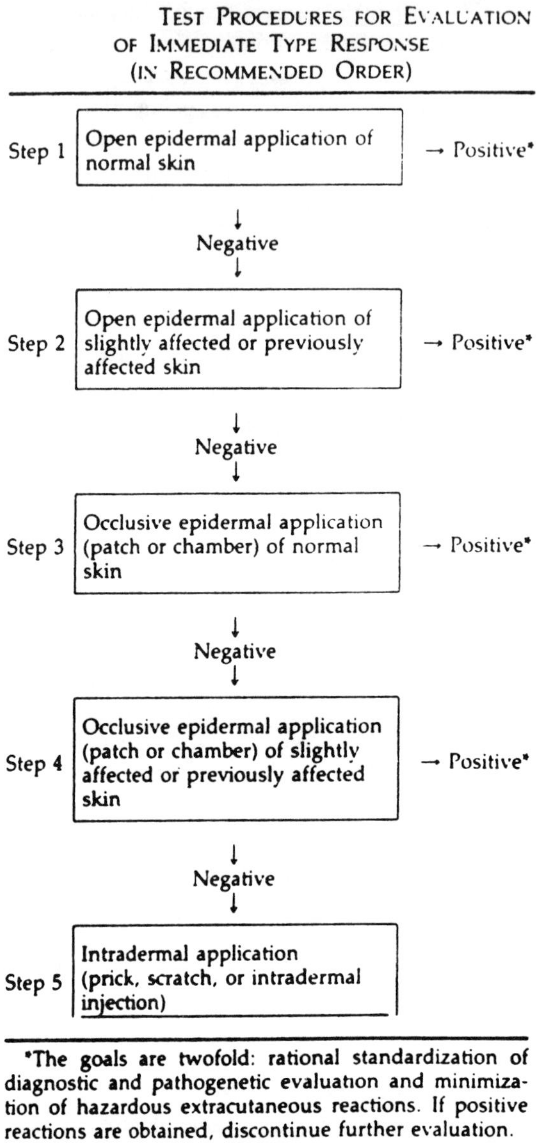

FIGURE 2. Test procedures for evaluation of immediate-type response (in recommended order). (From von Krogh, G. and Maibach, H. I., in *Dermatology*, Vol. 1, 2nd ed., Samuel, M. and Hurley, H., Eds., W. B. Saunders, Philadelphia, 1985, 323. With permission.)

VII. ASSOCIATION WITH BARRIER FUNCTION

Cases of immunologic contact urticaria frequently occur in patients with previously affected skin or eczematous skin. It can be hypothesized that this impaired barrier may allow more penetration and, thus, increase the susceptibility to contact urticaria. In support, Maibach[111] tested patients allergic to turkey, lamb, and flour. Those with intact normal skin had a negative patch; however, those with slightly affected skin demonstrated a positive wheal and flare. Hjorth and Roed-Peterson[36] tested kitchen workers with chronic eczema with complaints of immediate itching, erythema, and urticaria within 10 to 30 min of contact with food. The correlation of clinical symptoms to causative food products were verified with patch and scratch tests and RAST. They concluded that irritants may have started the dermatitis and facilitated the penetration of proteins. Contact with certain proteins caused the urticaria which facilitated the development of irritant or allergic contact dermatitis or worsening of preexisting atopic dermatitis. Finally, preexisting eczema predisposes to immediate skin reactions because of penetration of certain chemicals.

VIII. ASSOCIATION WITH ATOPIC DERMATITIS

The literature suggests a correlation between development of immunologic contact urticaria and history of atopy. In 1968 Malten[112] coined the term "eczema hybrids", referring to dermatitis caused by concurrent contact and atopic sensitivities. It has been proposed that atopic dermatitis might be aggravated by contact with a chemical whose penetration through the epidermis is facilitated by a coexistent contact dermatitis. Wraith[40] studied 119 patients with atopic dermatitis who developed immediate systemic reactions after contact with foods. Of 27 patients with symptoms that correlated with a positive RAST to food allergens, 20 developed swelling and irritation of the mouth and tongue and vomiting with contact. A total of 56% of his population had a positive RAST to food and symptoms of asthma and eczema. Furthermore, all of his patients with verified IgE, specifically to fish and nuts, had a clinical correlation of immediate symptoms. Finally, Wrangsjo[32] compared patients' allergy to rubber to their atopic history; 25 of 30 patients were atopics, and 15 of 30 had a preexisting hand dermatitis.[32]

In Elpern's retrospective study of 1020 patients with contact urticaria, atopy was found twice as often in patients with a history of the contact urticaria syndrome than those without this background.[3-5]

IX. PROPOSED ANIMAL MODEL

In order to test the potential of certain formulations to cause immunologic contact urticaria an animal model has been needed, and to date none exists. Lahti and Maibach,[119] in 1984, developed one for nonimmunologic contact urticaria. The animal used is the Hartley guinea pig. They are topically challenged with 50 μl on both sides of the ear. The ear thickness is measured with an engineer's micrometer, and each ear is examined for evidence of erythema or urticarial lesions.

A modification of this method is being proposed for the model for immunologic contact urticaria.[121] The Hartley guinea pig is sensitized by an intraperitoneal injection of an adjuvant (alum or carageenan) plus an antigen. A topical preparation of the antigen is applied to each ear on day 14. The ear thickness is measured with an engineer's micrometer, and each ear is examined for evidence of erythema or urticarial lesions. Blood is drawn on the same day for the passive cutaneous anaphylaxis test. In the original experiments, seminal plasma protein was used as the tested antigen; 50% of the guinea pigs produced the reaginic antibody, and 25% displayed an increased ear erythema.

X. CONCLUSION

Immunologic contact urticaria is reported with increasing frequency, presumably because of physician awareness. Diagnosing the etiology provides dramatic relief for the patient; this is especially important when laryngeal edema, asthma, food intolerance (gastrointestinal), and anaphylaxis are included.

REFERENCES

1. **Maibach, H. I. and Johnson, H.,** Contact urticaria syndrome: contact urticaria to diethyltoluamide (immediate-type hypersensitivity), *Arch. Dermatol.,* 111, 726, 1975.
2. **van Krogh, G. and Maibach, H. I.,** The contact urticaria syndrome — an updated review, *J. Am. Acad. Dermatol.,* 5, 328, 1981.
3. **Elpern, D. J.,** The syndrome of immediate reactivities (contact urticaria syndrome). An historical study from a dermatology practice. I. Age, sex, race and putative substances, *Hawaii Med. J.,* 44(11), 426, 1985.
4. **Elpern, D. J.,** The syndrome of immediate reactivities (contact urticaria syndrome). An historical study from a dermatology practice. II. The atopic diathesis and drug reactions, *Hawaii Med. J.,* 44(12), 10, 466, 1985.
5. **Elpern, D. J.,** The syndrome of immediate reactivities (contact urticaria syndrome). An historical study from a dermatology practice. III. General discussion and conclusion, *Hawaii Med. J.,* 45(1), 1986.
6. **Dahl, M. V.,** *Clinical Immunodermatology,* Year Book Medical Publishers, Chicago, 1988.
7. **Dahl, M. V.,** Basic immunology, in *Pediatric Dermatology,* Vol. 1, Schachner, L. A. and Hanson, R. C., Eds., Churchill Livingstone, New York, 1988, 85.
8. **Coombs, R. R. A. and Gell, P. G. H.,** Classification of allergic reactions responsible for clinical hypersensitivity and disease, in *Clinical Aspects of Immunology,* Gell, P. G. H. and Coombs, R. R. A., Eds., Blackwell Scientific, Oxford, 1968, 575.
9. **Landsteimer, K.,** *The Specificity of Serological Reactions,* Harvard University Press, Cambridge, MA, 1945.
10. **Cronin, E.,** Immediate-type hypersensitivity to henna, *Contact Dermatitis,* 5(3), 198, 1979.
11. **Temesvari, E.,** Contact urticaria from paraphenylenediamine, *Contact Dermatitis,* 11, 125, 1984.
12. **Goldberg, B. J., Herman, F. F., and Hirata, I.,** Systemic anaphylaxis due to an oxidatin product of *p*-phenylenediamine in a hair dye, *Ann. Allergy,* 58, 205, 1987.
13. **Alanko, K., Keskinen, H., Björkstén, F., and Ojansen, S.,** Immediate-type hypersensitivity to reactive dyes, *Clin. Allergy,* 8, 25, 1978.
14. **Agrup, G. and Sjostedt, L.,** Contact urticaria in laboratory technicians working with animals, *Acta Derm. Venereol.,* 65, 111, 1985.
15. **Botham, P. A., Davies, G. E., and Teasdale, E. L.,** Allergy to laboratory animals: a prospective study of its incidence and of the influence of atopy on its development, *Br. J. Ind. Med.,* 44, 627, 1987.
16. **Epstein, S.,** Milker's eczema: an analysis of forty-two cases, *J. Allergy,* 19, 333, 1948.
17. **Schmidt, H.,** Contact urticaria, *Contact Dermatitis,* 4, 230, 1978.
18. **Turjanmaa, K.,** Incidence of immediate allergy to latex gloves in hospital personnel, *Contact Dermatitis,* 17, 270, 1987.
19. **Kopman, A. and Hannuksela, M.,** Kumin aiheuttama kosketusurtikari, *Duodecim,* 99, 221, 1983.
20. **Axelsson, J. G. K., Johansson, S. G. O., and Wrangsjo, K.,** IgE-mediated anaphylactoid reactions to rubber, *Allergy,* 42, 46, 1987.
21. **Carrillo, T., Cuevas, M., Munos, T., Hinojosa, M., and Moneo, I.,** Contact urticaria and rhinitis from latex surgical gloves, *Contact Dermatitis,* 15, 69, 1986.
22. **Turjanmaa, K., Reunala, T., Tuimala, R., and Karkkainen, T.,** Severe IgE mediated allergy to surgical gloves, in Proc. 15th Nordic Congr. Allergology, Turku, 1984, *Allergy (Suppl),* 235 (Abstr.), 1984.
23. **Estlander, T., Jolanki, P., and Kanerva, L.,** Dermatitis and urticaria from rubber and plastic gloves, *Contact Dermatitis,* 14, 20, 1986.
24. **Fisher, A. A.,** Contact urticaria and anaphylactoid reaction due to cornstarch surgical glove powder, *Contact Dermatitis,* 16, 224, 1987.
25. **Forstrom, L.,** Contact urticaria from latex surgical gloves, *Contact Dermatitis,* 6, 33, 1980.
26. **Frosch, P. J., Wahl, R., Bahmer, F. A., and Maasch, H. J.,** Contact urticaria to rubber gloves is IgE-mediated, *Contact Dermatitis,* 14, 241, 1986.
27. **van Ketel, W. G.,** Contact urticaria from rubber gloves after dermatitis from thiurams, *Contact Dermatitis,* 11, 323, 1984.
28. **Kleinhans, D.,** Contact urticaria to rubber gloves, *Contact Dermatitis,* 10, 124, 1984.
29. **Meding, B. and Fregert, S.,** Contact urticaria from natural latex gloves, *Contact Dermatitis,* 10, 52, 1984.
30. **van der Meeren, H. L. M.,** Life threatening contact urticaria from glove powder, *Contact Dermatitis,* 14(3), 190, 1986.
31. **Nutter, A. F.,** Contact urticaria to rubber, *Br. J. Dermatol.,* 101, 597, 1979.
32. **Wrangsjo, K., Mellstrom, G., and Axelsson, G.,** Discomfort from rubber gloves indicating contact urticaria, *Contact Dermatitis,* 15, 79, 1986.
33. **Castells, M. C., Pascual, C., Esteban, M., and Ojeda, J. A.,** Allergy to white potato, *J. Allergy Clin. Immunol.,* 78, 1110, 1986.

34. **Hannulsela, M. and Lahti, A.,** Immediate reactions to fruits and vegetables, *Contact Dermatitis,* 3, 79, 1977.
35. **Herxheimer, H.,** Skin sensitivity to flour in bakers' apprentices, *Lancet,* 1, 83, 1967.
36. **Hjorth, N. and Roed-Peterson, J.,** Occupational protein contact dermatitis in food handlers, *Contact Dermatitis,* 2, 28, 1976.
37. **Kavli, G. and Moseng, D.,** Contact urticaria from mustard in fishstick production, *Contact Dermatitis,* 17, 153, 1987.
38. **Lahti, A. and Hannuksela, M.,** Hypersensitivity to apple and carrot can be reliably detected with fresh material, *Allergy,* 33, 143, 1978.
39. **Melino, M., Toni, F., and Riguzzi, G.,** Immunologic contact urticaria to fish, *Contact Dermatitis,* 17, 182, 1987.
40. **Wraith, D. G., Merrett, J., Roth, A., Yman, L., and Merrett, T. G.,** Recognition of food-allergic patients and their allergens by the RAST technique and clinical investigation, *Clin. Allergy,* 9, 25, 1979.
41. **Munoz, D., Leanizbarrutia, I., Lobera, T., and Fernandez de Corres, L.,** Anaphylaxis from contact with carrot, *Contact Dermatitis,* 13(5), 345, 1985.
42. **Pearson, R. S. B.,** Potato sensitivity, an occupational allergy in housewives, *Acta Allergol.,* 21, 507, 1966.
43. **Pigatto, P. D., Riva, F., Altomare, G. F., and Parotelli, R.,** Short-term anaphylactic antibodies in contact urticaria and generalized anaphylaxis to apple, *Contact Dermatitis,* 9, 511, 1983.
44. **Bergqvist-Karlsson, A.,** Delayed and immediate-type hypersensitivity to chlorhexidine, *Contact Dermatitis,* 18, 84, 1988.
45. **Bigby, M., Jick, S., Jick, H., and Arndt, K.,** Drug-induced cutaneous reactions: a report from the Boston Collaborative Drug Surveillance Program on 15,438 consecutive inpatients 1975-1982, *J. Am. Med. Assoc.,* 24, 3358, 1986.
46. **Comaish, J. S. and Cunliffe, W. J.,** Absorption of drugs from varicose ulcers: a cause of anaphylaxis, *Br. J. Clin. Pract.,* 21(2), 97, 1967.
47. **Grob, J. J., Pommier, G., Robaglia, A., Collet-Villette, M., and Bonerandi, J. J.,** Contact urticaria from rifamycin, *Contact Dermatitis,* 16, 284, 1987.
48. **Grunnet, E.,** Contact urticaria and anaphylactoid reaction induced by topical application of nitrogen mustard, *Br. J. Dermatol.,* 94, 101, 1976.
49. **Katz, B. E. and Fisher, A.,** Bacitracin: a unique topical antibiotic sensitizer, *J. Am. Acad. Dermatol.,* 17, 1016, 1987.
50. **Rosen, F.,** Sensitivity to streptomycin, *J. Am. Med. Assoc.,* 137, 1128, 1948.
51. **Roupe, G. and Strannegard, O.,** Anaphylactic shock elicited by topical administration of bacitracin, *Arch. Dermatol.,* 100, 450, 1969.
52. **Rudzki, E. and Rebandel, P.,** Occupational contact urticaria from penicillin, *Contact Dermatitis,* 13, 192, 1985.
53. **Schecter, J. F., Wilkinson, R. D., and Del Carpio, J.,** Anaphylaxis following the use of bacitracin ointment, *Arch. Dermatol.,* 120, 909, 1984.
54. **Okano, M., Masao, N., Nomura, M., Hata, S., Okada, N., Sato, K., Kitano, Y., Tashiro, M., Yoshimoto, Y., Hana, R., and Aoki, R.,** Anaphylactic symptoms due to chlorhexidine gluconate, *Arch. Dermatol.,* 125, 50, 1989.
55. **Horio, T.,** Chlorpromazine photoallergy: coexistence of immediate and delayed type, *Arch. Dermatol.,* 111, 1469, 1975.
56. **Lasthein, A. B. and Brandrup, F.,** Contact dermatitis from chlorhexidine, *Contact Dermatitis,* 13, 307, 1985.
57. **Nishioka, K., Doi, T., and Katayama, J.,** Histamine release in contact urticaria, *Contact Dermatitis,* 11, 191, 1984.
58. **Ohtoshi, T., Yamanchi, N., Tadokoro, K., Miyachi, S., Suzuki, S., Miyamoto, T., and Muranaka, M.,** IgE antibody mediated shock reaction caused by topical application of chlorhexidine, *Clin. Allergy,* 16, 155, 1986.
59. **Apted, J.,** Acute contact urticaria from *Grevillea juniperina, Contact Dermatitis,* 18, 126, 1988.
60. **Tanaka, T., Moriwaki, S., and Horio, T.,** Occupational dermatitis with simultaneous immediate and delayed allergy to chrysanthemum, *Contact Dermatitis,* 16, 152, 1987.
61. **Rivers, J. K. and Rycroft, R. J. G.,** Occupational allergic contact urticaria from colophony, *Contact Dermatitis,* 17, 181, 1987.
62. **Nakamura, T.,** Contact dermatitis to Japanese black pine, *Contact Dermatitis,* 14, 317, 1986.
63. **Lahti, A.,** Contact urticaria and respiratory symptoms from tulips and lilies, *Contact Dermatitis,* 14(5), 317, 1986.
64. **Frankland, A. W. and Parish, W. E.,** Anaphylactic sensitivity to human seminal fluid, *Clin. Allergy,* 4, 249, 1974.
65. **Freeman, S.,** Woman allergic to husband's sweat and semen, *Contact Dermatitis,* 14, 110, 1986.

66. **Halpern, B. N., Ky, T., and Robert, B.,** Clinical and immunological study of an exceptional case of reaginic type sensitization to human seminal fluid, *Immunology,* 12, 247, 1967.
67. **Levine, B., Siraganian, R. P., and Schenkein, I.,** Allergy to human seminal plasma, *N. Engl. J. Med.,* 288(17), 894, 1973.
68. **Mathias, C. G., Toby, F. O., Caldwell, T. M., Yunginger, J. W., and Maibach, H. I.,** Immediate hypersensitivity to seminal fluid and atopic dermatitis, *Arch. Dermatol.,* 116, 209, 1980.
69. **Mikkelsen, E. J., Henderson, L. L., Leiferman, K. M., and Gleich, G. J.,** Allergy to human seminal fluid, *Ann. Allergy,* 34, 239, 1975.
70. **Boso, E. B. and Brestel, E. P.,** Contact urticaria to cow milk, *Allergy,* 42, 151, 1987.
71. **Grattan, C. E. H. and Harman, R. R. M.,** Contact urticaria to strawberry, *Contact Dermatitis,* 13, 191, 1985.
72. **Halmepuro, L. and Lowenstein, H.,** Immunological investigation of possible structural similarities between pollen antigens and antigens in apple, carrot and celery tuber, *Allergy,* 40, 264, 1985.
73. **Lombardi, P., Camponmi, P., Giorgini, S., Spallanzani, P., and Sertoli, A.,** Contact urticaria from fish, honey and peach skin, *Contact Dermatitis,* 9, 422, 1983.
74. **Nater, J. P. and Zwartz, J. A.,** Atopic allergic reactions due to raw potato, *J. Allergy,* 40(4), 202, 1967.
75. **Rudzki, E. and Grzywa, Z.,** Contact urticaria from egg, *Contact Dermatitis,* 103, 1977.
76. **Salo, O. P., Makinen-Kiljunen, S., and Juntunen, K.,** Milk causes a rapid urticarial reaction on the skin of children with atopic dermatitis and milk allergy, *Acta Derm. Venereol.,* 66, 438, 1986.
77. **Tuft, L. and Blumstein, G. I.,** Studies in food allergy: sensitization to fresh fruits: clinical and experimental observations, *J. Allergy,* 13, 574, 1942.
78. **von Krogh, G. and Maibach, H. I.,** The contact urticaris syndrome and associated disease entities, in *Dermatology,* Vol. 1, 2nd ed., Moschella, S. and Hurley, H., Eds., W. B. Saunders, Philadelphia, 1985, 323.
79. **Bardana, E. J., Jr. and Andrach, R. H.,** Occupational asthma secondary to low molecular weight agents used in the plastic and resin industries, *Eur. J. Respir. Dis.,* 64, 241, 1983.
80. **Bernstein, D. I., Gallagher, J. S., D'Souza, L., and Bernstein, I. L.,** Heterogeneity of specific-IgE responses in workers sensitized to acid anhydride compounds, *J. Allergy Clin. Immunol.,* 74(6), 197, 1984.
81. **Bernstein, I. L.,** Isocyanate-induced pulmonary diseases: a current perspective, *J. Allergy Clin. Immunol.,* 70, 24, 1982.
82. **Chan-Yeung, M., Barton, G. M., MacLean, L., and Grzybowski, S.,** Occupational asthma and rhinitis due to western red cedar *(Thuja plicata), Am. Rev. Respir. Dis.,* 108, 1094, 1973.
83. **Cleare, M. J., Hughes, E. G., Jacoby, B., and Pepys, J.,** Immediate (type I) allergic responses to platinum compounds, *Clin. Allergy,* 6, 183, 1976.
84. **Cromwell, O., Pepys, J., Parish, E., and Hughes, E. G.,** Specific IgE antibodies to platinum salts in sensitized workers, *Clin. Allergy,* 9, 109, 1979.
85. **Dooms-Goossens, A., Deveylder, H., Duron, C., Dooms, M., and Degreef, H.,** Airborne contact urticaria due to cinchona, *Contact Dermatitis,* 15, 258, 1986.
86. **Dykewicz,** Occupational respiratory and dermatologic disease, *Primary Care,* 14(3), 559, 1987.
87. **Freedman, S. O. and Krupey, J.,** Respiratory allergy caused by platinum salts, *J. Allergy,* 42(4), 233, 1968.
88. **Howe, W., Venables, K. M., Topping, M. D., Dally, M. B., Hawkins, R., Law, S., and Newman, A. J.,** Tetrachlorophthalicanhydride asthma: evidence for specific IgE antibody, *J. Allergy Clin. Immunol.,* 71(1), 5, 1983.
89. **Korbee, L. and Bernstein, I. L.,** Workers risk occupational asthma by exposure to myriad of chemicals, *Occup. Health Safety,* 57(3), 30, 28, 1988.
90. **Lovell, C. R. and Rycroft, R. J. G.,** Contact urticaria from winged bean, *Contact Dermatitis,* 10, 314, 1984.
91. **Malo, J.-L., Cartier, A., Doepner, M., Nieborer, E., Evans, S., and Dolovich, J.,** Occupational asthma caused by nickel sulfate, *J. Allergy Clin. Immunol.,* 69(1), 55, 1982.
92. **Moller, D. R., Brooks, S. M., Bernstein, D. I., Cassedy, K., Enrione, M., and Bernstein, I. L.,** Delayed anaphylactoid reaction in a worker exposed to chromium, *J. Allergy Clin. Immunol.,* 77(3), 451, 1986.
93. **Moller, D. R., Gallagher, J. S., Bernstein, D. I., Wilcox, T. G., Burroughs, H. E., and Bernstein, I. L.,** Detection of IgE-mediated respiratory sensitization in workers exposed to hexahydrophthalic anhydride, *J. Allergy Clin. Immunol.,* 75(6), 663, 1985.
94. **Nieboer, E., Evans, S. L., and Dolovich, J.,** Occupational asthma from nickel sensitivity. II. Actors influencing the interaction of Ni^{2+}, HSA, and serum antibodies with nickel-related specificity, *Br. J. Ind. Med.,* 41, 56, 1984.
95. **Scheel, L. D., Killens, R., and Josephson, A.,** Immunochemical aspects of toluene diisocyanate toxicity, *Ind. Hyg. J.,* 25, 179, 1964.

96. **Zapp, J. A., Jr.,** Hazards of isocyanates in polyurethane foam plastic production, *Arch. Ind. Health,* 15, 324, 1957.
97. **Zeiss, C., Patterson, R., Pruzansky, R., Miller, J. J., Rosenberg, M., and Levitz, D. L.,** Trimellitic anhydride-induced airway syndromes: clinical and immunologic studies, *J. Allergy Clin. Immunol.,* 60(2), 96, 1977.
98. **Zeiss, C. R., Wolkonsky, P., Pruzansky, J. J., and Patterson, R.,** Clinical and immunologic evaluation of trimellitic anhydride workers in multiple industrial settings, *J. Allergy Clin. Immunol.,* 70(1), 15, 1982.
99. **Hunter, D., Milton, R., and Perry, K. M. A.,** Asthma caused by the complex salts of platinum, *Br. J. Ind. Med.,* 2, 92, 1945.
100. **Reitamo, S., Visa, K., Kahonen, K., Kayhko, K., Stubb, S., and Salo, O. P.,** Eczematous reactions in atopic patients caused by epicutaneous testing with inhalant allergens, *Br. J. Dermatol.,* 114, 303, 1986.
101. **Patterson, R., Zeiss, C. R., and Pruzansky, J. J.,** Immunology and immunopathology of trimellitic anhydride pulmonary reactions, *J. Allergy Clin. Immunol.,* 70(1), 19, 1982.
102. **Tuer, W. F., James, W. D., and Summers, R. J.,** Contact urticaria to *O*-phenylphenate, *Ann. Allergy,* 56, 19, 1986.
103. **Key, M. M.,** Some unusual allergic reactions in industry, *Arch. Dermatol.,* 83, 57, 1961.
104. **Marks, J. G., Jr., Trautlein, J. J., and Zwillich, C. W.,** Contact urticaria and airway obstruction from carbonless copy paper, *J. Am. Med. Assoc.,* 252(8), 1038, 1984.
105. **Burge, S. M. and Powell, S. M.,** Contact urticaria to phosphorous to sesquisulphide, *Contact Dermatitis,* 9, 424, 1983.
106. **Valsecchi, R. and Cainelli, T.,** Contact urticaria from nickel, *Contact Dermatitis,* 17, 187, 1987.
107. **Lowell, C. R., Cronin, E., and Rhodes, E. L.,** Photocontact urticaria from chlorpromazine, *Contact Dermatitis,* 14, 290, 1986.
108. **Matthews, K. P.,** Immediate type hypersensitivity to phenylmercuric compounds, *Am. J. Med.,* 44(2), 310, 1968.
109. **Rilliet, A., Hunziker, N., and Brun, R.,** Alcohol contact urticaria syndrome (immediate-type hypersensitivity), *Dermatologica,* 161, 361, 1980.
110. **Maurice, F., Rivory, J. P., Larsson, P. H., Johansson, S. G. O., and Bousquet, J.,** Anaphylactic shock caused by formaldehyde in a patient undergoing long-term hemodialysis, *J. Allergy Clin. Immunol.,* 77, 594, 1986.
111. **Maibach, H.,** Immediate hypersensitivity in hand dermatitis, *Arch. Dermatol.,* 112, 1289, 1976.
112. **Malten, K.,** Occurrence of hybrids between allergic eczema and atopic dermatitis (and vice-versa) and their significance, *Dermatologica,* 136, 404, 1968.
113. **von Krogh, G. and Maibach, H. I.,** The contact urticaria syndrome, *Semin. Dermatol.,* 1, 59, 1982.
114. **Lahti, A. and Maibach, H. I.,** Contact urticaria from diethyl fumarate, *Contact Dermatitis,* 12, 139, 1985.
115. **Winton, G. B. and Lewis, C. W.,** Contact urticaria, *Int. J. Dermatol.,* 21, 573, 1982.
116. **Odom, R. B. and Maibach, H. I.,** Contact urticaria: a different contact dermatitis, *Cutis,* 18, 672, 1976.
117. **Lahti, A., von Krogh, G., and Maibach, H. I.,** Contact urticaria syndrome: an expanding phenomenon, in *Dermatologic Immunology and Allergy,* C. V. Mosby, St. Louis, 1985, 379.
118. **Mathews, K. P.,** The urticarias: current concepts in pathogenesis and treatment, *Drugs,* 30, 552, 1985.
119. **Lahti, A. and Maibach, H. I.,** Species specificity of nonimmunologic contact urticaria: guinea pig, rat and mouse, *J. Am. Acad. Dermatol.,* 13, 66, 1985.
120. **Johansson, S. G. O., Bennich, H., and Berg, T.,** *In vitro* diagnosis of atopic allergy. III. Quantitative estimation of circulating IgE antibodies by radioallergosorbent test, *Int. Arch. Allergy Appl. Immunol.,* 41, 443, 1971.
121. **Katchen, B. and Maibach, H. I.,** unpublished data, 1988.

Chapter 5

SKIN COLOR AND PROCLIVITY TO IRRITATION

E. Berardesca and H. I. Maibach

TABLE OF CONTENTS

I. INTRODUCTION

The relationship between skin color and proclivity to develop irritant dermatitis has been investigated. It is commonly accepted that dark skin, presumably because erythema is less discernible, better resists chemical injury than white skin. The problem is open and the literature controversial. The minimal objective, qualitative and quantitative data, makes it difficult to evaluate and compare precisely racial differences in skin function and skin responses. Indeed, not only are there physical and structural differences between blacks and whites, but race seems to modulate the response of skin to chemical and environmental agents. If such a difference exists, the mechanism involved could be on an anatomical (differences in stratum corneum thickness, cell cohesion, barrier function) or biochemical (transcutaneous penetration, increased formation of mediators of inflammation, etc.) basis. This paper reviews the more important and recent aspects regarding racial differences in skin physiology and reactivity in relation to the development of irritant dermatitis.

II. PHYSIOLOGICAL DIFFERENCES

Although stratum corneum thickness is equal in blacks and whites,[1,2] more tape strippings are required to remove the stratum corneum in blacks. Weigand et al.[3] state that this is due to the increased number of corneocyte layers in black skin. Moreover, the same study revealed a great interindividual variability in black race, whereas data from white subjects were more homogeneous. A correlation between the number of stratum corneum layers and the degree of pigmentation has not yet been demonstrated. The increased number of cell layers in the stratum corneum and the increased resistance to stripping could be related to lipid molecules in the intercellular matrix that increase cell cohesion. Indeed, Rienertson and Wheatley,[4] investigating the stratum corneum lipid content, found higher values in blacks. Weigand et al.[3] confirmed this result. Other parameters investigated in different studies, such as skin electrical resistance,[5] were consistent with these findings.

III. CUTANEOUS ABSORPTION

The differences in anatomical structure reported above should account for a decreased transcutaneous penetration in black skin and, therefore, for a decreased proclivity of the black race to develop contact irritant dermatitis. Unfortunately, the problem is more complex. Indeed, *in vitro* penetration of fluocinolone acetonide through skin samples obtained from the legs of blacks and whites revealed an increased permeability in whites,[6] but *in vitro* water permeation did not show racial differences.[7] Furthermore, *in vivo* studies show different patterns of penetration depending on the molecules tested. Diflorasone acetate does not have different transcutaneous penetration,[8] while dipyrithione penetrates approximately 34% less efficiently through black skin than white.[9] Racial differences in methylnicotinate-induced penetration have been investigated,[10] and no important differences seem to exist between blacks and whites besides a decreased magnitude of peak responses in blacks.

IV. IRRITANT REACTIONS

The proclivity of black skin to develop irritant reactions has been investigated since early in this century. In 1919, Marshall et al.[11] applied a drop of 1% dichloroethylsulfide on the arm and elicited erythema in 58% of whites, compared to 15% of blacks. Weigand and Mershon[12] studied patch test reactions to *o*-chlorobenzylidene malonitrile. The chemical was applied with occlusive plastic patches for 24 h; black subjects were more resistant to developing irritant reactions. Weigand and Gaylor[13] measured erythema in blacks and whites after the application of dinitrochlorobenzene on the back in normal and stripped skin. Blacks were less reactive than whites, but the removal of the stratum corneum resulted in equivalent reactivity. All these data support the view that racial differences in skin reactivity do exist. These differences seem to be modulated by the stratum corneum. Stripped skin tends to yield equal responses.[13] The studies briefly reported here were based on subjective observations and visual assessment of irritation.

Recently, we investigated differences in racial proclivity to develop skin irritation using bioengineering techniques allowing the nonvisual detection and quantification of erythema, blood flow, and damage of the water barrier.[14,15] Irritant reactions were induced in whites, blacks, and Hispanics by applying two concentration of sodium lauryl sulfate (SLS) via the Finn Chamber® patch test for 24 h. SLS was applied on normal untreated skin, on skin pretreated with delipidization, and on skin preoccluded for 30 min. Irritant reactions were evaluated noninvasively by means of laser Doppler velocimetry (LDV), measurements of skin water content (WC), and measurements of transepidermal water loss (TEWL). The study reveals a different susceptibility in the racial groups investigated to SLS-induced irritation mainly related to a different response of the water-barrier function and of microcirculation after chemical injury (Table 1). Blacks revealed a more sensitive TEWL response with higher TEWL values and increased modulation of responses. On the other hand, blacks demonstrated a decreased reactivity of blood vessels with decreased levels of erythema, more noticeable at lower irritant concentrations. The data confirms previous visual findings described by other groups reporting decreased erythematous reactions in blacks.[11-13] Nevertheless, the statement that black skin is tougher than white cannot be confirmed. In blacks, the damage of the water barrier is clearly more evident.[14] *In vitro* TEWL differences between blacks and whites have been also reported.[16] TEWL is higher *in vitro* in the black race. The data are difficult to interpret in the light of the physicochemical differences between black and white skin reported above; a different responsiveness of black skin to thermal stimulation may be involved.[17]

Blood vessel reactivity is decreased in blacks, as demonstrated previously,[14] with decreased visible erythematous reaction after chemical irritation.[11-13] The data are not explainable only with a decreased transcutaneous penetration; a different mechanism determining vasodilative and vasoconstrictive reactions should be considered. In order to elucidate these questions, we noninvasively investigated the mechanism of cutaneous reactive hyperemia after the application of a constrictive stimulus[18] (corticosteroid ointment). Reactive hyperemia is the temporary increase in blood flow in a tissue when the circulation in this organ is released after a period of vascular occlusion. The mechanism involved in determining this response is myogenic or metabolic. The former reflects the view that arteries and smooth muscles react to stretching by contraction and to decreased tension by relaxation;[19] the latter is based on the release of a vasoactive mediator during occlusion.[20] Reactive hyperemia is decreased by topical application of corticoids, which induce a vasconstrictive reaction.[21] Before the application of the stimulus, no differences between blacks and whites were detected; on the contrary, different responses were recorded after the application of clobetasol diproprionate (0.05%). Blacks gave a significantly decreased response for the area under the curve response and the peak response. The slope of the blood-flow rise did not reveal differences between races, but the different reactivity of the blood vessels of blacks was confirmed by the decay of blood flow after the peak response, which was significantly decreased. This pattern of response could be explained on the basis of an increased transcutaneous penetration of the chemical in blacks, contrasting with previous findings and with the data reported in the literature. The data is also consistent with a decreased reactivity of blood vessels in blacks during the hyperemic reaction due to either an increased removal of a vasodilator metabolite which accumulates during occlusion time or to a different viscoelastic response of the arteriolar wall in the black race. The stiffness and the compliance of the

TABLE 1
Transepidermal Water Loss (TEWL) and Laser Doppler Velocimetry (LDV) Values in Blacks and Whites

	Not pretreated			Preoccluded			Predelipidized		
	0.5%		**2.0%**	**0.5%**[a]		**2.0%**	**0.5%**		**2.0%**
TEWL (g/m^2h)									
Whites	4.5 ± 4.4		8.3 ± 8.1	4.7 ± 3.9		14.7 ± 13.2	7.4 ± 4.7		8.5 ± 4.8
		N.S.			$p < 0.02$			N.S.	
Blacks	7.7 ± 6.5		16.11 ± 17.0	12.9 ± 9.9		20.3 ± 15.2	8.4 ± 5.8		24.1 ± 23
		$p < 0.04$			N.S.			$p < 0.03$	
LDV (mV)									
Whites	73.2 ± 35		174.2 ± 140	73.6 ± 34		179 ± 128	98.6 ± 74		122 ± 73
		$p < 0.05$			$p < 0.01$			N.S.	
Blacks	81 ± 63		151.6 ± 139	58.9 ± 25		234 ± 194	78.6 ± 40		183.6 ± 120
		N.S.			$p < 0.01$			N.S.	

Note: p values refer to the comparison between 0.5% and 2% sodium lauryl sulfate concentrations. In the black group, TEWL increases significantly ($p < 0.04$) for the 0.5% concentration in the occluded site when compared to the white. All 0.5% and 2.0% sites were significantly different from baseline control sites as recorded by LDV, except for 0.5% in non-pretreated black skin. N.S. is nonsignificant.

[a] $p \leq 0.04$ for blacks vs. whites.

From Berardesca, E. and Maibach, H. I., *Contact Dermatitis,* 18, 65, 1988. With permission.

ground substance of the tissues surrounding the vessels should be taken into account. Occlusion of the arterial supply has been demonstrated to increase the drainage of extravascular fluid across the vessel.[22]

V. CONCLUSIONS

A different proclivity of black and white skin to develop irritation is evident. This is of relevant importance for environmental and occupational dermatitis and should be taken into account by investigators and drug manufacturers. In 1973, the North American Contact Dermatitis Group reported no racial difference between blacks and whites in the frequency of cutaneous reactions to multiple antigens.[23] Further investigations are needed, and a different data base is required to evaluate the problem extensively in terms of different allergen exposure as well as social and environmental factors determining the responses. The availability of transdermal systems for drug delivery highlight the problem. The occlusive transdermal patch system for delivery of clonidine shows sensitization rates of 34% in white women, 18% in white men, 14% in black women, and 8% in black men.[24] No data are available regarding the Hispanic population. Recently, when investigating the proclivity of brown (Hispanic) skin to develop irritant reactions,[15] we found the same behavior of the water barrier in blacks and Hispanics with higher TEWL and water content responses compared to whites, but the microcirculatory response was similar in the white and Hispanic race. At present, we do not know the reason for such differences.

We need more experimental data to evaluate, compare, and understand racial differences in skin irritation. Further studies should be performed with irritants of different physicochemical properties, in different experimental circumstances, and comparing several anatomical sites with the ultimate goal being a better understanding of skin function.

REFERENCES

1. **Freeman, R. G., Cockerell, E. G., Armstrong, J., et al.,** Sunlight as a factor influencing the thickness of the epidermis, *J. Invest. Dermatol.*, 39, 295, 1962.
2. **Thomson, M. L.,** Relative efficiency of pigment and horny layer thickness in protecting the skin of Europeans and Africans against solar ultraviolet radiation, *J. Physiol. (London)*, 127, 236, 1955.
3. **Weigand, D. A., Haygood, C., and Gaylor, J. R.,** Cell layers and density of Negro and Caucasian stratum corneum, *J. Invest. Dermatol.*, 62, 563, 1974.
4. **Rienertson, R. P. and Wheatley, V. R.,** Studies on the chemical composition of human epidermal lipids, *J. Invest. Dermatol.*, 32, 49, 1959.
5. **Johnson, L. C. and Corah, N. L.,** Racial differences in skin resistance, *Science*, 139, 766, 1963.
6. **Stoughton, R. B.,** Bioassay methods for measuring percutaneous absorption, in *Pharmacology of the Skin*, Montagna, W., Stoughton, R. B., and Van Scott, E. J., Eds., Appleton-Century-Crofts, New York, 1969, 542.
7. **Bronaugh, R. L., Stewart, F. R., and Simon, M.,** Methods for *in vitro* percutaneous absorption studies. VII. Use of excised human skin, *J. Pharm. Sci.*, 75, 1094, 1986.
8. **Wickema-Sinha, W. J., Shaw, S. R., and Weber, O. J.,** Percutaneous absorption and excretion of tritium labelled diflorasone diacetate, a new topical corticosteroid in the rat, monkey and man, *J. Invest. Dermatol.*, 7, 373, 1978.
9. **Wedig, J. H. and Maibach, H. I.,** Percutaneous penetration of dipyrithione in man. Effect of skin color (race), *J. Am. Acad. Dermatol.*, 5, 433, 1981.
10. **Guy, R. H., Tur, E., Bierke, S., et al.,** Are there age and racial differences to methyl nicotinate-induced vasodilatation in human skin?, *J. Am. Acad. Dermatol.*, 12, 1001, 1985.
11. **Marshall, E. K., Lynch, V., and Smith, H. V.,** Variation in susceptibility of the skin to dichloroethylsulfide, *J. Pharmacol. Exp. Ther.*, 12, 291, 1919.
12. **Weigand, D. A. and Mershon, G.,** The Cutaneous Irritant Reaction to Agent *O*-Chlorobenzilidene Malonitrile (CS). Quantification and Racial Influence in Human Subjects, Tech. Rep. 4332, Edgewood Arsenal, February 1970.

13. **Weigand, D. A. and Gaylor, J. R.,** Irritant reaction in Negro and Caucasian Skin, *South. Med. J.,* 67, 548, 1974.
14. **Berardesca, E. and Maibach, H. I.,** Racial differences in sodium lauryl sulphate induced cutaneous irritation: black and white, *Contact Dermatitis,* 18, 65, 1988.
15. **Berardesca, E. and Maibach, H. I.,** Sodium lauryl sulphate induced cutaneous irritation: comparison of white and hispanic subjects, *Contact Dermatitis,* 19, 136, 1988.
16. **Wilson, D., Berardesca, E., and Maibach, H. I.,** *In vitro* transepidermal water loss: differences between black and white human skin, *Br. J. Dermatol.,* 119, 647, 1988.
17. **Robinson, S., Dill, D., Wilson, J., et al.,** Adaptations of white men and negroes to prolonged work in humid heat, *Am. J. Trop. Med.,* 21, 261, 1941.
18. **Berardesca, E. and Maibach, H. I.,** Cutaneous reactive hyperemia: racial differences induced by corticoid application, *Br. J. Dermatol.,* 120, 787, 1989.
19. **Bayliss, W. M.,** On the local reactions of the arterial wall to changes of internal pressure, *J. Physiol. (London),* 28, 220, 1902.
20. **Olsson, R. A.,** Myocardial reactive hyperemia, *Circ. Res.,* 37, 263, 1975.
21. **Bisgaard, H., Kristensen, J. K., and Sondergaard, J.,** A new technique for ranking vascular corticosteroid effect in human using laser Doppler velocimetry, *J. Invest. Dermatol.,* 86, 275, 1986.
22. **Rodbard, S., Takeda, Y., and Takacs, L.,** Post-occlusion hyperemia: a study on a model, *Q.J. Exp. Physiol.,* 54, 346, 1969.
23. **Rudner, J. et al.,** Epidemiology of contact dermatitis in North America, 1972, *Arch. Dermatol.,* 108, 537, 1973.
24. Catapres-TTS (Clonidine), package insert, Boeringer Ingelheim, Ridgefield, CT, 1987.

Section II. Diagnostic Methods

Chapter 6

PHYSICOCHEMICAL METHODS FOR DETECTION OF CONTACT ALLERGENS*

Sigfrid Fregert

TABLE OF CONTENTS

* This chapter was reprinted from Fregert, S., *Dermatol. Clin.*, 6(1), 97, 1988. With permission.

I. INTRODUCTION

Dermatologic examinations routinely include chemical and biological analyses of tissues and of the functions of the skin, liver, blood, kidneys, and other organ systems. However, very few dermatologists use chemical methods to analyze environmental factors, which often cause or worsen dermatoses. By chemical analysis of environmental or occupational products, we can often demonstrate the presence or absence of particular allergens. In turn, this can influence the treatment, prevention, and prognosis of contact eczema. Moreover, we can check the purity of a substance and isolate the ingredient that is the allergen. New allergens also can be traced.

To demonstrate the presence of some allergens in products, simple methods or spot tests can sometimes be used. More often, however, more complicated chemical methods must be applied. Such work requires cooperation between physicians and chemists. The former must have some knowledge of the chemistry of allergens to communicate with the chemists and laboratory assistants. This chapter provides a foundation of such knowledge.

II. NICKEL

Nickel metal can, to some extent, be dissolved by the skin. Sensitization and elicitation of eczema usually result from contact with the metal and, less often, from contact with oxides or salts of the metal. Therefore, spot testing is usually performed with nickel-plated objects or with alloys.

The spot test for nickel is qualitative, not quantitative. A few drops of 1% dimethylglyoxime in ethanol and of a 10% ammonium hydroxide solution are successively applied to a cotton-tipped applicator, which is rubbed against the metal object. A strawberry red color on the cotton indicates the presence of nickel.[1] The test also can be performed on solutions suspected of containing nickel.

Patients can perform this test at home or at work, or the dermatologist can do it on objects provided. Coins known to contain nickel can be used to test the reagents and to observe the color. One caution: because nickel alloys such as stainless steel have chromium oxide on the surface, little or no nickel may be released unless strong chemicals such as sulfuric acid are applied. Accordingly, a spot test can be falsely negative unless a worn area of the surface is checked.

Many objects such as water taps and eyeglasses have nickel plating under chromium plating. Nickel also is often present under gold plating in jewelry, and gold- and rhodium-plated spectacles often have an undercoat of nickel plating. Both gold and rhodium are eroded by sweat and wear, and therefore after considerable use, nickel may be released to sensitize or elicit eczema. These reactions can occur even before the surface is sufficiently eroded to give a positive spot test.

Nickel-sensitive patients often state that they can wear a gold-plated earring for only a certain time without developing dermatitis. Usage studies show that a gold-plated stud or clasp releasing only 0.05 μg of nickel for 1 week in synthetic sweat can elicit eczema in mature holes in the earlobes after 5 to 6 days use, even if the nickel spot test is negative.[2]

Silver and white-gold jewelry often contains as much as 10% nickel. In these cases, the spot test is positive.

Other methods are available when the spot test fails. Small amounts of dissolved nickel can be determined quantitatively by atomic absorption spectrometry (discussed in Section IX.D of this chapter). Less desirable is a quantitative colorimetric method using dimethylglyoxime and ultraviolet (UV) spectrophotometry. Special paper strips marketed for the detection of nickel are unreliable.[14]

III. CHROMIUM

Soluble chromates and dichromates are the principal sensitizers in chromium sensitivity. Trivalent chromium compounds, particularly basic chromium sulfate found in leather, are sensitizers, but they do not penetrate the skin so easily as does hexavalent chromium. Chromium metal, chromium-plated metal, and chromium alloys are not dissolved by skin sweat unless acted on by strong chemicals and, thus, are rarely sensitizing.

The chromium spot test is qualitative only and is valid only for the hexavalent form. To the solution thought to contain dissolved chromate or dichromate, a few drops of concentrated sulfuric acid and then of 0.5% *sym*-diphenylcarbazide in ethanol are added. A red color develops if chromates or dichromates are present.[3] To check the reagents and to observe the color, a match head is dipped in water to which a few drops of the reagents have been added.

To demonstrate chromate on a solid object, such as zinc-galvanized iron, the object should be dipped into water containing the reagents. Iron can give a similar false positive reaction.

Atomic absorption spectrometry can be used to analyze dissolved chromium compounds quantitatively. This method measures the total amount of chromium, not just the hexavalent form. To analyze the latter, it is chelated with ammonium pyrrolidine dithiocarbamate and extracted with methyl isobutyl ketone (isopropylacetone) before spectrometry. Hexavalent chromium also can be analyzed quantitatively by UV spectrophotometry with diphenylcarbazide as a reagent. Again, the special paper strips for detecting the metal are not reliable.[14]

IV. COBALT

Cobalt metal, like nickel, can be dissolved by the skin to some extent. There is no reliable spot test for cobalt.[4] In solution, cobalt can be determined by UV spectrometry with α-nitroso-β-naphthol as a reagent. However, atomic absorption spectrometry is the best method.

V. FORMALDEHYDE

Formaldehyde is a gas that dissolves easily in water-based products. High concentrations occur in formalin (maximum 38% formaldehyde in water) and in melamine/urea formaldehyde resins before hardening. Small amounts may be released from many materials used as preservatives, such as quaternium-15 and imidazolidinyl urea, and small amounts may contaminate cosmetic creams and other products.

A. CHROMOTROPIC ACID METHOD OF ANALYSIS

The glassware for this procedure should be carefully washed with detergent, rinsed in distilled water, and stored protected from dust, as organic matter gives sulfuric acid a brown color that can interfere with interpretation of the test results.

The test reagent contains 40 mg of chromotropic acid in 10 ml of concentrated sulfuric acid.[5] Although this reagent probably can be stored for weeks, freshly prepared solution is better. Standard solutions of formaldehyde (2.5 μg, 5 μg, and 10 μg/ml) are also prepared.

Approximately 0.5 to 1.0 g of the sample (ointment, cream, oil, rubber, textile) is placed in a 25- or 50-ml glass jar with a ground-glass stopper. Beside it is placed a 2- or 3-ml glass tube containing 0.5 ml of the reagent. The jar is kept in the dark and observed at intervals for 1 to 2 d for development of a violet color indicating formaldehyde. A brownish color does not exclude the presence of formaldehyde.

Although this method is quantitative, a rough impression of the concentration of formaldehyde can be obtained by comparing the color of the test material with that of the formaldehyde standards. Also, the violet color appears more quickly, within minutes or a few hours, if the concentration of formaldehyde in the specimen is high.

B. ACETYLACETONE (LUTIDINE) METHOD

Formaldehyde reacts with acetylacetone in the presence of ammonia to form a yellow compound.[6] This test may be either qualitative or quantitative.

1. Qualitative Method

The following reagent should be made fresh weekly and refrigerated: 15 g ammonium acetate, 0.2 ml acetylacetone, and 0.3 ml glacial acetic acid in distilled water to 100 ml. A stock solution of formaldehyde in distilled water (100 μg/ml) is used to prepare standards of 2.5 μg, 5.0 μg, and 10.0 μg/ml. These solutions also should be refrigerated and prepared fresh each week.

For the test, a sample of approximately 0.5 g is used. It is placed in a 25-ml glass jar with a ground-glass stopper. Solids (e.g., paper, textiles, leather, rubber) are used as such; ointments or other fat products should be mixed with a few drops of a formaldehyde-free emulsifier such as Triton® X-100. Each of the standard solutions (1 ml) is placed in a separate 10-ml volumetric flask; 1 ml of distilled water in a flask serves as a blank. The reagent (1.5 ml) is then added to each flask.

2. Quantitative Method

After a qualitative test has been performed to determine the approximate concentration of formaldehyde in the sample, a quantitative test can be performed. An aqueous solution or suspension of the sample thought to contain approximately 10 μg of formaldehyde is used. Standard solutions of formaldehyde (2.5 μg, 15.0 μg, and 20.0 μg/ml) are prepared. Again, 1 ml of each standard solution, 1 ml of distilled water, and 1 ml of the dissolved sample are placed in separate 10-ml volumetric flasks, and 1.5 ml of the color reagent is added to each flask. The flasks are heated at 60°C for 10 min in a cabinet, cooled to room temperature, and made up to 10 ml with distilled water. The absorption of the standards is measured against that of the blank at 412 nm, and a calibration curve is prepared. The absorption of the specimen solution is then measured and the formaldehyde concentration determined. The lowest detectable concentration of formaldehyde by this method is 1 μg per gram of sample.

If turbidities appear in the standard solutions after the color reaction, the solutions are extracted with l-butanol as described in the original article on this technique.[6]

C. PRACTICAL APPLICATIONS

Several types of products are not suitable for analysis by the chromotropic acid method. For example, those containing other aldehydes and those containing ketones produce a yellow or brownish discoloration. Most perfumes cannot be analyzed with the method for this reason. Some cosmetic and industrial products cannot be analyzed by the chromotropic method.[7]

Both the chromotropic acid and the acetylacetone method, especially the former, are simple, so that 20 to 30 samples can be analyzed simultaneously.

Some products to which formaldehyde has not been deliberately added, nevertheless, contain traces (5 to 50 ppm), probably as a result of contamination of some ingredient. If a patient's skin is already damaged, such as by hand eczema of any origin, such small amounts of formaldehyde may cause the patient already sensitive to the substance to suffer chronic eczema.

Gas chromatography and high-pressure liquid chromatography also can be used for quantitative determination of formaldehyde in water-based products and in air samples.

VI. EPOXY RESINS OF BISPHENOL A TYPE

Epoxy systems contain resin, hardener, and, sometimes, reactive diluents, dyes, fillers, tar, and even other resins such as phenol formaldehyde. The most common epoxy resin of the bisphenol A type is diglycidyl ether of bisphenol A (DGEBA). Other types are found in composite materials, special electronic circuits, and mounting materials for electron microscopy samples. In two-component epoxy systems, the resin and hardener are mixed when used. In high-molecular-weight, one-component systems, the hardener has been added to the resin during product formulation, and hardening occurs only at high temperature (e.g., in an oven).

Completely hardened resins are nonsensitizing. However, when low-molecular-weight resin is hardened at room temperature, 5 to 20% does not react and, thus, still contains sensitizers. Traces of nonhardened resin also have been found in heated high-molecular-weight products.

Epoxy resins of the bisphenol A type contain oligomers of various molecular weights. That of 340 Da is the sensitizer and so should be looked for in analyzing products for reactivity. The hardeners, particularly the aliphatic amines, are also sensitizers, but only in approximately 10% of patients with epoxy contact allergy. Even then, the patient often has resin sensitivity also.

There is no simple test for hardeners. The standard test for the oligomers of molecular weight 340, 624, and 908 Da in *uncured* epoxy resins of the bisphenol A type is thin-layer chromatography (TLC).[8] Chloroform:acetonitrile 90:10 (v/v) is used as the solvent system, 0.1 *M* sulfuric acid is used as the first spray reagent, and anisaldehyde in methanol 2.5% (v/v) in acetone is used as the second spray reagent. The standard is 1% (w/v) of resin.

Solid samples are extracted in ethanol at room temperature. Because of differences in the amount of nonhardened epoxy resin, the time required may range from a few minutes to 24 h. The ethanol is evaporated to a volume of a few milliliters before being applied to the plate.

The procedure is as follows. The standard solution (2 to 5 μl; 20 to 50 μg) is deposited with a capillary pipette on a TLC silica plate a few millimeters from the edge. A similar volume of the sample is applied beside the standard spot. If the concentration of the epoxy resin is unknown, or when a solid object is being analyzed, three or four samples of various dilutions are applied. The plate is allowed to dry for a few minutes until the solvent in the samples has evaporated. The chromatogram is then developed in the chloroform:acetonitrile in a glass tank, air dried, and examined first by UV light and then after spraying with the reagents specified. In UV light, the plate will show dark spots at the sites of low-molecular-weight oligomers. The plate is then sprayed with sulfuric acid until just moist and sprayed lightly with anisaldehyde. It is dried for about 10 min at 110°C. Violet spots will appear, with the 340-Da oligomer at the top, followed by the 624-Da one, etc., corresponding to the spots from the standard. The limit of detection on sprayed plates is approximately 0.5 μg of the 340-Da oligomer, about tenfold below the minimum amount detectable by UV. The technique of TLC is described in more detail later (Section IX.A).

These methods can be used to detect the presence of epoxy resins of the bisphenol A type in unidentified or unknown products. However, the most important application is in discovering whether cast or coated objects contain incompletely hardened resins. Such resins have been found in twist-off caps, film cassettes, furniture, metal pieces, signboards, textile labels, stoma pouches, polyvinylchloride plastic, nasal cannulas, hemodialysis sets, cardiac pacemakers, glass fibers, brass doorknobs, and tool handles, among other places.

Migration or development in the TLC technique can be disturbed by certain ingredients in the epoxy product such as phthallates used as plasticizers. In such cases, high-pressure liquid chromatography (discussed in Section IX.B) can be used.

VII. PRIMIN

Both spot tests and patch tests can be used to detect primin, the sensitizer found in *Primula obconica,* the predominant cause of plant dermatitis in Europe.

The spot test (Craven's test) is performed as follows.[9] Equal parts of ethanol and ammonium hydroxide (d = 0.880) are mixed to make 2 ml. Three drops of ethylcyanoacetate are added to the mixture. Two drops of this freshly prepared mixture are added to filter paper on which a leaf or stem of the suspected plant has been pressed or to filter paper prepared with plant extract. A blue color develops within seconds in the presence of primin.

For the patch test, approximately 6 g of *Primula* leaves harvested in the spring are dipped for a few seconds into 10 ml of ethyl ether and then immediately rinsed with an additional 10 ml of ethyl ether. The ether is evaporated to 5 ml at room temperature, and 25 μl of the concentrated solution is applied to patch test units. The patches are left for several minutes at room temperature to allow the ether to evaporate and then wrapped in aluminum foil and stored in the refrigerator.[10]

The allergenic and irritant properties of the patches should be tested on several individuals known to be sensitive to primin and with some control persons. If sensitive patients are not available, the presence of allergen and its approximate strength can be confirmed by the chemical spot test. Patches can be stored refrigerated for several years.

VIII. ACIDITY AND ALKALINITY DETERMINATIONS

Acid and, particularly, alkaline products play a significant role in the development of irritant contact dermatitis and in chemical burns. It is important to determine the degree of acidity or alkalinity in a product suspected of causing skin problems in order to avoid false positive diagnoses of contact dermatitis.

A universal pH paper is usually satisfactory for clinical work, although it is suitable only for water-based products, not those in organic solvents. Emulsions of oil in water, such as cutting fluids, also can be tested. The paper is simply dipped into the liquid and the resulting color compared with the color scale on the paper's package. When examining concentrated organic amines, the pH paper should first be dipped in water. Moistened paper also can be applied to solid objects, which might be coated with residual detergent or other product. The paper can be used to check the skin pH when artifactual dermatitis is suspected.

For precise determination of the pH, a pH meter is necessary. These devices are now available in pen size so that they can be carried in the pocket.

The most important use of pH measurement is in patch testing. Water solutions used in patch tests should have a pH of 4 to 9 to avoid creating false positive irritant reactions. Commercial products such as detergents, hand cleaners, hair preparations, and cutting oils often have a pH outside this range. To avoid irritant reactions caused by the extreme pH, it is possible to dilute such a solution, but avoidance of irritation might require use of a concentration of less than 1%, which could lead to false negative reactions to any sensitizers. To prevent this, buffer solutions can be used, permitting more concentrated solutions to be applied in the patch test. The product should be diluted gradually with buffer, with monitoring by pH paper.

The following buffer solutions are useful and do not reduce irritancy by mechanisms other than changes in pH. Thus, they do not interfere with the patch tests.

- *Acid buffer:* 1:1 mixture of 0.1 *N* sodium acetate and 0.1 *N* acetic acid (pH 4.7)
- *Alkaline buffer:* 1:1 mixture of 0.1 *M* sodium carbonate (anhydrous) and 0.1 *M* sodium bicarbonate (pH 10.0)[11]

Both solutions can be stored refrigerated for at least 6 months.

IX. PHYSIOCOCHEMICAL METHODS FOR DETECTION OF OTHER SENSITIZERS

A number of the most common sensitizers cannot be identified by simple spot tests: thiuramsulfides, dithiocarbamates, mercaptobenzthiazole, *p*-phenylenediamine derivatives and other rubber chemicals, ethylenediamine, neomycin, hydroxyquinolines, benzocaine and other local anesthetics, parabens and other preservatives, and single fragrance substances. However, several of these can be demonstrated with TLC or high-pressure liquid chromatography. At present, the necessary equipment is available in only a few dermatology clinics. However, as interest in contact dermatitis has increased during the past decade, greater interest has been expressed in these techniques. Just as we have had to master histologic examination, determination of serum IgE, and liver function tests, so it may become necessary to gain ready access to TLC and high-pressure liquid chromatography. The former will be described in detail, as it is simple and the equipment is not necessarily complex. Several other potentially useful methods will be described briefly to give the reader an idea of the principles behind them and of their uses.

Another problem awaiting solution is the difficulty of analyzing complex natural or synthetic products containing many ingredients. Examples of such products are lanolin, colophony (rosin), balsams, phenol formaldehyde resins, and acrylates. As a result, only a few of the sensitizers in many of these products have been identified.

A. THIN-LAYER CHROMATOGRAPHY

Thin-layer chromatography takes advantage of differential solubility of chemicals in solvent mixtures. It is performed on special TLC plates made of glass, plastic, or aluminum and coated with an inert absorbent such as silica gel, aluminum oxide, or cellulose. Such plates are available ready for use from many companies.

Several solvent mixtures (eluents) are in common use. If one suspects the identity of the sensitizer in a sample, a specific eluent may be recommended, as described earlier for TLC of epoxy resins. If the sensitizer is not known, or if the best eluent has not been established, it is best to run several samples in different eluents simultaneously. Six popular eluent systems, with all proportions given in volumes, are

- Methanol:l-butanol 60:40
- Methanol:acetic acid 90:10
- Methanol:methylal:ammonia 50:50:1
- Toluene:acetic acid 100:15
- Chloroform:acetonitrile 90:10
- Chloroform:acetonitrile:methanol:acetic acid 90:10:10:2

The sample is dissolved or extracted in an organic solvent such as ethanol, acetone, or ethyl ether, usually to an estimated concentration of 1 to 10%. A small amount (often 10 μl) of the extract is applied to an indelibly marked site approximately 1 cm from the bottom of the TLC plate with a capillary micropipette. If a particular substance is suspected of being present, a small volume of a standard solution of that substance may be applied at the same level.

The plate is then placed in a covered flat-bottomed chamber containing the chosen eluent. The surface level of the eluent should be below the spot(s) of sample and is marked indelibly on the plate. Filter paper soaked in the eluent may be deposited to saturate the air in the chamber.

The chamber is covered and left undisturbed, with occasional monitoring to note the position of the eluent front. The solvents rise by capillary action. As they move, they pick up those components of the sample that are soluble in them. Because the solvents in the eluent mixture differ in their affinity for the absorbent coating of the TLC plate, they move at different speeds, thus separating the components of the sample. When the solvent front has almost reached (but has not gone as far as!) the top of the plate, the plate is removed from the chamber and dried.

When the test substance is colored, such as a textile dye, its position on the TLC plate is easily seen. If the absorbent coating of the plate has been prepared with a fluorescent ink, the specimen spots will be visible in UV light (approximately 254 nm) as dark spots on a bright background. Still other compounds can be made visible by spraying with developing reagents, such as in the TLC procedure for epoxy resins.

From the position of the spots, the migration of the ingredients in the sample can be calculated. This value, the distance from the starting point to the endpoint divided by the distance traveled by the solvent front, is called Rf and is specific for each substance on the same type of plate in a particular eluent. When two spots, one from the test substance and one from a known standard, have the same color and the same Rf, it is an indication that they are the same chemical. However, to be confident of the identity, similarity should be apparent in at least three solvent systems.

If only one spot appears from a test substance, this is an indication that the compound is pure. More often, a trail of spots is produced, making comparison with a known standard particularly helpful. Permanent records can be made of the developed plates with a camera or photocopier.

Special TLC plates are available with an extra-thick absorbent coating. With these plates, spots can be scraped off and used for patch testing. The best results are usually gained by obtaining several spots and extracting the chemical from the absorbent with a suitable solvent before preparing the patches.

B. HIGH-PRESSURE LIQUID CHROMATOGRAPHY

This method, also called high-performance liquid chromatography or simply HPLC, can be used for identification and isolation of compounds. The sample is dissolved in an organic solvent and forced through a column by a high-pressure pump. The speed of travel of the ingredients is determined by their chemical properties, the solvent system, and the kind of column, and thus the various ingredients in complex mixtures separate. By analyzing a known substance in the same way, it is often possible to identify product ingredients. This technique has been used to identify allergens in such products as epoxy resins, acrylics, phenol formaldehyde resins, and isothiazolones (e.g., Kathon® CG). Moreover, with special semipreparative columns, it is often possible to isolate individual ingredients for chemical testing, patch tests, or animal predictive tests.

C. GAS CHROMATOGRAPHY

In this method, the sample, dissolved in an organic solvent, is injected into the system, where it is evaporated and transported by the flow of gas (often nitrogen or helium) through a column containing a static phase. Because of interaction with the static phase, the time the substance takes to traverse the column (the retention time) is characteristic. Different substances thus are separated. Substances are detected as they leave the column, commonly by a flame ionization detector, and signals are registered by a recorder to produce a profile of the sample ingredients that is compared with profiles of known standards. The substances are not collected, as they are in high-pressure liquid chromatography.

D. ATOMIC ABSORPTION SPECTROMETRY

Analysis by this method depends on the fact that free atoms absorb light of specific wavelengths. The absorption can be detected, and its extent is proportional to the number of atoms.

In flame atomic absorption spectrometry, a solution of the sample is sprayed by a pneumatic nebulizer into a flame, commonly of air and acetylene. Alternatively, the sample can be dissociated into free atoms by heating it in a graphite furnace. Absorption of light from the light source by the freed atoms is then measured.

Approximately 70 elements can be identified at concentrations of as little as 0.005 ppm. In occupational dermatology, atomic absorption spectrometry is most often used to determine chromium, nickel, and cobalt.

E. ULTRAVIOLET AND VISIBLE-LIGHT SPECTROPHOTOMETRY

For this method, the sample is dissolved in an organic solvent with low UV absorbance. The absorbance of light at different wavelengths is measured and plotted. Many substances have a characteristic absorption curve. This method can be used for quantitative determination of some materials such as formaldehyde, as described earlier (see Section V.B).[6]

F. INFRARED SPECTROPHOTOMETRY

This method demonstrates the presence of functional groups in molecules, such as –OH (alcohol), C=O (ketone, aldehyde, or acid), $-NH_2$ (amine), and $-NO_2$ (nitro). Each molecule thus has a characteristic infrared spectrum or fingerprint. Transparent samples such as plastic films can be analyzed without processing. Otherwise, the sample is dissolved or pressed into a tablet with potassium bromide.

G. MASS SPECTROMETRY

For this method, one needs a pure sample or one separated by gas chromatography. In the mass spectrometer, the molecules are bombarded by an electron stream that breaks them into fragments. Various types of chemical bonds break in characteristic ways. With this technique, it is possible to measure the molecular weight exactly. Computer programs containing information on many substances can be used to identify the sample.

H. NUCLEAR MAGNETIC RESONANCE SPECTROSCOPY

In nuclear magnetic resonance spectroscopy, a sample is subjected to a strong magnetic field that is varied in controlled ways. As the field changes, absorption and re-emission occur at characteristic frequencies, causing oscillations that are detected by the instrument. From this information, it is possible to determine the position in a molecule of an element with a nonzero nuclear spin (i.e., one with an odd number of protons, neutrons, or both). Hydrogen-1 has been studied most, but increasing attention is being given in medical applications to carbon-13, fluorine-19, and phosphorus-31, which are making it possible to do noninvasive chemical analyses and monitor certain kinds of metabolism *in vivo*.

X. SPECIAL PROBLEMS: BATCH VARIATIONS

Variations in chemical composition of seemingly identical objects are often noted in daily life — for example, when a vegetable or fruit is more (or less) bitter or sweet than usual. Technicians in industry are well aware of the chemical differences that may occur between batches of chemicals or of a product. Sometimes, variation is manifest as differences in color, scent, viscosity, or stickiness; other times, it is more subtle. The batch or product will not always be rejected; indeed, the manufacturer may not be aware of it, especially in industries in which chemicals manufactured by other firms are used.

The practical importance of batch variation in occupational dermatology is the difficulty it may create in identifying a contact-sensitizing substance. Chemicals may be present that are not listed on the label and are not known to be present, even in materials labeled "hypoallergenic". Some examples of dermatologic relevance are: products of incomplete hardening of epoxy resins; the 340-Da oligomer in epoxy resins of the bisphenol A type; antimicrobial agents, nickel, or cobalt in cutting oils; chromate in welding fumes or cement; formaldehyde in cosmetics and toiletries; cobalt in nickel; ethylene thiourea in the pesticides Maneb® and Zineb®; contaminants in rubber; and trace substances in lanolin, wood, balsams, colophony, and other natural products.[13]

Therefore, when trying to identify a sensitizer, one must take batch variation into consideration. In patch testing with nonstandard preparations, one should whenever possible use a sample from the batch believed to be the source of the problem. The same precaution is appropriate when performing chemical analyses. When cross-reactivity patterns are studied, only pure chemicals should be used, and the method used in purity control should be specified in the report.

In several countries, attempts are now being made to computerize information on the ingredients in chemical products. Helpful as this will be when available, it will not take batch variations into account, so dermatologists will still need to be batch conscious if they are to be of the greatest help to their patients.

XI. SUMMARY

In the same way that it often is necessary to perform chemical and biological analysis of skin and body fluids in the management of skin disease, it may be necessary to analyze the chemical environment. Some analyses, such as those for nickel, chromate, and formaldehyde, require only simple tests. Others require complex and expensive equipment and specialized skills. In the future, it may become necessary to install some of this equipment for routine use in dermatology clinics. In the meantime, the dermatologist needs to be aware of the possibilities.

REFERENCES

1. **Fisher, A. A.,** *Contact Dermatitis,* 3rd ed., Lea & Febiger, Philadelphia, 1986, 752.
2. **Fischer, T., Fregert, S., Gruvberger, B., and Rystedt, J.,** Nickel release from ear piercing T kits and earrings, *Contact Dermatitis,* 10, 39, 1984.
3. **Grogan, C. H., Cahmann, H. I., and Lethco, E.,** Microdetermination of chromium in small samples of various biological media, *Anal. Chem.,* 27, 983, 1955.
4. **Feigl, F.,** *Spot Tests in Inorganic Analysis,* 5th ed., Elsevier, Amsterdam, 1958, 394.
5. **Blohm, G.,** Formaldehyde contact dermatitis, *Acta Dermatol.,* 39, 450, 1959.
6. **Fregert, S., Dahlquist, I., and Gruvberger, B.,** A simple method for the detection of formaldehyde, *Contact Dermatitis,* 10, 132, 1984.
7. **Dahlquist, I., Fregert, S., and Gruvberger, B.,** Reliability of the chromotropic acid method for qualitative formaldehyde determination, *Contact Dermatitis,* 6, 357, 1980.
8. **Fregert, S. and Trulson, L.,** Simple methods for demonstration of epoxy resin of bisphenol A type, *Contact Dermatitis,* 4, 69, 1978.
9. **Craven, R.,** A sensitive colour reaction for certain quinones, *J. Chem. Soc. Trans.,* 1, 1605, 1931.
10. **Hjorth, N., Fregert, S., and Schildknecht, H.,** Cross-sensitization between synthetic primin and related quinones, *Acta Dermatol.,* 49, 552, 1969.

11. **Bruze, M.,** Use of buffer solutions for patch testing, *Contact Dermatitis,* 10, 267, 1984.
12. **Bruze, M., Fregert, S., and Ljunggren, B.,** Effect of ultraviolet irridation of photo patch test substances *in vitro, Photodermatology,* 15, 24, 1985.
13. **Fregert, S.,** Batch consciousness in dermatologic management, *Acta Dermatol.,* 59 (Suppl. 85), 63, 1979.
14. **Fregert, S.,** unpublished data.

Chapter 7

PATCH TESTING IN ALLERGIC CONTACT DERMATITIS

Torkel Fischer and Howard I. Maibach

TABLE OF CONTENTS

ABSTRACT

The history of patch testing is an almost 100-year development, from short exposure to raw materials or pieces of fabric soaked with solutions to the present more sophisticated and exact methods. The present patch test methods are good tools in the hands of a skilled dermatologist. A new technique using a thin-layer film vehicle to incorporate the allergen suggests a future developmental line to improve and simplify the patch test technique. The computer registration of allergens and products and computerized studies of the chemical background of sensitizers promise to increase the efficiency and accuracy of both patch testing and patient counseling.[1-3]

I. INTRODUCTION

Eczema accounts for about one quarter of dermatologic patients, and one half of these patients suffer from contact dermatitis. Besides the clinical evaluation, the patch test is our most important tool for a correct diagnosis. Accurate patch testing indicates that 30 to 50% of the patients with contact dermatitis have positive reactions, often unexpected by history.[4-14] Most cases of allergic contact dermatitis have a background of a type four sensitivity which can be verified by patch testing. Suspicion of allergic background in a contact dermatitis is the principal indication for patch testing, but patch testing is also valuable in other types of eczema which do not heal properly and in hand psoriasis where part of the background may be contact allergy.[15] Details about different allergens and how to test with them are given in many excellent textbooks.[16-21] Testing for irritants, contact urticaria, and photopatch testing are not included in this chapter.

II. HISTORICAL NOTES

The patch test technique has developed during the last 100 years. Sporadic experiments have been mentioned in earlier literature. Individual intolerance to wood and other natural products was recognized in antiquity.[22] In 1847, Staedler[23] described a blotting paper strip method to test idiosyncrasy. This was forgotten for almost 50 years. In 1895, Jadahsson[25] introduced his application method; this was the start of modern patch testing.[22,24]

III. DEMANDS ON A GOOD PATCH TEST

A good patch test indicates contact sensitization and produces no false positive reactions. This apparently easy method of testing holds many difficulties and pitfalls, both technical and of interpretation. Ideally, allergens should be chemically defined and have high purity. The optimal amount of allergen applied per square area produces a patch test reaction that is high in quality and easy to evaluate, both in weakly and strongly sensitized individuals. The risk of sensitizing the patient must be kept at a low level, and the test should produce a minimum of irritant reactions. The test should be easy to apply, have good stability, be easy to read, be suitable to the patient, and not unnecessarily limit his activities.

IV. PRESENT PATCH TEST METHODS

Present standard methods involve a device and patch test material. The Al-Test® was the standard method of applying antigens, and it is still widely used.[20] This method uses filter paper discs fixed to a strip of plastic-coated aluminum foil. In recent years, it has been surpassed by chambers and cups made from aluminum or polypropene plastic.[26-29]

Most allergens are presently incorporated in a petrolatum vehicle.[30] This material is applied on the patches or in the cups and fixed with tape to the upper part of the back for 48 h. The precision of patch testing is insufficient.[31] One major reason is that the pharmaceutical preparation of patch test material is unsatisfactory.[30-35] However, not even with optimal material and technique can perfect tests be expected. Individuals with recent hand eczema and with low thresholds of irritation will sometimes present false positive reactions to allergens which are also minor irritants.

The Pharmacia TRUE Test™ is a new, innovative, ready-to-apply test method which uses polyester patches coated with allergen in hydrophilic vehicles.[36] Other ready-to-use test methods are currently announced.[37,38]

Uncommon test methods concern mucosal test technique, testing with volatile substances (such as Freon), and the technique of buffering strongly acid and alkaline test material.[39-42]

A. THE TAPE

Old-fashioned tape is made from colored cotton fabric covered with a sticky mass composed of crepe rubber, colophony rosin, zinc oxide, lanolin, antioxidants, and preservatives. Tape allergy is common with this type of material. The most frequent background is colophony sensitization. Most of the ingredients of this old tape have been described as sensitizers.[43-45]

Modern tapes have adequate tack and keep the patch tests in close contact with the skin. Excessive exercise or sweating during the test period may remove tape corners or even the whole test strip. Greasy skin decreases the adhesiveness of the tape.

Modern tapes are made from nonwoven textile and acrylate adhesive. Allergy is rare, but may occur to uncured acrylate[46,47] or to formaldehyde, which is present in low concentration in such tapes.[48] Some modern tapes have polyvinylchloride (PVC) backing, and additives to this plastic may cause allergic reactions.[49]

Many tape glues are still prepared from colophony derivatives, which sensitize. Patch testing with ordinary colophony does not always identify this type of allergy.[50,51] However, mild irritant, acneiform, and miliarial reactions to all kinds of tapes are frequent.[52] Furthermore, individuals with a tendency to dermographism often present an urticarial reaction for a short period of time after the removal of the tape.[53] Vitiligo of the taped area is a rare side effect of tape application.

B. THE PATCH

The first patches were pieces of cotton fabric soaked with allergen solutions; later, filter paper discs were used in a similar way and taped to the skin under a water-impermeable cover. The Al-Test® is a development of this method.[20] Presently, most dermatologists use aluminum or plastic chambers, their diameter being 10 to 15 mm and the volume 15 to 100 μl.[26-29] The chambers give the same accuracy as the Al-Test,® but use a smaller test area.[54-56]

The patch test device should keep the test material in close contact with the skin in a small, well-delineated area. With accurate application technique, most chambers fulfill these criteria; however, even with correct technique, minimal amounts of test material diffuse 1 mm around the patch.[57] A rare disadvantage of Al-chambers are cases of Al-sensitization, most commonly seen in individuals earlier desensitized with Al-prolonged, intradermal allergen extracts.[58-61]

Metal salts may interact with aluminum. Mercury has a high positive redox potential and reacts rapidly with aluminum, which has strong negative charge.[62-64] Cobalt and nickel salts also interact to a moderate degree with aluminum.[65]

Aluminum may interfere with organic substances. Methylene blue together with metallic aluminum fixes strongly to stratum corneum, probably by forming a double salt of low solubility.[57] Aluminum increases the rate of acrylate polymerization, probably a catalytic reaction.[66] With allergens mixed in petrolatum, these interactions have minimal clinical significance. Plastic-coated Finn Chambers® are now available and will eliminate this side effect.

C. THE VEHICLE

The vehicles of patch test allergens are currently subject to investigations and reevaluations.[32,67-74] Several different vehicles can be used. Aqueous solutions and solvents, such as ethanol, acetone, methylethylketone (MEK), and ethyl ether, have been tried and are still recommended for special allergens.[18-21] The stability of solutions is low, and they are difficult to handle and preserve. These problems are partly overcome with semisolid vehicles, e.g., lanolin and petrolatum. Lanolin, the first tried, was soon abandoned because of its sensitizing properties and exchanged for yellow petrolatum.[75-79] This vehicle was recently changed to pure white petrolatum (snow white), presently our standard vehicle.[80-82] The petrolatum vehicle has both advantages and disadvantages and has been questioned regularly.[83,84] Petrolatum is highly lipophilic, whereas most allergens are hydrophilic. The bioavailability of such allergens from petrolatum is usually low.[85-88] Softisan®, a hydrophilic lanolin substitute, is a promising but little-tested alternative.[89] Rare cases of petrolatum allergy have been reported to be caused by phenantrene derivatives normally present in petrolatum.[80-82]

Hydrated gels, used dry as a vehicle for the TRUE Test™, may be an alternative for chamber tests.[90]

D. THE ANTIGENS

More than 6 million substances are chemically defined at this time. At least 2800 of them have contact sensitizing properties.[21] Moreover, a large number of less well-defined biological materials also sensitize. As a rule, antigens in biological materials are not well defined.[76,78,91,92] Attempts have been made to describe the chemical basis of sensitization, so far with limited success.[93,94]

The selection of proper chemical for patch testing is demanding and difficult. Soluble metal compounds are, almost without exception, marginally irritant. Mild salts seldom present the antigen well.[90,95] The *p*-phenylenediamine base has recently been reelected for the European standard series instead of the salt because it seems to give a higher percentage of positive reactions.[96]

A small number of substances account for most contact sensitivity. National and international groups of patch test specialists suggest that series of allergens and allergen mixes be used for standard testing. Such series, which comprise 20 to 25 materials, include 50 to 60 different allergens. Such series detect 70 to 80% of allergic sensitivity in a normal population.[4-12,96]

Test series composed for special purposes include plastics, fragrances, cosmetics, sunscreens, preservatives, rubber chemicals, shoe chemicals, and textile dyes and cover such professions as engineering, hairdressing, baking, and dentistry.[19,97,98] Airborne allergens deserve their own series.[99,100]

Standard patch test material available from different manufacturers consists of finely ground chemicals or other substances incorporated in petrolatum.[32,34] The major exception is formaldehyde, which has to be kept in aqueous solution. Many allergens outside the standard series can be obtained standardized, most often in petrolatum.[97,98]

Purification of antigens and their preparation require skill. The purity of the test material is of utmost importance. Declaration of quality and purity of commercial patch test material is vague. However, a change is expected since several countries plan to set standards of quality, stability, and labeling of patch test materials.

The stability of patch test material depends largely on the allergen. Metal salts are stable.

Contrary to expectations, formaldehyde in an aqueous solution is quite stable.[101] *p*-Phenylenediamine patch test material is colorless when fresh, but is rapidly degraded as indicated by a blue-black discoloration. Aging of fragrances will decrease their antigenicity.[102] Photopatch test material kept in sunlight deteriorates rapidly.[103] Complex antigens may lose antigenicity with increased purity.[104]

E. THE ANTIGEN DOSE FOR PATCH TESTING

Textbooks recommend test concentrations and vehicles for several allergens.[16-21] The dose of allergen is then dependent on the amount of test material applied. The TRUE Test™ is declared as dose per patch.[36]

Unstandardized substances should be tested with some care. Positive patch test reactions with such material should not be accepted as accurate unless adequate controls have been made to rule out irritant test reactions. The structure and properties of chemical compounds may indicate risks of irritancy. Extremes of pH, high polarity, and solvent properties increase this risk and suggest a test concentration below 1%. Most other chemicals can be tested using three concentrations (aqueous or in petrolatum) of 10, 1.0, and 0.1% in an attempt to find a nonirritant concentration. A positive reaction should not be accepted as allergic unless at least ten controls are negative.

Fine grading of the dose for patch testing is done by the serial dilution technique.[105-108] Common allergens ought to be optimized this way. The sensitivities of allergic individuals distribute geometrically with a standard deviation of $+3$ and a mean of $-{}^{1}/_{3}$. For some allergens, especially metal salts, irritant reactions and weak allergic reactions will appear together within a certain range of dosage. The choice of test dose for such allergens is a balance between demands of sensitivity vs. specificity. The dose should be high enough to detect weak allergies, but should allow only an insignificant number of irritant reactions. Sometimes an increased test concentration reduces the number of positive reactions.[91] Care also must be taken not to sensitize.[109]

The optimal dose for an allergen is related to the test method, as the bioavailability for different allergens differs between methods.[110] An example is formaldehyde, where 2% (recommended for the Al-Test®) will cause irritancy with the Finn Chamber® test method; 1% is better. For most allergens, the optimal TRUE Test™ allergen dose is one third of what is used with chamber methods.

F. THE ANTIGEN PREPARATION

The present method of preparing test material from solids involves grinding and mixing the allergen with the vehicle. It is difficult to obtain a sufficiently fine particle size. Most materials present have a particle size of 2 to 200 μm. In some commercially available preparations the particles may be as large as 0.5 mm.[32,35] The petrolatum vehicle will keep the particles in closed compartments against the skin. Large crystals dissolve slowly. This can be studied with tape-fixed glass chambers which have a central window in the tape. Colored antigens, such as cobalt chloride crystals, can be observed to persist undissolved in the petrolatum for a test period of 48 h, and large crystals in close contact with the skin dissolve slowly, forming spots of strong color. This proves the presence of small areas with high allergen concentration.[32,34] The follicular openings are particularly sensitive to irritant reactions because many substances easily enter living epidermis in these areas. Therefore, large particles of an allergen with irritant properties in contact follicular areas may cause patchy irritant reactions. Large particles will also cause significant variation in the amount of allergen applied.[33]

Liquid allergens are difficult to incorporate in petrolatum unless they are highly lipophilic. Some preparations of fragrance mix include a detergent. This solves a pharmaceutical problem, but may change the irritant or sensitizing properties of the material. As a rule, microscopic inspection of balsam of Peru or fragrance test material discloses large droplets of allergen within and outside the petrolatum unless a detergent is added.[32,98]

G. MIXES

Patch testing with single allergens gives the best and most exact information, but the use of allergen mixes saves time and space.[111] Therefore, much effort has been made to include several allergens in one test.[111-116] This involves problems of concentration, interference, stability, and formulation of the material.[116] The major problem, however, is that mixes often become irritants because of a high total concentration of allergen. To avoid this, the participating allergens are often incorporated in suboptimal amounts, resulting in a significant number of weak allergies being missed.[111,113] Thus, tests with mercaptomix often give false negatives.[111,117] An interesting example of interference is our recent finding that the three thiuramdisulfides in the thiuram mix change to an equilibrium of six different thiuramdisulfides, while the thiuram monosulfide remains stable.[118] The choice of vehicle and the optimization of allergens with different chemical reactivity, solubility, and stability in a mix may be a real challenge, shown in the caine-mix standardization of the TRUE Test™.[116]

Many of the six mixes in our present standard series are close to the irritant concentration. Therefore, it is advisable, when getting a doubtful or weakly positive mix reaction, to repeat the test after some time with the separate mix ingredients.[96] Test substances such as balsam of Peru, lanolin, and colophony are, in fact, natural mixes. Balsam of Peru contains several sensitizing fragrances and will suggest about 50% of fragrance allergy; the fragrance mix, which is composed of eight perfumes — seven chemically defined and one natural product (oak moss) — will indicate about 70 to 80% of this sensitivity.[119-125]

The fragrance petrolatum mix is a borderline irritant. Most of the fragrance mix ingredients are incorporated in suboptimal concentrations, and a test with the perfumes separately will increase the reliability of this test and detect weak allergies otherwise missed. The TRUE Test™ has incorporated the fragrance mix in β-cyclodextrin and is less irritant than the petrolatum mix.

H. THE PACKING OF THE TEST MATERIAL

The packing of the test material serves to protect it from environmental influences. Previously used plastic bottles exposed the allergens to atmospheric influences, and some allergens even penetrated the bottles with the risk of sensitizing the technician.[32] Polypropene syringes, presently replacing the bottles, are highly resistant to chemicals and protect the test material well against air and humidity. These syringes also protect against microbials and other contamination.[126] At room temperature, mercury in petrolatum sediments and homogeneity of the preparation are rapidly lost. This also happens with nickel and chromium salts at increased temperature.[127] Heat, light, and UV radiation penetrate the syringes; therefore, the allergens are best kept in a cool and dark place.[103]

V. TEST TECHNIQUE

Patch test materials are standardized for application to the sides of the upper back. If insufficient or inappropriate, the outsides of the upper arms may be substituted. The amount of test material applied is critical to obtain a good test result. The dose should be sufficient to cover the test area completely without extrusion. With Finn Chambers®, which have a volume of 25 μl, application of 15 μl of test material will give the best results. Volatile test material, also in aqueous solution, must be applied without a time lag to reduce evaporation of the vehicle and, thus, reduce skin contact.[30] The test strip should be applied from below with mild pressure to remove air pouches, followed by some moderate strokes with the back of the hand to improve adherence. Extra tape to secure corners most often has the opposite effect and should be avoided. Extreme exercise should be avoided during the test period. The most frequent technical cause of tape loosening is the extrusion of excess test material; the next most frequent cause is extremely sebaceous skin. A reduction of the skin lipids by ethanol washing to degrease the test area is a method used to improve the tack of the tape. The ethanol, however, must evaporate completely before test application; otherwise, irritant reactions from both tape and patches may appear.

A. MARKING THE TEST AREA

The test strips may be marked with either colored or fluorescent ink. Different water-resistant skin markers have been used. We used an aqueous solution of methylrosaniline (1%) and silver nitrate (20%) in ethanol (70%) dispensed in a marking pen. To avoid discoloration of the patient's clothes, UV-fluorescent stains such as sodium fluorescein can be used.[128] Application of transparent tape on top of ink markings is another method to reduce the contamination of clothing.[129] Skin markers may cause toxic skin reactions.[130,131] The edge of the tape may be stripped off and utilized as an indicator. The limiting factor of this good idea is that the test strips must be removed at the test clinic and not by the patients themselves.

B. PATIENT INSTRUCTION

The test application should include good patient instruction about the test. A handout explaining that the test is applied to detect contact allergy, that it must remain in contact with the skin for 48 h, and with appointment times for the test reading facilitates and completes the oral explanations to the patients. This handout should also instruct the patient to avoid sweat-producing exercise, showers, and baths of the test area and to avoid scratching and rubbing, even if the area is itching. Loose tape should be reaffixed and such losses reported in the test record. The test area should not be exposed to sunlight or sunlamps.

VI. BIOAVAILABILITY OF THE ALLERGENS AND THE TEST TIME

In the future, skin permeation studies will be an integral part of the optimization of patch test material. It has been found to be valuable in the evaluation of test concentration, choice of test substance, vehicles, and test system.[132,133] Only small amounts of metal salts enter the living epidermis from a patch test.[86,87,134] Although there already is (at least with chromium) an influx equilibrium from the test within 5 h, and some individuals present positive reactions after less than 1 test exposure to nickel salts,[88,134] there is an increased frequency of positive patch test reactions with increased exposure time for at least 48 h. Sensitized patients who have had no signs of contact dermatitis for a long period of time need this continuous, long exposure to react.[87] Neomycin and *p*-phenylenediamine often have their maximal reaction 4 to 7 d after the application, and this also happens with other allergens.[135,136] The reason may be a slow penetration of allergen into the skin. Another possibility is that there may be a tardy immune response, slow coupling, or propagation of the antigen. Since the rate of antigen flux into the epidermis seems to be only a minor exposure time limiting factor, the 48-h patch test will probably remain our standard procedure to avoid significant loss of positive reactions.

VII. THE TRUE TEST

A new test, the Pharmacia TRUE Test™, represents a new generation of patch tests which solves many of the problems indicated above.[36,137] It is an easy, time-saving, ready-to-use test method. The allergens are incorporated in hydrophilic gels, preferably cellulose derivatives. The allergen in gel preparation is printed on an impermeable backing of polyester and dried to a thin film. The coated sheet is cut into 9 × 9 mm squares — test patches — which are mounted on a tape, covered with a protective sheet, and packed in an airtight, light-impermeable envelope.

Cellulose derivatives have low sensitizing potential, stick adequately to the plastic backing, and are compatible with most allergens. Both carboxymethylcellulose and hydroxypropylcellulose have been reported to sensitize.[138,139] The surface distribution and the amount of allergen can be dosed with high accuracy. When the test patches come in contact with the skin, perspired water hydrates the film to a gel. The gel covered with plastic will produce optimal contact with the skin and good skin permeation opportunities, resulting in high bioavailability of the allergen.

The test has been subjected to a comprehensive control program. Each allergen is standardized with a serial dilution technique on patients known to be sensitive to the allergen.[108] The intensity of the test reactions is also evaluated against tests with the Finn Chamber® method. For most allergens, equivalent reactions will be obtained from the Finn Chamber® test and the new test at an application of 20 to 40% of the amount of allergen normally applied in Finn Chambers® (range, 5 to 150%). Analytic methods have been worked out for the different allergens. The amount of allergen on the test patches has been controlled by qualitative and quantitative analysis and *in vivo* on patients at different temperatures at different time intervals. The test will be manufactured in a control area according to guidelines for good manufacturing of pharmaceuticals and must be validated according to GMP rules. The product is also controlled microbiologically. All this guarantees a final product of high and uniform quality. In special *in vitro* experiments, the permeation profiles of several allergens through skin and different membranes has been studied both with ordinary standard patch test material in Finn Chambers® and the new test patches. There was good agreement between the clinical and the diffusion chamber experiments. Multicenter studies have verified excellent performance of the test.[140,141]

VIII. THE READING AND EVALUATION OF THE PATCH TEST REACTION

The patch test reading evaluates contact sensitization and also gives some information about its intensity. The optimal time for allergen exposure seems to be 48 h.[142-147] The patch test is best evaluated at 72 to 96 h, when immediate irritant reactions have faded and most reactions of slow allergens are fully developed.[135,136,148-150]

With small variations, modern textbooks and the American Academy of Dermatology use the following evaluation model:

? or +?	doubtful reaction; faint erythema only
+	weak positive reaction; erythema, infiltration, discrete papules
++	strong positive reaction; erythema, infiltration, papules, vesicles
+++	extreme positive reaction; intense erythema, infiltration and coalescing vesicles
−	negative reaction
IR	irritant reaction of different types
NT	not tested

A. READING PARAMETERS

The following parameters of a patch test reaction can be discerned and evaluated: erythema, infiltration (edema), fine structure, the surface distribution of the reaction, and the area involved. These five variables have separate qualities in the discrimination of the test reaction.

The erythema is an intensity parameter which does not discriminate between allergic and irritant reactions.

The infiltration is essentially an intensity parameter, but infiltration should be present in the allergic reaction. Reactions without infiltration are almost always irritant. Infiltration, however, is no proof of allergy.

The fine structure is the most important discriminating parameter between allergic and irritant reactions. Allergic reactions present the following structural panorama in order of increasing strength: discrete papules, distinct papules, papulovesicles, and coalescing vesicles. Homogeneous reactions of such types are usually allergic, but irritant reactions have been reported which are impossible to discriminate from allergies.[151,152]

Irritant reactions present many different types of fine structure: faint homogeneous erythema without infiltration, fine wrinkling (silk paper structure), erythema and papules in follicular distribution, petechiae, pustules, bullae, and necrosis.[153-158] Petechial reactions are rarely allergic.[157,158] Patch testing with chambers and aqueous vehicle may result in ring-shaped allergic as well as irritant reactions.[57,159,160]

If a reaction presents a mixture of allergic and irritant fine-structure components, a retest is recommended. The cause may be skin disturbances in the test area such as scars, acne, miliaria, folliculitis, or others.

In general, allergic reactions cover the test area homogenously, while many irritant reactions are irregular, patchy, ring-shaped, or follicular in distribution. Exceptions of this rule are weak allergic reactions, which may present a patchy or follicular character, best demonstrated in the most dilute steps of a serial dilution test.[31,105-107]

The surface involvement of the test reaction is a variable of quality rather than of discrimination. Extrusion of the reaction beyond the size of the patch is less common in irritant than in allergic reactions. Strong irritant reactions, however, also may present this characteristic.

Multiple reactions of the allergic type should raise the suspicion of spillover effects from other test reactions or irritable skin, "the angry back".[161-165]

Squared test chambers such as van der Bent chambers produce squared test reactions. Distinct corners used to be regarded as a sign of irritancy, while allergic reactions should have diffuse borders. Such allergic reactions are probably caused by efflux of allergen. Allergic reactions from the squared TRUE Test® patches are well delineated. With this test system the diffusion of allergen from the test area is low.

B. HISTOPATHOLOGY OF THE PATCH TEST REACTION

There are no distinct distinguishing criteria between allergic and irritant patch test reactions.[166] In the guinea pig, the proportion of basophilic cells in the allergic reaction is higher than in an irritant reaction. Human experiments have not verified this difference.[167,168]

The monoclonal antibody technique has provided new possibilities for subgrouping lymphoid cells. No distinct differences between the cell infiltrates of allergic and irritant reactions has been reported.[169-175] Recent research has verified a higher concentration of interleukin-1 in suction blister fluid of an allergic patch test reaction than in an irritant, which is close to normal.[175]

C. OBJECTIVE EVALUATION METHODS

Several instruments and principles have been attempted to devise an objective evaluation method of patch test reactions which discriminates between allergic and irritant reactions. However, no reliable method has been developed yet.[176]

Reflectance measurements with different wavelength regions give no discrimination, only a gradation of the intensity of the reaction.[177] Registration of infrared radiation from the skin using termography gives an image without differences between allergic and irritant reactions.[178] Skin conductivity and skin impedance measurements give variable results.[176] Evaporimetry is of no differential diagnostic value.[179-181] The methods of carbon dioxide diffusion and transcutaneous oxygen resistance measurement need technical improvements.[181,182] Skin edema measurements with caliper and ultrasound reflectance are the most promising methods, but the whole time course of the test development has to be followed to differentiate the two types of reactions, and the accuracy of the results is lower than unaided eye observations.[183-185] Determination of epidermal water content by microwave resonance is a promising new method.[186] These methods primarily grade the intensity of test reactions. Blood flow measurements with the laser Doppler technique have the same limitations as the edema evaluations.[187-189]

IX. RISKS WITH PATCH TESTING

General toxic and allergic reactions with patch testing are extremely rare. Formaldehyde has been suspected of causing background reactions in a few cases of anaphylaxis.[190] In 1 to 2 per 1000 tests, there is a general flare-up of dermatitis which may last for 1 or 2 weeks. In a low percentage of tests, patients experience worsening of their dermatitis and increased itching for a few days or sometimes 1 week.

The normal disappearance time of a patch test reaction is 1 to 2 weeks. Test reactions may remain longer than 1 month, and persistence times of 8 months have been reported. Depigmented, pseudolymphomatous, and keloid reactions may appear as rare late sequelae.[191-193]

Flare-up of patch test reactions later than day 10 to 14 after the test application is regarded as a sign of active sensitization. Such reactions are common with dinitrochlorobenzene (DNCB) and natural primin, which have high sensitizing potencies. Synthetic primin implies a relatively lower risk of sensitization. In the standard test series, cobalt and *p*-phenylenediamine give the highest sensitizing risk.[194] Kathon® CG, now included in the European standard series as Cl+/−Me-isothiazolinon,[96] will sensitize strongly when used in a 0.03% aqueous solution for tests, whereas the risk with a 0.01% test concentration is insignificant.[195-197] The background may be due to strong binding to the horny layer where most of the substance is stopped before reaching the living epidermis. When applied in the 0.01% solution, the risk of sensitization is small, but this concentration will suggest only 50% of the sensitization.[197]

X. SENSITIVITY, SPECIFICITY, ACCURACY, AND RELEVANCE OF PATCH TESTS

Sensitivity is the ability of a test to detect contact allergy. Specificity is the capability of the test to discriminate between true allergic and nonallergic test reactions. Thus, sensitivity indicates the risk of false negative test reactions, whereas specificity indicates the risk of false positives. Sensitivity and specificity taken together describe the accuracy of the test, and this integer of the variables is an expression of the quality of the test which considers both false positive and false negative test reactions.

Relevance is the relationship between an accurate test reaction and the patient's disease. The accuracy of ? or +? reactions is low; one or just a few percent have an allergic background. The accuracy of + reactions depends somewhat on the allergen and the patch test material. For metal allergens in petrolatum, it is about 20%; for other allergens, it is between 20 and 50%. The accuracy of + + reactions is 80 to 90% and of + + + reactions, 95 to 100%.[31] If the result of a test is important to the patient and the reaction is questionable, retest after 4 to 6 weeks with the same test material, with another test technique, or with serial dilution tests.[105-107] Intracutaneous testing may be of value to confirm dubious test reactions, especially with metals.[198,199]

The background of false positive reactions is usually irritant and has already been discussed. The frequency of false negative reactions is uncertain.[200-204] Even with correct patch test material there may be several technical reasons for false negatives, most often insufficient penetration of allergen. The reaction also may be suppressed. UV radiation has both local and systemic suppressive effects on patch test reactions,[205] but there seems to be no seasonal influence of UV sunlight on patch tests.[206,207] Systemic and topical corticosteroids are well-known suppressors.[167,208] Antihistamines have earlier been regarded as inert, but cinnarizine seems to influence patch test reactions.[209] Sodium chromoglicate also inhibits allergic patch test reactions.[210] Both systemic and topical cyclosporine can be expected to influence the contact allergic reaction.[211] Unexpected factors like dietary zinc may contribute to the intensity of the allergic response.[212]

The evaluation of relevance is highly dependent on the skill of the physician. Unknown or hidden relevance is, from a practical point of view, the same thing as negative relevance. Thus, the only possible relevance evaluations are present, past, or no/unknown relevance.

A dermatologist well experienced in patch testing can predict 80% of nickel sensitivity, 50% of rubber, colophony, and fragrance sensitivity, but only 10 to 20% of reactions to other allergens. A complementary history after the test shows that about 80% of positive patch test reactions are relevant and provide information to the patient. Remaining positive reactions are of unknown relevance.[198,213-217] Contact sensitization develops after environmental contact with allergens. Our knowledge about products is limited, as is the memory of the patients. This may be the background of these unknown/nonrelevant patch test reactions.[218,219]

REFERENCES

1. **Edman, B.,** The usefulness of detailed information to patients with contact allergy, *Contact Dermatitis,* 19, 42, 1988.
2. **Gailhofer, G.,** Computerunterstutzte Erfassung und Bearbeitung von Epikutantestdaten, *Dermatosen,* 36, 10, 1988.
3. **MacFarlane, H. A. F. and Ross, J. A.,** Use of a computer program for contact clinic results, *Contact Dermatitis,* 14, 162, 1986.
4. **Hammershöy, O.,** Standard patch test results in 3,225 consecutive Danish patients from 1973 to 1977, *Contact Dermatitis,* 6, 263, 1980.
5. **Hirano, S. and Yoshikawa, K.,** Patch testing with European and American standard allergens in Japanese patients, *Contact Dermatitis,* 8, 48, 1982.
6. **Hjorth, N.,** Routine patch tests, *Trans. St. Johns Hosp. Dermatol. Soc.,* 49, 99, 1963.
7. **Lepine, E. M.,** Results of routine office patch testing, *Contact Dermatitis,* 2, 89, 1976.
8. **Gailhofer, G. and Ludvan, M.,** Zur Änderungen der Allergenspectrums bei Kontaktekzem in der Jahren 1975-1984, *Dermatosen,* 35, 12, 1987.
9. **Magnusson, B., Blohm, S.-G., Fregert, S., Hjorth, N., Hövding, G., Pirilä, V., and Skog, E.,** Routine patch testing. III. Frequency of contact allergy at six Scandinavian clinics, *Acta Derm. Venereol.,* 46, 396, 1966.
10. **Young, E. and Honwing, R. H.,** Patch test results with standard allergens over a decade, *Contact Dermatitis,* 17, 64, 1987.
11. **Mitchell, J. C., Adams, R. M., Glendenning, W. E., Fisher, A., Kanof, N., Larsen, W., Maibach, H. I., Rudner, E., Schorr, W., Storrs, F., and Taylor, J. S.,** Results of patch tests with substances abandoned, *Contact Dermatitis,* 8, 336, 1982.
12. **North American Contact Dermatitis Group,** The frequency of contact sensitivity in North America 1972-1974, *Contact Dermatitis,* 1, 277, 1975.

13. **Pevny, I., Brennenstuhl, M., and Raxinskas, G.,** Patch testing in children. II. Results and case reports, *Contact Dermatitis,* 11, 302, 1984.
14. **Goh, C. L.,** Epidemiology of contact dermatitis in Singapore, *Int. J. Dermatol.,* 27, 308, 1988.
15. **Fransson, J., Storgårds, K., and Hammar, H.,** Palmoplantar lesions in patients and their relation to inverse psoriasis, tinea infection and contact allergy, *Acta Derm. Venereol.,* 65, 218, 1985.
16. **Adams, R. M.,** *Occupational Skin Disease,* Grune & Stratton, New York, 1983.
17. **Cronin, E.,** *Contact Dermatitis,* Churchill Livingstone, Edinburgh, 1980.
18. **Fisher, A. A.,** *Contact Dermatitis,* 3rd ed., Lea & Febiger, Philadelphia, 1986.
19. **Fossereau, J., Benezra, C., and Maibach, H. I.,** *Occupational Contact Dermatitis, Clinical and Chemical Aspects,* Munksgaard, Copenhagen, 1982.
20. **Fregert, S.,** *Manual of Contact Dermatitis,* 2nd ed., Munksgaard, Copenhagen, 1981.
21. **de Grooth, A. C.,** *Patch Testing. Test Concentrations and Vehicles for 2800 Allergens,* Elsevier, Amsterdam, 1986.
22. **Foussereau, J.,** History of epicutaneous testing: the blotting paper and other methods, *Contact Dermatitis,* 11, 219, 1984.
23. **Staedeler,** Über die eigentumlichen Bestandteile der Anacardium Fruchte. *Ann. Chem. Pharm.,* 63, 117, 1847.
24. **Sulzberger, M. B.,** The patch test — who should and should not use it and why, *Contact Dermatitis,* 1, 117, 1975.
25. **Jadassohn, J.,** Zur Kenntnis der medikamentose Dermatosen, in *Verhandlungern der Deutschen Dermatologischen Gesellschaft,* Funfter Kongress, Graz, 1895, W. Braunmuller, Vienna, 1896, 106.
26. **Pirilä, V.,** Chamber test versus patch test for epicutaneous testing, *Contact Dermatitis,* 1, 48, 1975.
27. **Frosch, P. J. and Kligman, A. M.,** The Duhring chamber. An improved technique for testing of irritant and allergic reactions, *Contact Dermatitis,* 5, 73, 1979.
28. **Dooms-Goossens, A.,** Van der Bend Chamber Test, *Nieuwsbrief Contactdermatol.,* December 1984.
29. **Quisno, R. A. and Doyle, R. L.,** A new occlusive patch test system with a plastic chamber, *J. Soc. Cosmet. Chem.,* 34, 13, 1983.
30. **Fischer, T. and Maibach, H. I.,** The Finn chamber patch test technique, *Contact Dermatitis,* 11, 137, 1984.
31. **Fischer, T. and Rystedt, I.,** False positive, follicular and irritant patch test reactions to metal salts, *Contact Dermatitis,* 12, 93, 1985.
32. **Fischer, T. and Maibach, H. I.,** Patch test allergens in petrolatum: a reappraisal, *Contact Dermatitis,* 11, 224, 1984.
33. **Fischer, T. and Maibach, H. I.,** The amount of nickel applied with a standard patch test, *Contact Dermatitis,* 11, 285, 1984.
34. **Vanneste, D., Martin, P., and Lachapelle, J. M.,** Comparative study of the density of particles in suspensions for patch testing, *Contact Dermatitis,* 6, 197, 1980.
35. **Puschmann, M.,** Galenische qualität von testsubstansen, *Z. Hautkr.,* 62, 1710, 1987.
36. **Fischer, T. and Maibach, H. I.,** Easier patch testing; TRUE Test, *J. Am. Assoc. Dermatol.,* 20, 447, 1989.
37. **Marks, J. G., Westheim, A., and Snipes, W.,** PA Patch — a new patch testing device, *J. Am. Assoc. Dermatol.,* 16, 961, 1987.
38. **Hornstein, M.,** Anvendung eines neuen Fertigpflasters in der Praxis (Epiquick), *Z. Hautkr.,* 62, 1719, 1987.
39. **van Loon, L. A. J., van Elsas, P. W., van Joost, Th., and Davidson, C. L.,** Contact stomatitis and dermatitis to nickel and palladium, *Contact Dermatitis,* 11, 294, 1984.
40. **Axwll, T., Spiechowiez, E., Glantz, P.-O., Andersson, G., and Larsson, Å.,** A new method for intraoral patch testing, *Contact Dermatitis,* 15, 58, 1986.
41. **van Ketel, W. G.,** Allergic contact dermatitis from propellants in deodorant sprays in combination with allergy to ethyl chloride, *Contact Dermatitis,* 2, 115, 1976.
42. **Bruze, M.,**. Use of buffer solutions for patch testing, *Contact Dermatitis,* 10, 267, 1984.
43. **Calnan, C. D.,** Diethyldithiocarbambate in adhesive tape, *Contact Dermatitis,* 4, 61, 1978.
44. **Cronin, E. and Calnan, C. D.,** Allergy to hydroabietic alcohol in adhesive tape, *Contact Dermatitis,* 4, 57, 1978.
45. **Rasmussen, J. E. and Fisher, A. A.,** Allergic contact dermatitis to a salicylic acid plaster, *Contact Dermatitis,* 2, 237, 1976.
46. **Jordan, W. P.,** Cross-sensitization pattern in acrylate allergies, *Contact Dermatitis,* 1, 13, 1975.
47. **Waegemaekers, T. H. and van der Walle, H. B.,** The sensitizing potential of 2-ethylhexyl acrylate in guinea pig, *Contact Dermatitis,* 9, 372, 1983.
48. **Andersen, K. E., Hjorth, N., Bundgaard, H., and Johansen, M.,** Formaldehyde in a hypoallergenic non-woven textile acrylate tape, *Contact Dermatitis,* 9, 228, 1983.

49. **Fregert, S., Trulson, L., and Zimerson, E.,** Contact allergic reactions to diphenylthiourea and phenylisothiocyanate in PVC adhesive tape, *Contact Dermatitis,* 8, 38, 1982.
50. **Sjöborg, S. and Fregert, S.,** Allergic contact dermatitis from a colophony derivative in a tape skin closure, *Contact Dermatitis,* 10, 114, 1984.
51. **Hausen, B. M., Jensen, S., and Mohnert, J.,** Contact allergy to colophony. IV. The sensitizing potency of commercial products. An investigation of French and American modified colophony derivatives, *Contact Dermatitis,* 20, 133, 1989.
52. **Marks, J. G. and Rainey, M. A.,** Cutaneous reactions to surgical preparations and dressings, *Contact Dermatitis,* 10, 1, 1984.
53. **Mathias, C. G. T.,** Dermographism and patch test reactions, *Contact Dermatitis,* 10, 110, 1984.
54. **Cronin, E.,** Comparison of Al-test and Finn chamber, *Contact Dermatitis,* 4, 301, 1978.
55. **Hammershöy, O.,** Contradictory results following patch testing with Finn chambers, *Contact Dermatitis,* 6, 216, 1980.
56. **Peltonen, L.,** Comparison of Al-test and Finn chamber test, *Contact Dermatitis,* 7, 192, 1981.
57. **Fischer, T. and Maibach, H. I.,** The Finn chamber methylene blue ring, *Contact Dermatitis,* 12, 12, 1985.
58. **Clemmensen, O. and Knudsen, H. E.,** Contact sensitivity to aluminum in a patient hyposensitized with aluminum precipitated grass pollen, *Contact Dermatitis,* 6, 305, 1980.
59. **Fischer, T. and Rystedt, I.,** A case of contact sensitivity to aluminum, *Contact Dermatitis,* 8, 343, 1982.
60. **Cox, N. H., Moss, C., and Forsyth, A.,** Allergy to non-toxoid constituents of vaccines and implications for patch testing, *Contact Dermatitis,* 18, 143, 1988.
61. **Veien, N. K., Hattel, T., Justesen, O., and Nörholm, A.,** Aluminum allergy, *Contact Dermatitis,* 15, 295, 1986.
62. **Kalveram, K.-J., Rapp-Frick, C., and Sorck, G.,** Misleading patch test results with aluminum Finn chambers and mercury salts, *Contact Dermatitis,* 6, 507, 1980.
63. **Lachapelle, J. M. and Douka, M. A.,** An evaluation of the compatibility between aluminum Finn chambers and various mercurials dissolved in water of dispensed in petrolatum, *Dermatosen,* 33, 12, 1985.
64. **Lindemayr, H. and Becerano, St.,** Interaction of mercury compounds and aluminum, *Contact Dermatitis,* 13, 274, 1985.
65. **Fischer, T. and Maibach, H. I.,** Aluminum in Finn chambers reacts with cobalt and nickel salts in patch test materials, *Contact Dermatitis,* 12, 200, 1985.
66. **Björkner, B. and Niklasson, B.,** Influence of the vehicle on the elicitation of contact allergic reactions to acrylic compounds in guinea pig, *Contact Dermatitis,* 11, 268, 1984.
67. **Atkinson, J. C. and Rodi, S. B.,** Effect of vehicles and elicitation concentration in contact dermatitis testing. II. Statistical analysis of data, *Contact Dermatitis,* 2, 330, 1976.
68. **Cohen, H. A.,** The role of carrier in sensitivity to chromium and cobalt, *Arch. Dermatol.,* 112, 37, 1976.
69. **van Ketel, W. G.,** Patch testing with nickel sulphate in DMSO, *Contact Dermatitis,* 4, 167, 1978.
70. **Marzulli, F. N. and Maibach, H. I.,** Effects of vehicles and elicitation concentration in contact dermatitis testing. I. Experimental contact sensitization in humans, *Contact Dermatitis,* 2, 325, 1976.
71. **Marzulli, F. N. and Maibach, H. I.,** Further studies of the effects of vehicles and the elicitation concentration in experimental contact sensitization testing in humans, *Contact Dermatitis,* 6, 131, 1980.
72. **Samsoen, M., Stampf, J. L., Lelievre, g., and Fossereau, J.,** Patch testing with hexavalent chromium salts in different vehicles and with nickel and cobalt in petrolatum, *Dermatosen,* 30, 181, 1982.
73. **Skog, E. and Wahlberg, J. E.,** Patch testing with potassium dichromate in different vehicles, *Arch. Dermatol.,* 99, 697, 1969.
74. **Suhonen, R.,** Photoepicutaneous testing. Influence of the vehicle, occlusion time and concentration of the test substances on the results, *Contact Dermatitis,* 2, 218, 1976.
75. **Clark, E. M., Blondeel, A., Cronin, E., Oleffe, J. A., and Wilkinson, D. S.,** Lanolin of reduced sensitizing potential, *Contact Dermatitis,* 7, 80, 1981.
76. **Fregert, S., Dahlquist, I., and Trulsson, L.,** An attempt to isolate and identify allergens in lanolin, *Contact Dermatitis,* 10, 16, 1984.
77. **Kligman, A. M.,** Lanolin allergy: crisis or comedy, *Contact Dermatitis,* 9, 99, 1983.
78. **Sugai, G. and Higashi, J.,** Hypersensitivity to hydrogenated lanolin, *Contact Dermatitis,* 1, 146, 1975.
79. **Schlossman, M. L. and McCarthy, J. P.,** Lanolin and derivatives chemistry: relationship to allergic contact dermatitis, *Contact Dermatitis,* 5, 65, 1979.
80. **Dooms-Goossens, A. and Degraff, H.,** Contact allergy to petrolatums. I. Sensitizing capacity of different brands of yellow and white petrolatums, *Contact Dermatitis,* 9, 175, 1983.
81. **Dooms-Goossens, A. and Degraff, H.,** Contact allergy to petrolatums. II. Attempts to identify the nature of the allergens, *Contact Dermatitis,* 9, 247, 1983.
82. **Dooms-Goossens, A. and Dooms, M.,** Contact allergy to petrolatums. III. Allergenicity prediction and pharmacopoeial requirements, *Contact Dermatitis,* 9, 352, 1983.

83. **van Ketel, W. G.**, Petrolatum again: an adequate vehicle in cases of metal allergy?, *Contact Dermatitis*, 5, 192, 1979.
84. **Wahlberg, J. E.**, The vehicle role of petrolatum, *Acta Derm. Venereol.*, 51, 129, 1971.
85. **Foussereau, J., Benezra, C., and Maibach, H.**, *Occupational Contact Dermatitis*, Munksgaard, Copenhagen, 1982, 45.
86. **Fischer, T. and Maibach, H. I.**, Recovery of nickel sulphate from a standard patch test, *Contact Dermatitis*, 11, 134, 1984.
87. **Kalimo, K., Lammintausta, K., Mäki, J., Teuho, J., and Jansén, C.**, Nickel penetration in allergic individuals: bioavailability versus X-ray microanalysis dection, *Contact Dermatitis*, 12, 255, 1985.
88. **Mendelow, A. Y., Forsyth, A., Florence, A. T., and Baillie, A. J.**, Patch testing for nickel allergy. The influence of the vehicle on the response rate to topical nickel sulphate, *Contact Dermatitis*, 13, 29, 1985.
89. **Väänänen, A. and Hannuksela, M.**, Softisan — a new vehicle for patch testing, *Contact Dermatitis*, 14, 215, 1986.
90. **Fullerton, A., Menné, T., and Hoelgaard, A.**, Patch testing with nickel chloride in a hydrogel, *Contact Dermatitis*, 20, 17, 1989.
91. **Karlberg, A. T. and Lidén, C.**, Comparison of colophony patch test preparations, *Contact Dermatitis*, 18, 158, 1988.
92. **Matthies, C., Dooms-Goosens, A., Lachapelle, J.-M., Lahti, A., Menné, T., White, I. R., and Wilkinson, J.**, Patch testing with fractionated balsam of Peru, *Contact Dermatitis*, 19, 384, 1988.
93. **Benezra, C. and Maibach, H.**, True cross-sensitization, false cross-sensitization and otherwise, *Contact Dermatitis*, 11, 65, 1984.
94. **Depuis, G. and Benezra, C.**, *Allergic Contact Dermatitis to Simple Chemicals: A Molecular Approach*, Marcel Dekker, New York, 1982.
95. **Fowler, J. F.**, Selection of patch test materials for gold allergy, *Contact Dermatitis*, 17, 23, 1987.
96. **Anon.**, Revised European standard series, *Contact Dermatitis*, 19, 391, 1988.
97. **Anon.**, *The Trolab Guide for Patch Testing*, Hermal, Kurt Herrmann, Reinbek, West Germany, 1987.
98. **Anon.**, Patch Test Allergens. Product Catalogue. Chemotechnique Diagnostics AB, Malmö, Sweden, 1988.
99. **Lachapelle, J.-M.**, Industrial airborne irritant or allergic contact dermatitis, *Contact Dermatitis*, 14, 137, 1986.
100. **Dooms-Goosens, A., Debusschere, K. M., Gevers, D. M., Depré, K. M., Degreef, H. J., Loncke, J. P., and Snauwaert, J. E.**, Contact dermatitis caused by airborne agents, *J. Amer. Assoc. Dermatol.*, 15, 1, 1986.
101. **Wahlberg, J. E. and Kartus, E.**, Stability of formalin test solutions, *Contact Dermatitis*, 7, 43, 1981.
102. **Fisher, A. A. and Dooms-Goossens, A.**, The effect of perfume "ageing" on the allergenicity of individual perfume ingredients, *Contact Dermatitis*, 2, 155, 1976.
103. **Bruze, M. and Fregert, S.**, Studies on purity and stability of photopatch test substances, *Contact Dermatitis*, 9, 33, 1983.
104. **Karlberg, A. T., Boman, A., Holmbom, B., and Liden, C.**, Contact allergy to acid and neutral fractions of rosins. Sensitization experiments in guinea pigs and patch testing in patients, *Derm. Beruf. Umwelt.*, 34, 31, 1986.
105. **Wahlberg, J. E.**, Thresholds of sensitivity in metal contact allergy, *Berufs-Dermatosen*, 21, 22, 1973.
106. **Marzulli, F. N. and Maibach, H. I.**, The use of graded concentrations in studying skin sensitizers: experimental contact sensitization in man, *Cosmet. Toxicol.*, 12, 219, 1974.
107. **Fischer, T. and Rystedt, I.**, Cobalt allergy in hard metal workers, *Contact Dermatitis*, 9, 115, 1983.
108. **Fischer, T. and Maibach, H. I.**, The science of patch test standardization, *Immunol. Allergy Clin.*, 9, 417, 1989.
109. **Bruze, M., Dahnquist, I., Fregert, S., Gruvberger, B., and Persson, K.**, Contact allergy to the active ingredients of Kathon CG, *Contact Dermatitis*, 16, 183, 1987.
110. **Kim, H. O., Wester, R. C., McMaster, J. A., Bucks, D., and Maibach, H. I. M.**, Skin absorption from patch test systems, *Contact Dermatitis*, 17, 178, 1987.
111. **Mitchell, J. C.**, Patch testing with mixes. Note on mercaptobenzothiazole mix, *Contact Dermatitis*, 7, 98, 1981.
112. **Lynde, C. W., Mitchell, J. C., Adams, R. M., Maibach, H. I., Schorr, W. J., Storrs, F. J., and Taylor, J.**, Patch testing with mercaptobenzothiazole and mercapto-mixes, *Contact Dermatitis*, 8, 273, 1982.
113. **van Ketel, W. G.**, Thiuram mix, *Contact Dermatitis*, 2, 232, 1976.
114. **Menné, T. and Hjorth, N.**, Routine patch testing with paraben esters, *Contact Dermatitis*, 19, 189, 1988.
115. **Beck, M. H. and Holden, A.**, Benzocaine — an unsatisfactory indicator of topical local anaesthetic sensitization for the U.K., *Br. J. Dermatol.*, 118, 91, 1988.

116. **Kreilgård, B., Hansen, J., and Fischer, T.**, Chemical pharmaceutical and clinical standardization of the TRUE Test® caine mix, *Contact Dermatitis*, 21, 23, 1989.
117. **Lammintausta, K. and Kalimo, K.**, Sensitivity to rubber. Study with rubber mixes and individual rubber chemicals, *Dermatosen*, 33, 204, 1985.
118. **Fransson, D. and Fischer, T.**, The thiuram mix, submitted.
119. **Adams, R. M. and Maibach, H. I. M.**, A five year study of cosmetic reactions, *J. Am. Assoc. Dermatol.*, 13, 1062, 1985.
120. **Fisher, A. A.**, Patch testing with perfume ingredients, *Contact Dermatitis*, 1, 166, 1975.
121. **de Groot, A. C., Bruynzeel, D. P., Bos, J. D., van der Meeren, H. L. M., van Joost, T., Jagtman, B. A., and Weyland, J. W.**, Allergens in cosmetics, *Arch. Dermatol.*, 124, 1525, 1988.
122. **Larsen, W. G.**, Perfume dermatitis, *J. Am. Assoc. Dermatol.*, 12, 1, 1985.
123. **Malten, K. E., van Ketel, W. G., Nater, J. P., and Liem, D. H.**, Reactions in selected patients to 22 fragrance materials, *Contact Dermatitis*, 11, 1, 1984.
124. **Santucci, B., Cristaudo, A., Camnistraci, C., and Picardo, M.**, Contact dermatitis to fragrances, *Contact Dermatitis*, 16, 93, 1987.
125. **Mitchell, J. C., Calnan, C. D., Clendenning, W. E., Cronin, E., Hjorth, N., Magnusson, B., Maibach, H. I., Meneghini, C. L., and Wilkinson, D. S.**, Patch testing with some components of balsam of Peru, *Contact Dermatitis*, 2, 57, 1976.
126. **Hjorth, N. and Trolle-Lassen, C.**, Quick and easy method for application of patch tests and storage of test substances, *Acta Derm. Venereol.*, 43, 324, 1963.
127. **Goh, C. L. and Kwok, S. F.**, The influence of temperature on the concentration homogeneity of patch test materials, *Contact Dermatitis*, 15, 231, 1986.
128. **Jordan, W. P.**, Fluorescent marking ink for patch test site identification, *Contact Dermatitis Newsl.*, 10, 229, 1971.
129. **McKee, S. and Burrows, D.**, Marking of patch tests, *Contact Dermatitis*, 9, 40, 1983.
130. **Björnberg, A.**, Toxic reactions to a patch test skin marker containing fuchsin-silver nitrate, *Contact Dermatitis*, 3, 101, 1977.
131. **Dooms-Goossens, A.**, Toxic reactions to skin markers, *Contact Dermatitis*, 3, 280, 1977.
132. **Kreilgård, B. and Hansen, J.**, Aspects of pharmaceutical and chemical standardization of patch test material, *J. Am. Assoc. Dermatol.*, 21, 836, 1989.
133. **Hansen, J., Albrechtsen, S., and Hedegaard, K.**, Proallergens of formaldehyde applied in patch testing of formaldehyde allergy, *J. Am. Assoc. Dermatol.*, 21, 838, 1989.
134. **Liden, S. and Lundberg, E.**, Penetration of chromium in intact human skin *in vivo*, *J. Invest. Dermatol.*, 72, 42, 1979.
135. **Paramsothy, Y., Collins, M., and Smith, A. G.**, Contact dermatitis in patients with leg ulcers. The prevalence of late positive reactions and evidence against systemic ampliative allergy, *Contact Dermatitis*, 18, 30, 1988.
136. **Macfarlane, A. W., Curley, R. K., Graham, R. M., Lewis-Jones, M. S., and King, C. M.**, Delayed patch test reactions at days 7 and 9, *Contact Dermatitis*, 20, 127, 1989.
137. **Fischer, T. and Maibach, H. I.**, The thin layer rapid use epicutaneous test (TRUE-test), a new patch test method with high accuracy, *Br. J. Dermatol.*, 112, 63, 1985.
138. **Hamada, T. and Horiguchi, S.**, Allergic contact dermatitis due to sodium carboxymethyl cellulose, *Contact Dermatitis*, 4, 244, 1978.
139. **Schwartz, B. K. and Clendenning, W. E.**, Allergic contact dermatitis from hydroxypropyl cellulose in a transdermal estradiol patch, *Contact Dermatitis*, 18, 106, 1988.
140. **Lachapelle, J. M., Bruynzeel, D. P., Ducombs, G., Hannuksela, M., Ring, J., White, I. R., Wilkinson, J., Fischer, T., and Billberg, K.**, European multicenter study of the TRUE Test™, *Contact Dermatitis*, 19, 91, 1988.
141. **Stenberg, B., Billberg, K., Fischer, T., Nordin, L., Pettersson, L., Runek-Forsbeck, M., Sundberg, K., Swanbeck, G., Svensson, L., Wahlberg, J. E., Widström, L., and Wrangsjö, K.**, Swedish multicenter study with TRUE Test™ panel 2, in *Current Topics in Contact Dermatitis*, Frosch, P., Dooms-Godsens, A., Lachapelle, J.-H., Rycroft, R. J. G., and Scheper, R. J., Eds., Springer-Verlag, Berlin, 1989, 518.
142. **Jordan, W. P.**, 24-, 48-, and 48/48 hour patch tests, *Contact Dermatitis*, 6, 151, 1980.
143. **Kalimo, K. and Lammintausta, K.**, 24 and 48 h allergen exposure in patch testing. Comparative study with 11 common contact allergens and $NiCl_2$, *Contact Dermatitis*, 10, 25, 1984.
144. **Menne, T.**, Nickel allergy — reliability of patch test. Evaluated in female twins, *Dermatosen*, 29, 156, 1981.
145. **Bruze, M.**, Patch testing with nickel sulphate under occlusion for 5 hours, *Acta Derm. Venereol.*, 68, 361, 1988.
146. **Rudzki, E., Zakrzewski, Z., Prokopczyk, G., and Kozlowska, A.**, Patch tests with potassium dichromate removed after 24 and 48 hours, *Contact Dermatitis*, 2, 309, 1976.

147. **Skog, E. and Forsbeck, M.,** Comparison between 24- and 48-hour exposure time in patch testing, *Contact Dermatitis,* 4, 362, 1978.
148. **Mathias, C. G. T. and Maibach, H. I.,** When to read a patch test?, *Int. J. Dermatol.,* 18, 127, 1979.
149. **Mitchell, J. C.,** Day 7 (D7) patch test reading — valuable or not?, *Contact Dermatitis,* 4, 139, 1978.
150. **Rietschel, R. L., Adams, R. M., Maibach, H. I. M., Storrs, F. J., and Rosenthal, L. E.,** The case for patch test readings beyond day 2, *J. Am. Assoc. Dermatol.,* 18, 42, 1988.
151. **Hannuksela, M., Pirilä, V., and Salo, O. P.,** Skin reactions to propylene glycol, *Contact Dermatitis,* 1, 112, 1975.
152. **Trancik, R. J. and Maibach, H. I.,** Propylene glycol: irritation or sensitization?, *Contact Dermatitis,* 8, 185, 1982.
153. **Romaguera, C. and Grimalt, F.,** PPPP syndrome, *Contact Dermatitis,* 3, 102, 1977.
154. **Schmidt, H., Schultz-Larsen, F., Ölholm, Larsen, P., and Sörgaard, H.,** Petechial reaction following patch testing with cobalt, *Contact Dermatitis,* 6, 91, 1980.
155. **Fischer, T. and Rystedt, I.,** Patch testing with sodium tungstate, *Contact Dermatitis,* 9, 69, 1983.
156. **Wahlberg, J. E. and Maibach, H. I.,** Sterile cutaneous pustules: a manifestation of primary irritancy? Identification of contact pustulogens, *J. Invest. Dermatol.,* 76, 381, 1981.
157. **Bruynzeel, D. P., van den Hoogenband, H. M., and Koedijk, F.,** Purpuric vasculitis-like eruption in a patient sensitive to balsam of Peru, *Contact Dermatitis,* 11, 207, 1984.
158. **Roed-Petersen, J., Clemmensen, O. J., Menné, T., and Larsen, E.,** Purpuric contact dermatitis from black rubber chemicals, *Contact Dermatitis,* 18, 166, 1988.
159. **Fyad, A., Masmuodi, M. L., and Lachapelle, J. M.,** The "edge effect" with patch test materials, *Contact Dermatitis,* 16, 147, 1987.
160. **Lachapelle, J. M., Tennstedt, D., Fyad, A., Masmoudi, M. L., and Nouaigui, H.,** Ring-shaped positive allergic patch test reactions to allergens in liquid vehicles, *Contact Dermatitis,* 18, 234, 1988.
161. **Bandmann, H.-J. and Agathos, M.,** New results and some remarks to the "Angry Back Syndrome", *Contact Dermatitis,* 7, 23, 1981.
162. **Bruynzeel, D. P. and Maibach, H. I.,** The excited skin syndrome (Angry Back), *Arch. Dermatol.,* 122, 323, 1986.
163. **Maibach, H. I., Fregert, S., Magnusson, B., Pirilä, V., Hjorth, N., Wilkinson, D., Malten, K., Meneghini, C., Lachapelle, J. M., Calnan, C., and Cronin, E.,** Quantification of the excited skin syndrome (the "angry back"), *Contact Dermatitis,* 8, 78, 1982.
164. **Mitchell, J. C.,** The angry back syndrome: eczema creates eczema, *Contact Dermatitis,* 1, 193, 1975.
165. **Mitchell, J. C.,** Multiple concomitant positive patch test reactions, *Contact Dermatitis,* 3, 315, 1977.
166. **Lindberg, M.,** Studies of the cellular and subcellular reactions in the epidermis at irritant and allergic dermatitis, *Acta Derm. Venereol. Suppl.,* Vol. 105, 1982.
167. **Skoog, M.-L. and Groth, O.,** Cellular infiltrate in experimental allergic and toxic contact dermatitis. Effects of systemically administered corticosteroids, *Int. Arch. Allergy Appl. Immunol.,* 65, 168, 1981.
168. **Anderson, C.,** The effect of selected immunomodulating agents on experimental contact reactions, *Acta Derm. Venereol. Suppl.,* 116, 32, 1985.
169. **Christensen, O. B., Daniels, T. E., and Maibach, H. I.,** Expression of OKT 6 antigen by Langerhans cell in patch test reactions, *Contact Dermatitis,* 14, 26, 1986.
170. **Kanerva, L., Ranki, A., and Lauharanta, J.,** Lymphocytes and Langerhans cells in patch tests, *Contact Dermatitis,* 11, 150, 1984.
171. **Scheynius, A., Fischer, T., Forssum, U., and Klareskog, L.,** Phenotypic characterization *in situ* of the cells in allergic and irritant contact dermatitis in man, *Clin. Exp. Immunol.,* 55, 81, 1984.
172. **Scheynius, A. and Fischer, T.,** Phenotypic differentiation between allergic and irritant patch test reactions in man, *Contact Dermatitis,* 14, 297, 1986.
173. **Gawkrodger, D. J., Carr, M. M., McVittie, E., Guy, K., and Hunter, J. A.,** Keratinocyte expression of MHC class II antigens in allergic sensitization and challenge reactions and in irritant contact dermatitis, *J. Invest. Dermatol.,* 88, 11, 1987.
174. **Avnstorp, C., Ralfkier, E., Jörgensen, J., Lange, and Wantzin, G.,** Sequential immunophenotypic study of lymphoid infiltrate in allergic and irritant reactions, *Contact Dermatitis,* 16, 239, 1987.
175. **Thestrup-Pedersen, K., Grönhöj-Larsen, C., Rönnevig, J.,** The immunology of contact dermatitis. A review with special reference to the pathophysiology of eczema, *Contact Dermatitis,* 80, 81, 1989.
176. **Berardesca, E. and Maibach, H. I.,** Review article. Bioengineering and the patch test, *Contact Dermatitis,* 18, 3, 1988.
177. **Mendelow, A. Y., Forsyth, A., Feather, J. W., Baille, A. J., and Florence, A. T.,** Skin reflectance measurements of patch test responses, *Contact Dermatitis,* 15, 83, 1986.
178. **Raskin, M. M. and Zies, P. M.,** Role of thermography in allergic contact dermatitis, *Contact Dermatitis,* 3, 206, 1977.

179. **van der Valk, P. G. M., Kruis-de Vries, M. H., Nater, J. P., Bleumink, E., and de Jong, M. C. J. M.,** Eczematous (irritant and allergic) reactions of the skin and barrier function as determined by vapour loss, *Clin. Exp. Dermatol.,* 10, 185, 1985.
180. **Serup, J. and Staberg, B.,** Differentiation of allergic and irritant reactions by transepidermal water loss, *Contact Dermatitis,* 16, 129, 1987.
181. **Malten, K. E. and Thiele, F. A. J.,** Evaluation of skin damage. II. water loss and carbon dioxide release measurements related to skin resistance measurements, *Br. J. Dermatol.,* 89, 565, 1973.
182. **Prens, E. P., van Joost, T. H., and Steketee, J.,** Quantification of patch test reactions by transcutaneous PO_2 measurement, *Contact Dermatitis,* 16, 142, 1987.
183. **Wahlberg, J. E.,** Assessment of skin irritancy: measurement of skin fold thickness, *Contact Dermatitis,* 9, 12, 1983.
184. **Serup, J. and Staberg, B.,** Ultrasound for assessment of allergic and irritant patch test reactions, *Contact Dermatitis,* 17, 80, 1987.
185. **Brazier, S. and Shaw, S.,** High-frequency ultrasound measurements of patch test reactions, *Contact Dermatitis,* 15, 199, 1986.
186. **Jacques, S. L., Maibach, H. I., and Suskind, C.,** Water content in stratum corneum measured by a focused microwave probe: normal and psoriatic, *Bioeng. Skin,* 3, 118, 1981.
187. **Staberg, B. and Serup, J.,** Allergic and irritant skin reactions evaluated by laser Doppler flowmetry, *Contact Dermatitis,* 18, 40, 1988.
188. **Nilsson, G., Otto, U., and Wahlberg, J. E.,** Assessment of skin irritancy in man by laser Doppler flowmetry, *Contact Dermatitis,* 8, 401, 1982.
189. **Wahlberg, J. E. and Wahlberg, E. N. G.,** Quantification of skin blood flow at patch test sites, *Contact Dermatitis,* 17, 229, 1987.
190. **Orlandini, A., Viotti, G., and Magno, L.,** Anaphylactoid reaction induced by patch testing with formaldehyde in an asthmatic, *Contact Dermatitis,* 19, 383, 1988.
191. **Björkner, B. E.,** Contact allergy and depigmentation from Alstroemeria, *Contact Dermatitis,* 8, 178, 1982.
192. **Monti, M., Berti, E., Caviccini, S., and Sala, F.,** Unusual cutaneous reaction after gold chloride patch test, *Contact Dermatitis,* 9, 150, 1983.
193. **Calnan, C.,** Keloid formation after patch test, *Contact Dermatitis,* 7, 279, 1981.
194. **Agrup, G.,** Sensitization induced by patch testing, *Br. J. Dermatol.,* 80, 631, 1968.
195. **De Groot, A. C., Bos, J. D., Jagtman, B. A., Bruynzeel, D. P., Van Jost, T., and Weyland, J. W.,** Contact allergy to preservatives. II, *Contact Dermatitis,* 15, 218, 1986.
196. **Weaver, J. E., Cardin, C. W., and Maibach, H. I.,** Dose-response assessments of Kathon biocide. I. Diagnostic use and diagnostic threshold patch testing with sensitized humans, *Contact Dermatitis,* 12, 141, 1985.
197. **Björkner, B., Bruze, M., Dahlquist, I., Gruvberger, B., and Persson, K.,** Contact allergy to the preservative Kathon CG, *Contact Dermatitis,* 14, 85, 1986.
198. **Möller, H.,** Intradermal testing in doubtful cases of contact allergy to metals, *Contact Dermatitis,* 20, 120, 1989.
199. **Ruzicka, T., Gerstmeier, M., Przybilla, B., and Ring, J.,** Allergy to local anesthetics: comparison of patch test with prick and intradermal test results, *J. Am. Assoc. Dermatol.,* 16, 1202, 1987.
200. **Podmore, P., Burrows, D., and Bingham, E. A.,** Prediction of patch test results, *Contact Dermatitis,* 11, 283, 1984.
201. **Cronin, E.,** Patch testing with nickel, *Contact Dermatitis,* 1, 56, 1975.
202. **Dooms, Goossens, A., Naret, C., Chrispeels, M. T., and Degreff, H.,** Is 5% nickel sulphate patch test concentration adequate?, *Contact Dermatitis,* 6, 232, 1980.
203. **Epstein, E.,** Contact dermatitis, to neomycine with false negative patch tests: allergy established by intradermal and usage tests, *Contact Dermatitis,* 6, 219, 1980.
204. **Epstein, E.,** Contact dermatitis to 5-fluoracil with false negative patch tests, *Contact Dermatitis,* 6, 220, 1980.
205. **Sjöwall, P. and Christensen, O. B.,** Local and systemic effect of ultraviolet irradiation (UV B and UV A) on human allergic contact dermatitis, *Arch. Dermatol. Res.,* 66, 290, 1986.
206. **Bruze, M.,** Seasonal influence on routine patch test results, *Contact Dermatitis,* 14, 184, 1986.
207. **Dooms-Goosens, A., Lesaffre, E., Heidbuchel, M., Dooms, M., and Degreef, H.,** UV sunlight and patch test reactions in humans, *Contact Dermatitis,* 19, 36, 1988.
208. **Sukanto, H., Nater, J. P., and Bleumink, E.,** Influence of topically applied corticosteroids on patch test reactions, *Contact Dermatitis,* 7, 180, 1981.
209. **Lembo, G., Presti, M. L., Balato, N., Ayala, F., and Santoianni, P.,** Influence of cinnarizine on patch test reactions, *Contact Dermatitis,* 13, 341, 1985.
210. **Meffert, H., Wischnewski, G. G., and Gunther, W.,** Disodium cromoglycate inhibits allergic patch test reactions, *Contact Dermatitis,* 12, 18, 1985.

211. **Biren, C. A., Barr, R. J., Ganderup, G. S., Lemus, L. L., and McCullough, J. L.,** Topical cyclosporine: effects on allergic contact dermatitis in guinea pigs, *Contact Dermatitis,* 20, 10, 1989.
212. **Warner, R. D., Dorn, C. R., Blakeslee, J. R., Gerken, D. F., Gordon, J. C., and Angrick, E. J.,** Zinc effects on nickel dermatitis in the guinea pig, *Contact Dermatitis,* 19, 98, 1988.
213. **Agrup, G., Dahlquist, I., Fregert, S., and Rorsman, H.,** Value of history and testing in suspected contact dermatitis, *Arch. Dermatol.,* 101, 212, 1970.
214. **Cronin, E.,** Clinical prediction of patch test results, *Trans. St. Johns Dermatol. Soc.,* 58, 153, 1972.
215. **Kiefer, M.,** Nickel sensitivity: relationship between history and patch test reaction, *Contact Dermatitis,* 5, 398, 1979.
216. **Rystedt, I.,** Evaluation and relevance of isolated test reactions to cobalt, *Contact Dermatitis,* 5, 233, 1979.
217. **Rystedt, I. and Fischer, T.,** Relationship between cobalt and nickel sensitization in hard metal workers, *Contact Dermatitis,* 9, 195, 1983.
218. **Magnusson, B. and Möller, H.,** Contact allergy without skin disease, *Acta Derm. Venereol.,* 59 (Suppl.85), 113, 1979.
219. **Fischer, T. and Rystedt, I.,** Hand eczema among hard metal workers, *Am. J. Ind. Med.,* 8, 381, 1985.

Chapter 8

SKIN TESTING FOR IMMEDIATE HYPERSENSITIVITY IN OCCUPATIONAL ALLERGOLOGY

Lasse Kanerva, Tuula Estlander, and Riitta Jolanki

TABLE OF CONTENTS

I. INTRODUCTION

Blackley,[1] in 1873, was probably the first to suggest that allergens could be introduced into the skin to detect sensitization. Schloss[2] used a scratch technique in studies of food allergy between 1910 and 1920. Later allergologists found the more sensitive intradermal technique,[3] but this technique led to more anaphylactic reactions, which had already been described in 1902.[4] In their classic experiments, Prausnitz and Küstner[5] then discovered "supersensitivity". They transferred serum from an allergic individual into the skin of a normal subject and then provoked local whealing by letting the test subject eat the allergen. They provided evidence that a humoral factor is necessary for the production of the cutaneous wheal, but it took 45 years to discover this factor, immunoglobulin E (IgE).[6,7] Now it is well known that a high number of the immediate whealing reactions are IgE-mediated, and specific *in vitro* tests have been developed to detect the allergen, but skin tests are still an extremely valuable diagnostic tool in allergology.

By far the most important skin test in occupational *dermatology* is the patch test because the majority of occupational dermatoses are contact dermatitis.[8] On the other hand, in occupational *allergology,* skin testing for immediate hypersensitivity has long been considered a valuable tool[9] because a large number of industrial inhalant allergens can be screened. The final diagnosis can then be confirmed by provocation (nasal, bronchial, oral, conjunctival) and *in vitro* tests.[10] Patients with dermatitis also need immediate allergy skin testing (IAST). First, type IV allergy can be associated with immediate hypersensitivity.[11-17] Indeed, of 67 patients tested for both types of responses, 22 (33%) developed a positive delayed response subsequent to the initial wheal and flare reaction.[18] Second, repetitive contact with type I allergens may result in dermatitis, known as protein contact dermatitis,[19] or atopic contact dermatitis.[20] Contact urticaria is an increasing problem in occupational dermatology, and IAST is needed for an accurate diagnosis.[18,21-23] IAST is also used when we need information on the atopic background of patients with dermatitis. It should also be remembered that skin tests are more sensitive than radioallergosorbent tests (RAST) for detection of IgE-mediated allergy.[23-25] Furthermore, even non-IgE-mediated reactions can be reproduced, and this further increases the value of IAST.

Therefore, we feel that it is important that an occupational dermatology clinic which wants to keep its diagnostic capacity at a high level needs to do IAST. In Finland, dermatologists have traditionally done type I allergy skin testing to help specialists in many other disciplines. At our clinic, we dermatologists do the skin tests for the pulmonologists, otorhinolaryngologists, ophtalmologists, industrial physicians, and internists. Currently, there are quite a number of industrial materials that give positive results in IAST systems. Specific IgE has been detected in serum, and these industrial substances also cause asthma, which can be verified with bronchial provocation tests. These materials are summarized in Table 1.

TABLE 1
Industrial Materials with Allergic Pathogenesis. Bronchial Provocation and Type I Skin Tests Have Been Positive in All Cases[9,86]

		Diagnostic tests			
Material	Industry, exposure, or occupation	Bronchial provocation tests	Skin tests prick or intracutaneous	Passive transfer tests to man or monkey	Specific IgE in serum
Animal, bird, fish, and insect serum, dander, secretions, excreta, contaminated water	Veterinarians, animal and poultry breeders, laboratory workers, fishermen, sericulture	+	+	+	+
Castor bean	Vegetable oil manufacture, coffee processors	+	+		+
Cotton dust	Textile and cotton seed oil	+	+		+
Complex salts of platinum	Metal refining	+	+	+	+
Diisocyanates (TDI, HDI, MDI)	Polyurethane, printers, adhesives	+	+		+
Enzymes from *Bacillus subtilis*	Detergent	+	+	+	+
Ethylenediamine	Plastics, rubber, and resin	+	+	+	
Flour	Bakers, farmers, and grain handlers	+	+		+
Formalin, formaldehyde	Medical, pharmaceutical	+	+		+
Grain, including insect and related grain	Farmers, grain elevator operators	+	+		
Green coffee bean	Coffee processors, plantation workers	+	+		+
Hog trypsin		+	+	+	+
Latex	Medical personnel, cleaners	+	+	+	+
Methyl hexahydrophthalic anhydride	Synthetic resin industry	+	+		+
Molds	Farmers, grain workers	+	+		
Organic dusts	Farmers	+	+		
Pancreatic enzymes	Distributors of enzymes		+	+	
Papain	Oil and food	+	+		+
Phenylglycine acid chlorine		+	+	+	
Phthalic anhydride	Synthetic resin industry	+	+		+
Reactive dyes	Dye house, textile plant	+	+		+
Salts of nickel	Metal plating	+	+		+
Sulfone chloramides	Pharmaceutical	+	+	+	
Trimellitic anhydride	Synthetic resin industry	+	+	+	+
Vegetable gums (acacia, karaya) natural resins	Printers	+	+	+	
Wood dusts	Wood mills, carpenters	+	+		

In the present chapter we report our IAST techniques and review our recent observations obtained using IAST regarding allergy to natural rubber (latex), reactive dyes, metals, epoxy resins, isocyanates, vegetable gums, and castor bean.

II. SKIN TESTING TECHNIQUES FOR IMMEDIATE HYPERSENSITIVITY

The techniques for the open and closed epicutaneous test, the ''rub'' test, the intracutaneous test, the prick test, the scratch (chamber) test, and the Prausnitz-Küstner test are reviewed.

A. OPEN TEST (OPEN APPLICATION TEST, OPEN PATCH TEST)

This can be performed on intact skin. Lahti[26] used 3 × 3 cm areas on the back or volar side of the forearm and recorded the results every 15 min for 6 h. Smaller or larger areas can also be used.[27] Sometimes a test on intact skin is negative, while a positive test reaction can be provoked at the sites of previous dermatitis.[19] A positive reaction is comprised of wheal and flare reactions or sometimes a vesicular eczematous reaction.[19]

B. CHAMBER TEST (CLOSED PATCH TEST)

The routine patch testing procedure is used, but readings are made earlier. Lahti[26] removed the chamber tests after 20 min and read results 10 min later. Readings should also be taken at 6 to 8 h.[25]

C. RUB TEST

The test is performed by rubbing the putative allergen with a firm object into the skin of the forearm. Hannuksela[27] suggested that the substance to be investigated could be rubbed gently into slightly affected skin. Penetration of the substance occurs mainly through the hair follicles, resulting in a follicular pattern of the whealing reaction.[25] The rub test is considered 10,000 times less sensitive than the intradermal test.[25]

D. INTRADERMAL TEST

The injection is made into the superficial layer of the dermis with a fine (26 or 27 gauge) needle. Because of the risk of anaphylaxis we have abandoned this test method. It is considered 100 times more sensitive than the prick and scratch tests.[25]

E. PRICK TESTS

Drops of allergen solutions are placed on the skin of the volar side of the forearm. When a large number of substances are tested, the additional allergens are tested on the volar upper arm or upper back 3 to 5 cm apart. A sterile disposable needle (e.g., Neolus® No. 20, Terumo Corporation, Japan) is then passed through the drop at a 45° angle and lifted, creating a small break in the epidermis. A minimal amount (3 μl)[28] of the extract penetrates the epidermis. Alternatively, a lancet with a tiny tip especially made for prick testing (e.g., the Dome-Hollister-Stier prick test lancet) is touched to the skin at a 90° angle. When properly done, no bleeding occurs. It has been reported that the needle or the lancet can be wiped off with a wet sponge between each prick and can then be reused without causing significant false positive results;[25] however, we routinely use a separate needle/lancet for each prick, although this is slower and more expensive. Standard allergens can nowadays also be tested with allergen-coated lancets (Phazet®, Pharmacia, Uppsala, Sweden), which have been reported to be equally effective in the diagnosis of immediate-type allergy as using skin prick tests with standardized extracts.[29]

If the industrial material is in a pulverized form and is tested ''undiluted'', we place a drop of water on the skin and then add the powder to the drop with a wooden toothpick. Thin solid substances (e.g., latex gloves) can be tested by placing the substance on the skin and then piercing it as described above with a needle or a lancet. Thicker materials can be tested by pricking the skin as described above and then applying the solid material tightly against the small break in the epidermis. In our clinic the tests are performed by specially trained nurses, and the test results are read by a dermatologist.

F. SCRATCH AND SCRATCH CHAMBER TESTS

A 5- to 10-mm-long superficial scratch is made with a sterile disposable needle in the epidermis of the forearm, upper arm, or upper back. Bleeding should be avoided. We use the upper back and a thicker needle (Neolus® No. 1, Terumo Corporation, Japan) than for the prick tests in order to get a somewhat wider scratch. The putative allergen is then applied to the scratch, and in the scratch chamber method[30] an 8-mm aluminum cup (Finn Chamber®, Epitest Ltd., Hyrylä, Finland) is applied to the scratch for 30 min with Scanpor® tape (Norgeplaster, Norway). Hannuksela and Lahti,[30] Lahti,[26] and Turjanmaa[23] have used a shorter challenge time (15 to 20 min). When we perform multiple tests with the scratch chamber method, we remove every second aluminum chamber from the Finn Chamber® kit of ten chambers, i.e., the second and fourth chambers from the left vertical row and the first, third, and fifth from the right. In this way, the midpoints of the chambers (scratches) are vertically 4 cm apart and horizontally about 3 cm. We make the scratches with the aid of the Finn Chamber® test reading plate and, accordingly, the chambers fit the scratches. We have a special reading plate with 10-mm holes, which enables us to make 10-mm scratches. It has been suggested that the test sites should be 5 cm apart from each other in order to prevent the test reactions from influencing each other,[31] but with our test system we very seldom encounter any problems.

G. THE PRAUSNITZ-KÜSTNER TEST[5]

In this test the patient's serum is injected into a healthy volunteer. Because of the risk of transmissible viruses (hepatitis, human immunodeficiency) we do not use this test. It has been used in occupational allergology, e.g., to confirm allergy to potato,[32] spider mite,[33] and latex.[23]

H. INTERPRETATION OF PRICK AND SCRATCH (CHAMBER) TEST REACTIONS

The wheal, and not the flare, is the test reaction to be evaluated. Wheals can be measured with a planimeter, but the product of two diameters at right angles yields equally good results.[28] The result is usually expressed as the mean of the longest diameter (D) and the diameter perpendicular to it (d): $(D + d)/2$. In the scratch chamber method the longest diameter perpendicular to the scratch is recorded. Histamine hydrochloride (10 mg/ml) and a negative control, consisting of the test solution without an antigen, are used as controls. At this concentration, histamine mostly gives a 4- to 8-mm wheal on the volar forearm. In the scratch chamber method on the upper back the thickness of the wheal is often 30 to 50% less. The other test reactions are evaluated semiquantitatively in relation to the histamine wheal.[25] The reactions equal to those of histamine are graded 3+, those with one half the histamine reaction 2+, and smaller wheals that are larger than the control solution wheals 1+ (wheals smaller than 2 mm are difficult to interpret and their significance is questionable). Twice the size of histamine reaction is called a 4+ reaction. The presence of pseudopodia is also recorded.

The skin test results should always be correlated with anamnesis and clinical symptoms. A positive test result does not necessarily show that the patient is clinically allergic to the tested substance. Neither does a negative test result always exclude an allergy; additional types of tests may be needed. With nonstandardized industrial putative allergens, there are no definite rules as to the size of the wheal for it to be positive. When a wheal is 3 mm or larger, further tests such as provocation tests and RAST are indicated.

I. LATE-PHASE SKIN TEST REACTION

In an IgE-dependent skin test reaction, the wheal and flare at 15 min can be followed by a late-phase reaction (LPR). Cutaneous LPR begins at 3 to 4 h following challenge, reaches maximal intensity in 6 to 10 h, and resolves by 24 to 48 h. LPR can be mediated by IgE antibody, but histamine alone is not responsible for the tissue inflammation observed in LPR. LPR may be involved in asthma, allergic rhinitis, urticaria, and atopic dermatitis. Pollen, molds, animal dander, housedust and dust mite, *Bacillus subtilis* enzyme, and insulin are agents that have been reported to cause LPR in the skin. The clinical significance of LPR in occupational dermatology is so far poorly understood, but more information could be obtained by recording IAST reactions at 6 and 24 h. LPR has recently been reviewed by Lemanske and Kaliner[34] and by Dorsch.[35]

III. FACTORS TO BE CONSIDERED WHEN PERFORMING IAST

A. ANAPHYLACTIC REACTIONS

When IAST is performed, the risk of anaphylactic reactions must always be taken into consideration, and readiness for the treatment of acute anaphylaxis must be maintained. If any anamnestic data of anaphylaxis exist and a skin test is being done, the procedures suggested by von Krogh and Maibach[18] are proposed (Table 2). In our and Hannuksea's[10] experience, generalized reactions are very rare, however. We perform about 25,000 prick or scratch tests per year; for example, in 1986 we performed 1170 prick test series, each consisting of 20 substances (i.e., 23,400 "pricks"), and 326 scratch chamber series, each consisting of 10 "scratches", which makes a total of 26,600 individual pricks or scratches during 1 year. In the last 15 years we have had less than ten generalized urticarial reactions. We have never needed to resuscitate a patient. In our clinic, generalized effects have been caused by reactive dye, electrocardiographic ink, ammonium persulfate, melamine formaldehyde resin, birch pollen, latex (use test), and castor bean. Therefore, we usually do not perform an open test (Table 2), but proceed directly to prick or scratch tests. We do not perform intracutaneous tests because of the risk of anaphylactic reactions. Anaphylactic/anaphylactoid reactions may occur even with open testing on normal skin.[37] Therefore, a worker could get an acute anaphylactic reaction in his work from an industrial substance. Accordingly, when the worker is exposed to industrial chemicals in a hospital environment, he has a better chance of getting immediate help if an anaphylactic reaction were to develop. Therefore, we find it ethically indicated to perform IAST also with industrial substances.

B. FACTORS INFLUENCING SIZE OF SKIN REACTIONS

There are several factors that might modulate skin reactivity. This has been reviewed by Czarnetzki[25] and is summarized in Table 3. In our testing system the patients are not allowed to take H_1-antihistamines or more than 20 mg of prednisone for 3 d before IAST. Astemizol therapy has to be stopped 4 weeks before IAST.

C. CONTRAINDICATIONS TO SKIN TESTING

Czarnetzki[25] has listed the following contraindications to skin testing: (1) severe systemic disease such as pneumonia and cardiac insufficiency, (2) first 4 months of pregnancy, (3) thrombocytopenia, (4) intrinsic clotting defects (not therapy), (5) acute allergic symptoms, (6) impetigo, and (7) generalized skin disease. An anamnesis of anaphylactic reactions should of course be taken into consideration, as mentioned in Section III.A.

TABLE 2
Suggested Test Procedures for Evaluation of Immediate-Type Responses[a]

1. Epidermal
   - A. Open application
     Nonaffected (normal) skin
     ↓
     negative
     ↓
     Slightly affected (or previously affected) skin
     ↓
     negative → Positive[b]
     ↓
   - B. Occlusive application (patch or chamber)
     Nonaffected (normal) skin
     ↓
     negative
     ↓
     Slightly affected (or previously affected) skin
     ↓
     negative
     ↓
2. Intradermal (prick, scratch, or intradermal injection)

[a] According to von Krogh and Maibach.[18]
[b] If positive reactions are obtained, discontinue further evaluation.

TABLE 3
Factors Influencing Size of Skin Test Reactions

Amount of allergen
Responsiveness of mast cells
Responsiveness of target tissue
Depth of injection
Body region (↑ on upper back)
Time of day (↑ in the evening)
Menstrual cycle (↑ prior to menstruation)
Age (↑ in third decade)
Associated diseases
(↑ with dermographic urticaria and erythroderma, ↓ with ichthyosis and atopic eczema)
Immunotherapy (↓)
Refractory state of the skin (after recent urtication at that site)
H1 antihistamines (↓)
Corticosteroids (with long, high-dose therapy)
Sympathomimetic drugs (↓)

From Czarnetzki, B. M., *Urticaria,* Springer-Verlag, Berlin, 1986. With permission.

IV. CHALLENGE TESTS

Conjunctival, nasal, and bronchial challenge tests are needed when occupational dermatological problems are associated with conjunctivitis, rhinitis, or asthma or when the shock organ is different from the skin. The oral mucosal challenge test may be useful in cases of food allergy. Peroral challenge tests may be needed in the differential diagnosis of generalized urticaria which may be of occupational origin. The above challenge tests have recently been reviewed by Hannuksela.[10]

TABLE 4
Some Occupational Allergens Commercially Now Available in RAST

Castor bean
Green coffee bean
Silk
Latex
Isocyanates
TDI
MDI
HDI
Phthalic anhydride
Trimellitic anhydride
Ethylene oxide
Chloramine T
Formalin

From Johansson, S. G. O., in *European Academy of Allergology and Clinical Immunology, 20 Years with IgE — New Prospects,* Debelic, M., Ed., Medicom, 1987, 23.

V. *IN VITRO* TESTS

The specific IgE level measured with the RAST[7,36] is of great importance in occupational allergology. The allergen is in most cases a protein, and chemicals have not usually been regarded as potential IgE-stimulating allergens. However, recent studies have shown that reactive chemicals can act as haptens and, after conjugation *in vivo* to a carrier, may stimulate an IgE antibody response.[36] Fortunately, new RASTs have been developed for industrial allergens (Table 4), and IgE-mediated occupational allergies can thus be verified with the RAST technique.[36] However, only IgE-mediated reactions will be revealed. Furthermore, skin tests are more sensitive for the detection of IgE-mediated allergy than RAST.[23,24] Other *in vitro* tests (e.g., the basophil degranulation test and histamine release from basophils) have so far been of little value in clinical occupational allergology,[10] but it seems evident that *in vitro* tests for allergy will most probably gain in importance in the future.

VI. IAST WITH VARIOUS TYPES OF ALLERGENS

In our clinic all patients with hints of their own or family atopy are prick tested with 20 common *inhalant allergens* (ALK, Copenhagen, Denmark) including epithelia, pollens, and house dust mite.[38] This is done to detect respiratory allergy and atopy and can be useful in the diagnosis of irritant contact dermatitis. However, Rystedt[39] and Lammintausta[40] did not consider a history of respiratory allergy without eczema as a risk factor for irritant hand eczema, but the opposite was found in another study.[41]

Patients with possible allergy to *spices* are prick tested with a "routine" series of ten spices (white pepper, allspice, cinnamon, ginger, paprika, cardamum, clove, coriander, garlic, and mustard).[30,42] We use powders without additives in a small drop of water. To detect allergy to *roots and vegetables,*[30,42] we have another "routine" series of nine fresh or freeze-dried substances: apple, potato, carrot, swede, celery, parsnip, parsley, onion, and tomato. Hannuksela[10] stated that edible roots and fruits such as potato, carrot, apple, and

TABLE 5
Foods and Beverages Causing Contact Urticaria; Probably of the Allergic Type

Beverages	Seafood
Alcohol (ethyl)	Fish
Beer	Lobster
Dairy	Prawns
Cheese	Shrimp
Egg	Vegetables, spices, and tobacco
Milk	Bamboo shoots
Fruits and berries	Beans
Apple	Cabbage
Apricot	Caraway seed
Banana	Carrot
Kiwi	Celery
Mango	Chives
Orange	Coriander
Peach	Cucumber
Plum	Curry
Strawberry	Endive
Grains	Lettuce
Corn	Mustard
Flour	Onion
Malt	Parsley
Wheat bran	Parsnip
Honey	Potato
Nuts and seeds	Rape
Almonds	Rutabaga (swede)
Peanut butter	Tobacco
Sesame seed	Tomato
Sunflower seed	Soybean
Meats and liver	Winged bean
Beef	
Chicken	
Lamb	
Turkey	

Note: The lists of Fisher,[13] Lahti and Maibach,[22] and Andersen et al.[21] is supplemented with more recent reports.[44-47]

celery loose their allergenicity when frozen, but we have had good experiences with freeze-dried test substances. There is a great need for commercial standardized allergens with proven efficacy.

Flours are a common cause of occupational rhinitis/asthma in bakers and farmers, and this may be associated with contact urticaria. Flours (wheat, barley, rye, and oats) are prick tested with noncommercial, freeze-dried flours, since our experience is that prick tests with the commercial flour test substances that we have used (Soluprick®, ALK, Copenhagen, Denmark) give false negative results. We believe that part of the allergens from the commercial flour test substances have disappeared during manufacturing of the flour prick test substances. There may be considerable variation between commercial test substances from different sources.[43] Flour RASTs (Pharmacia, Uppsala, Sweden) have given results with good clinical correlation, in agreement with our own prick test results.[93]

Foods are one of the most common causes of immediate-type hypersensitivity in occupations where raw foodstuffs are handled.[19] Foods that have been reported to cause contact urticaria are listed in Table 5.[13,21,22,42,44-47] In our test system, the patient brings in the raw food that he suspects to cause the allergy. We then apply the raw food to the skin and use the scratch chamber method for IAST. During the last few years we have increasingly used the prick test, which is easier to interpret. It has also been suggested that tests could be performed by applying the raw food to eczematized skin,[13,18,19] but we find it difficult to interpret the test result, since raw food often irritates eczematized skin even without allergy.

There is now an increasing number of *industrial substances* that have caused contact urticaria and allergic asthma (Table 1), but there is little information available on how to prepare test substances for IAST. Water-soluble substances are mixed in water, and allergens can easily be extracted (e.g., latex used for manufacturing of natural rubber gloves).[23] Chemicals mixed in petrolatum and used for patch testing sometimes give positive reactions (e.g., with reactive dyes[48] and with colophony[49]), but many chemicals (e.g., isocyanates in petrolatum) probably give false negative results.[50] Chemicals conjugated with human serum albumin have been reported to give more positive skin reactions,[51,52] but the conjugates are not very easily manufactured[52] and, therefore, are not good for screening high numbers of putative allergens.

In all cases of positive IAST 10 to 20 control subjects should be tested with the same test procedure, and these should be negative. If a substance is able to cause both nonimmunologic and immunologic contact urticaria, the mechanism cannot be revealed by IAST.

VII. OCCUPATIONAL ALLERGY TO SELECTED SUBSTANCES VERIFIED BY IAST

A. NATURAL RUBBER (LATEX)

Already in 1927 Stern[53] described a female patient who suffered from generalized urticaria associated with attacks of angioedema, caused by an immediate-type allergy from her rubber dentures. However, only during the last decade has latex allergy become a well-known entity.[23,54,55] Today, latex allergy is one of the most important occupational immediate hypersensitivity skin diseases. Latex allergy may manifest itself as contact urticaria, rhinitis, facial and eyelid edema, asthma, generalized urticaria, and anaphylactic reactions.[23] We have earlier reported on our first four cases up to 1985,[56,57] and since then we have had several new cases. Recently we have had patients who have had an asthmatic attack from exposure (use test) to only one finger of a latex glove. However, prick tests have been safe and have not caused generalized reactions. The prick test with latex is more reliable and reveals more allergy cases than the commercially available RAST.[23] Latex allergy has recently been reviewed by Turjanmaa.[23]

B. REACTIVE DYES

Reactive dyes have been used for textile dyeing since 1956[48] and are now used extensively in the textile industry because they produce bright colors; their application is fast, and their fixation is permanent in both natural and synthetic fibers. Of the dyes used in the textile industry in Finland in 1977, 29.7% were reactive dyes.[58] The first cases of asthma were reported in 1978,[59] and we reported the first cases of allergic contact dermatitis from reactive dyes in 1986.[60] So far we have detected five cases of occupational allergic dermatoses from reactive dyes (Table 6). The patients had been working in dye houses or textile plants and had been exposed to reactive dyes for 8 months to 4 years before symptoms developed. Four of the patients had eczema and one only urticaria as a skin symptom. The patient with urticaria also had rhinitis and asthma, and two others had rhinitis and/or asthma in addition to eczema.

The four patients with eczema were patch test positive to reactive dyes, but the patient with urticaria was not. On scratch and/or prick testing, the two patients who also had respiratory symptoms and/or urticaria reacted positively to the same dyes as on patch testing.

TABLE 6
Clinical Characteristics, Scratch, Scratch Chamber, Prick and Challenge Tests of Five Patients with Allergic Dermatoses from Reactive Dyes[a]

Patient no.	1	2	3	4	5
Age (years)	51	24	47	47	52
Sex	M	M	M	M	F
Occupation	Dye mixer	Dyer	Storeman/ machinist	Dyer	Dyer
Employer	Textile plant	Dye house	Dye house	Dye house	Textile plant
Skin symptoms					
Eczema	Yes	Yes	Yes	Yes	No
Urticaria	No	No	Yes	No	Yes
Respiratory symptoms					
Rhinitis	Yes	No	Yes	No	Yes
Asthma	No	No	Yes	No	Yes
Results of patch tests to reactive dyes					
Remazol dyes 1—2% pet.	Pos	ND	ND	Pos	Neg
Cibacron dyes 1—2% pet.	ND	Pos	ND	ND	ND
Drimaren dyes 1% pet.	ND	Neg	Pos	ND	ND
Results of scratch, scratch chamber, or prick tests					
Reactive dyes					
Remazol dyes 1—2% pet.	Pos	ND	ND	Neg	ND
Remazol dyes 1% aq.	ND	ND	ND	ND	Pos
Drimaren dyes 1% pet.	ND	ND	Pos	ND	ND
Standard series	Neg	ND	Neg	Neg	Neg
Results of RAST tests	ND	ND	Pos	Neg	ND
Results of challenge tests					
Nasal					
Remazol Schwarz B	Pos	ND	ND	ND	ND
Bronchial					
Drimaren Scharlach K 2G	ND	ND	Pos	ND	ND
Remazol Gold Gelb RNL	ND	ND	ND	ND	Pos

Note: ND = not done; Neg = negative; Pos = positive; M = male; F = female; aq = aqueous; pet. = petrolatum.

[a] Adapted from Reference 48.

Patient 3 (Table 6) was RAST positive to five reactive dyes: Drimaren Scharlach K 2G (score 3), Drimaren BR Yellow K 3GL (3), Drimaren BR Blue K BL (1), Levafix BR Yellow E 3G (3), and Cibacron BR Scarlet 3B (3). The RASTs were prepared by us and are not commercially available. Patient 1 was scratch chamber positive with undiluted commercial dye powders, and patients 3 and 4 with dyes at 1 to 2% (w/w). Patient 5 was prick tested with 1% dye in distilled water. Ten nonexposed controls were negative. The potential risk of anaphylactic reactions with reactive dyes is emphasized. One patient with reactive-dye asthma had a generalized urticarial reaction from a scratch chamber test with undiluted commercial reactive dye powder (Figure 1); therefore, we now prick test with 1% dye in distilled water. This has not caused side effects. Extreme care is necessary during bronchial challenge tests, and it should be remembered that even a lethal asthma attack has been described from reactive dye at work.[61] One of the patients was patch tested with fabric dyed with a reactive dye; he was negative, and there were no indications that dyed fabrics should act as allergens. Reactive dyes are good examples of chemicals causing both type I and type IV allergy, but the actual allergen(s) in the reactive dyes is/are not known because pure substances cannot be obtained for testing.[48] Type I and type IV allergy could be caused by different substances present in the reactive dye "mixture".

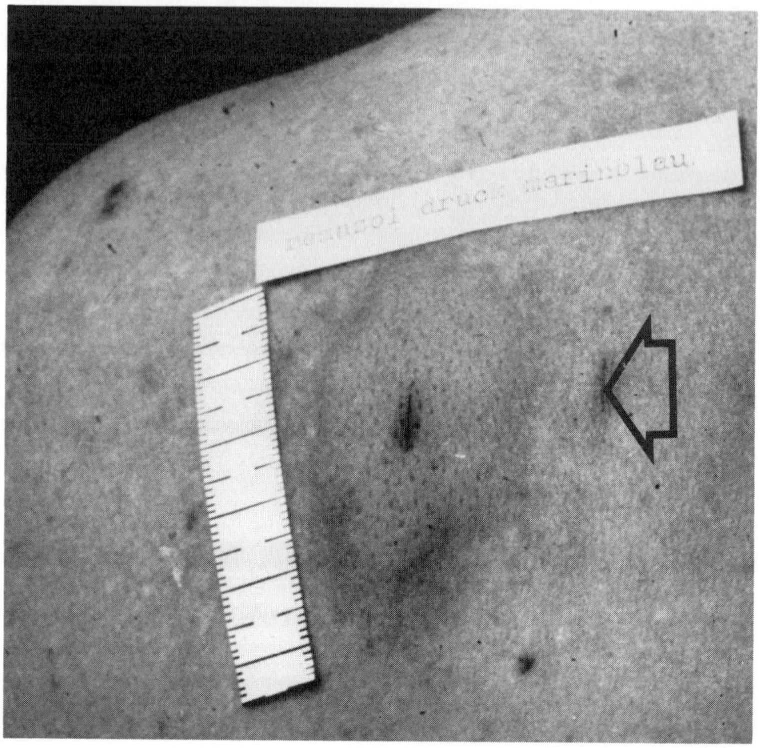

FIGURE 1. Abnormally strong scratch chamber test reaction caused by undiluted commercial reactive dye powder (Remazol® druck mainblau). The test was performed on the upper back, and the patient developed generalized urticaria after the test. Note the negative control scratch (arrow).

C. METALS

The following metals have been associated with asthma: cobalt, nickel, tungsten, cadmium, beryllium, platinum, vanadium, chromium, and zinc;[62] some of them are well-known contact allergens. Here we review some aspects of cobalt, chromium, and nickel which are best known to dermatologists.

1. Cobalt

The largest working population exposed to cobalt-containing dusts is found in the hardmetal industry. Several studies have detected wheezing and chest tightness after exposure to hard metal dust.[62] The prevalence of asthma in such workers has been estimated to be 5%.[63] It is not known whether respiratory symptoms following cobalt inhalation are caused by irritant or immunological events.[62] Cobalt has been listed as an agent causing nonallergic contact urticaria,[13] and there are no certain cases of allergic contact urticaria.[22]

2. Chromium

Chromium salts are well-known contact sensitizers; long ago chromium salts were reported to be associated with asthma in the chrome-plating industry,[64] and intracutaneous skin reactions to chrome salts were demonstrated in affected workers. Chromium or nickel in stainless steel welders has been considered an inciting factor of occupational asthma.[65] Asthma and IgE antibodies have been induced by chromium and nickel salts.[66]

3. Nickel

The first case of asthma due to inhalation and sensitization to nickel sulfate ($NiSO_4$) was reported in 1973,[67] and since then additional cases have been detected.[68,69] Skin testing revealed an immediate wheal and flare reaction, and IgE antibodies to a $NiSO_4$-human serum albumin conjugate were demonstrated. An asthma reaction to nickel sulfate without antibodies to nickel has also been reported.[70]

4. Scratch Chamber and Prick Tests to Metals

We have performed scratch chamber tests with metal salts on about 50 patients per year during the last 10 years. The following test concentrations (aqueous) have been used: six-valency chromate (1 mg/ml, 0.1 mg/ml, and 0.01 mg/ml), three-valency chromate (1 mg/ml and 0.1 mg/ml), $NiSO_4$ (10 mg/ml and 1 mg/ml) and cobalt chloride (1 mg/ml and 0.1 mg/ml). Very few patients have been skin test positive, and we do not consider this test method valuable in diagnosing chromate-, cobalt, or nickel-induced rhinitis/asthma. In most cases, rhinitis and asthma caused by these metals may not be of the conventional IgE-mediated type. One case of contact allergy, contact urticaria, and asthma due to nickel has been reported from our clinic.[71] Our patient was a 27-year-old woman who for 2 years had performed manual machine grinding of metal castings that contained 9% nickel. After 2 months of employment, she developed a longstanding cough and paroxysmal dyspnoe followed by occasional urticarial eruptions on both arms. During the second year of employment her cough and dyspnoe worsened markedly. On examination the patient had only a mild cough, and her skin was symptomless. Neither she nor her family was atopic. Inexpensive jewelry had caused dermatitis since she was 14 years old. A standard patch test series confirmed her allergic contact dermatitis to $NiSO_4$, but was otherwise negative. Prick tests to common inhalant allergens were negative. A scratch chamber test elicited a pseudopodial reaction to $NiSO_4$ (10 mg/ml). An open test on the volar aspect of the right arm with the same $NiSO_4$ solution elicited contact urticaria in 15 min. Bronchial and nasal provocation tests with the same $NiSO_4$ solution were positive. RAST to nickel was also positive, indicating that the immediate allergy was IgE-mediated. Nickel and cobalt may cause mast cell depletion on a nonimmunological mechanism,[13,18] but prick or scratch chamber tests with nickel and cobalt salts are negative in most patients, according to our experience. Cases of immediate nickel skin allergy have previously been reported, and they have been combined with a delayed allergy.[72-76] Scratch chamber tests to the above metals are often positive 24 to 96 h after challenge in patients with allergic contact dermatitis to these allergens, displaying an eczema reaction around the scratch and corresponding to the results obtained after conventional patch testing.

D. EPOXY RESINS

In a 10-year material (1974 to 1983) we had 1082 cases of occupational skin disease, of which 542 cases were allergic contact dermatitis.[8] Epoxy resins caused 71 cases (13.1%). During the same period we had 12 cases of occupational contact urticaria (1.1% of all occupational skin disease cases) and 9 (0.8%) cases of protein contact dermatitis. Only one case of occupational contact urticaria was caused by the epoxy resin system.[77] The patient was a worker in an electrical factory who had to fill condensers with a slightly warmed uncured mixture of an epoxy resin, a hardener, an accelerator, and a colorant. Within 2 months she developed urticarial eruptions on her face, neck, breast, and arms. The skin eruptions disappeared 2 weeks after she changed jobs. Open tests with the filling mixture and the hardener tested separately were positive, causing small urticas to develop on the test area. Tests with all other components were negative. Open tests with the undiluted hardener in 12 control persons were negative. Prick tests to common inhalant allergens were negative. The epicutaneous tests with standard series, series of plastics and glues, and the patient's own epoxy resin (1, 0.5, and 0.1% petrolatum [pet.]), hardener (1% pet.), accelerator (0.5% pet.), and colorant (1% pet.) were all negative. The causative hardener, according to the material safety sheet, was methylhexahydrophthalic acid anhydride (MHHPA).

At that time we stopped the test procedure after a positive open test and did not proceed to the prick tests according to the suggestions of von Krogh and Maibach (Table 2),[18] but today, in most cases, we perform prick and/or scratch chamber tests to industrial materials. Recently, we had cases of occupational asthma and rhinitis from MHHPA, and these patients had had positive prick tests to MHHPA and methyltetrahydrophthalic anhydride (MHTPA), respectively.[78,94]

Earlier, another anhydride used as an epoxy hardner (dodecenyl succinic anhydride) was reported to cause allergic contact dermatitis.[79] Phthalic anhydride, trimellitic anhydride, and other anhydrides are well-known industrial inhalant allergens (Table 1).[80] The aliphatic polyamine hardeners (e.g., triethylenetetramine), which are strong contact sensitizers,[13] can also cause occupational asthma[81] and contact urticaria.[13] In an older report, a patch test with epoxy resin (containing Epidian 1 to 3 and 5, Araldite® 175 and 185, Epoxyd 201, and ERL 4221) caused a generalized urticaria and bronchial asthma.[82] Romaguera et al.[83] reported a case of hand eczema, rhinitis, and conjunctivitis which was caused by epoxy resin. The product contained, in addition to Bisphenol A, a diglycidyl ether oligomer with a molecular weight of 340. We have recently had two patients with epoxy asthma who displayed a positive prick test to epoxy resin,[95] and we believe that more cases of immediate allergy to epoxy resin can be diagnosed by IAST.

E. ISOCYANATES

Isocyanates are used extensively in the manufacture of foams, films, elastomeric fibers, and surface coatings. Exposure to isocyanate resins occurs among spray painters, manufacturing workers, lacquer and adhesive workers, insulation installers, and those accidentally exposed to the pyrolysis products. Pulmonary irritation and sensitization are the chief hazards. Obstructive lung disease may result from short-term or prolonged exposure to isocyanates. The chief pulmonary sensitizers are toluene 2,4-diisocyanate (TDI), 4,4′-diphenylmethane diisocyanate (MDI), and 1,6-hexamethylene diisocyanate (HDI). Isocyanates are skin irritants, and only about 20 cases of allergic contact dermatitis from isocyanates have been reported to be caused by MDI, TDI, dicyclohexylmethane 4,4′-diisocyanate (DMDI), isophorone diisocyanate (IPDI), polymethylene polyphenyl isocyanate (PAPI), and trimethyl hexamethylene diisocyanate (TMDI).[50] We have recently described the first case of allergic contact dermatitis from HDI.[50] The patient was a car painter exposed to HDI-containing paints. He had HDI-induced asthma as well, but scratch chamber tests and RASTs to TDI, MDI, and HDI were negative. In order to obtain information about the diagnostic value of using patch test substances for scratch chamber tests (in petrolatum) and whether patients exposed to isocyanates would display allergic contact dermatitis, we skin tested 83 consecutive patients suspected of having occupational inhalant or skin diseases from isocyanates. The patch-test concentrations in petrolatum were: TDI and MDI, 5, 2, and 1%; HDI, 0.5, 0.2, and 0.1%. A total of 20 unexposed control persons, tested with these isocyanates at the above-mentioned concentrations, were negative. Scratch chamber tests were performed with 1% TDI, MDI, and HDI in pet.

Based on specific provocation tests, 34 of the 83 patients turned out to have an isocyanate-induced asthma and/or rhinitis. Only one of these patients was positive to isocyanates in scratch chamber tests, with TDI (3 mm) and HDI (3mm). This patient also had positive RASTs to TDI and HDI. He did not have dermatitis and was patch test negative. The study showed that scratch chamber tests with isocyanate test substances in petrolatum are not useful in the diagnosis of isocyanate-induced respiratory diseases. Allergic contact dermatitis was also rare. In addition to the HDI-allergic patient, we had a 28-year-old nonatopic female who, after 2 months of polyurethane spraying, developed a strongly itching dermatitis on her hands and respiratory symptoms soon thereafter. Patch tests with MDI (5, 2, and 1%) and TDI (5 and 2%) gave a 2 to 3+ positive allergic reaction; 1% TDI and HDI were

TABLE 7
Allergens of Vegetable Origin Causing Occupational Inhalant Allergy[a]

Allergen	Occupational environment
Bromelain (pineapple enzyme)	Pineapple processors, meat tenderizers
Castor oil bean	Vegetable oil manufacture, dock workers
Coffee (green coffee dust)	Plantation workers, coffee processors
Colophony (pine resin)	Tin-lead solderers (used as flux), hot melt glue workers
Cotton, flax, hemp dust	Textile workers
Devil's tongue root	Producers of the Japanese food, konnyaku
Grain/flour (barley, corn, oats, rye, wheat)	Farmers, grain handlers, bakers
Gums (acacia, arabic, guar, karaya)	Printing industry, paper industry, power cable laboratory
Mushrooms	Workers processing mushroom soup powder
Latex	Hospital and medical personnel, cleaners
Papain (proteolytic enzyme)	Pharmaceutical industry, meat tenderizer
Psyllium	Producers of the laxative Metamucil
Soya bean oil	Additive in polyvinylchloride; has been shown to be contributory to ''meat wrappers' asthma''
Soya bean flour	Food industry
Tea	Plantation workers, tea packagers
Tobacco	Tobacco harvesters, cigarette and tobacco workers
Woods	Carpenters, joiners, lumberworkers

[a] Adapted from Reference 86.

negative, while MDI was positive in an open test. Inhalation provocation tests proved that she also had MDI-induced asthma.

It is evident that isocyanates cause asthma by several mechanisms, and a rather small proportion (15 to 20%) have been estimated to be IgE-mediated.[84,85] We are now using human serum albumin-conjugated isocyanates for prick or scratch chamber tests, and these test materials[51] seem to give more positive skin tests than isocyanates in petrolatum. Our study indicates that HDI, MDI, and TDI are weak contact allergens in humans.

F. VEGETABLE GUMS

Many allergens causing occupational allergic contact urticaria or inhalant allergy are of plant origin (Table 7).[86] Vegetable gums are resinous plant exudates, and they can cause both type I and type IV allergy.[13] They occur as exudates from trees and shrubs in tropical areas. Guar gum is used in paper manufacture, foods, cosmetics, pharmaceuticals, interior coating of firehose nozzles, fracturing aid in oil wells, textiles, printing, polishing, atomic metal processing, and as a thickener and emulsifier. World production of guar gum is increasing, since new applications have been discovered (e.g., in medicine and the food industry). In Finland, guar gum has been used as a peroral medication for diabetes and hypercholesterolemia since 1982.[38]

Recently we reported three cases of occupational allergic rhinitis from one vegetable gum, guar gum.[38] The data and test results are summarized in Table 8. Patients 1 and 2 suffered from rhinitis and dermatitis. They were from the same workplace and were exposed to several chemicals, but they suspected that Emco Gum 563 (Meyhall Chemical AG, Switzerland), i.e., pure guar gum, caused their symptoms. After power cable experiments they had to open the cables and were then exposed to the fine guar gum powder, which was used as an insulator in rubber cables (Figure 2). Scratch chamber tests (Table 9), a nasal provocation test, nasal eosinophilia, and a RAST proved their allergy.

It is well known that nonstandardized vegetable allergens show considerable differences in quality. In order to obtain more information about the allergenic nature of guar gum, we used different commercially available guar gum products for skin testing in patients 1 and

TABLE 8
Characteristics of Three Patients with Allergic Occupational Rhinitis from Guar Gum[a]

	Patient 1	Patient 2	Patient 3
Age (years)	27	49	42
Sex	Male	Male	Male
Causative agent	Emco gum (guar gum)	Emco gum (guar gum)	Meyproid 5306 (guar gum)
Scratch chamber test with guar gum (see Table 9)	Positive	Positive	Positive
Nasal provocation test with guar gum	Positive	Positive	Positive
Amnestic atopy/family atopy	+/+	−/−	−/−
IgE	370 kU/ml	37 kU/ml	60 kU/ml
RAST tests to guar gum (score in parentheses)	4.7 PRU/ml (3)	1.1 PRU/ml (2)	ND
Spirometry with bronchodilation test and histamine provocation test	Normal	ND	Normal

Note: ND = not done, PRU = Phadebas RAST units.

[a] Adapted from Reference 38.

2 (Table 9). Scratch chamber tests revealed an allergic reaction to all substances containing guar gum. The commercially available products of guar gum contain mainly galactomannans (85 to 94%), and also protein (3 to 7%), raw fiber, ash, fat, and natural impurities. Because proteins are usually responsible for type I allergy it seems probable that the protein part acted as allergen, even though heat-treated guar gum (heat often destroys the allergenicity of a protein) gave positive skin reactions (Table 9). Another feature shown is encountered occasionally; i.e., a stronger scratch test response is noted with a more diluted substance. The depth of the scratch might influence the wheal/flare response, but with guar gum the undiluted test substance formed a thick hydrocolloid and the allergenic molecules might not be optimally apposed to the scratch, resulting in a weaker response.

Vegetable gum-induced allergic contact dermatitis has been reported from agar, gum arabic, karaya gum, and gum traganth.[13] Because patients 1 and 2 had suffered from dermatitis, they were patch tested with the allergens to which they had been exposed, including guar gum. They did not show type IV allergy to the vegetable gums, although on first testing patient 1 showed patch test irritation to guar gum (10% and 100%). Retesting revealed the irritative nature of the first patch test, and an airborne contact dermatitis from these substances was excluded.

G. CASTOR BEAN — LONG-LASTING CONTACT URTICARIA

The castor bean plant is known to have several components with toxic and/or allergenic properties. The residue left after extracting the oil can cause asthma, urticaria, and dermatitis among castor oil extractors, fertilizer workers, and farmers. The castor bean plant is cultivated in the tropics and in southern countries, but occupational rhinitis, asthma, and dermatoses from castor bean dust can also be found in the northern countries among workers in the coffee industry from contaminated burlap bags used to transport the coffee. We have recently reported a patient who attained both type I and type IV allergy and whose skin test reactions were quite unique and interesting from many points of view.[87,88] The contact urticarial reaction persisted for more than 48 h and was therefore called "long-lasting contact urticaria".[88]

A 50-year-old male stockroom worker in a coffee-roasting plant had been in daily airborne

FIGURE 2. Two patients developed allergic rhinitis from guar gum. After power cable experiments they opened the cables and were exposed to the fine guar gum powder (arrow), which was used as an insulator in rubber cables. The diameter of the cable was about 3 cm.

contact with sacks of green coffee beans, thus being exposed to the dust of the coffee sacks. He himself had not handled sacks, coffee beans, or castor beans. He developed rhinitis, facial dermatitis, and periodic dermatitis on the backs of his hands on work days after 22 years of employment. Scratch chamber tests with crushed green coffee bean gave a histamine-sized (3 mm) reaction. Crushed castor bean, as well as dust from the sacks of the patients' work place, gave a very strong cutaneous reaction (20- to 35-mm wheal with pseudopodia). A scratch chamber test with crushed castor bean was negative in ten controls (five atopics and five nonatopics). A nasal provocation test (crushed castor bean diluted 1:10,000 in physiological saline) was positive. A RAST for castor bean was positive (score 3). Crushed castor bean tested as such with the Finn Chamber® closed epicutaneous test method gave a 30 × 40 mm urticarial reaction at 5 h and a 55 × 125 mm urticarial reaction at 24 h, when the occlusion was removed. The urticarial component had diminished slightly at 48 h, measuring 55 × 110 mm, and had disappeared at the 3-d reading. At 4 d a classical strong vesicular (3+) reaction (diameter 10 mm) was observed. A total of 20 control persons (10 atopics and 10 nonatopics) were tested with the same crushed castor bean and were all negative both for an urticarial and a delayed patch test reaction. Although our observations were interesting, it should be stressed that skin tests should not be performed with castor bean because of its toxicity.[88]

Light and electron microscopy in the 5-h reaction showed lymphocytes, macrophages, neutrophils, eosinophils, and mast cells. In the 4-d reaction eosinophils were still present, in contrast to normal allergic patch tests.[89-91] Eosinophils were degenerated and partly degranulated; granules were also detected free in the interstitial tissue (Figure 3). The electron microscopic observations indicate that eosinophils and mast cells are activated and participate in patch test reactions where both type I and type IV allergy reactions occur. Recently, the role of eosinophils in this type of patch test reaction has been emphasized.[16,17]

TABLE 9
Scratch-Chamber Tests with Vegetable Gums

		Patient 1	Patient 2
Scratch-chamber test with Emco gum			
10% Emco gum (guar gum)		2/30 + p.s.	4/15
100% Emco gum (guar gum)[a]		2/30 + p.s.	4/20
Control		—	—
Histamine 10 mg/ml		2/30	5/20
Scratch-chamber tests with other vegetable gums			
Vidocrem® (heat-treated guar gum)	Undilated[a]	2/20	10/30 + p.s.
	1%	8/25 + p.s.	5/30 + p.s.
Vidogum FS 21® (mixture of guar gum and locust bean gum)	Undilated[a]	7/30 + p.s.	3/20
	1%	2/15	3/20
Vidogel NR 20® (mixture of galactomannans and hydrocolloids)	Undilated[a]	2/10	2/10
	1%	2/20	—
Guarem® (pure guar gum)	Undilated[a]	2/20	7/30 + p.s.
	1%	12/30 + p.s.	5/30 + p.s.
Glupec® (glucomannan-psyllium compound)	1% aqua	—	—
Gum arabic	10% aqua	—	—
Control		—	—
Histamine 10 mg/ml		3/25	7/30

Note: Numbers show wheal/flare in millimeters; p.s. = pseudopodia.

[a] Mixed in a drop of water.

From Kanerva, L., Tupasela, O., Jolanki, R., Vaheri, E., Estlander, T., and Keskinen, H., *Clin. Allergy,* 18, 245, 1988. With permission.

VIII. MECHANISM OF CONCOMITANT TYPE I AND TYPE IV ALLERGIC SKIN REACTIONS

There is now increasing evidence that type I allergic reactions might be associated with type IV allergic reactions more often than was previously believed.[11-18] Bruynzeel-Koomen and Brynzeel[17] have recently presented a new hypothesis linking type I and type IV hypersensitivity reactions when patients with atopic dermatitis are patch tested with inhalant allergens (Figure 4). Having penetrated the horny layer, ''type I allergens'' (e.g., common inhalant allergens) could bind to IgE molecules on epidermal Langerhans' cells (LC)[15] or dermal mast cells. This could be responsible for the immediate skin reactions observed 20 to 360 min after challenge. Simultaneously, eosinophil chemotactic factors may be generated, resulting in an early influx of eosinophils into the epidermis and the dermis. LCs may simultaneously induce a T-lymphocyte proliferative response. LC-eosinophil apposition[16,17] suggests interaction between these cells, although its significance is not clear. This hypothesis would explain how inhalant allergens via skin contact exacerbate the eczema in atopic dermatitis. However, as shown in Figure 4, there are still numerous questions to be clarified, and at the moment the significance of IgE on LCs is not clear. Interestingly, in the above studies, ''type I allergens'' were used for type IV allergy skin testing. This ought to encourage clinicians to explore new skin testing systems that may give us new insight into allergy.

IX. HYPOTHETICAL MODEL EXPLAINING ALLERGIC DERMATITIS FROM SYSTEMICALLY ADMINISTERED FOODSTUFFS

The scheme of Bruynzeel-Koomen and Bruynzeel[17] could be extended to explain the clinically well-known dermatitis reaction in atopics after intake of allergenic foods (Figure

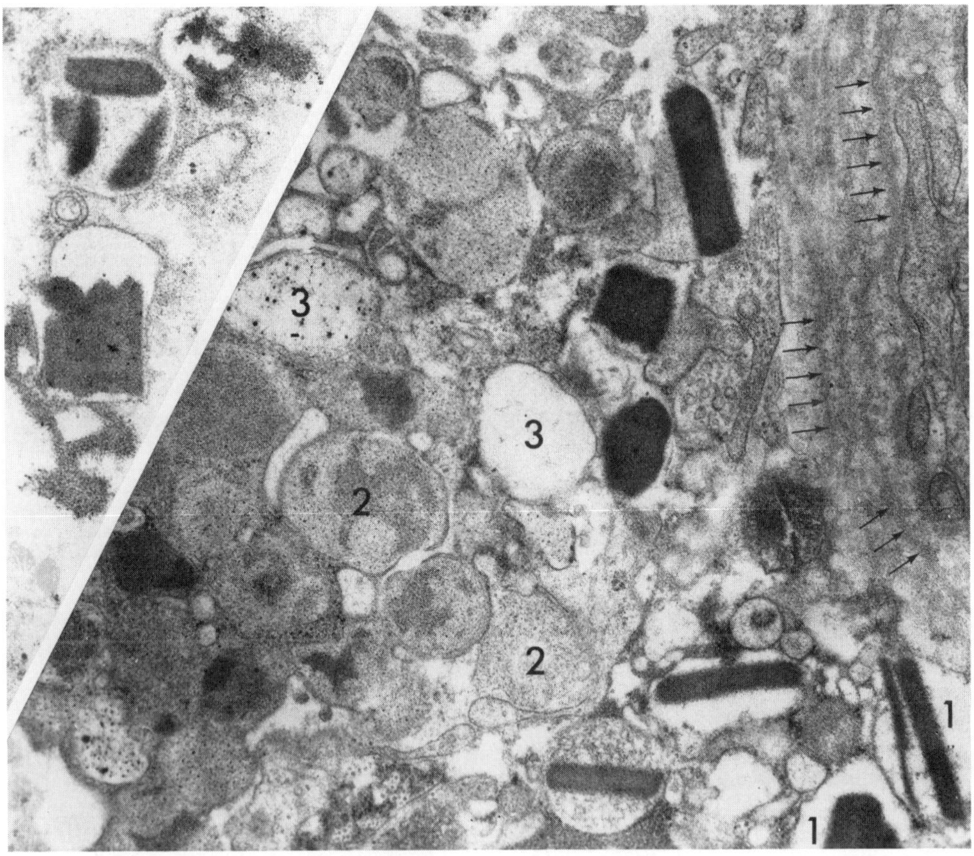

FIGURE 3. Electron micrograph showing partly degranulated perivascular eosinophils in the upper dermis. The biopsy was taken from an urticarial reaction 5 h after closed patch test with castor bean. The matrix (1), the core (2), or both (3) have dissolved from the granules. Granules from eosinophils were also detected free in the dermis (inset). Arrows show multiple arrays of basal lamina of dermal vessel. (Magnification × 24,300; inset × 31,500.)

4). The hypothesis would also explain allergic skin reactions caused by inhalant allergens (e.g., environmental airborne chemicals) reaching the circulation and the skin via the respiratory tract. In patients with skin allergy from foods or chemicals, the allergens would reach the dermis and epidermis via the circulation and bind to the specific IgE molecules of epidermal and dermal LCs or mast cells (Figure 4). This could cause an immediate (minutes, hours) or a delayed (hours, days) skin reaction, or both types of reactions, concomitantly. The diagnosis of a delayed food allergy reaction could, however, be difficult to make. First, the actual allergen in the circulation reaching the skin may be different from the substance the patient was exposed to because it may have been metabolically modified in the body. A chemical (hapten) could bind to a different protein when entering the skin via the circulation (Figure 4, right) than when reaching the immunocompetent cells through the horny layer and the epidermis (Figure 4, left), in this way becoming immunologically different. This would explain why food allergens and chemicals could be negative in IAST or patch testing, although exposure to the allergen *per os* causes dermatitis. Because the allergen would be modified, a RAST would also be negative. Second, it may take several days for a delayed allergic reaction to develop. An exacerbated dermatitis several days after challenge is, in most instances, very difficult to connect to the exposure. Therefore, even with this model, researchers face the challenge of developing methods for helping clinicians make a correct diagnosis of allergic dermatitis caused by peroral or respiratory intake of

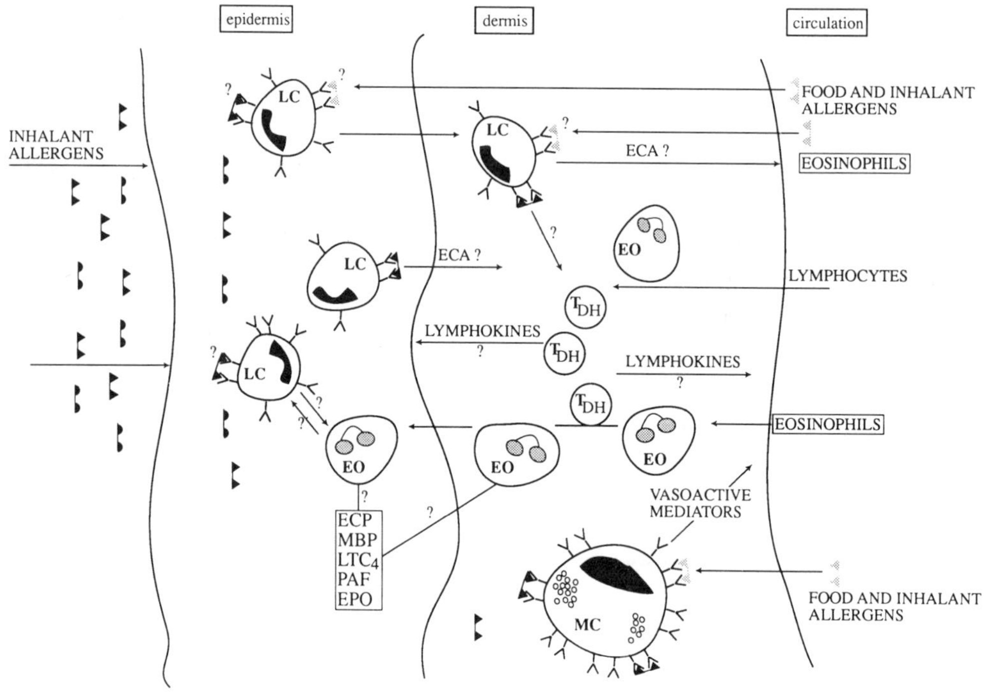

FIGURE 4. Hypothetical scheme of the reaction sequence taking place when patients with atopic dermatitis are patch tested with inhalant allergens based on the scheme of Bruynzeel-Koomen, and Bruynzeel[17] (left) or when challenged *per os* or via the respiratory tract with allergens causing allergic dermatitis via circulation (right). Abbreviations used: LC = Langerhans' cell; EO = eosinophil; TDH = T-lymphocyte participating in the delayed-type hypersensitivity reaction; MC = mast cell; ECA = eosinophil chemotactic activity; ECP = eosinophil cationic protein; MBP = major basic protein; LTC_4 = Leukotriene C4; PAF = platelet activating factor; EPO = eosinophil peroxidase.

foodstuffs and inhalant allergens, including chemicals. Another type of skin allergy, systemic contact dermatitis, which has to be patch test positive before this term can be used, has been reviewed by Fisher[13] and Menné and Hjorth.[92]

X. CONCLUDING REMARKS

In the present chapter the common way of doing skin testing for immediate hypersensitivity in occupational allergology, as we at an occupational dermatology clinic do it, has been reviewed. The main emphasis has been directed toward immediate hypersensitivity reactions with an immunological background. In most cases we prefer prick tests and scratch chamber tests. A number of examples on how the diagnosis of an occupational allergic disease has been made and confirmed are given. Currently, a great number of new agents causing immediate hypersensitivity reactions are detected. Because skin tests are an extremely valuable diagnostic tool, we feel that dermatologists need to become acquainted with these techniques — and perform the tests. For an occupational dermatology clinic that wants to keep its diagnostic capacity at a high level, immediate skin testing ought to be a routine procedure.

REFERENCES

1. **Blackley, C. H.,** *Experimental Researches on the Causes and Nature of Catarrhus Aestivus,* Bailliere, Tindall and Cox, London, 1873.
2. **Schloss, O. M.,** Allergy in infants and children, *Am. J. Dis. Child.,* 19, 433, 1920.
3. **Cooke, R. A.,** Cutaneous reactions in human hypersensitiveness, *Proc. N.Y. Pathol. Soc.,* 21, 8, 1921.
4. **Portier, P. G. and Richet, C.,** De l'action anaphylactique de certain venins, *C.R. Soc. Biol.,* 54, 170, 1902.
5. **Prausnitz, K., and Küstner, H.,** Studien über die Überempfindlichkeit, *Zentralbl. Bakteriol.,* 86, 160, 1921.
6. **Ishizaka, K., Ishizaka, T., and Hornbrook, M. M.,** Physicochemical properties of reaginic antibody. V. Correlation of reaginic activity with E globulin antibody, *J. Immunol.,* 97, 840, 1966.
7. **Wide, L., Bennich, H., and Johansson, S. G. O.,** Diagnosis of allergy by an *in vitro* test for allergen antibodies, *Lancet,* 2, 1105, 1967.
8. **Kanerva, L., Estlander, T., and Jolanki, R.,** Occupational skin disease in Finland. An analysis of 10 years of statistics from an occupational dermatology clinic, *Int. Arch. Occup. Environ. Health.,* 60, 89, 1988.
9. **Goldstein, R. A., Sogn, D. D., and Ayres, J.,** Occupational and environmental immunologic lung disease: a perspective, in *Immunotoxicology and Immunopharmacology,* Dean, J. et al., Eds., Raven Press, New York, 1985.
10. **Hannuksela, M.,** Tests for immediate hypersensitivity, in *Occupational and Industrial Dermatology,* 2nd ed., Maibach, H. I., Ed., Year Book Medical Publishers, Chicago, 1987, 168.
11. **Mitchell, E. B., Crow, J., Chapman, M. D., Jouhal, S. S., Pope, F. M., and Platts-Mills, T. A. E.,** Basophils in allergen-induced patch test sites in atopic dermatitis, *Lancet,* 1, 127, 1983.
12. **Platts-Mills, T. A. E., Mitchell, E. B., Rowntree, S., Chapman, M. D., and Wilkins, S. R.,** The role of dust mite allergens in atopic dermatitis, *Clin. Exp. Dermatol.,* 8, 233, 1983.
13. **Fisher, A. A.,** *Contact Dermatitis,* Lea & Febiger, Philadelphia, 1986.
14. **Reitamo, S., Visa, K., Kähkönen, K., Käyhkö, K., Stubb, S., and Salo, O. P.,** Eczematous reactions in atopic patients caused by epicutaneous testing with inhalant allergens, *Br. J. Dermatol.,* 114, 303, 1986.
15. **Bruynzeel-Koomen, C. A. F. M., van Wichen, D. F., Toonsra, J., Berrens, L., and Bruynzeel, P. L. B.,** The presence of IgE molecules on epidermal Langerhans cells in patients with atopic dermatitis, *Arch. Dermatol. Res.,* 278, 199, 1986.
16. **Bruynzeel-Koomen, C. A., van Wichen, D. F., Spry, C. J., Venge, P., and Bruynzeel, P. L.,** Active participation of eosinophils in patch test reactions to inhalant allergens in patients with atopic dermatitis, *Br. J. Dermatol.,* 118, 229, 1988.
17. **Bruynzeel-Koomen, C. A. and Bruynzeel, P. L.,** A role for IgE in patch test reactions to inhalant allergens in patients with atopic dermatitis, *Allergy,* 43 (Suppl. 5), 15, 1988.
18. **von Krogh, G. and Maibach, H. I.,** The contact urticaria syndrome — an updated review, *J. Am. Acad. Dermatol.,* 5, 328, 1981.
19. **Hjorth, N. and Roed-Petersen, J.,** Occupational protein contact dermatitis in food handlers, *Contact Dermatitis,* 2, 28, 1976.
20. **Hannuksela, M.,** Atopic contact dermatitis, *Contact Dermatitis,* 6, 30, 1980.
21. **Andersen, K. E., Benezra, C., Burrows, D., Camarasa, J., Dooms-Goossens, A., Ducombs, G., Frosch, P., Lachapelle, J. M., Lahti, A., Menné, T., Rycroft, R., Scheper, R., White, I., and Wilkinson, J.,** Contact Dermatitis. A Review, *Contact Dermatitis,* 16, 55, 1987.
22. **Lahti, A. and Maibach, H. I.,** Immediate contact reactions (contact urticaria syndrome), in *Occupational and Industrial Dermatology,* 2nd ed., Maibach, H. I., Ed., Year Book Medical Publishers, Chicago, 1987, 32.
23. **Turjanmaa, K.,** Latex glove contact urticaria, *Acta Univ. Tamperensis Ser. A,* 254, 1, 1988.
24. **van der Zee, J. S., de Groot, H., van Swieten, P., Jansen, H. M., and Aalberse, R. C.,** Discrepancies between the skin test and IgE antibody assays: study of histamine release, complement activation *in vitro,* and occurrence of allergen-specific IgG, *J. Allergy Clin. Immunol.,* 82, 270, 1988.
25. **Czarnetzki, B. M.,** *Urticaria,* Springer-Verlag, Berlin, 1986.
26. **Lahti, A.,** Nonimmunologic contact urticaria, *Acta Derm. Venereol.,* 60 (Suppl. 91), 1, 1980.
27. **Hannuksela, M.,** Contact urticaria from foods, in *Nutrition and the Skin,* Roe, D. A., Ed., Alan R. Liss, New York, 1986, 153.
28. **Nelson, H. S.,** Diagnostic procedures in allergy. I. Allergy skin testing, *Ann. Allergy,* 51, 411, 1983.
29. **Chanal, I., Horst, M., Segalen, C., Dreborg, S., Michel, F. B., and Bousquet, J.,** Comparison between modified skin prick test with standardized allergen extracts and Phazet, *J. Allergy Clin. Immunol.,* 82, 878, 1988.
30. **Hannuksela, M. and Lahti, A.,** Immediate reactions to fruits and vegetables, *Contact Dermatitis,* 3, 79, 1977.

31. **Voorhorst, R. and van Krieken, H.**, Atopic skin test reevaluated. II. Variability in results of skin testing done in duplicate, *Ann. Allergy*, 31, 195, 1973.
32. **Tuft, L. and Blumstein, G. I.**, Studies in food allergy. II. Sensitization to fresh fruits: clinical and experimental observations, *J. Allergol.*, 13, 574, 1942.
33. **Reunala, T., Björksten, F., Förström, L., and Kanerva, L.**, IgE-mediated occupational allergy to the spider mite, *Clin. Allergy*, 13, 383, 1983.
34. **Lemanske, R. F. and Kaliner, M.**, Late-phase IgE-mediated reactions, *J. Clin. Immunol.*, 8, 1, 1988.
35. **Dorsch, W.**, IgE and common allergic diseases. Studies on pathogenesis and pharmacotherapy of late phase reactions, *Allergy*, 43 (Suppl. 5), 38, 1988.
36. **Johansson, S. G. O.**, *In vitro* diagnosis of IgE-mediated allergy, in *European Academy of Allergology and Clinical Immunology, 20 years with IgE — New Prospects*, Debelic, M., Ed., Medicom, 1987, 23.
37. **Daughters, D., Zackheim, H., and Maibach, H. I.**, Urticaria and analphylactoid reactions after topical application of mechlorethamine, *Arch. Dermatol.*, 107, 429, 1973.
38. **Kanerva, L., Tupasela, O., Jolanki, R., Vaheri, E., Estlander, T., and Keskinen, H.**, Occupational allergic rhinitis from guar gum, *Clin. Allergy*, 18, 245, 1988.
39. **Rystedt, I.**, Hand eczema and long-term prognosis in atopic dermatitis, *Acta Derm. Venereol., Suppl.*, 117, 1, 1985.
40. **Lammintausta, K.**, Risk Factors for Hand Dermatitis in Wet Work. Atopy and Contact Sensitivity in Hospital Workers, Ph.D. thesis, University of Turku, Turku, Finland, 1982.
41. **Shmunes, E. and Keil, J. E.**, The role of atopy in occupational dermatoses, *Contact Dermatitis*, 11, 174, 1984.
42. **Niinimäki, A.**, Scratch-chamber tests in food handler dermatitis, *Contact Dermatitis*, 16, 11, 1987.
43. **Sampson, H. A.**, Comparative study of commercial food antigen extracts for the diagnosis of food hypersensitivity, *J. Allergy Clin. Immunol.*, 82, 718, 1988.
44. **Fisher, A. A.**, Contact urticaria and anaphylactoid reaction due to corn starch surgical glove powder, *Contact Dermatitis*, 16, 224, 1987.
45. **Kitajima, T.**, Contact allergy caused by bamboo shoots, *Contact Dermatitis*, 15, 100, 1986.
46. **Meding, B.**, Immediate hypersensitivity to mustard and rape, *Contact Dermatitis*, 13, 121, 1985.
47. **Tosti, A., Melino, M., and Veronesi, S.**, Contact urticaria to tobacco, *Contact Dermatitis*, 16, 225, 1987.
48. **Estlander, T.**, Allergic dermatoses and respiratory diseases from reactive dyes, *Contact Dermatitis*, 18, 290, 1988.
49. **Rivers, J. K. and Rycroft, R. J. G.**, Occupational allergic contact urticaria from colophony, *Contact Dermatitis*, 17, 181, 1987.
50. **Kanerva, L., Lähteenmäki, M. T., Estlander, T., Jolanki, R., and Keskinen, H.**, Allergic contact dermatitis from isocyanates, in *Current Topics in Contact Dermatitis*, Frosch, P. J., Dooms-Goossens, A., Lachapelle, J.-M., Rycroft, R. J. G., and Scheper, R. J., Eds., Springer-Verlag, Berlin, 1989, 368.
51. **Baur, X.**, Immunologic cross-reactivity between different albumin-bound isocyanates, *J. Allergy Clin. Immunol.*, 71, 197, 1983.
52. **Wass, U., Belin, L., and Delin, K.**, Longitudinal study of specific IgE and IgG antibodies in a patient sensitized to ethylene oxide through dialysis, *J. Allergy Clin. Immunol.*, 82, 679, 1988.
53. **Stern, G.**, Überempfindlichkeit gegen Kautschuk als Ursache von Urticaria und quinckeschem Oedem, *Klin. Wschr.*, 6, 1096, 1927.
54. **Nutter, A. F.**, Contact urticaria to rubber, *Br. J. Dermatol.*, 101, 597, 1979.
55. **Wrangsjö, K., Wahlberg, J. E., and Axelsson, G. K.**, IgE-mediated allergy to natural rubber in 30 patients with contact urticaria, *Contact Dermatitis*, 19, 264, 1988.
56. **Estlander, T., Jolanki, R., and Kanerva, L.**, Dermatitis and urticaria from rubber and plastic gloves, *Contact Dermatitis*, 14, 20, 1986.
57. **Estlander, T., Jolanki, R., and Kanerva, L.**, Contact urticaria from rubber gloves: a detailed description of four cases, *Acta Derm. Venereol. Suppl.*, 134, 98, 1987.
58. **Estlander, T. and Jolanki, R.**, Tekstili- ja vaatetusteollisuus, kemialliset aineet ja ihottumat, Työterveyslaitos, Helsinki, 1980 (in Finnish).
59. **Alanko, K., Keskinen, H., Björksten, F., and Ojanen, S.**, Immediate-type hypersensitivity to reactive dyes, *Clin. Allergy*, 8, 25, 1978.
60. **Estlander, T. and Kanerva, L.**, Dermatitis and urticaria from reactive dyes, in *8th Int. Symp. Contact Dermatitis*, Cambridge, March 20 to 22, 1986, 48.
61. **Stern, M. A.**, Occupational asthma from a reactive dye, *Ann. Allergy*, 55, 264, 1985.
62. **Fink, J. N.**, Occupational asthma from inorganic chemicals, in *Clinics in Immunology and Allergy. Occupational Respiratory Allergy*, Vol. 4, Pepys, J., Ed., W. B. Saunders, Philadelphia, 1984, 125.
63. **Key, M. M.**, Some unusual allergic reactions in industry, *Arch. Dermatol.*, 83, 57, 1961.
64. **Joules, H.**, Asthma from sensitization to chromium, *Lancet*, 2, 182, 1932.
65. **Keskinen, H., Kalliomäki, P. L., and Alanko, K.**, Occupational asthma due to stainless steel welding fumes, *Clin. Allergy*, 10, 151, 1980.

66. **Novey, H. S., Habib, M., and Wells, I. D.,** Asthma and IgE antibodies induced by chromium and nickel salts, *J. Allergy Clin. Immunol.*, 72, 407, 1983.
67. **McConnel, L., Fink, J. N., Schlueter, D. P., and Schmidt, M. G.,** Asthma caused by nickel sensitivity, *Ann. Intern. Med.*, 78, 888, 1973.
68. **Block, G. T. and Yeung, M.,** Asthma caused by nickel, *J. Am. Med. Assoc.*, 247, 1600, 1982.
69. **Malo, J. Cartier, A., Doepner, M., Nieboer, E., and Dolovich, J.,** Occupational asthma caused by nickel sulfate, *J. Allergy Clin. Immunol.*, 69, 55, 1982.
70. **Malo, J. L., Cartier, A., Gagnon, G., Evans, S., and Dolovich, J.,** Isolated late asthmatic reaction due to nickel sulphate without antibodies to nickel, *Clin. Allergy,* 15, 95, 1985.
71. **Estlander, T., Keskinen, H., Vaheri, E., and Kuuliala, O.,** Contact allergy, contact urticaria and asthma due to nickel, in *7th Int. Symp. Contact Dermatitis,* Erfurt, G.D.R., June 3 to 5, 1983, 34.
72. **Stoddart, J. C.,** Nickel sensitivity as a cause of infusion reactions, *Lancet,* 2, 741, 1960.
73. **Holti, G.,** Immediate and Arthus-type hypersensitivity to nickel, *Clin. Allergy,* 4, 437, 1974.
74. **Osmundsen, P. E.,** Contact urticaria from nickel and plastic additive (butylhydroxytoluene, oleylamide), *Contact Dermatitis,* 6, 452, 1980.
75. **Tosti, A., Melino, M., Labanca, M., and Ragazzi, R.,** Immediate hypersensitivity to nickel, *Contact Dermatitis,* 15, 95, 1986.
76. **Valsecchi, R. and Cainelli, T.,** Contact urticaria from nickel, *Contact Dermatitis,* 17, 187, 1987.
77. **Jolanki, R., Estlander, T., and Kanerva, L.,** Occupational contact dermatitis and contact urticaria caused by epoxy resins, *Acta Derm. Venereol., Suppl.*, 134, 90, 1987.
78. **Keskinen, H., Nordman, H., Tupasela, O., Vaheri, E., Pfäffli, P., and Sarjanen, M.,** Methyl hexahydrophthalic anhydride (MHHPA) induced asthma and rhinitis, in Int. Congr. Allergology and Clinical Immunology, Montreal, Switzerland, October 16 to 21, 1988; N. Engl. Reg. Allergy Proc., 9, 397, 1988.
79. **Göransson, K.,** Allergic contact dermatitis to an epoxy hardener: dodecenyl-succinic anhydride, *Contact Dermatitis,* 5, 277, 1977.
80. **Bernstein, I. L. and Bernstein, D. I.,** Respiratory allergy to synthetic resins, in *Clinics in Immunology and Allergy. Occupational Respiratory Allergy,* Vol. 4, Pepys, J., Ed., W. B. Saunders, Philadelphia, 1984, 483.
81. **Newman-Taylor, A. J. and Venables, K. M.,** Clinical and epidemiological methods in investigating occupational asthma, in *Clinics in Immunology and Allergy. Occupational Respiratory Allergy,* Vol. 4, Pepys, J., Ed., W. B. Saunders, Philadelphia, 1984, 43.
82. **Woytoń, Wąsik, F., and Bliżanowska, A.,** An uncommon anaphylactic and eczematous reaction after contact with epoxide resins (in Polish), *Przeg Derm,* LXI, 3, 303, 1974.
83. **Romaguera, C., Grimalt, F., and Vilaplana, Y. J.,** Dermatitis de contacto profesional y por resina epoxi, *Med. Segur. Trab.*, 31, 63, 1984.
84. **Davies, J. R.,** Respiratory hypersensitivity to diisocyanates, in *Clinics in Immunology and Allergy. Occupational Respiratory Allergy,* Vol. 4, Pepys, J., Ed., W. B. Saunders, Philadelphia, 1984, 103.
85. **Keskinen, H., Tupasela, O., Tiikkainen, U., and Nordman, H.,** Experiences of specific IgE in asthma due to diisocyanates, *Clin. Allergy,* 18, 597, 1988.
86. **Hendrick, D. J., Butcher, B. T., and Connolly, M. J.,** Occupational asthma due to allergens of organic origin, in *Clinics in Immunology and Allergy. Occupational Respiratory Allergy,* Vol. 4, Pepys, J., Ed., W. B. Saunders, Philadelphia, 1984, 437.
87. **Kanerva, L., Jolanki, R., Estlander, T., and Niemi, K. M.,** Light and electron microscopic observations of an epicutaneous patch test showing both immediate (type I) and delayed (type IV) allergy, in *Immunodermatology,* Caputo, R., Ed., CIC Editioni Internationali, Rome, 1987, 255.
88. **Kanerva, L., Estlander, T., and Jolanki, R.,** Long-lasting contact urticaria from castor bean, *J. Am. Acad. Derm.*, in press.
89. **Ranki, A., Kanerva, L., Förström, L., Konttinen, Y., and Mustakallio, K. K.,** T and B lymphocytes, macrophages and Langerhans cells during the course of contact allergic and irritant skin reactions in man. An immunohistochemical and electron microscopic analysis, *Acta Derm. Venereol.*, 63, 376, 1983.
90. **Kanerva, L., Ranki, A., and Lauharanta, A.,** Lymphocytes and Langerhans cells in patch tests. An immunohistological and electron microscopic study, *Contact Dermatitis,* 11, 150, 1984.
91. **Kanerva, L., Estlander, T., and Ranki, A. M.,** Lymphocytes and Langerhans cells in allergic patch tests, *Dermatosen,* 35, 16, 1987.
92. **Menne, T. and Hjorth, N.,** Reactions from systemic exposure to contact allergens, Semin. Dermatol., 1, 15, 1982.
93. **Kanerva, L., Estlander, T., and Jolanki, R.,** unpublished results.

94. **Kanerva, L., Jolanki, R., Estlander, T., Tarvainen, K., Tupasela, O., and Niemi, K.-M.,** Long-lasting IgE-mediated skin prick test reaction from methyltetrahydrophthalic anhydride (MHTPA), in 17th Annu. Meet. Society for Cutaneous Ultrastructure Research, Rome, May 4 to 5, 1990, 18.
95. **Kanerva, L., Jolanki, R., Halmepuro, L., Keskinen, H., and Estlander, T.,** Immediate and delayed allergy from diglycidyl ether bisphenol a epoxy resin, in 9th Int. Symp. Contact Dermatitis, Stockholm, May 17 to 19, 1990, 53.

Chapter 9

DIAGNOSTIC PROCEDURES FOR ECZEMA PATIENTS

Niels K. Veien

TABLE OF CONTENTS

I. INTRODUCTION

A. DEFINITION

Eczema is defined as an inflammatory reaction in the dermis and epidermis. Clinically speaking, eczema usually presents with pruritus, erythema, edema, infiltration, and various forms of epidermal reaction, from papules and vesicles to scales, fissures, and crusts.

B. TYPES OF ECZEMA

The diagnosis of eczema will normally fall within one of the following groups:

- Seborrheic dermatitis
- Atopic dermatitis
- Contact dermatitis
  - Allergic contact dermatitis
  - Irritant contact dermatitis
  - Mechanical contact dermatitis
  - Photo contact dermatitis
    - Allergic photo contact dermatitis
    - Other types of photo contact dermatitis
  - Protein contact dermatitis
- Lichen simplex chronicus (neurodermatitis)
- Nummular eczema
- Hyperkeratotic hand and/or foot eczema
- Pompholyx
- Gravitational eczema (stasis eczema)
- Eczema due to malnutrition and metabolic diseases
- Asteatotic eczema
- Pityriasis alba

Certain clinical characteristics can often be used to identify some types of eczema. The diagnoses of seborrheic dermatitis and atopic dermatitis are usually based, for example, on the pattern of the dermatitis.[1,2] The morphology and anatomical site are used to diagnose pompholyx, hyperkeratotic hand and/or foot eczema, and stasis eczema.

A diagnosis of allergic contact dermatitis usually requires confirmatory immunological testing, while a diagnosis of irritant contact dermatitis is usually made when the history of the patient reveals relevant exposures and no sensitization has been demonstrated. It is necessary to make a specific diagnosis in order to provide the patient with correct information regarding the prognosis of the disease and to treat it properly. It is particularly important to identify the type of eczema when dealing with occupational dermatoses.

C. AGGRAVATING FACTORS

Some diagnostic tests are carried out in order to place the diagnosis of eczema in one or more of the categories mentioned in Section I.B. In addition, it is important to carry out tests to determine which factors aggravate the eczema in question. This is of particular importance in dealing with some chronic eczemas where the cause or causative agents are ill defined and, therefore, difficult to remove.

Seborrheic dermatitis may be aggravated in patients suffering from acquired immunodeficiency syndrome (AIDS) and by the growth of *Pityrosporum ovale,* which has been considered to be a part of the etiology of seborrheic dermatitis.[1] Immediate-type sensitization to this same microorganism may be an aggravating factor in some patients with atopic dermatitis of the head and neck.[3]

Hyperkeratotic eczema of the palms has no well-defined cause, but physical trauma will aggravate the eczema.[4] The pattern of some cases of juvenile plantar dermatoses indicates that friction or pressure can be important aggravating factors. A number of factors may aggravate chronic hand eczema, regardless of its original cause.[5-9] The diagnostic workup of a patient with chronic hand eczema should, therefore, include consideration of the following:

Physical trauma
Extreme climatic conditions
Chemical irritants
Secondary contact sensitizations
Contact urticaria
  Immunological
  Nonimmunological
Psychological stress
Sweating
Smoking
Ingested foods
Drug intake (e.g., acetylsalicylic acid and contraceptive pills)
Infections
  Dermatophytes and yeasts
  Bacteria
  Herpes simplex virus

In typical cases, the diagnostic workup presents no difficulty. In other cases, however, a specific diagnosis may not be obvious and may not, in fact, be made until extensive investigations have been carried out.

II. THE DIAGNOSTIC WORKUP

In carrying out the diagnostic workup, the following parameters should be considered:

1. The history of the patient
2. The results of a clinical examination, including the patterns of dermatitis
3. The results of diagnostic tests such as:
    a. A histological examination
    b. Tests for delayed-type hypersensitivity
        i. Patch tests
        ii. Scratch chamber tests
        iii. Patch testing after tape stripping
        iv. Intradermal tests
        v. Photopatch tests
        vi. *In vitro* tests
    c. Oral challenge
    d. Use tests
    e. Diagnostic tests for irritancy
    f. Tests for immediate-type hypersensitivity
        i. Prick tests
        ii. RAST tests
4. The patient's response to the elimination of causative substances and reexposure to these substances

A. THE MEDICAL HISTORY OF THE PATIENT

It is important to ask a patient with eczema whether he or she has a personal history of atopic diseases and if parents and/or siblings have suffered from atopy in any form. It can also be of significance to determine whether there is a family history of psoriasis or ichthyosis.

The duration of the eczema and the activities of the patient at the time of onset may provide diagnostic clues. An abrupt onset 1 to 2 d after the patient was engaged in a specific activity suggests contact dermatitis, while life-long dermatitis is more likely to have a connection to atopy.

Occupational dermatoses often fade during week-long periods away from the work site only to become aggravated again a few days after work is resumed. Such dermatoses rarely improve significantly in the course of just a weekend.

The course of the dermatitis can also have diagnostic significance. Long-term dermatitis which improves during the summer months and is aggravated in the winter could be seborrheic or atopic dermatitis.

Aggravation occurring in the summer months may be related to sun exposure. When this aspect is discussed with the patient, it should be stressed that the offending ultraviolet (UV) irradiation can penetrate window glass in the home and in cars. Aggravation during the summer months may also be related to sweating or to specific outdoor activities, to direct contact with plants, or to exposure to airborne substances such as dust particles, pollen, and other plant particles.[10] A record of drug intake and contact with potential photoreactants other than plants can also contribute significantly to a patient's medical history.

The history of a patient suffering from eczema should also include a careful determination of possible aggravating factors such as those listed in Section I.C. In cases of chronic eczema, aggravating factors may be as important in keeping the eczema active as the fundamental cause of the dermatitis. It is also important to remember that many aggravating factors can be operative at the same time. The patient should be informed that, while the improvement of eczema is often slow and inconspicuous, aggravation of eczema usually occurs within a few days of exposure to the aggravating factor.

Patients with recurrent, vesicular hand eczema can often supply particularly helpful information. Their flares of pruritic vesicles are usually so characteristic that the patient himself can often register aggravation and determine whether he was in contact with a particular aggravating factor or ingested specific food items 1 to 3 d before the aggravation occurred. Aggravating factors for this type of eczema can also include smoking, contraceptive pills, the intake of analgesics containing acetylsalicylic acid, and psychological factors.[8,11]

For other eczemas, the specific location may provide an important clue as to the diagnosis. Contact dermatitis of the eyelid is often caused by substances touched by the hands which are transferred to the eyelids by rubbing them. Eczema of the face and genitals may be connubial dermatitis, and it can be of diagnostic significance to note whether aggravation occurs after sexual or other contact with a partner. Aggravation of an existing dermatitis of the ears or anogenital area or of stasis dermatitis is commonly due to sensitization to a component of a topic medicament.[12]

A history of stinging, burning, and itching seconds to minutes after contact with proteinaceous substances, possibly followed by aggravation of eczema, is characteristic of contact urticaria.[13] This condition is commonly caused by raw food items and should be suspected if aggravation frequently occurs following the preparation of meals.[7,14] Vesicles can appear on the fingers as soon as 20 min after contact with offending foods.[15] These symptoms of contact urticaria also characteristically appear on the lips as well as on the oral mucosa after contact with foods. Mayonnaise preserved with sorbic acid caused an epidemic of perioral contact urticaria in a kindergarten.[16] In this instance, the carefully taken history was the most important tool in arriving at the correct diagnosis.

If hazel nuts and apples cause symptoms of contact urticaria on the skin or oral mucosa and the patient has a history of hay fever caused by birch pollen, no further testing should be necessary. The common antigen present in all three items and the history of the patient provide the basis for a firm diagnosis of contact urticaria.[17,18]

B. THE RESULTS OF A CLINICAL EXAMINATION, INCLUDING THE PATTERNS OF DERMATITIS

The clinical examination of a patient with eczema includes taking an overview of the distribution of the eczema and the symmetry and patterns of the lesions. This is then followed by a careful inspection of the individual lesions to determine whether there are vesicles or other indications of acute dermatitis, whether foreign bodies or other substances are imbedded in the skin, and whether any apparent scaling is characteristic of psoriasis, Darier's disease, or some other noneczematous dermatosis.

Seborrheic dermatitis is easily diagnosed if, as in typical cases, there is pityriasiform scaling, a history of seasonal variation, and a typical distribution of the dermatitis. A diagnosis of atopic dermatitis is equally straightforward in typical cases; when less characteristic eruptions are seen, Hanifin and Rajka's diagnostic criteria may be applied.[2]

Contact dermatitis may present with a typical contact pattern, thus facilitating the diagnosis, especially when the contactant can be readily identified. A statistical correlation has been shown between the location of dermatitis and a number of specific contact allergens.[19] Some allergens are irritants, however, can give rise to dermatitis at locations and in patterns which do not immediately arouse suspicion. In such cases, a thorough history and a physical examination will be necessary before a correct diagnosis can be made.

Nummular eczema is a descriptive diagnosis based on the clinical pattern of the eruptions.[20]

Erosions and ulcerations of the knees and lower legs caused by irritancy from wet cement is another typical pattern of dermatitis,[21] as is the erosion of areas of skin which have been in contact with plants of *Dieffenbachia* spp.[22]

Excoriated nodules and papules at the sites of childhood immunizations or sites of hyposensitization therapy are characteristically seen in patients with aluminum hypersensitivity.[23,24]

Striped patterns of acute dermatitis are characteristic of plant contact dermatitis or dermatitis caused by the poison found in the tentacles of jellyfish or hydrozoans such as the Portuguese man-of-war.[25] If when striped acute dermatitis fades it is followed by marked hyperpigmentation, it is likely to have been a phototoxic dermatitis caused by plants which contain psoralens. Berloque dermatitis is a similar type of hyperpigmented dermatitis. It is seen when perfumes containing psoralens are applied to skin sites which are then exposed to the sun.

There have been reports of pigmented allergic contact dermatitis caused by naphthol AS used in dyeing fabric and by optic brighteners in detergents.[26,27] Such cases present clinically as pigmented clothing dermatitis.

Mechanical self-induced dermatitis caused by rubbing or picking the skin may present as eczematous lesions. Such lesions are usually asymmetric. Trauma associated with physical exercise led to cases of dermatitis graphically described as "rower's rump" and "jogger's nipples".[28,29] Occupational friction dermatitis may present as a pattern of dry hyperkeratotic eczema on the fingertips of persons who handle large amounts of paper. Other well-recognized causes of fingertip eczema include the allergic contact dermatitis caused by garlic seen among chefs.

Some patterns of dermatitis are thought to have endogenous causes; these include certain morphological types of hand and/or foot eczema, such as symmetrical recurrent vesicular eczema of the central portion of the palms and/or soles and the sides of the fingers and/or toes.[9,30] Symmetrical hyperkeratotic eczema at the same sites with no vesicles and no evidence of psoriasis is another example of an endogenous pattern of dermatitis.[4] Other endogenous patterns of dermatitis include the "baboon syndrome", an intensely inflamed eruption, particularly in the anogenital area,[31] and "fixed eczema", an eczematous patch which recurs repeatedly in the same area following the ingestion of a certain substance.[7] Hyperkeratotic psoriasiform lesions seen on the elbows of nickel- and chromate-allergic patients have been described as having an endogenous cause.[32]

The presence of white dermographism is of clinical significance in adults with nonspecific eczemas and indicates that a search for atopic diseases is warranted. Hyperlinearity of the palms is another useful clinical marker in atopic hand eczema.[33]

III. DIAGNOSTIC TESTS

A. THE HISTOLOGICAL EXAMINATION

Taking a biopsy from what appears to be an eczematous lesion is sometimes warranted. Even though there are no specific histological features, spongiotic dermatitis is highly suggestive of a diagnosis of eczema.[34] Chronic eczematous lesions tend to develop a psoriasiform epidermis, and a biopsy taken from an acute lesion may be helpful in distinguishing between psoriasis and eczema. Spongiotic dermatoses include pityriasis rosea and guttate parapsoriasis; therefore, a diagnosis of eczema cannot be based solely on the histological features.

Lichen planus may present with a dyshidrosiform eruption, and it may be necessary to carry out a histological examination to confirm this diagnosis.[35]

B. TESTS FOR DELAYED-TYPE HYPERSENSITIVITY

1. Patch Tests

The patch test is the major specific diagnostic test when an eczematous eruption involves a cellular immune reaction. If the substance suspected to cause the eruption is applied to the skin under controlled conditions, acute allergic contact dermatitis will be seen at the site of application after 2 to 3 d.

Various collections of substances are commercially available for battery testing. Important allergens may vary throughout the world due to cultural differences and differences in climate and geography; in spite of this, certain substances are considered to be ubiquitous allergens. These include the metals nickel, cobalt, and chromium, fragrances, rubber chemicals, and formaldehyde.[36-38]

In addition to the trays of substances used for general screening, various other collections have been assembled for testing those engaged in specific occupations or those who have daily contact with, for example, plastics, preservatives, etc. Cronin has recently provided detailed suggestions for assembling trays which could be used for particular occupations.[39] There are also trays of substances to use when the eczema is found in a specific area of the body; a shoe tray can be helpful in diagnosing foot eczema, for example.

Much research has gone into the determination of the correct concentration of the substances in various standard collections or series as well as the determination of the vehicle best suited for a particular substance. Therefore, it is wise to use these standardized test substances.

Occasionally, however, testing with nonstandardized substances, often substances brought in by the patients themselves, may be warranted.[40] Before testing with nonstandardized substances is attempted, one should always consult the literature to determine the correct test concentration and vehicle.[41-43] If no information can be found, an open test should be carried out prior to occlusive testing.

It can be difficult to avoid false negative and false positive reactions when using nonstandardized substances for patch testing; therefore, such test results should be interpreted cautiously.[44] It should also be borne in mind that, in cases of occupational dermatoses, patch test results have legal consequences.

Patch tests are best read 72 h after application.[1] Neomycin patch tests, however, typically do not show a positive reaction until 4 to 7 d after application. In some situations false negative reactions can be avoided if a longer application time is allowed.[45] The late occurrence of some positive patch test reactions may indicate that patch test sensitization has taken place. Positive patch test reactions which appear after 1 week should not automatically be interpreted as the explanation for an eczematous eruption, as the test procedure itself may have caused the sensitization. Experimental sensitization with strong sensitizers has shown that the sensitization period can vary from 10 to 14 d.[41]

Patch tests with simple chemical compounds, such as nickel, may become positive after 30 min and show an urticarial reaction.[46,47] Similar but presumably nonimmunological reactions have been seen after testing with balsam of Peru.[48,49] Open as well as occluded patch tests applied on intact skin and read after 25 to 30 min are an established test procedure for identifying contact urticaria[13,50] (protein contact dermatitis).[15] This type of test has also been used for the diagnosis of chronic urticaria.[51] It has been suggested that the diagnosis of protein contact dermatitis could be made by applying the test to the area where dermatitis normally occurs during a quiescent phase.[15]

2. Scratch Chamber Tests

The haptens used for patch testing all have low molecular weights. Tests with more complex chemical compounds such as proteins may require scarification of the skin prior to the test procedure in order to facilitate the penetration of the test substance. This type of test is called the scratch chamber test if the test site is occluded with, for example, a Finn Chamber® (Epitest 04, Helsinki, Finland). Since both immediate-type reactions and late reactions may occur when this test is carried out, the test site should be evaluated after 15 to 20 min and again after 48 to 72 h.

The scratch chamber method is particularly useful when testing with labile allergens such as those found in fruits and vegetables.[52] The test procedure itself may present some difficulties of interpretation.[17] The test is carried out with nonstandardized materials, and the erythema and wheal which appear after 15 to 20 min may be due to irritants in the test substance. Therefore, several controls should be tested at the same time as the patient suspected to have an allergic reaction to the substance.

In spite of these difficulties, good agreement has been seen between the results of scratch chamber tests with fruits and vegetables and clinical symptoms of itching and tingling of the hands and/or lips upon contact with uncooked fruits and vegetables.[53]

3. Patch Testing after Tape Stripping

Patch tests may be applied to skin which has been stripped with adhesive tape. Like scarification, this procedure facilitates hapten penetration of the skin. Stripping is indicated when it is necessary to test with haptens which have a complex chemical structure or which bind to the stratum corneum.

4. Intradermal Testing

If a patch test with nickel is negative in spite of a history of nickel dermatitis, intradermal testing may reveal the hypersensitivity.[54,55] Four patients who experienced dermatitis at the sites of subcutaneous heparin injections all had negative patch tests to heparin, while intradermal tests were positive in all the patients.[56] Both patch testing after stripping and intradermal testing represent alternative methods of presenting the hapten to the immunocompetent cells of the skin and the peripheral blood.

5. Photopatch Testing

If photoallergic dermatitis is suspected, photopatch testing can be of help in making the diagnosis. The procedure involves the application of two sets of test substances to the skin; one set is then subjected to 5 J/cm^2 of UV-A irradiation 24 h later, and the tests are read after a total application time of 48 and 72 h. Series of photoallergens have been proposed by, among others, the Scandinavian Photodermatological Research Group.[57]

6. *In Vitro* Tests for Cellular Immune Reactions

The use of an *in vitro* test to detect cellular immune reactions in the skin is indicated when there is extensive, severe eczema which makes patch testing impossible. In one study in which the results of patch tests were compared with the results of *in vitro* tests, it was shown that some patients reacted in the *in vitro* system, but not *in vivo. In vitro* testing, therefore, may also be useful when a false negative patch test to nickel is suspected. This form of testing has also been used to detect some cases of drug allergy.[58,59]

The diagnosis of allergic contact dermatitis by means of *in vitro* tests has not yet become practical. The leukocyte migration inhibition test has been applied to groups of patients known to be sensitive to nickel or chromium. While it is possible to distinguish between groups of patients and controls, there is so much overlap between the results among the two groups that the value of the test for the individual patient remains low.[60-62]

The lymphocyte transformation test has been used more extensively than the leukocyte migration inhibition test, and Al-Tawil[59] has attempted to standardize the test for the allergens nickel, cobalt, and chromium. This test, however, has the same limitations as the leukocyte migration inhibition test with respect to its use for screening purposes.

C. ORAL CHALLENGE

The hapten may also be presented to the immunocompetent cells of the skin via the gastrointestinal tract. Oral challenge with haptens has been utilized in patients with positive patch tests to nickel, cobalt, chromate, balsam of Peru, and medicaments such as neomycin and the hydroxyquinolines.[9] Flares of the actual dermatitis following oral challenge have been seen in an appreciable number of cases, but the clinical relevance of this type of positive reaction has not yet been established. An effective depletion in the body of the substance in question and subsequent improvement of the dermatitis would help to establish a causative relationship.

Some patients with negative patch tests to nickel, cobalt, or chromium react to oral challenge with salts of the metals.[9] This may indicate that the patch test results were false negative and/or that the route of presentation of the hapten may itself be of importance for the reaction. False negative patch tests to nickel have been seen.[55,63]

In a dermatology clinic where oral challenge procedures were used routinely as a last resort in making a diagnosis, the procedure led to a specific diagnosis in 122 of 7887 patients with eczema.[64]

D. USE TESTS

Low-grade cellular immune reactions also can be demonstrated by the repeated open application test or use test.[65,66] A small amount of the substance in question is rubbed into the skin twice daily, usually in an area where the skin is delicate such as the antecubital folds. If dermatitis appears within 1 week, the test is considered to be positive. This test can also be used to detect low-grade irritants in cosmetic preparations.

E. DIAGNOSTIC TESTS FOR IRRITANCY

The diagnosis of irritant contact dermatitis is usually one of exclusion. If a substance with which the skin comes in contact causes dermatitis and there is no evidence of a cellular or immediate-type immune reaction, the dermatitis is considered to be caused by an irritant reaction.

Although there are various animal models for determining the irritancy potential of substances prior to their use on human skin, there are no similar reliable test models for detecting irritancy on human skin.[1,67] One reason for this lack of a test system is that irritant reactions are complex and involve many means of triggering inflammatory reactions on a nonimmunological basis.[68,69] The interindividual threshold for irritancy reactions is great, and the threshold for irritant reactions may vary with the age of the patient.[70,71] At present, clinical test systems for irritancy are complex and require the use of several substances and several different types of tests to provide reliable results.[69]

F. TESTS FOR IMMEDIATE-TYPE HYPERSENSITIVITY

1. The Prick Test

The introduction into the skin of standardized allergens using a standard prick test needle represents the best routine method of detecting immediate-type hypersensitivity.[72,73] For control purposes, each patient should be given a prick test with both a saline and a histamine solution. The tests can be read after 15 to 20 min, and there is evidence supporting the importance of also reading late reactions (after 12 to 48 h). This late reading could lead to the detection of cellular immune reactions to proteins or other types of immunological reactions.

Test substances are usually available as trays of inhalant allergens or food allergens. Both types of allergen are useful in diagnosing an eczema patient as they can help to identify a specific allergic reaction or, if a number of inhalant allergies are found, provide evidence of a diagnosis of atopic dermatitis. The prick test can also be used to determine specific aggravating factors, such as house-dust mites or *Pityrosporum ovale,* in patients with atopic dermatitis.

Atopics have been seen to react to a scratch test with spices more often than nonatopics, and a correlation has been seen between a reaction to spices and reactions to tree pollen, fruits, and vegetables.[74]

The scratch chamber test is useful for the detection of immediate-type hypersensitivity to certain substances, as described in Section III.B.2.

2. The RAST Test

If an immediate-type reaction is suspected and the eczema of a patient is so widespread that prick tests cannot be applied, allergy to an individual substance may be determined by means of *in vitro* tests such as the RAST (radioallergosorbent test). This test should be considered as a supplement, not as a substitute, for prick tests, and because of its cost, its use is not feasible when carrying out battery testing.[75,76]

3. Histamine Release

In vitro test systems based on histamine release are currently being developed. They can be used for battery testing and may in due time become proper substitutes for the time-consuming *in vivo* skin tests.[77,78]

G. THE PATIENT'S RESPONSE TO ELIMINATION AND REEXPOSURE

For immediate as well as delayed reactions it is sometimes possible to see an obvious relationship between the eczema and the substance causing it. In such cases the dermatitis is likely to clear if the causative substance can be avoided. Unfortunately, such an obvious relationship is rarely seen, and neither is it always possible to avoid a causative substance entirely. In such cases it is necessary to evaluate the activity of the dermatitis, following as complete an avoidance of the causative substance as possible, as well as the activity of the dermatitis when the patient is reexposed to the suspected substance. Such an evaluation often makes it possible to distinguish between relevant and nonrelevant reactions resulting from the various test systems. For example, one should not consider that a positive test has identified the cause of the dermatitis unless there is satisfactory clearing of the dermatitis after avoidance of the causative substance and recurrence of the dermatitis upon reexposure.

1. Chemical Analysis

It may be helpful to use simple chemical analysis in situations where it is difficult to determine if the patient has actually been exposed to the suspected substance. The dimethylglyoxime test for nickel is easy to carry out in the office setting. There are similar, slightly more complicated tests for determining the presence of formaldehyde and hexavalent chromate.

IV. DIFFERENTIAL DIAGNOSIS

If a patient is suspected of suffering from a dermatosis which is morphologically similar to eczema, it may be useful to carry out diagnostic procedures which exclude the diagnosis of eczema. Eczematous lesions associated with systemic disease meet the criteria for a diagnosis of eczema, but the cutaneous lesions are often minor symptoms of a more serious fundamental problem. Examples of such lesions include those seen in untreated cases of phenylketonuria, acrodermatitis enteropathica, and the similar zinc-deficiency dermatitis seen in malnourished patients.

The clinical indicators which make it possible to distinguish between eczema and psoriasis, for example, include the type of scaling and the presence of Auspitz' sign. This is of particular significance in the palms, where the differential diagnosis between psoriasis and hyperkeratotic eczema may be difficult.

The distinction between psoriasis and seborrheic dermatitis is particularly difficult when eruptions are seen on the face and scalp. The term ''sebopsoriasis'' has been proposed for such cases. Infants may have eczematous lesions on the face and scalp for which it is impossible to make a distinction between seborrheic and atopic dermatitis. Facial discoid lupus erythematosus and seborrheic dermatitis have many common features, but seasonal variations in intensity and a histological examination will in most cases make a specific diagnosis possible. Polymorphic light eruption may also be eczematous.

Because of the vesicles and pruritus which can occur in connection with palmo-plantar pustulosis, this disease can mimic recurrent vesicular hand and foot eczema on the palms and soles. Acrodermatitis Hallopeau may also mimic eczematous lesions.

Intertriginous eczemas may have a striking resemblance to Hailey and Hailey's disease and to inverse psoriasis. In such cases a biopsy is the best diagnostic procedure.

Superficial basal cell carcinomas, Bowen's disease, and early stages of mycosis fungoides may have the clinical appearance of eczema, but histological examination will usually make a specific diagnosis possible.

A herpetic whitlow may simulate vesicular hand eczema, but isolation of the virus or viral antigen and the predominance of pain rather than itching are usually characteristic. Superficial impetigo may be difficult to distinguish from eczema unless a bacterial culture is made. Dermatophytosis of the scalp, *Trichophyton rubrum* infections of the hands, and widespread dermatophytosis may simulate eczematous lesions, but microscopy and culture make an exact diagnosis possible. The diagnosis of Norwegian scabies requires a high level of suspicion, and this entity may be mistaken for eczema.

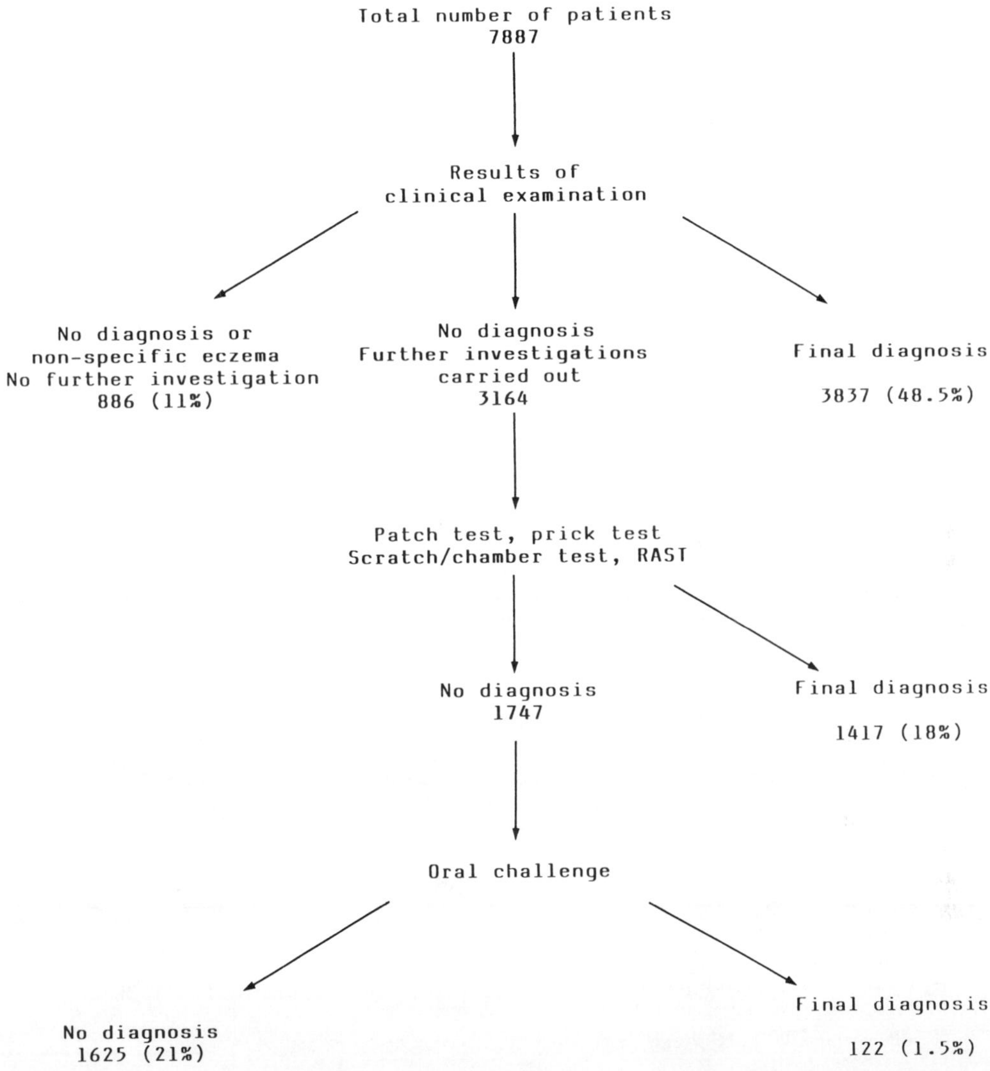

FIGURE 1. Results of the diagnostic work-up of 7887 patients with eczema. (From Veien, N. K., Hattel, T., Justeson, O., and Nørholm, A., *Contact Dermatitis,* 17, 35, 1987. With permission.)

V. CONCLUSION

The medical history of the patient and the results of a clinical examination will often be sufficient to enable a specific diagnosis of eczema to be made.[64] (Figure 1) If diagnostic tests are carried out, these should be cost effective.

Histological examinations may be useful for the differentiation between eczema and other dermatoses, while different types of eczemas can be more difficult to distinguish from one another on the basis of histology.

When it is not possible to make the specific diagnosis of contact dermatitis on the basis of the history and a clinical examination, patch testing is an effective method of determining the diagnosis, while *in vitro* tests are both time-consuming and expensive. Prick tests are relatively easy to perform and are inexpensive. These tests may help to substantiate a diagnosis of atopic dermatitis and to determine the factors which aggravate it.[64] RAST tests are expensive to carry out and should be used to supplement skin tests. Oral challenge procedures should be used primarily in treating refractory eczemas of long standing or prior to tedious diet treatments. Such procedures are difficult to perform and have not yet been well standardized. The same applies to the various skin tests for irritancy.

REFERENCES

1. **Rook, A., Wilkinson, D. S., Ebling, F. J. G., Champion, R. H., and Burton, J. L., Eds.,** *Textbook of Dermatology,* 4th ed., Blackwell Scientific, Oxford, 1986, 367.
2. **Hanifin, J. M. and Rajka, G.,** Diagnostic features of atopic dermatitis, *Acta Derm. Venereol. Suppl.,* 92, 44, 1980.
3. **Waersted, A. and Hjorth, N.,** Pityrosporum orbiculare — a pathogenic factor in atopic dermatitis of the face, scalp and neck?, *Acta Derm. Venereol. Suppl.,* 114, 146, 1985.
4. **Hersle, K. and Mobacken, H.,** Hyperkeratotic dermatitis of the palms, *Br. J. Dermatol.,* 107, 195, 1982.
5. **Rietschel, R. L. and Ray, M. C.,** Nonatopic eczemas, *J. Am. Acad. Dermatol.,* 18, 569, 1988.
6. **Kavli, G. and Førde, O. H.,** Hand dermatoses in Tromsø, *Contact Dermatitis,* 10, 174, 1984.
7. **Veien, N. K., Hattel, T., Justesen, O., and Nørholm, A.,** Dermatitis induced or aggravated by selected foodstuffs, *Acta Derm. Venereol.,* 67, 133, 1987.
8. **Edman, B.,** Computerized Patch Data in Contact Allergy, Doctoral thesis, Lund University, Malmö, Sweden, 1988.
9. **Menné, T.,** Reactions to systemic exposure to contact allergens, in *Dermatotoxicology,* 2nd ed., Marzulli, F. N. and Maibach, H. I., Eds., Hemisphere, Washington, D.C., 1983, 483.
10. **Dooms-Goossens, E., Debusschere, K. M., Gevers, D. M., Dupré, K. M., Degreef, H. J., Loncke, J. P., and Snauwaert, J. E.,** Contact dermatitis caused by airborne agents, *J. Am. Acad. Dermatol.,* 15, 1, 1986.
11. **Hansen, O., Küchler, T., Lotz, G.-R., Richter, R., and Wilckens, A.,** Es juckt mich in den Fingern, aber mir sind die Hände gebunden, *Z. Psychosom. Med. Psychoanal.,* 27, 275, 1981.
12. **Wilkinson, J. D., Hambly, E. M., and Wilkinson, D. S.,** Comparison of patch test results in two adjacent areas of England. II. Medicaments, *Acta Derm. Venereol.,* 60, 245, 1980.
13. **von Krogh, G. and Maibach, H. I.,** The contact urticaria syndrome — an updated review, *J. Am. Acad. Dermatol.,* 5, 328, 1981.
14. **Rystedt, I.,** Hand Eczema and Long-Term Prognosis in Atopic Dermatitis, Doctoral thesis, Karolinska Institute, Stockholm, 1985.
15. **Hjorth, N. and Roed-Petersen, J.,** Occupational protein contact dermatitis in food handlers, *Contact Dermatitis,* 2, 28, 1976.
16. **Clemmensen, O. J., Hjorth, N., and Gram, H. J.,** "Mayonnaise-eksem" i en vuggestue, *Ugeskr. Laeg.,* 143, 2725, 1981.
17. **Eriksson, N. E., Formgren, H., and Svenonius, E.,** Food hypersensitivity in patients with pollen allergy, *Allergy,* 37, 437, 1982.
18. **Kremser, M. and Lindemayr, W.** Zur Häufigkeit sogenannter "Apfelallergie" ("Apfelkontakt-Urticariasyndrom") bei Patienten mit Birkenpollenallergie, *Z. Hautkr.,* 58, 543, 1983.
19. **Edman, B.,** Sites of contact dermatitis in relationship to particular allergens, *Contact Dermatitis,* 13, 129, 1985.
20. **Hellgren, L. and Mobacken, H.,** Nummular eczema — clinical and statistical data, *Acta Derm. Venereol.,* 40, 189, 1969.
21. **Vickers, H. R. and Edwards, D. H.,** Cement burns, *Contact Dermatitis,* 2, 73, 1976.
22. **Ippen, H., Wereta-Kubek, M., and Rose, U.,** Haut- und Schleimhautreaktionen durch Zimmerpflanzen der Gattung Dieffenbachia, *Dermatosen,* 34, 93, 1986.
23. **Clemmensen, O. and Knudsen, H. E.,** Contact sensitivity to aluminium in a patient hypsensitized with aluminium precipitated grass pollen, *Contact Dermatitis,* 6, 305, 1980.
24. **Veien, N. K., Hattel, T., Justesen, O., and Nørholm, A.,** Aluminum allergy, *Contact Dermatitis,* 15, 295, 1986.
25. **Burnett, J. W. and Calton, G. J.,** Jellyfish envenomation syndromes updated, *Ann. Emerg. Med.,* 16, 1000, 1987.
26. **Hayakawa, R., Matsunaga, K., Kojima, S., Kaniwa, M., and Nakamura, A.,** Naphthol AS as a cause of pigmented contact dermatitis, *Contact Dermatitis,* 13, 20, 1985.
27. **Osmundsen, P. E.,** Pigmented contact dermatitis, *Br. J. Dermatol.,* 83, 296, 1970.
28. **Tomecki, K. J. and Mikesell, J. F.,** Rower's rump, *J. Am. Acad. Dermatol.,* 16, 890, 1987.

29. **Nequin, N. D.,** More on jogger's ailments, *N. Engl. J. Med.*, 298, 405, 1978.
30. **Schwanitz, H. J.,** *Das Atopische Palmoplantareksem*, Springer-Verlag, Berlin, 1986.
31. **Andersen, K. E., Hjorth, N., and Menné, T.,** The baboon syndrome: systemically-induced allergic contact dermatitis, *Contact Dermatitis*, 10, 97, 1984.
32. **Kaaber, K., Sjølin, K. E., and Menné, T.,** Elbow eruptions in nickel and chromate dermatitis, *Contact Dermatitis*, 9, 213, 1983.
33. **Svensson, Å.,** Hand eczema: an evaluation of the frequency of atopic background and the difference in clinical pattern between patients with and without atopic dermatitis, *Acta Derm. Venereol.*, 68, 509, 1988.
34. **Ackerman, A. B.,** *Histologic Diagnosis of Inflammatory Skin Diseases*, Lea & Febiger, Philadelphia, 1978.
35. **Feuerman, E. J., Ingber, A., David, M., and Weissman-Katzenelson, V.,** Lichen ruber planus beginning as a dyshidrosiform eruption, *Cutis*, 30, 401, 1982.
36. **Hammershøy, O.,** Standard patch test results in 3,225 consecutive Danish patients from 1973 to 1977, *Contact Dermatitis*, 6, 263, 1980.
37. **Lämmer, D.,** Testergebnisse von 1008 Patienten mit Kontaktallergie, *Z. Hautkr.*, 54, 571, 1979.
38. **Van Hecke, E.,** Frequency of contact allergy in Belgium, *Arch. Belg. Dermatol.*, 30, 235, 1974.
39. **Cronin, E.,** Some practical supplementary trays for special occupations, *Semin. Dermatol.*, 5, 243, 1986.
40. **Veien, N. K.,** Why test with environmental agents? A review of recent studies concerning the value of routine testing in dermatologic practice, *Semin. Dermatol.*, 5, 231, 1986.
41. **Cronin, E.,** *Contact Dermatitis*, Churchill Livingstone, London, 1980.
42. **Adams, R. M.,** *Occupational Skin Disease*, Grune & Stratton, New York, 1983.
43. **Fisher, A. A.,** *Contact Dermatitis*, 3rd ed., Lea & Febiger, Philadelphia, 1986.
44. **Rycroft, R. J. G.,** False reactions to nonstandard patch tests, *Semin. Dermatol.*, 5, 225, 1986.
45. **Fisher, T. and Maibach, H. I.,** Patch testing in allergic contact dermatitis: an update, *Semin. Dermatol.*, 5, 214, 1986.
46. **Osmundsen, P. E.,** Contact urticaria from nickel and plastic additives (butylhydroxytoluene, oleylamide), *Contact Dermatitis*, 6, 452, 1980.
47. **Valsecchi, R. and Cainelli, T.,** Contact urticaria from nickel, *Contact Dermatitis*, 17, 187, 1987.
48. **Forsbeck, M. and Skog, E.,** Immediate reactions to patch tests with balsam of Peru, *Contact Dermatitis*, 3, 201, 1977.
49. **Temesvári, E., Soos, G., Podányi, B., Kovács, I., and Nemeth, I.,** Contact urticaria provoked by balsam of Peru, *Contact Dermatitis*, 4, 65, 1978.
50. **Winton, G. B. and Lewis, C. W.,** Contact urticaria, *Int. J. Dermatol.*, 21, 573, 1982.
51. **Warin, R. P. and Smith, R. J.,** Chronic urticaria investigations with patch and challenge tests, *Contact Dermatitis*, 8, 117, 1982.
52. **Lahti, A. and Hannuksela, M.,** Hypersensitivity to apple and carrot can be reliably detected with fresh material, *Allergy*, 33, 143, 1978.
53. **Hannuksela, M. and Lahti, A.,** Immediate reactions to fruits and vegetables, *Contact Dermatitis*, 3, 79, 1977.
54. **Meneghini, C. and Angelini, G.,** Intradermal test in contact allergy to metals, *Acta Derm. Venereol. Suppl.*, 85, 1979.
55. **Marcussen, P. V.,** Comparison of intradermal test and patch test using nickel sulphate and formaldehyde, *J. Invest. Dermatol.*, 40, 263, 1963.
56. **Meissner, K. and Schulz, K.-H.,** Kutane Allergie auf Heparin, *Allergologie*, 9, 295, 1986.
57. **Wennersten, G., Thune, P., Jansén, C. T., and Brodthagen, H.,** Photocontact dermatitis: current status with emphasis on allergic contact photosensitivity (CPS) occurrence, allergens, and practical phototesting, *Semin. Dermatol.*, 5, 277, 1986.
58. **Stejskal, V. D. M., Olin, R. G., and Forsbeck, M.,** The lymphocyte transformation test for diagnosis of drug-induced occupational allergy, *J. Allergy Clin. Immunol.*, 77, 411, 1986.
59. **Al-Tawil, N. G.,** Immunological Studies in Patients with Nickel, Cobalt and Chromium Sensitivity, Doctoral thesis, Karolinska Institute Medical School, Stockholm, 1985.
60. **Thulin, H.,** The Leucocyte Migration Test in Dermato-Venereological Diseases, Doctoral thesis, University of Aarhus, Aarhus, Denmark, 1978.
61. **Nordlind, K. and Sandberg, G.,** Leukocytes from patients allergic to chromium and nickel examined by the sealed capillary migration technique, *Int. Arch. Allergy Appl. Immunol.*, 70, 30, 1983.
62. **Czernielewski, A., Libiszowski, T., and Dudek, H.,** Die Anwendung des Leukozytenmigrationshemmtestes (LMH) bei der Diagnostik der Chromatsensibilisierung, *Dermatol. Monatsschr.*, 163, 889, 1977.
63. **Kieffer, M.,** Nickel sensitivity: relationship between history and patch test reaction, *Contact Dermatitis*, 5, 398, 1979.
64. **Veien, N. K., Hattel, T., Justesen, O., and Nørholm, A.,** Diagnostic procedures for eczema patients, *Contact Dermatitis*, 17, 35, 1987.

65. **Epstein, W. L.,** The use test for contact hypersensitivity, *Arch. Dermatol. Res.*, 272, 279, 1982.
66. **Hannuksela, M. and Salo, H.,** The repeated open application test (ROAT), *Contact Dermatitis,* 14, 221, 1986.
67. **Andersen, K. E., Benezra, C., Burrows, D., Camarasa, J., Dooms-Goosens, A., Ducombs, G., Frosch, P., Lachapelle, J.-M., Lahti, A., Menné, T., Rycroft, R., Scheper, R., White, I., and Wilkinson, J.,** Contact dermatitis. A review, *Contact Dermatitis,* 16, 55, 1987.
68. **Malten, K. E.,** Thoughts on irritant contact dermatitis, *Contact Dermatitis,* 7, 238, 1981.
69. **Björnberg, A.,** Irritant dermatitis, in *Occupational and Industrial Dermatology,* 2nd ed., Maibach, H. I., Ed., Year Book Medical Publishers, Chicago, 1987.
70. **Lammintausta, K., Maibach, H. I., and Wilson, D.,** Human cutaneous irritation: induced hyporeactivity, *Contact Dermatitis,* 17, 193, 1987.
71. **Coenraads, P. J., Bleumink, E., and Nater, J. P.,** Susceptibility to primary irritants, *Contact Dermatitis,* 1, 377, 1975.
72. **Dreborg, S.,** The Skin Prick Test, Thesis, Linköping University, Linköping, Sweden, 1987.
73. **Østerballe, O., Dirksen, A., Weeke, B., and Weeke, E. R.,** Forekomst af medicinsk-allergiske sygdomme i speciallaegeregie, *Ugeskr. Laeg.,* 143, 3211, 1981.
74. **Niinimäki, A. and Hannuksela, M.,** Immediate skin test reactions to spices, *Allergy,* 36, 487, 1981.
75. **Norman, P. S., Lichtenstein, L. M., and Ishizaka, K.,** Diagnostic tests in ragweed hay fever, *J. Allergy Clin. Immunol.,* 52, 210, 1973.
76. **Panzani, R. and Johansson, S. G. O.,** Results of skin test and RAST in allergy to a clinically potent allergen (castor bean), *Clin. Allergy,* 16, 259, 1986.
77. **Siraganian, R. P.,** Automated histamine analysis for *in vitro* allergy testing, *J. Allergy Clin. Immunol.,* 59, 214, 1977.
78. **Diamant, B. and Parkar, S.,** Histamine release from washed whole blood, *Int. Arch. Allergy Appl. Immunol.,* 67, 13, 1982.

Chapter 10

EXCITED SKIN SYNDROME AND THE HYPOREACTIVE STATE: CURRENT STATUS

D. P. Bruynzeel and H. I. Maibach

TABLE OF CONTENTS

I. INTRODUCTION

Since the introduction of the diagnostic patch test in 1895 by Jadassohn, it has been recognized that patients with eczema may show cutaneous hyperirritability.[1-7] Hyperirritability was mentioned by Von Zumbusch[8] in 1921 and since then has frequently been described under different names (see Table 1). A seminal publication is the thesis of Björnberg[9] in 1968 about hyperirritability in patients with eczema. Apart from interindividual variation in skin reactivity, as was shown again recently,[10] enhanced reactions to irritants were seen in patients with hand eczema. In 1975 Mitchell[11,12] emphasized that hyperirritability was a serious problem in the patch test clinic, responsible for the induction of false positive reactions. He called this phenomenon the "Angry Back Syndrome" (ABS). As it is not restricted to the back but involves the entire skin, we changed the name to "Excited Skin Syndrome" (ESS).[13,14]

II. CLINICAL SETTING

When assuming that the skin (i.e., of the back) can be hyperirritable, then even minor injuries (e.g., due to a marginal irritant or weak immunological reaction) may cause an enhanced inflammatory reaction. Therefore, performing patch tests in such circumstances may elicit reactions to test substances that will not give any reaction under normal conditions. When the hyperirritability has subsided and the substances are retested, we presume that they will be negative. In such cases we speak of a false positive reaction due to the ESS or, if preferred, skin hyperirritability. It is only possible to denote false positive reactions if the retested allergens are negative at subsequent testing ("lost reactions"). Supposing that this is not due to a faulty technique or other technical peculiarities (e.g., patch test systems that may influence the outcome of test results[15], we assume that the reason might be hyperirritable skin (ESS, or an "angry back"). Retesting is not absolute proof.

In case of low-grade contact sensitization, allergic patch test reactions may not always reach the level of a clinically visible reaction. With the available methods, it is still far from easy to make a distinction between a weak positive immunological test reaction and a real false positive reaction (that is, skin irritation).

The ESS remains controversial in the dermatological literature.[16-18] To what extent is the syndrome responsible for false positive patch test reactions? The nonreproducibility of patch test reactions has been reported to be in the range of 40%,[12,19-21] but could be as high as 60%[22] or, on the other hand, as low as 8.6%.[23] Utilization of different protocols makes comparison between these studies difficult.

Recently, Gollhausen et al.[24] suggested that there is a low reproducibility of weak positive patch test reactions, as expected from earlier studies.[16,21,25] A total of 24 patients who 1 week previously had shown a positive reaction to one or more allergens of a standard series were retested on the other side of the back. Of the total 70 reactions, 40% were positive only on the first or the second occasion. In addition, they patch tested in duplicate on different sides of the back; 52% of the reactions were only positive on one side. Weak positive reactions were nonreproducible far more often than strong reactions (see Table 2). Comparable results are reported in investigations into the reproducibility of patch test reactions comparing the TRUE Test™ and Finn Chamber® technique. In these studies, the test series were applied at the same time on different sides of the back. An overall concordance of 67% was found; however, one half of the weak reactions were disconcordant.[26] This implies that we may expect that about one half of the weaker reactions are not reproducible at retesting, but on the other hand, we might miss an equal amount of positive reactions on the first test session. One could argue that this nonreproducibility has nothing to do with hyper- or hypoirritiability. It is critical that we realize that the interpretation of a weak

TABLE 1
Expressions Used to Describe Skin Hyperreactivity[71]

Expression	Year	Ref.
Reflektorisches Ekzem	1921	8
Crazy back	1979	73
Metallergic and parallergic reactions	1939	74
Status eczematicus	1959	59
Skin fatigue	1967	47
Multiple nonspecific reactions	1970	75
Conditioned hyperirritability	1975	76
Angry back syndrome	1975	11
Spillover	1977	12
Rogue positive reactions	1980	77
Excited skin syndrome	1981	13

TABLE 2
Reproducibility of Positive Patch Test Reactions in Patients Tested with a Standard Series (39 Allergens)[24]

	Number of:		
Test[a]	Patients with positive reactions	Total positive reactions	Reactions positive in one test session or one side
Sequential	24	70	40%
Duplicate	13	25	52%

[a] Tests were performed sequentially (41 patients) or in duplicate (28 patients).

positive reaction (+: erythema, induration, and papules[27]) is difficult. The question remains: are these reactions false positive? What is even more troublesome, did we miss reactions (i.e., false negatives)? What is the importance of a weak positive reaction if reproducibility is so low?

III. MECHANISMS

Another important question is: which mechanism is responsible for the fluctuations in reactivity? The following hypothetical mechanisms might be involved:

- Skin reactivity fluctuations may be regulated by humoral inflammation mediators; the irritability state depends on the balance between mediators that promote and depress inflammation. Dermatitis is presumed to be a condition involved in the release of such mediators. Depending on the balance of such factors, a (marginal) irritating test substance may elicit a positive reaction or no reaction at all.
- Fluctuations in immune reactivity do occur; the underlying causes are imperfectly known.[13,28,29] Distribution or compartmentalization of lymphocytes[30,31] probably plays a role in immune reactivity of the patient, among other factors such as the influence of infectious diseases (e.g., viral), stress, and circadian rhythm (cortisol levels). If there are such fluctuations, it will be the weak sensitizers which are most likely to evoke weak positive reactions or elicit no reaction (false negatives).

TABLE 3
Skin Hyperreactivity Studies in Guinea Pigs

Dermatitis (all, irr, FCA)	Test	Year	Ref.
irr	irr	1959	32
irr	all	1970	33
FCA	irr	1980	40
irr	irr	1981	34[a]
FCA	all	1983	37
FCA	irr	1983	38
irr, FCA	all, irr	1983	39
all	irr	1984	35
all, irr	irr	1985	36

Note: The state of hyperreactivity, induced by an allergic (all) or irritant (irr) dermatitis or Freunds' complete adjuvant (FCA), was shown by epicutaneous allergic (all) or irritant (irr) challenge tests.

[a] Experiments performed in mice.

The cascade of reactions involved in the development of inflammation is induced by irritation as well as by means of an immunological reaction. In the end the result is the same: a dermatitis. This means that the immunological reaction is dependent not only on fluctuations in the immunological system, but also on nonimmunological phenomena which regulate the irritability state of the skin.

IV. ANIMAL STUDIES

Several experiments in animals, predominantly guinea pigs, demonstrate the existence of skin hyperirritability (see Table 3). Haeberlin and Fox[32] observed enhanced reactions to an irritant, croton oil, in guinea pigs at the height of an induced dermatitis compared with the reactions in animals without dermatitis. Magnussen and Kligman[33] observed similar results for allergic reactions. In the presence of a dermatitis, the reaction to a challenge with an allergen was enhanced in number and severity in these previously sensitized guinea pigs. Van der Lugt and Venninga,[34] who used albino hairless mice, reported the enhancing effect of a concomitant irritant dermatitis on skin test reactions elicited by the same irritant several days later. Others[35,36] demonstrated in guinea pig models that an acute irritant and an allergic contact dermatitis lowered the irritation threshold of the skin. Enhancement of skin reactivity was noted in animals pretreated with an intradermal injection of Freund's complete adjuvant (FCA).[37-39] Andersen and Maibach[40] used this phenomenon in a cumulative irritancy assay in guinea pigs by repeated application of low-grade irritant products. Several investigators noted that this altered response due to FCA should lead to careful evaluations of the results of challenge tests in the guinea pig maximization test (GMT), where FCA is used to enhance sensitization.[37,38]

V. HYPOREACTIVITY

Bruynzeel et al.[39] showed similar results in FCA-treated guinea pigs for skin tests to both an irritant and an allergen. Less marked hyperreactivity could be evoked by a localized chronic croton-oil-induced dermatitis. In this experiment, hyperirritability was preceded by

TABLE 4
Skin Hyporeactivity in Relation to Inflammatory Processes (e.g., Dermatitis, Freunds' Complete Adjuvant, Skin Tests) in the Skin in Experimental Settings

Species	Test	Year	Ref.
Guinea pig	ec, all	1951	41
Guinea pig	ic, all	1973	43
Man	ic, all	1975	44
Guinea pig	ic, all	1977	46
Guinea pig	ic, all	1978	45
Guinea pig	ec, all, irr	1983	39
Man	ec, irr	1987	42

Note: ec = epicutaneous; ic = intracutaneous; all = allergic; irr = irritant.

a short period of hyporeactivity of the skin. This depression of skin reactivity had already been observed by Frey[41] in dinitrochlorobenzene-sensitized guinea pigs (see Table 4). Lammintausta et al.[42] showed that repeated open applications with sodium lauryl sulfate (SLS) in females induced a hyporeactive state on simultaneously performed tests with SLS elsewhere. Depressed skin reactivity has also been described for intracutaneous skin test reactions.[43-46]

In summary, these animal experiments show that the reactivity of the skin is influenced by a concomitant dermatitis or an inflammation process caused by FCA. The influence on the skin can run a biphasic course: hyporeactivity in the early phase of a severe acute inflammation followed by hyperreactivity in the more chronic phase. Little is known of the humoral and cellular events in these experiments.

In humans this hyporeactivity or even anergy is seen as well in connection with an acute dermatitis.[47-49] Skin hyporeactivity could be a result of compartmentalization of lymphocytes; i.e., some cells leave the circulation and enter the skin, whereas others remain in the draining lymph nodes or are attracted to a site of acute inflammation[30,31,50] and are thus not available for the patch test reaction, allergic or irritant. On the other hand, extra inflammatory cells might be released in the circulation due to a stimulating effect by a (chronic) inflammatory process as induced by FCA.[51-53] These activated cells might contribute to the inflammatory reaction of otherwise weak or even subclinical irritant and allergic reactions.

VI. SPILLOVER

Inhibiting as well as inflammation-promoting humoral mediators are probably important. The balance between these mediators might be responsible for the skin irritant state.[54,55] Recent investigations showed that the activity of the epidermal inflammation mediator epidermal thymocyte activating factor (ETAF) or interleukin-1 (IL-1) is increased in skin overlying tuberculin skin tests and positive patch test reactions.[56] Also, the newly described epidermal lymphocyte chemotactic factor (ELCF) was increased.[57] Both mediators are involved in the development of skin-test reactions (immunologic and irritant).

ETAF/IL-1 was increased in normal-appearing skin. ELCF is not present in the skin of healthy persons. However, ELCF was found in nontested normal-appearing skin in three of seven patients with eczema prior to patch testing.[56] This shows that the entire skin seems to be activated during patch testing or in the case of active eczema. Similar experiments by

TABLE 5
Observations on Lymphokines, Skin Temperature, Blood Flow, and Cellular Infiltrate Suggesting that Inflammatory Responses in the Skin Are Not Confined to the Test Site But Also Involve Apparently (Nearby) Clinically Normal Skin

	ETAF[a] IL-1	ELCF[b]	Skin Temperature	Blood flow	Cellular infiltrate
Allergic reaction					
Patch test site	▲	▲			
Nontested skin	▲	▲	▲	▲	▲
Irritation reaction					
Patch test site	=	▲			
Nontested skin	=	—	▲*	▲*	
Normal skin					
Eczematous person	▲	▲			
Noneczematous person	=	—			

Note: ▲ increase; = no increase; — absent; * not significant.

[a] ETAF = epidermal thymocyte activating factor; IL-1 = interleukin-1.
[b] ELCF = epidermal lymphocyte chemotactic factor.

Modified from References 56—58, 63.

these authors[58] with irritants indicated that ELCF, but not ETAF/IL-1, was increased by patch testing with irritants.

These observations, together with the availability of inflammatory cells, might explain the hyperirritability of the skin in patients with eczema or even just a strong skin test reaction (see Table 5). It might also explain the mechanism of spillover reactions, i.e., induction of positive reactions due to a runover effect from one test site to another neighboring test site.[12,21,59] Although spillover reactions clearly occur in the patch test clinic, in experimental settings it is more difficult to demonstrate this phenomenon. Bruynzeel et al.[60] showed enhancement of SLS reactions which were situated close to allergic patch test reactions. No increase was seen in the number of inflammatory cells in the dermis of the uninvolved skin adjacent to the area exhibiting a strong patch test reaction (3-cm distance). Therefore, the hyperirritability was supposed to mediated by humoral factors. The study of Larsen[56] with ETAF/IL-1 and ELCF indicates at least the importance of such factors in spillover. Bandmann and Agathos[61] and later Kligman and Gollhausen[62] however, did not succeed in demonstrating a spillover effect. A reason for this could be the difference in volunteers; here they were patients with nonactive eczema, while Kligman and Gollhausen used volunteers who did not have eczema.

Controversy on these topics can only be overcome if experiments like these are repeated under well-standardized conditions; in addition to histological investigation, assays in the presence or absence of inflammation mediators in the skin should be done. Hamani and Marks[63] reported interesting results concerning different measurements at a distance of 2, 5, and 10 cm from small areas with irritation and an allergic patch test reaction. There was an increase in skin temperature and blood flow 2 cm from the allergic patch test reaction compared to control sites. The histology of ''normal skin'' 1 cm away revealed signs of inflammation (spongiosis and/or increased inflammatory cells). Their findings indicate that normal appearing skin near foci of inflamed sites on the skin may be abnormal regarding blood flow and histology.

Furthermore, previously inflamed skin has been reported to have an abnormal barrier

TABLE 6
Dealing with the Excited Skin Syndrome Resulting from Battery Testing[13]

1. When more than one positive reaction is noted, retake the clinical history.
2. It may not be necessary to retest positive reactions individually if
   a. The material is readily avoided (e.g., ammoniated mercury)
   b. History strongly negates the positive patch-test results
3. It is important to retest positive reactions individually if the material is ubiquitous:
   a. Preservatives (paraben mixture)
   b. Vehicle (wool wax alcohol)
   c. Medicament (neomycin)
   d. Fragrance (fragrance mixture)
   e. Rubber chemical
   f. Formaldehyde solution
   g. Possible job change
   h. Medicolegal situation
   i. Regulatory situation
   j. Public health issues
4. With mixtures, testing with components is often helpful (especially the fragrance and paraben mixtures).
5. Consider the role of the Provocation Use Test (PUT) or Repeated Open Application Test (ROAT) in such an evaluation.

function and a lower threshold to irritants near areas with eczema.[64] Spillover reactions could be explained by these observations.

Another form of enhanced skin reactivity of apparently normal skin is seen in recently inflamed sites. Many inflammatory cells remain in the dermis due to the recent dermatitis. The studies of Sjöborg et al.,[65] Kanerva,[66] and Scheper[67] indicate that, in previously inflamed skin, increased numbers of Langerhans' cells and T-lymphocytes may persist in the skin for a long time. Therefore, enhanced skin reactivity can be expected as long as an increased amount of these cells is present. This may result in false positive reactions if patch testing is performed in such areas.

All these facts together indicate that normal-looking skin can react more strongly to irritants than expected and that is likely to play a role in spillover and the ESS. Humoral as well as cellular elements seem to be important and play a role in hyper- and hyporeactivity of the skin. Immunohistological differences between allergic and irritant (''false positives'') reactions seem to be absent or hardly detectable.[60,68-71] Humoral factors, although difficult to estimate, might give us better insight into the mechanisms involved in skin reactivity and, thus, in the ESS.

VII. PRACTICAL IMPLICATIONS

How does the dermatologist handle the ESS, for example, in a case involving many positive reactions, the existence of even minor eczema elsewhere, a positive reaction to marginal irritant test material, or several positive reactions in the proximity of a strong reaction?[13,72] (See Table 6.) We stress that, especially in a case of job change and when medicolegal problems are involved, retesting or provocative use tests (repeated open application tests) might be required before a well-balanced, scientifically rational answer will be possible to the question whether we are dealing with a true allergic reaction.

The performance of patch tests can be so easy; newer test materials like the TRUE Test™ will simplify it even more. However, interpretation of patch test results is far more complicated. There is a significant risk that the results will be misinterpretted, especially when phenomena of hypo- and hyperreactivity are involved. With regard to skin hyperirritability, it is not so important which name we give it as long as we recognize it. The term ESS is, however, easy and practical to use.

REFERENCES

1. **Jaeger, H.,** De la nature de l'eczéma: recherces expérimentales: hypersensibilité, idio syncrasie et anaphylaxie cutanées chez les eczémateux, *Ann. Dermatol.,* 6, 109, 1923.
2. **Geiger, R.,** Unspezifische Desensibilisierungsversuche an Ekzematikern mit Terpentinöl auf percutanen Wege, *Arch. Dermatol.,* 158, 76, 1929.
3. **Rostenberg, A., Jr. and Sulzberger, M. B.,** Some results of patch tests, *Arch. Dermatol.,* 35, 433, 1937.
4. **Beek, C. H.,** The sensibility of the skin against soap among patients with eczema, *Dermatological,* 93, 167, 1946.
5. **Grolnick, M.,** Studies in contact dermatitis: the response of healed specific sites to stimulation with another contactant, *J. Allergy Clin. Immunol.,* 29, 298, 1948.
6. **Wilson, H. T. H.,** Standard patch tests in eczema and dermatitis, *Br. J. Dermatol.,* 67, 291, 1955.
7. **Bettley, F. R.,** Non-specific irritant reactions in eczematous subjects, *Br. J. Dermatol.,* 76, 116, 1964.
8. **Von Zumbusch, L.,** Ueber die Behandlung des Ekzems, *Muench. Med. Wochenschr.,* 68, 401, 1921.
9. **Björnberg, A.,** Skin Reactions to Primary Irritants in Patients With Hand Eczema, Oscar Isacsons Tryckeri AB, Göteborg, Sweden, 1968.
10. **Lammintausta, K., Maibach, H. I., and Wilson, D.,** Irritant reactivity in males and females, *Contact Dermatitis,* 17, 276, 1987.
11. **Mitchell, J. C.,** The angry back syndrome: eczema creates eczema, *Contact Dermatitis,* 1, 193, 1975.
12. **Mitchell, J. C.,** Multiple concomitant positive patch test reactions, *Contact Dermatitis,* 3, 315, 1977.
13. **Maibach, H. I.,** The ESS: excited skin syndrome, in *New Trends in Allergy,* Ring, J. and Burg, G., Eds.,Springer-Verlag, New York, 208, 1981.
14. **Mitchell, J. C. and Maibach, H. I.,** The angry back syndrome: the excited skin syndrome, *Semin. Dermatol.,* 1, 9, 1982.
15. **Kim, H. O., Wester, R. C., McMaster, J. A., Bucks, D. A., and Maibach, H. I.,** Skin absorption from patch test systems, *Contact Dermatitis,* 17, 178, 1987.
16. **Mitchell, J. C.,** Angry back syndrome, *Contact Dermatitis,* 7, 359, 1981.
17. **Bandmann, H.-J.,** Angry back syndrome, *Contact Dermatitis,* 7, 360, 1981.
18. **Kligman, A. and Gollhausen, R.,** The "angry back": a new concept or old confusion?, *Br. J. Dermatol.,* 115(Suppl 31), 93, 1986.
19. **Bruynzeel, D. P., Van Ketel, W. G., Scheper, R. J., and Von Blomberg-Van der Flier, B. M. E.,** The angry back syndrome: a retrospective study, *Contact Dermatitis,* 7, 293, 1981.
20. **Maibach, H. I., Fregert, S., Magnusson, B., Pirilä, V., Hjorth, N., Willemson, D. S., Malten, K. E., Meneghini, C. L., Lachapells J. M., Calnan, C. D., and Cronin,E.,** Quantification of the excited skin syndrome (the "angry back"): retesting one patch at a time, *Contact Dermatitis,* 8, 78, 1982.
21. **Bruynzeel, D. P., Van Ketel, W. G., Scheper, R. J., and Von Blomberg-Van der Flier, B. M. E.,** Angry back or the excited skin syndrome: a prospective study, *J. Am. Acad. Dermatol.,* 8, 392, 1983.
22. **Von Luderschmidt, C., Heilgemeir, R., Ring, J., and Burg, G.,** Polyvalente Kontaktallergie versus "Angry Back". Zur Problematik falsch positiver Epikutantestreaktionen, *Allergie,* 5, 262, 1982.
23. **Bandmann, H.-J., and Agathos, M.,** New results and some remarks to the "angry back syndrome", *Contact Dermatitis,* 7, 23, 1981.
24. **Gollhausen, R., Przybilla, B., and Ring, J.,** Reproducibility of patch testing, in *Abstr. 17th World Congr. Dermatology, Part II,* 1987, 59.
25. **Fischer, T. and Rystedt, I.,** Cobalt allergy in hard metal workers, *Contact Dermatitis,* 9, 115, 1983.
26. **Lachapelle, J.-M., Bruynzeel, D. P., Ducombs, G., Hannuksela, M., Ring, J., White, I. R., Wilkinson, J., Fischer, T., and Billberg, K.,** European multicenter study of the True Test®, *Contact Dermatitis,* 19, 91, 1988.
27. **Wilkinson, D. S., Fregert, S., Magnusson, B., Bandmann, H. J., Calnan, C. D., Cronin, E., Hjorth, N., Maibach, H. I., Maltem, K. E., Meneghini, C. L., Pirilä, V.,** Terminology of contact dermatitis, *Acta Derm. Venereol.,* 50, 287, 1970.
28. **Bruynzeel, D. P.,** Angry Back or Excited Skin Syndrome, M.D. thesis, Free University, Amsterdam, 35, 1983.
29. **Von Blomberg-Van der Flier, B. M. E.,** Recent Advances in Contact Allergy Research, Ph.D. thesis, Free University, Amsterdam, 1989.
30. **Rook, G. A. W., Carswell, J. W., and Stanford, J. L.,** Preliminary evidence for the trapping of antigen-specific lymphocytes in the lymphoid tissue of "anergic" tuberculosis patients, *Clin. Exp. Immunol.,* 26, 129, 1976.
31. **Ford, D. K., Da Roza, D. M., and Schulzer, M.,** Lymphocytes from the site of disease but not blood lymphocytes indicate the cause of arthritis, *Ann. Rheum. Dis.,* 44, 701, 1985.
32. **Haeberlin, J. B. and Fox, J. M.,** Primary irritation of the skin, *Arch. Dermatol.,* 80, 690, 1959.
33. **Magnusson, B. and Kligman, A. M.,** *Allergic Contact Dermatitis in the Guinea Pig,* Charles C Thomas, Springfield, IL, 1970, 100.

34. **Van der Lugt, L., and Veninga, T. S.,** Nonspecific hypersensitivity of the skin, *Dermatologica,* 162, 438, 1981.
35. **Arlette, J. P. and Fritzler, M. J.,** Reduced skin threshold to irritation in the presence of allergic contact dermatitis in the guinea pig, *Contact Dermatitis,* 11, 31, 1984.
36. **Roper, S. S. and Jones, H. E.,** An animal model for altering the irritability threshold of normal skin, *Contact Dermatitis,* 13, 91, 1985.
37. **Klecak, G.,** Identification of Contact Allergens: Predictive Tests in Animals, in *Dermatoxicology,* Marzulli, F. N. and Maibach, H. I., Eds., Hemisphere, Washington, D.C., 1983, 193.
38. **Nethercot, J. R. and Lawrence, M.,** The effect of the guinea pig maximization protocol on the irritant response to deodorized kerosene. The excited skin syndrome, *Contact Dermatitis,* 9, 439, 1983.
39. **Bruynzeel, D. P., Von Blomberg-Van der Flier, B. M. E., Van Ketel, W. G., and Scheper, R. J.,** Depression or enhancement of skin reactivity by inflammatory processes in the guinea pig, *Int. Arch. Allergy Appl. Immunol.,* 72, 67, 1983.
40. **Andersen, K. E. and Maibach, H. I.,** Cumulative irritancy in the guinea pig from low-grade irritant vehicles and the angry skin syndrome, *Contact Dermatitis,* 6, 430, 1980.
41. **Frey, J. R.,** Quantitative Untersuchungen bei epidermaler Sensibilisierung von Meerschweinchen mit Dinitrochlorobenzol, *Dermatologica,* 102, 1, 1951.
42. **Lammintausta, K., Maibach, H. I., and Wilson, D.,** Human cutaneous irritation: induced hypo reactivity, *Contact Dermatitis,* 17, 193, 1987.
43. **Jokipii, A. M. M. and Jokipii, L.,** Dose-response relationships and interference of simultaneous skin tests in delayed hypersensitivity, *J. Clin. Pathol.,* 26, 687, 1973.
44. **Thestrup-Pedersen, K.,** Suppression of tuberculin skin reactivity by prior tuberculin skin testing, *Immunology,* 28, 343, 1975.
45. **Von Blomberg, M., Boerrigter, G. H., and Scheper, R. J.,** Interference of simultaneous skin tests in delayed hypersensitivity, *Immunology,* 35, 361, 1978.
46. **Uehara, M. and Ofuji, S.,** Suppressed cell-mediated immunity associated with eczematous inflammation, *Acta Derm. Venereol.,* 57, 137, 1977.
47. **Fisher, A. A.,** *Contact Dermatitis,* Lea & Febiger, Philadelphia, 1967, 18.
48. **Hering, H.,** Zur Beurteilung berufsbedingter Hauterkrankungen unter Berücksichtigung von Fehlerquellen bei epikutanen Testproben, *Dermatol. Monatsschr.,* 130, 1129, 1954.
49. **Kligman, A. M.,** Poison ivy (Rhus) dermatitis, *Arch. Dermatol.,* 77, 149, 1958.
50. **McGuire, R. and Fox, R. A.,** Suppression of delayed hypersensitivity by the depletion of circulating monocytes, *Immunology,* 38, 157, 1979.
51. **Meyer-Bloch, J. C. H., Grundmann, H.-P., Weis, H., Trachsel, H., and Storck, H.,** Skin test and lymphocyte stimulation in delayed hypersensitivity against Staphylococcus aureus antigens, *Arch. Dermatol. Res.,* 267, 291, 1980.
52. **Asherson, G. L., Allwood, G. G., and Mayhew, B.,** Contact sensitivity in the mouse: movement of T blasts in the draining lymph nodes to sites of inflammation, *Immunology,* 25, 485, 1973.
53. **Van Dinther-Janssen, A. C. H. M., Van Maarsseveen, A. C. M. Th., De Groot, J., and Scheper, R. J.,** Comparative migration of T- and B-lymphocyte subpopulations into the skin inflammatory sites, *Immunology,* 48, 519, 1983.
54. **Van Gool, J., Schreuder, J., and Ladiges, N. C. J. J.,** Inhibitory effect of foetal α_2-globulin, an acute phase protein, on carrageenin oedema in the rat, *J. Pathol.,* 112, 245, 1974.
55. **Van Gool, J.,** Acute-fase-eiwitten, betekenis voor de ontstekingsreactie, *Ned. Tijdschr. Geneeskde.,* 124, 869, 1980.
56. **Larsen, C. G., Ternowitz, T., Larsen, F. G., and Thestrup-Pedersen, K.,** Epidermis and lymphocyte interactions during an allergic patch test reaction, *J. Invest. Dermatol.,* 90, 230, 1988.
57. **Ternowitz, T. and Thestrup-Pedersen, K.,** Epidermis and lymphocyte interaction during a tuberculin skin reaction. II Epdiermis contains specific lymphocyte chemotactic factors, *J. Invest. Dermatol.,* 87, 613, 1986.
58. **Larsen, C. G., Ternowitz, T., Larsen, F. G., Zachariae, C., and Thestrup-Pedersen, K.,** Etaf/Interleukin-1 and epidermal lymphocyte chemotactic factor in epidermis overlying an irritant patch test, submitted.
59. **Rostenberg, A.,** Predictive procedures for eczematous hypersensitivity, *Arch. Indust. Health,* 20, 181, 1959.
60. **Bruynzeel, D. P., Nieboer, C., Boorsma, D. M., Scheper, R. J., and Van Ketel, W. G.,** Allergic reactions, "spillover" reactions and T-cell subsets, *Arch. Dermatol. Res.,* 275, 80, 1983.
61. **Bandmann, H.-J., and Agathos, M.,** Das "Angry Back Syndrome" — Untersuchungsergebnisse mit Sequenztestungen, Wiederholungstestungen und dem Cocarden-(Target-) Test, Hautarzt, 32, (Suppl 5), 97, 1981.
62. **Kligman, A. and Gollhausen, R.,** The "angry back": a new concept or old confusion?, *Br. J. Dermatol,* 115 (Suppl 31), 93, 1986.

63. **Hamani, I. and Marks, R.,** Abnormalities in clinically normal skin — a possible explanation of the "angry back syndrome", *Clin. Exp. Dermatol.*, 13, 328, 1988.
64. **Roper, S. S. and Jones, H. E.,** A new look at conditioned hyperirritability, *J. Am. Acad. Dermatol.*, 7, 643, 1982.
65. **Sjöborg, S., Andersson, A., and Christensen, O. B.,** The Langerhans cells in healed patch test reactions, before and after oral administration of nickel, *Acta Derm. Venereol. Suppl.*, Vol. 111, 1984.
66. **Kanerva, L., Estlander, T., and Jolanki, R.,** Immunohistochemistry of lymphocytes and Langerhans' cells in longlasting allergic patch tests, *Acta Derm. Venereol.*, 68, 116, 1988.
67. **Scheper, R. J., Von Blomberg-Van der Flier, B. M. E., Boerrigter, G. H., Bruynzeel, D. P., Van Dinther, A., and Vos, A.,** Induction of immunological memory in the skin: role of local T cell retention, *Clin. Exp. Immunol.*, 51, 141, 1983.
68. **Ranki, A., Kanerva, L., Förström, L., Konttinen, Y., and Mustikalallio, K. K.,** T and B lymphocytes, macrophages and Langerhans' cells during the course of contact allergic and irritant skin reactions in man, *Acta Derm. Venereol.*, 63, 376, 1983.
69. **Scheynius, A., Fischer, T., Forsum, U., and Klaveskog, L.,** Phenotypic characterization in sites of inflammatory cells in allergic and irritant contact dermatitis in man, *Clin. Exp. Immunol.*, 55, 81, 1984.
70. **Kanerva, L., Estlander, T., and Ranki, A.,** Lymphocytes and Langerhans' cells in allergic patch tests, *Dermatosen*, 35, 16, 1987.
71. **Willis, C. M., Young, E., Brandon, D. R., and Wilkinson, J. D.,** Immunopathological and ultrastructural findings in human allergic and irritant contact dermatitis, *Br. J. Dermatol.*, 115, 305, 1986.
72. **Bruynzeel, D. P. and Maibach, H. I.,** Excited skin syndrome (angry back), *Arch. Dermatol.*, 122, 323, 1986.
73. **Jessner, M., quoted by Fisher, A. A.,** Occupational, industrial and plant dermatology symposium, San Francisco, *Cutis*, 24, 143, 1979.
74. **Shelmire, B.,** Contact dermatitis from weeds: patch testing with their oleoresins, *J. Am. Med. Assoc.*, 113, 1085, 1939.
75. **Wilkinson, D. S., Fregert, S., Magnusson, B., et al.,** Terminology of contact dermatitis, *Acta Derm. Venereol.*, 50, 287, 1970.
76. **Kligman, A. M. and Epstein, W.,** Updating the maximization test for identifying contact allergens, *Contact Dermatitis*, 1, 231, 1975.
77. **Fisher, A. A.,** quoted by Cronin, E., *Contact Dermatitis*, Churchill Livingstone, Edinburgh, 1980, 12.

Chapter 11

TECHNOLOGY OF ALLERGEN DELIVERY

Jens Hansen

TABLE OF CONTENTS

I. INTRODUCTION

Optimization of allergen bioavailability and, consequently, an improvement of the possibility to diagnose allergic contact dermatitis imply both an efficient and selective delivery and transport of the allergen to the epidermal antigen-presenting cells and an optimal rate of delivery. The recognition of the importance of allergen bioavailability in the diagnosis of contact allergy has increased in recent years. Novel approaches to improve antigen delivery have been developed. Until recently petrolatum has been the vehicle of choice for patch test materials. Problems encountered with the petrolatum preparations include (1) inhomogeneous mixtures, (2) the necessity of several steps before the preparation can be applied to the skin, and (3) varying amounts applied. Recently, new ready-to-use patch test systems with controlled amounts of allergen have been published.[1-2] However, the only product marketed so far is the TRUE Test™, in which the allergen is incorporated as dissolved or micronized in a dry vehicle.

In the development of allergen delivery systems it is useful to understand the diffusional characteristics of allergen both in the vehicle as well as into the skin and, most importantly, the stability of the allergen in the vehicle.

Contact allergens are low-molecular-weight compounds with a molecular weight under about 1000 Da. The bioavailability varies from chemical to chemical (e.g., stability in the vehicle, dissolution from the vehicle, and passage through the skin). One possible approach for managing some of these problems appears to be the prohapten concept. A prohapten denotes a per se inactive derivative of a known allergen. It may improve stability in the patch test material; when applied to the skin, the derivative enhances the delivery characteristics of the allergen and, subsequently, the prohapten is converted to the allergen by virtue of enzymatic and/or chemical activity.

II. ALLERGEN DELIVERY FROM PATCH TEST VEHICLES

Stratum corneum is the rate-determining barrier for percutaneous absorption. It is important to identify and understand the factors which influence the absorption process. In this section some of these factors will be discussed, with special emphasis on nickel, *p*-phenylenediamine, and chlorocresol.

The influence of nickel counterion concentration and the amount of preparation applied has been studied *in vitro* by Fullerton et al.[3-5] by means of excised human skin. Their results show that the transport of nickel ion from petrolatum was nearly unaffected by the counterion. In Table 1 is shown the effect of the concentration of nickel chloride in petrolatum and the amount of ointment applied on the recovery of nickel from epidermis and dermis after application for 96 h. The investigation showed that the amount of allergen applied significantly affected the nickel content in the skin. This result is of practical relevance clinically, as Fisher and Maibach[6] have reported a fivefold variation in the amount (8 to 43 mg) of test preparation used in ordinary Finn Chambers®. This observation partly explains the problems with reproducibility of the present standard patch test procedure (allergen in petrolatum) compared to the TRUE Test reported by Gollhausen.[7]

The influence on the permeability of human skin to nickel chloride from different hydrogels (e.g., methyl cellulose, Methocels®, polyvinyl alcohol, and polyoxyethylene-polyoxypropylene copolymer 30% has been studied *in vitro* and compared to the standard patch test in petrolatum by Fullerton et al.[4] The authors concluded that the nickel concentration can be reduced at least fivefold (compared to the concentration required in the petrolatum vehicle) if Methocel® E4M in water is used as a vehicle. Recently, Kim et al.[8] showed that not only the vehicle but also the patch test system influences the bioavailability of the allergen. The results showed that there was a sixfold difference in skin absorption of

TABLE 1
The Effect of Nickel Chloride Concentration and the Amount of Petrolatum Applied on the Recovery of Nickel in the Epidermis and Dermis at 96 h

$NiCl_2$ conc. (% w/w)	Applied amount (mg)	Epidermis (μg Ni/1.8 cm^2)	Dermis (μg Ni/1.8 cm^2)
2.5	27	9.2, 4.9	2.7, 3.1
2.5	67.5	10.6, 6.8	4.2, 3.9
5	27	9.9, 10.6	3.0, 3.3
5	67.5	16.3, 12.9	5.1, 4.0

1% ^{14}C-phenylenediamine in petrolatum through guinea pig skin *in vivo* which was dependent on the patch test system at a concentration normalized to equal surface area. The skin absorption was determined in the guinea pig by urinary excretion of ^{14}C. The systems, in decreasing order of skin absorption, were Hill Top Chamber > Teflon® control patch > small Finn Chamber® > large Finn Chamber® > Al-test chamber.

The contact sensitivity and bioavailability of chlorocresol from five topical preparations was determined by Andersen et al.[9] in guinea pigs using the cumulative contact enhancement test. Chlorocresol (5%) in olive oil/acetone (4:1) and in 5% aqueous suspension stabilized with carbomer 941 were more sensitizing (55% and 60% of the animals were positive, respectively) than chlorocresol (5%) in propylene glycol with or without carbomer 941 (20% positive). The sensitization from a saturated aqueous chlorocresol solution (about 0.38% w/v) was comparable to that of a 5% propylene glycol solution containing 13 times more chlorocresol. Of the applied doses, 0.2 to 1.6% remained at the patch test skin sites as free chlorocresol. Of the chlorocresol in aqueous suspension, 75% permeated the skin in contrast to 34% and 35% of the chlorocresol in olive oil/acetone (4:1) and propylene glycol, respectively. In spite of the equality of chlorocresol absorption the two latter preparations, they showed significantly different sensitizing capacities. No simple relationship between the sensitization rates and the calculated bioavailability was observed with the preparations tested.

In conclusion, these investigations stress the importance of a standardization of patch test materials with the aim of minimizing the risk of false negative and false positive reactions. One promising approach to overcome these problems seems to be the ready-to-use patch test systems.

III. READY-TO-USE PATCH TEST SYSTEM

The ready-to-use patch test systems can be classified into two categories: one in which the allergen is mixed in petrolatum or in a gel in a sealed reservoir, the PA-Patch, and one in which the allergen is prepared in a dry matrix. The TRUE Test system is of the latter category, where the allergens are dissolved or micronized in a matrix at a fixed concentration (in milligrams per square centimeter).

In the laboratory the TRUE Test patches are made, as described by Fischer and Maibach,[2] by dissolving or suspending the allergen in water and/or ethanol. A gel of a water-soluble polymer (e.g., hydroxypropylcellulose, methylcellulose, or polyvidone) is incorporated into the solution/suspension. Subsequently, the gel is distributed uniformly on polyester sheets and dried. The coated sheets are cut into small patches (9 × 9 mm); 12 patches, each containing specific allergens, are attached to the tape to form a panel. A protective sheet covers the TRUE Test panel, which is finally packed in laminate foil (Figure 1).

The PA test is shown in Figure 2. The cylindrical reservoirs into which the allergens

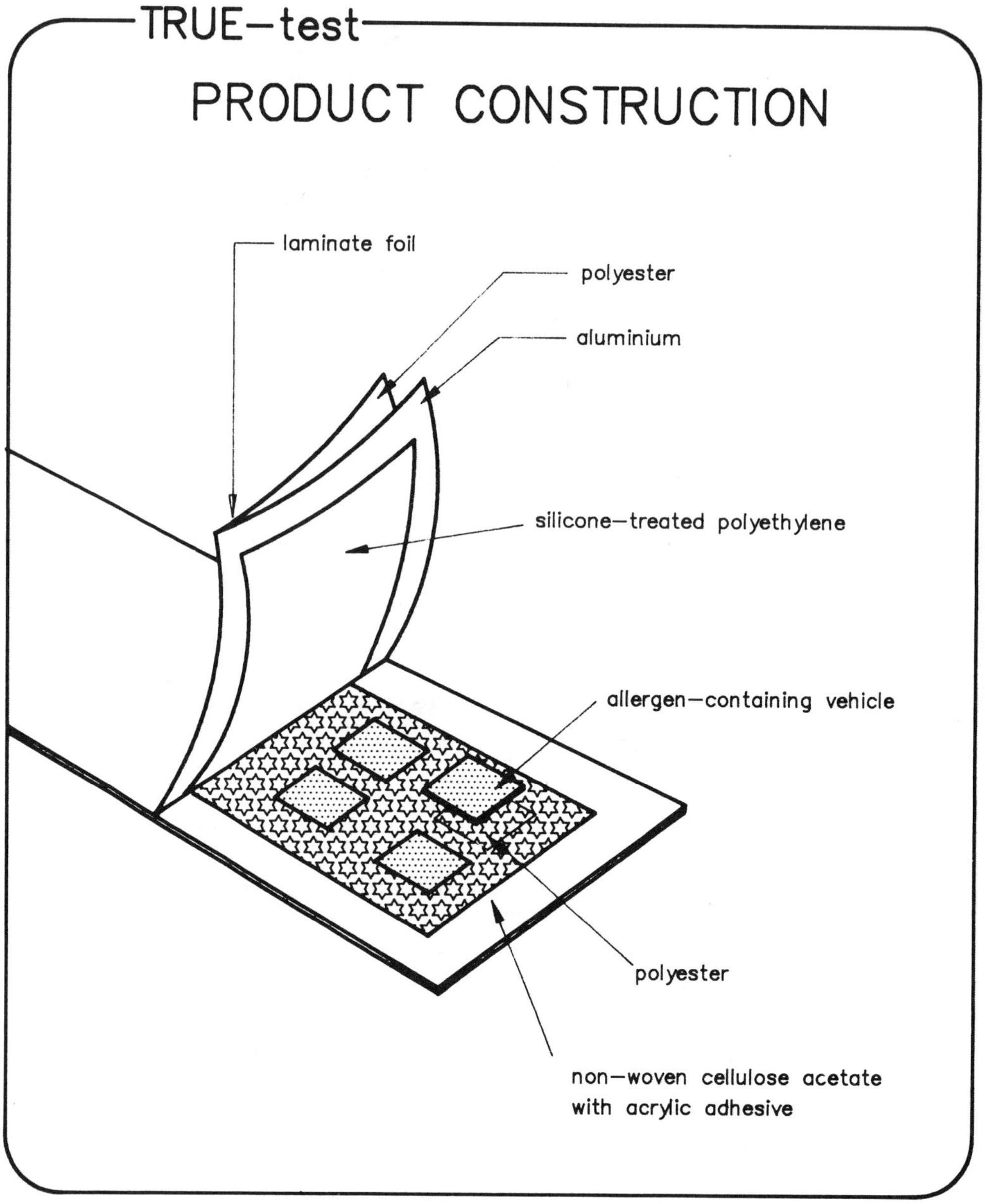

FIGURE 1. Construction of the TRUE Test™

are deposited, are stamped into a closed cell foam with adhesive on both sides. A circular piece of loosely woven fabric covers the bottom of each reservoir. The woven fabric must be tight enough to hold the petrolatum in place as well as open enough to allow petrolatum containing the antigen to seep through onto the patient's skin. A release line without holes is placed over the bottom surface of the foam covering the loosely woven fabric. An impermeable adhesive backing, covering the holes to create sealed reservoirs, is placed on the top surface of the foam.

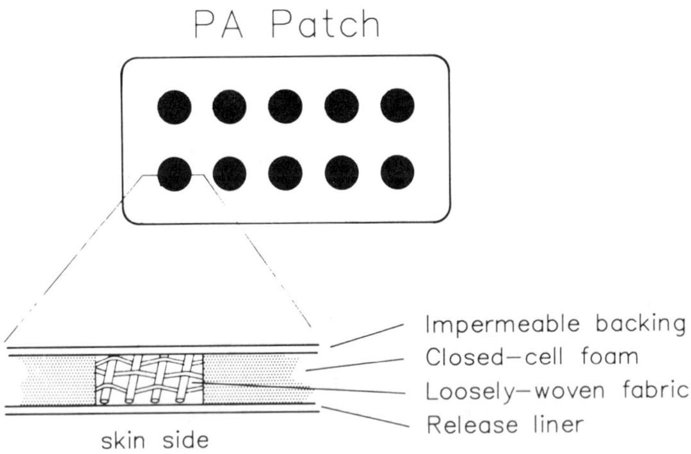

FIGURE 2. Construction of the PA Patch.[1] The top drawing is the top view and the bottom drawing is the side view. (From Merks, J., Westheim, A., and Snipes, W., *J. Am. Acad. Dermatol.*, 16, 960, 1987. With permission.)

IV. BIOPHARMACEUTICAL OPTIMIZATION

Various steps are used to optimize the bioavailability of each standard allergen in the TRUE Test. Formaldehyde and perfumes cause special problems due to evaporation and chemical reactivity.

A. NICKEL

We studied the release of nickel sulfate hexahydrate from the TRUE Test patch by *in vtiro* penetration studies using human cadaver skin. From a theoretical point of view, the degree of allergen saturation (thermodynamic activity) on the patch determines the partitioning of the allergen between the patch and the skin surface. In the solid-phase matrix system (TRUE Test), the allergen is dispersed as micronized particles in the matrix. It was therefore unexpected that an increased nickel sulfate hexahydrate concentration (0.038 to 0.322 mg/cm^2) raised the flux of nickel through the skin by a factor of ten, as shown in Figure 3. In contact with the human skin, the allergen is dissolved by the perspiration water or the lipid in the skin, giving patches with different degrees of saturation. Consequently, nickel from the patch containing the highest amount of that metal gave the highest flux.

In vitro penetration studies of nickel sulfate into human skin were carried out for 6, 24, and 48 h under occlusion. As expected, results (Figure 4) showed that more nickel was absorbed after 48 h than after 6 h of occlusion. Kalimo[10] tested 20 nickel-sensitive patients and found that 8 reacted to a nickel sulfate hexahydrate patch (0.200 mg/cm^2) after 6 h of application and 18 patients reacted after 48 h of application.

The insufficient clinical response after 6 h of application may be overcome by using nickel chloride in the patch instead of nickel sulfate hexahydrate. Fullerton et al.[3] have recently shown that more nickel is delivered to the skin *in vitro* from nickel chloride in a hydrogel compared to nickel sulfate in the same hydrogel. By using nickel chloried in the TRUE Test, Kalimo[10] observed that 6 to 12 h of application gave positive results comparable to those obtained for nickel sulfate tests applied for 48 h. Nickel chloride may, therefore, be useful in short-application patch testing.

B. BENZOCAINE

Benzocaine crystalized when hydroxypropyl cellulose was used as a vechicle. Consequently, the release of benzocaine from the formulation into water was insufficient, as seen

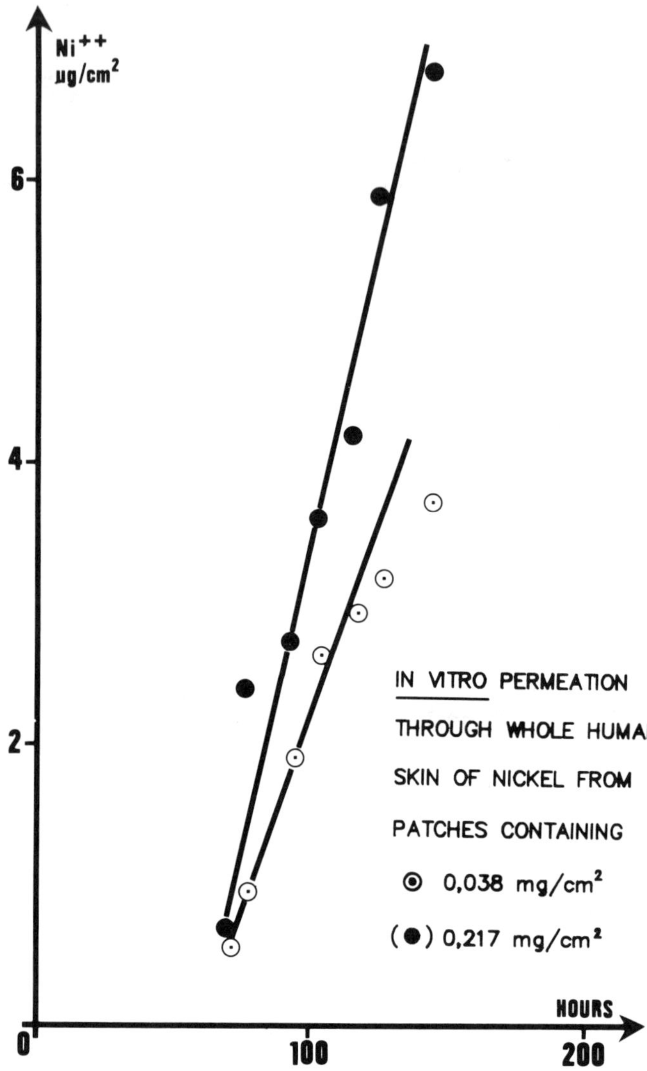

FIGURE 3. *In vitro* permeation through excised human skin of nickel from patches containing 0.038 mg/cm², 0.217 mg/cm², and 0.319 mg/cm² nickel sulfate hexahydrate.

in Figure 5. This was confirmed in six patients with known allergy to benzocaine. Change of the vehicle to polyvidone increased the release of benzocaine significantly (Figure 6). This was attributed, in part, to a solid dispersion being formed between benzocaine and polyvidone. The improved benzocaine release from the preparation was confirmed in a clinical study on benzocaine-allergic patients.[11]

V. PROHAPTENS

A. FORMALDEHYDE

A prohapten should decompose into the allergen and the carrier after contact with sweat and/or transepidermal water on the surface of the skin. The allergen should then penetrate through the stratum corneum and possibly provoke an allergic reaction. Alternatively, the prohapten is absorbed and converted into the allergen in the skin by virtue of enzymatic and/or chemical reactivity.

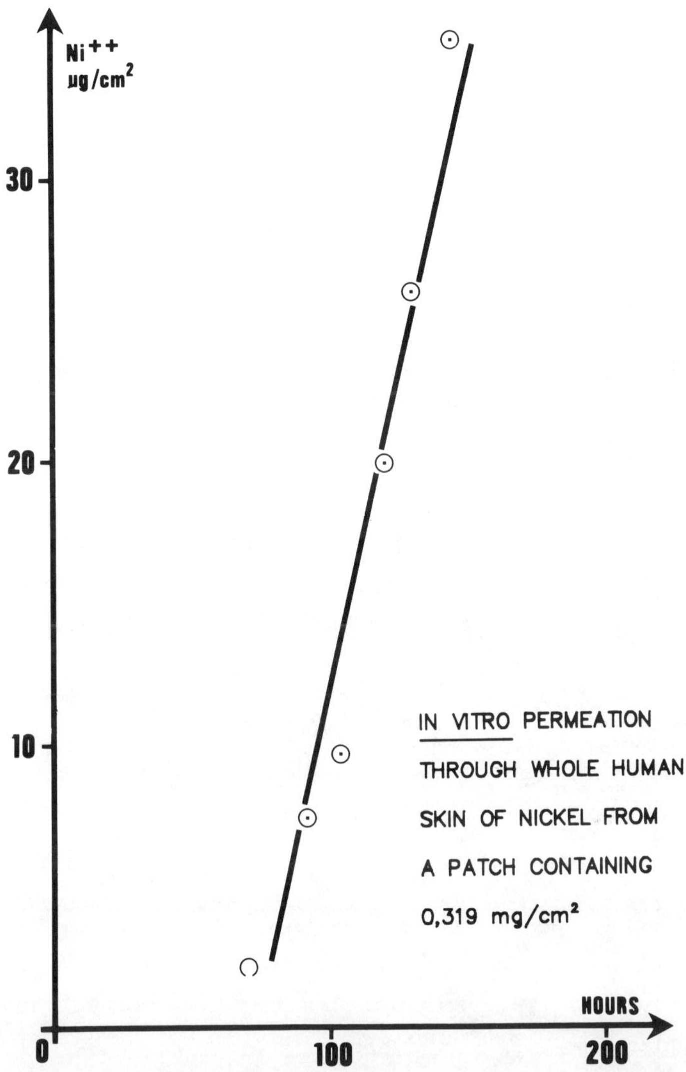

FIGURE 3 (continued).

Aqueous formaldehyde is used in the standard series. It is a volatile allergen, a fact which created problems in the development of the TRUE Test formaldehyde patch. We evaluated *N*-hydroxymethyl[12] derivatives as possible prohapten candidates for formaldehyde.

A number of reports[13-15] have indicated that *N*-hydroxymethylated derivaties of amides and imides fulfill the requirements for a prohapten of formaldehyde, as illustrated in Figure 7.

The effect of concentration, pH, and carrier on the degradation of a series of *N*-hydroxymethylated amides and imides to formaldehyde and carrier was studied in aqueous solution. The results showed that at low concentration, less than 1×10^{-3} *M* of prohapten (e.g., hydroxymethyl succinimide), formaldehyde was found to be formed in stoichiometric amounts. However, by increasing the concentration of prohapten, nonstoichiometric amounts of formaldehyde were formed, indicating that the decomposition of *N*-hydroxymethyl succinimide into succinimide and formaldehyde was a reversible process, in accordance with earlier studies by Johansen and Bundgaard.[13]

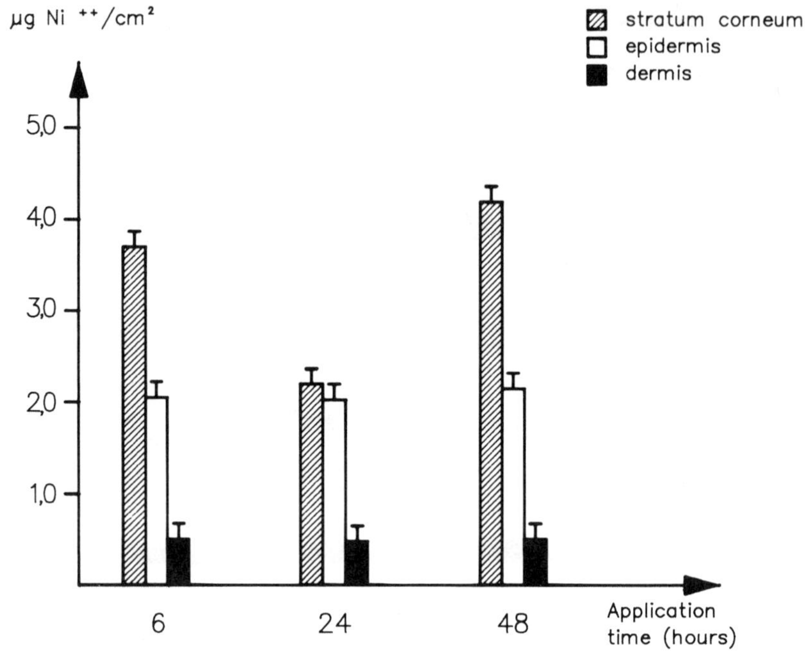

FIGURE 4. Recovery of nickel in the stratum cornuem (5th to 15th strippings), epidermis, and dermis after application of 0.200 mg/cm² nickel sulfate hexahydrate at different times.

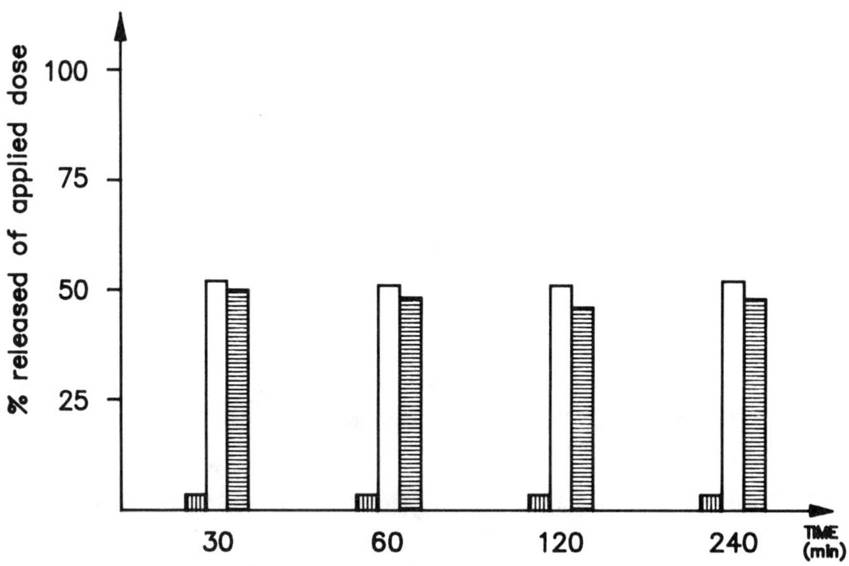

FIGURE 5. *In vitro* release at 30°C of the caine mix from the hydroxypropyl cellulose matirx. Key: benzocaine (▥), tetracaine hydrochloride (□), dibucaine hydrochloride (▤).

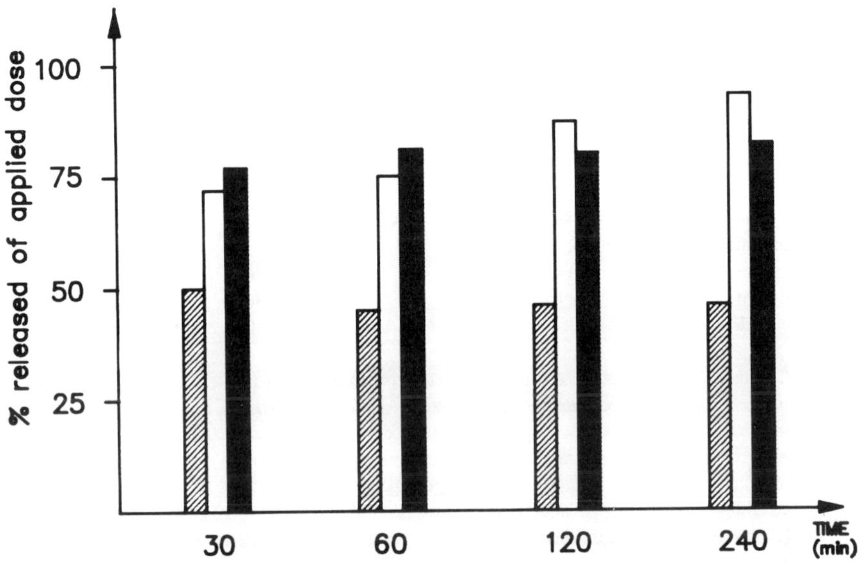

FIGURE 6. *In vitro* release at 30°C of the caine mix from the polyvidone matrix. Key: benzocaine (▨), tetracaine hydrochloride (□), dibucaine hydrochloride (■).

$$R\text{-}CONH\text{-}CH_2OH \rightleftharpoons R\text{-}CONH_2 + HCHO$$

FIGURE 7. Decomposition of *N*-hydroxymethylated amides into amide and formaldehyde.

TABLE 2
Half-Life ($T^{1/2}$) for the Decomposition of 10^{-4}*M* *N*-Hydroxymethyl Succinimide in Aqueous Solution at Different pH Values

pH	$T^{1/2}$
9.00	ca. 0.03
7.36	1.03
6.00	16
4.00	20.80

From Table 2 it is seen that the pH of the solution influences the half-life of cleavage of the prohapten to formaldehyde. From the table it can be calculated that by increasing the pH one unit the half-life of the prohapten decreases by a factor of ten. Similar results were obtained for the other prohaptens of formaldehyde. Table 3 shows that the carrier component of the *N*-hydroxymethyl derivatives has a pronounced effect on the decomposition rates expressed as the half-life for the cleavage to formaldehyde at pH 9.00. Based on this information a basic formulation for the different prohaptens of formaldehyde was designed. The polymer polyvidone was selected as a matrix, and a buffer system was incorporated to ensure that the pH in the patch was kept at 9 after contact with the perspiration water.

In guinea pigs sensitized to formaldehyde by the methods described for the guinea pig

TABLE 3
***In Vivo* Release of Formaldehyde Determined by the GMP Test**

Application		$T^{1/2}$ (pH 9.0)	Response (% of total)			
			0	1	2	3
N-hydroxymethyl						
Phatalimide	(0.85 mg/cm²)	⩾0.03 min	40	45	10	5
Succinimide	(0.50 mg/cm²)	0.03 min	0	5	0	95
Nicotinamide	(0.62 mg/cm²)	0.7 h	10	45	10	35
Benzamide	(2.07 mg/cm²)	6.7 h	90	10	0	0
Formaldehyde solution	(0.78 mg/cm²)	—	0	0	0	100

Note: Challenge response read 72 h after application of proallergen in guinea pigs sensitized to formaldehyde by the guinea pig maximization test (GMP test) method, and the corresponding *in vitro* half-life for the degradation of the proallergens to formaldehyde and carrier to pH 9.0. The evaluation scale which has been used: 0 = no visible reaction; 1 = discrete or patchy erythema; 2 = moderate or confluent erythema; 3 = intense erythema and swelling. The number of animals in the control and test group is 10 and 20, respectively. (From Hansen, J., Albrectsen, S., and Hedegaard, K., *J. Am. Acad. Dermatol.*, 21, 838, 1989. With permission.)

maximization test (GPM test),[16] the contact allergic response toward the prohaptens of formaldehyde was compared to the present standard test (1.0% formaldehyde in water). The *in vivo* release of formaldehyde determined by the GPM test is shown in Table 3. The results show a correlation between the challenge reaction and half-lives for the cleavages of the prohapten measured *in vitro*. This correlation is not strictly linear, possibly due to differences in solubility and diffusibility properties of the guinea pig skin. Based on these results, *N*-hydroxymethyl succinimide was selected as the prohapten for formaldehyde (Figure 8). In the TRUE Test, succinimide is nonallergenic and nonirritating in animals.

Preliminary experiments in healthy volunteers showed that neither *N*-hydroxymethyl succinimide (0.5 mg/cm²), equivalent to 0.12 mg/cm² formaldehyde, nor succinimide (1.6 mg/cm²) caused irritant reactions. Patch testing on a limited number of formaldehyde-allergic patients showed that *N*-hydroxymethyl succinimide elicited allergic reactions. The appropriate patch test concentration for hydroxymethyl succinimide in the TRUE Test seems to be around 0.12 mg/cm² with respect to formaldehyde. Further tests are required.

B. PERFUMES

To overcome the problem with volatile or unstable allergens, the concept of molecular encapsulation by β-cyclodextrin, a cyclic oligosaccharide containing seven glucose units, has been used. Generally, the monomolecular inclusion complex involves the entrapment of a single guest molecule in the cavity of one β-cyclodextrin molecule without any covalent bonds being formed, in contrast to the prohapten concept.

Toxicity studies have shown that orally administered β-cyclodextrin is harmless. In addition, animal experiments have shown that β-cyclodextrin is a nonallergic and nonirritating substance.

Table 4 illustrates the volatility of the perfume mix (geraniol, cinnamaldehyde, hydroxycitronellall, cinnamyl alcohol, eugenol, isoeugenol, α-amylcinnamaldehyde, and oak moss in equal parts) incorporated in a hydroxypropyl cellulose matrix and the perfume-cyclodestrin complex in the same matrix, stored at 40°C.

Of perfume mix incorporated in the hydroxypropyl cellulose matrix, 95% evaporated after 1 month of storage at 40°C while only 5% evaporated from the β-cyclodextrin complex. The formation of a complex between the fragrance-mix allergens results in a 1:1 (molar ratio complex. The release of cinnamaldehyde from such a complex *in vitro* through human

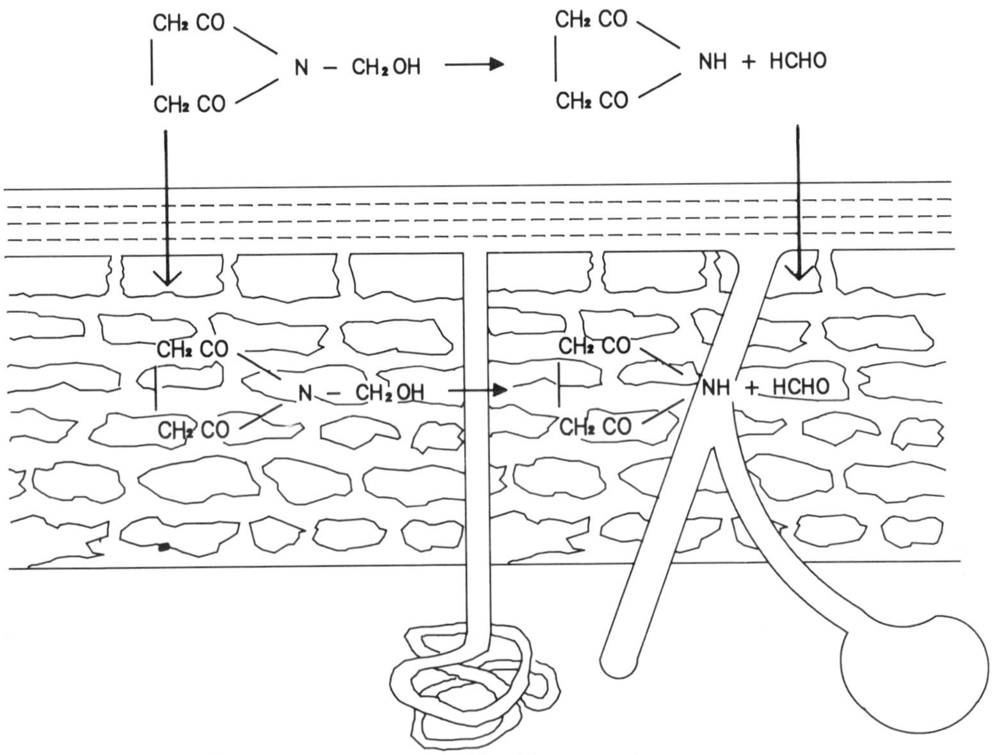

FIGURE 8. Absorption and degradation of *N*-hydroxymethyl succinimide to succinimide and formaldehyde on the surface of the skin and after absorption of the prohapten.

TABLE 4
Degradation of the Perfume Mix 0.37 mg/cm² in Different Vehicles at 40°C (% Residual)

Storage time (months)	Vehicle: Hydroxypropyl cellulose matrix	Vehicle: β-Cyclodextrin
0	100	100
1	5	92
2	0	87

skin is shown in Figure 9. Application of β-cyclodextrin-cinnamaldehyde (1:1 complex) resulted in the appearance of cinnamaldehyde, cinnamyl alcohol, and a lesser amount of cinnamic acid in the receptor phase. This is probably due to skin metabolism, in accordance with results reported by Weibel et al.[17]

The *in vivo* penetration and release of cinnamaldehyde from the complex was investigated in guinea pigs sensitized to cinnamaldehyde by the GPM test method. The animals were challenged by means of pure cinnamaldehyde and by an equimolar concentration of cinnamaldehyde-β-cyclodextrin complex. The results showed that 21 out of 24 animals responded to pure cinnamaldehyde, while 7 out of 24 animals responded to the β-cinnamaldehyde-β-cyclodextrin complex (1:1).

When the cyclodextrin complex is administered topically it must dissociate before absorption, as it might be expected that the hydrophilic complex can only diffuse across the stratum corneum with considerable difficulty (if at all) because the molecular weight of β-

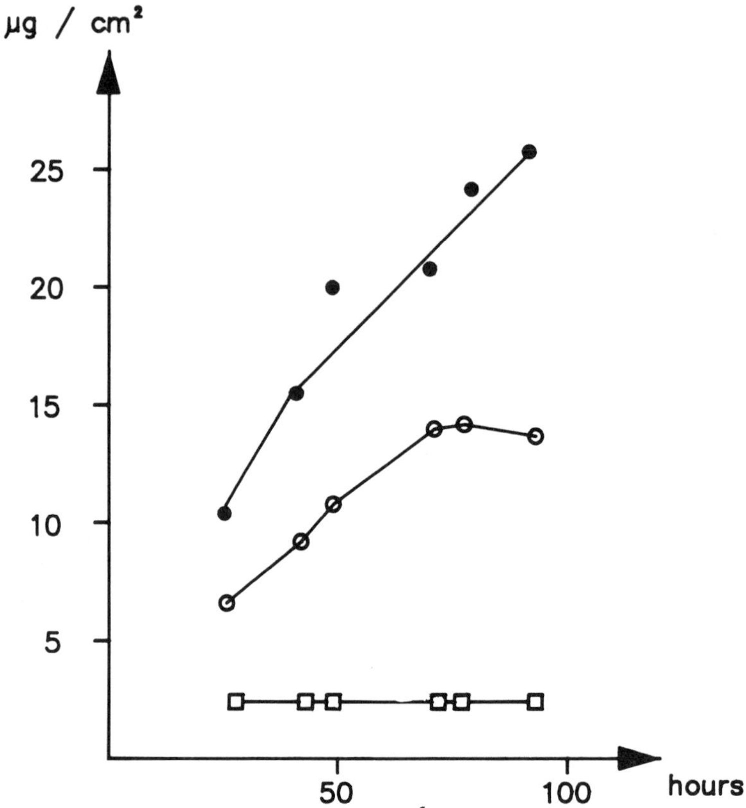

FIGURE 9. *In vitro* percutaneous penetration of cinnamaldehyde from a β-cyclodextrin-cinnamaldehyde (1:1 molar complex equal to 62 µg/cm² cinnamaldehyde). Key: cinnamaldehyde (□), cinnamyl alcohol (●), cinnamic acid (○).

cyclodextrin is 1135. This overall process is visualized in Figure 10. It is obvious that if complexes with a very low stability constant (K) are applied to the skin a quick dissociation of the complex to allergen and β-cyclodextrin will result, with little or no effect on the overall absorption process of the allergen in contrast to a complex with a higher complexation constant.

It is therefore reasonable that equimolar concentrations of cinnamaldehyde-β-cyclodextrin complex and pure cinnamaldehyde applied to guinea pigs allergic to cinnamaldehyde result in fewer allergic reactions to the complex compared to the pure allergen. However, it is possible to increase the response rate in the GPM test by applying higher concentrations of cinnamaldehyde-β-cyclodextrin complex to the guinea pig skin. An inclusion complex between allergens and β-cyclodextrin seems to be a useful concept in the delivery of volatile allergens.

VI. CONCLUSION

A new series of patch test material, the ready-to-use patch test, has been developed to provide a better standardization of patch test materials with regard to bioavailability and stability of the allergen. The systems described improve the diagnosis of contact allergy by simplifying the application and deliver a more reproducible amount of allergen to the skin than is possible with petrolatum preparations. As these dosage forms are products of a new technology, they require development of new approaches for allergen deliver, i.e., the prohapten concept, which denotes a per se inactive derivative of a known allergen.

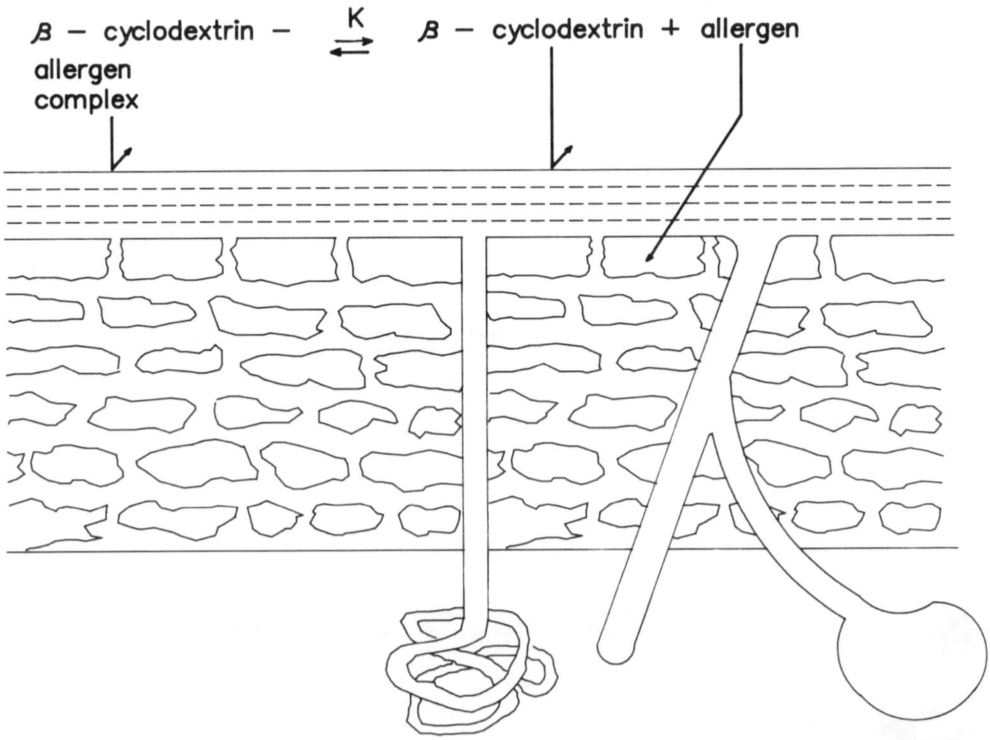

FIGURE 10. Absorption of allergen from a β-cyclodextrin-allergen complex.

REFERENCES

1. **Marks, J., Westheim, A., and Snipes, W.,** PA-Patch — a new patch testing device, *J. Am. Acad. Dermatol.,* 16, 961, 1987.
2. **Fischer, T. and Maibach, H. I.,** The thin layer rapid use epicutaneous test (TRUE test), a new patch test method with high accuracy, *Br. J. Dermatol.,* 112, 63, 1985.
3. **Fullerton, A., Andersen, J. R., Hoelgaard, A., and Menne, T.,** Permeation of nickel salts through human skin *in vitro, Contact Dermatitis,* 15, 173, 1986.
4. **Fullerton, A., Andersen, J. R., and Hoelgaard, A.,** Permeation of nickel through human skin *in vitro* — effect of vehicles, *Br. J. Dermatol.,*118, 509, 1988.
5. **Fullerton, A. and Hoelgaard, A.,** Nickel patch test material: effect of counter ion and vehicle on skin permeation *in vitro, FARMACI Sci. Ed.,* 16, 53, 1988.
6. **Fischer, T. and Maibach, H.,** Patch test allergens in petrolatum: a reappraisal, *Contact Dermatitis,* 11, 285, 1984.
7. **Gollhausen, R., Przybilia, B., and Ring, J.,** in *Current Topics in Contact Dermatitis,* Springer-Verlag, Berlin, 1989, 524.
8. **Kim, H. O., Wester, R. C., McMaster, J. A., Bucks, D. A. W., and Maibach, H. J.,** Skin absorption from patch test systems, *Contact Dermatitis,* 17, 178, 1987.
9. **Andersen, K. E., Carlsen, L., Egsgaard, H., and Larsen, E.,** Contact sensitivity and bioavailability of chlorocresol, *Contact Dermatitis,* 13, 246, 1985.
10. **Kalimo, K., and Lammintausta, K.,** in *Current Topics in Contact Dermatitis,* Springer-Verlag, Berlin, 1989, 540.
11. **Kreilgård, B., Hansen, J., and Fischer, T.,** Chemical, pharmaceutical and clinical standardization of the TRUE Test™ caine mix, *Contact Dermatitis,* 21, 23, 1989.
12. **Hansen, J., Albrechtsen, S., and Hedegaard, K.,** Proallergens of formaldehyde applied in patch testing of formaldehyde contact allergy, *J. Am. Acad. Dermatol.,* 21, 838, 1989.

13. **Johansen, M. and Bundgaard, H.,** Pro-drugs as delivery systems: VI kinetics and mechanism of the decomposition of N-hydroxy-methylated amides and imides in aqueous solution and assessment of their suitability as pro-drugs, *Arch. Pharm. Chem. Sci. Ed.*, 7, 175, 1979.
14. **Johansen, M. and Bundgaard, H.,** Decomposition of rolitetracycline and other N-Mannich bases and of N-hydroxymethylated derivatives in the presence of plasma, *Arch. Pharm. Chemi. Sci. Ed.*, 9, 43, 1981.
15. **Bundgaard, H. and Johansen, M.,** Pro-drug as drug delivery systems IV. N-Mannich bases as potential novel pro-drugs for amindes, ureides, amides and other NH-acidic compounds, *J. Pharm. Sci.*, 69, 44, 1980.
16. **Andersen, K. E., Boman, A., Vølund, A., and Wahlberg, J. E.,** Induction of formaldehyde contact sensitivity: dose reponse relationship on the guinea pig maximization test, *Acta Derm. Venereol.*, 65, 472, 1985.
17. **Weibel, H., Hansen, J., and Andersen, K. E.,** Cross-sensitization patterns in guinea pigs between cinnamaldehyde, cinnamylalcohol and cinnamic acid, *Acta Derm. Venereol.*, 69, 302, 1989.

Chapter 12

IN VITRO MODELS OF HUMAN SKIN: KERATINOCYTE AND ORGAN CULTURE RESPONSES

Ernest Bloom, Howard I. Maibach, Raija Tammi, Donald J. Fauss, and Jon R. Polansky

TABLE OF CONTENTS

I. INTRODUCTION

Irritant effects have been examined in experimental animal systems using physiological and histological approaches (e.g., see References 1 and 2). Knowledge gained from these studies is limited by the complexity of the *in vivo* system, and it is not known to what extent different irritants mediate their effects on the skin by influencing keratinocytes, dermal fibroblasts, or other cell types. Another concern is that tissues from experimental animal systems could have different responses than human tissues.

In vitro studies of human tissues provide a possible means to overcome some of these difficulties, especially when studies comparing isolated cells propagated in culture can be compared to intact skin organ culture preparations. Propagation of keratinocytes in serial culture permits experiments to be performed on reproducible populations of cells; human skin organ culture permits examination of *in situ* responses without potential alterations due to cell proliferation and loss of the spatial arrangements of tissues *in vivo*.

Previous studies of irritant effects on keratinocytes and fibroblasts have examined membrane leakage of cytosolic enzymes and penetration of vital dyes, as well as effects on cell viability/cell division.[3-6] Irritant effects on the release of arachidonic acid and phospholipid cleavages[7-10] and changes in specific protein synthesis and mRNAs have also been examined following tetradecanoylphorbolacetate (TPA) treatment of keratinocytes.[11,12] Studies in our laboratory comparing TPA and ethylphenyl propiolate (EPP) have examined eicosanoid changes at various doses and times[13,14] and effects on specific protein synthesis.

In this report we present data from our initial evaluations of irritant effects on human keratinocytes and human skin organ culture using TPA, EPP, and hydrogen peroxide (H_2O_2). In addition, the effects of dexamethasone on keratinocyte and skin organ culture models were examined because of the known anti-inflammatory effects of glucocorticoids in reducing irritation in human skin. Observations of different irritant responses may help in the classification of irritant effects on specific cell types and to reveal underlying mechanisms of inflammatory (and anti-inflammatory) agents in the skin.

II. MATERIALS AND METHODS

A. CELL CULTURE AND ORGAN CULTURE

Human epidermal keratinocytes were grown from normal skin specimens using collagenase digestion methods.[15-17] The culture medium for the cells was keratinocyte growth media without hydrocortisone added (KGM, Clonetics Corporation, San Diego, CA). All experiments were performed on second or third passage cells. Experiments were started when the cells were at 75 to 85% of confluence.

Studies on human skin organ culture were conducted using the method of Tammi et al.[18-20] Briefly, skin specimens were obtained from patients undergoing mastectomy operations. Immediately after the operation, subcutaneous fat was dissected from the skin, and the skin was then transferred into culture medium. Much of the dermal layer of the specimens was then removed using microsurgical techniques and the skin cut into 2 × 2 mm pieces. The skin pieces were incubated, supported by wire grids, in organ culture dishes. Explants had an area of about 3 to 4 mm^2 and had a thickness of about 0.5 mm. They were cultured in Trowell-type organ culture dishes. The culture medium was Eagle's minimum essential medium (MEM) with Earle's salts, 20 m*M* HEPES, 12 m*M* bicarbonate, 50 IU/ml penicillin G, and 50 μg/ml streptomycin sulfate added. No serum was used. The culture dishes were kept in airtight chambers containing a mixture of 40% oxygen, 5% carbon dioxide, and 55% nitrogen. The chambers were regassed daily, and the medium was renewed every other day.

B. EXPOSURE TO IRRITATING CHEMICALS

Both TPA and EPP are strong cutaneous irritants. Both irritants, if given in sufficient amounts, are capable of causing similar maximal levels of inflammation.[21] Exposure of the skin *in vivo* to these irritants, however, causes different patterns of physiological[1,21] and histological[22] inflammatory changes to occur.

Cultured keratinocytes and human skin in organ culture were exposed to 10^{-6} *M* TPA or 10^{-4} *M* EPP for 16 to 24 h. After exposure to the irritants, the media was collected for prostaglandin E_2 (PGE_2) analysis, and the cells were labeled with ^{35}S-methionine as described in Section II.C. For examination of the effects of H_2O_2, cultured keratinocytes were exposed to 10^{-4} or 3×10^{-4} *M* H_2O_2 for 1 h. After 1 h the cells were changed back to growth media without H_2O_2, and ^{35}S-methionine labeling was performed the next day. PGE_2 samples were taken the day after the H_2Q_2 exposure. The concentrations of all irritants used were chosen on the basis of prior dose-response studies in which effects on cell viability/growth, PGE_2 production, and morphological criteria were examined. PGE_2 was measured by radioimmunoassay.

C. ^{35}S-METHIONINE LABELING

Following exposure to the irritants or dexamethasone, the tissue pieces or cells were changed into methionine-free medium with 100 μCi/ml translation-grade ^{35}S-methionine added. Short labeling periods (15 min to 1 h) best reflect regulation of mRNA production and specific protein synthesis, the pathway which appears to be responsible for most clinically important glucocorticoid effects as well as other receptor-mediated events.[23,24] Following 1 h of incubation with the ^{35}S-methionine, the cultures were washed three times in serum-free media. A lysis buffer (20 m*M* Tris [pH 7.6], 10 m*M* $MgSO_4$, 0.1% Triton® X-100, and 1 m*M* phenylmethane sulfonyl fluoride [PMSF]) was then added to the dishes for 14 min at 0 to 4°C. The cell lysates were collected and trichloro acetic acid (TCA)-precipitable radioactivity determined.

D. GEL ELECTROPHORESIS

For one-dimensional (1-D) sodium dodecyl sulfate polyacrylamide gel electrophoresis (SDS-PAGE), lysate samples were adjusted based on the TCA-precipitable radioactivity to give equal counts by volume. The addition of a concentrated sample buffer resulted in the following final concentrations: 10% glycerol, 5% 2-mercaptoethanol, 2.3% SDS, and 0.063 *M* Tris HCl, pH 6.8. Samples were then boiled for 2 min and spun at high speed in a microcentrifuge for 2 min. The supernatants were loaded onto the gels, and the proteins were separated using a 4% stacking gel and either a 7 or 12% SDS polyacrylamide running gel at 35 mA per gel for approximately 3 h. The proteins were stained with Coomassie blue and molecular weights estimated by standards run on the same gel. The gels were fixed in a solution containing 10% ethanol and 7% acetic acid. After drying, the gels were loaded in X-ray film cassettes and exposed at −80°C.

For two-dimensional (2-D) gel analyses, we used the method of Ivarie and O'Farrell.[24] Cell lysates were prepared in the same buffer as had been used for the 1-D analyses. Gels were run with equal counts loaded on each gel. Isoelectric focusing was done using a pH 3 to 10 range, with molecular weight separations performed on 10% slab gels.

III. RESULTS

A. IRRITANT EFFECTS ON SPECIFIC PROTEIN SYNTHESIS AND CELL MORPHOLOGY

The responses to different irritants were examined by pulse labeling with ^{35}S-methionine in the cultured keratinocytes as presented in Figure 1. TPA (10^{-6} *M*) treatment, a dose

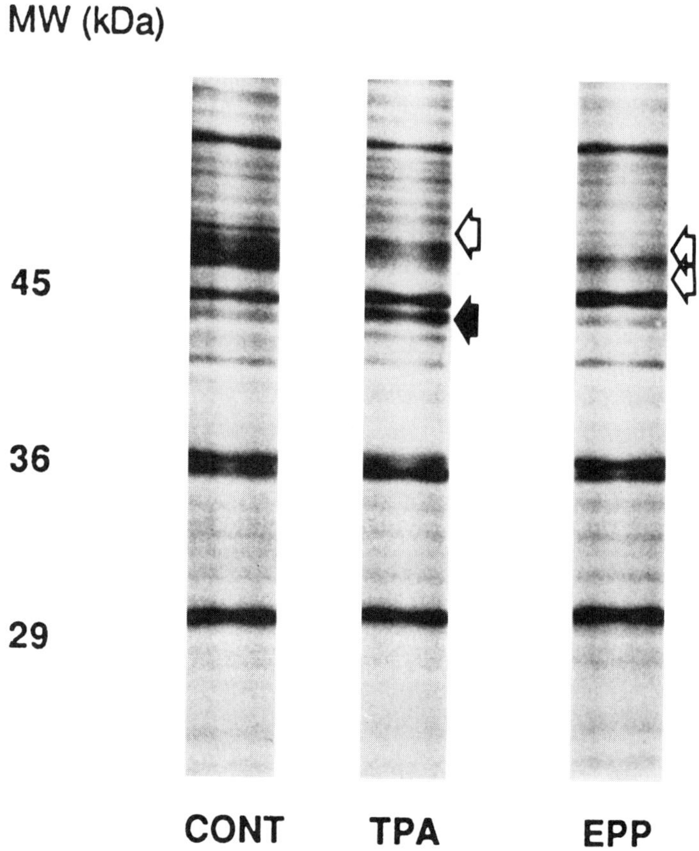

FIGURE 1. Effects of irritants on cultured human keratinocytes. The cells were exposed to 10^{-6} *M* TPA or 10^{-4} *M* EPP for 24 h followed by ^{35}S-methionine labeling, electrophoresis, and autoradiography. Concentrations of the irritants were chosen based on relative potencies of inflammation in animal studies (see text). TPA induced a protein band at approximately 44 kDa and deinduced a protein band at approximately 48 kDa. EPP showed no major inductions, but deinduced protein bands at approximately 48 kDa and 46 kDa.

which maximally stimulates PGE_2 in the cultured human cells,[25] produced a major induction of a protein band at approximately 44 kDa and a clear deinduction at approximately 48 kDa. As described below, these changes in specific protein synthesis occur with relatively little alteration of cell morphology. In contrast to the results obtained with TPA, 10^{-4} *M* EPP (which is also a maximal PGE_2-stimulating dose)[25] showed no major inductions under the conditions tested and produced deinductions of protein bands at approximately 48 and 45 kDa. One deinduction (48 kDa) appears at a molecular weight similar to a deinduction observed with TPA treatment. Although only minor changes are observed with EPP on the pattern of specific protein synthesis, there are marked morphological changes with this agent (as described below).

H_2O_2 effects on cultured human keratinocytes were also investigated for possible similarities to and/or differences from TPA with EPP effects. As shown in Figure 2, exposure of keratinocytes to 10^{-4} and 3×10^{-4} *M* H_2O_2 also induced specific proteins. A major induction was observed at approximately 44 kDa, similar to that seen with TPA. In addition, a partially induced protein band was also observed at approximately 49 kDa, and an induction also occurred at 32 kDa (not shown in photograph). In full dose-response studies of H_2O_2 damage, the gels show changes in protein synthesis with H_2O_2 treatment at lower concen-

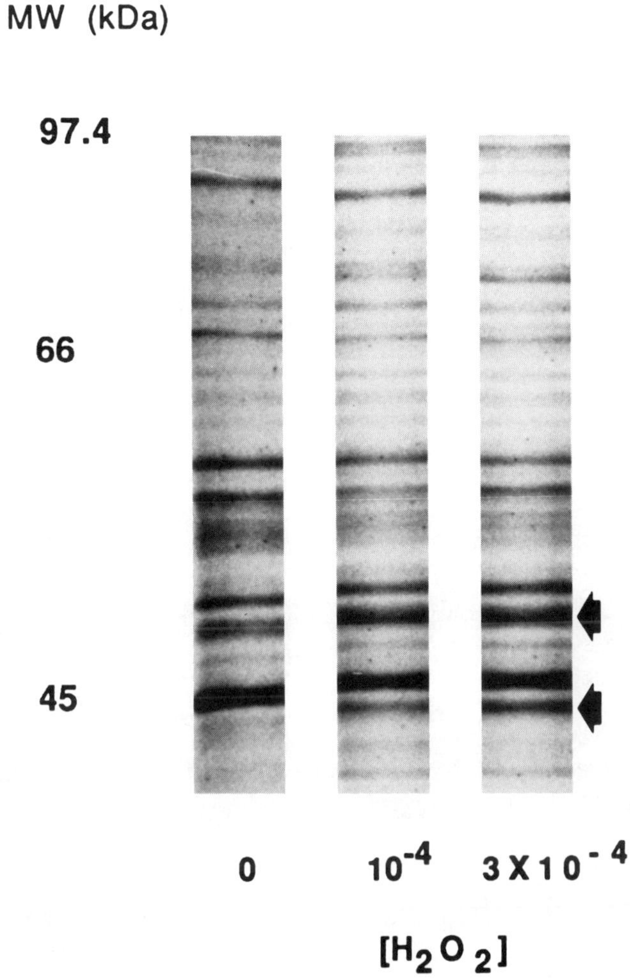

FIGURE 2. Effect of 10^{-4} and 3×10^{-4} *M* H_2O_2 on cultured normal human keratinocytes. Cells were exposed to H_2O_2 for 1 h and ^{35}S-methionine labeled 24 h later. H_2O_2 caused induction in protein bands at approximately 49 and 44 kDa.

trations than those required to produce morphological changes. For example, 10^{-4} *M* H_2O_2 treatment, which results in only subtle changes in cellular morphology, shows clear changes in specific protein synthesis. It is also possible that other changes are taking place which are obscured by overlying heavy bands of similar molecular weight but different isoelectric point. Identification of the proteins will require 2-D gel electrophoresis studies, which separate proteins on the basis of both size and isoelectric point.

The effects observed on cultured keratinocyte morphology after exposure to the irritants are shown in Figure 3. TPA at 10^{-6} *M* produced only subtle alterations in morphology, with slightly increased definition of the cell borders and the possibility of slightly decreased cell packing. EPP at 10^{-4} *M* produced major changes in the morphological appearance of the cells, causing an increase in vacuolization, membrane blebbing, granular-appearing cytoplasm, and cell loss. At 10^{-5} *M* EPP, only subtle morphological changes could be observed (not shown). H_2O_2 (3×10^{-4} *M*) treatment of the cells also caused morphological changes which included vacuolization, membrane blebbing, "rounding up" of some cells (probably due to partial detachment of the cells), and cell loss. Treatment of the cells with

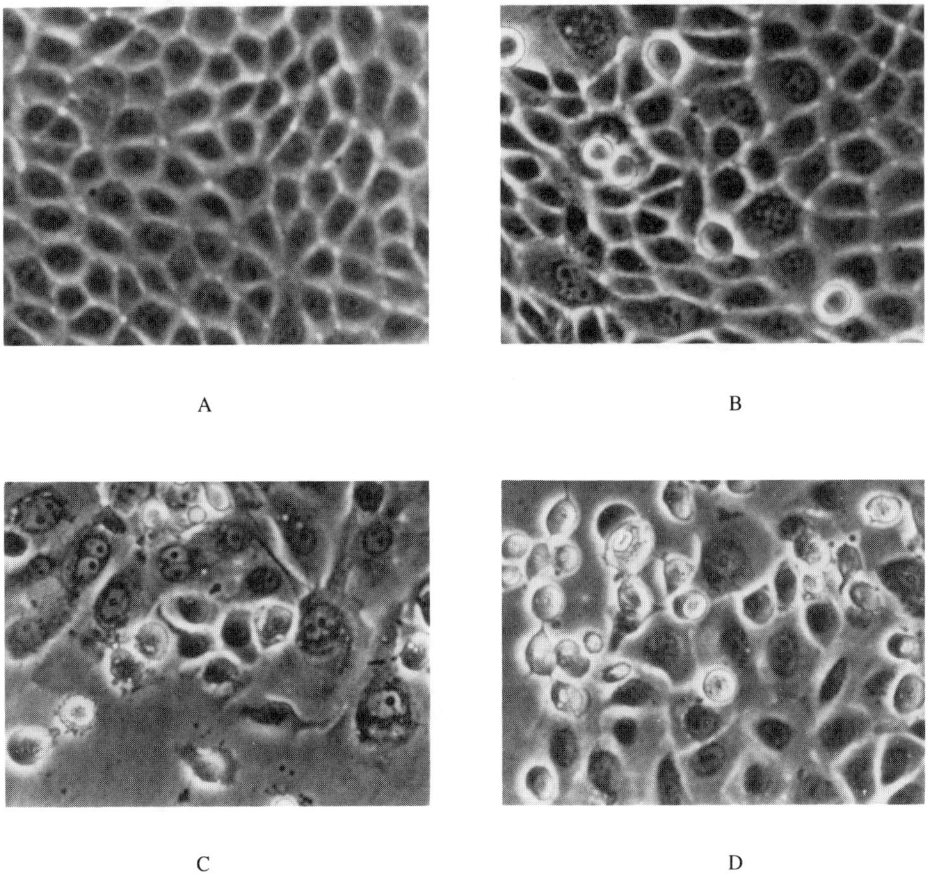

FIGURE 3. Morphological effects of irritants on cultured keratinocytes. (A) Control; (B) 10^{-6} *M* TPA produced a small increase in cell-cell border definition; (C) 10^{-4} *M* EPP produced major morphological changes including vacuolization, membrane blebbing, granular-appearing cytoplasm, and cell loss; (D) 3×10^{-4} *M* H_2O_2 also produced overt morphological changes, including increased vacuolization, membrane blebbing, and cell rounding. Changes due to dexamethasone and 10^{-5} *M* EPP treatment are not shown because the morphological alterations are difficult to appreciate in photographs.

10^{-6} *M* dexamethasone did not produce changes easily visualized in a photograph, but did result in a major change in protein synthesis pattern, as described below in Section III.B.

B. GLUCOCORTICOID EFFECTS ON SPECIFIC PROTEIN SYNTHESIS

Glucocorticoid effects on ^{35}S-methionine-labeled protein profiles were measured by both 1-D (SDS-PAGE) and 2-D (isoelectric focusing/SDS-PAGE) methods on human keratinocytes. Treatment of the cultured keratinocytes with 10^{-6} *M* dexamethasone for 5 d produced a major induction at approximately 57 kDa, as shown on the 1-D SDS-PAGE in Figure 4. Although the actual degree of induction of this band varied slightly from experiment to experiment, the induction itself has been a consistent finding in a large series of experiments. In initial experiments in the skin organ system, a similar induction has also been noted. In contrast, in preliminary experiments, treatment of cultured fibroblasts with dexamethasone resulted in a different set of changes in protein synthesis which did not show the induction at 57 kDa. Figure 5 shows a 2-D gel electrophoresis study of the proteins induced by 10^{-6} *M* dexamethasone treatment of cultured keratinocytes. The 2-D system improves resolution to approximately 1000 spots, allowing additional inductions to be seen in areas that normally would not be visualized in a 1-D gel system due to the presence of overlying proteins of

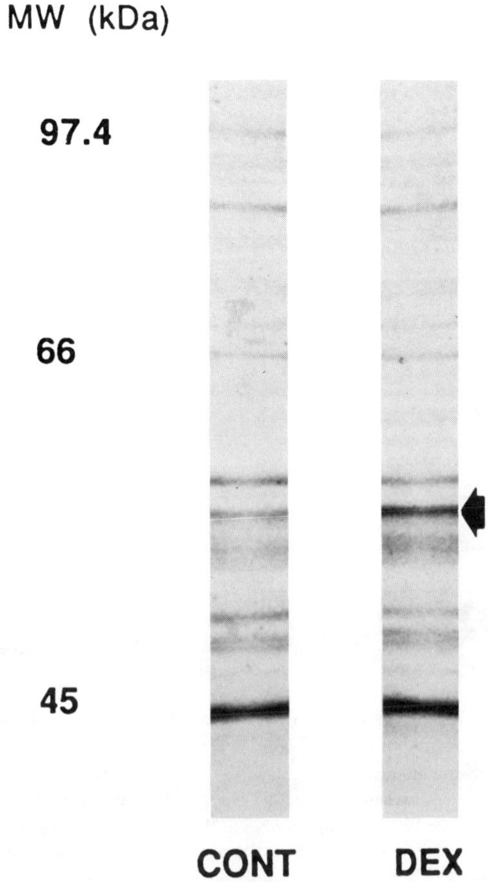

FIGURE 4. Autoradiograph of 1-D gel of cultured keratinocytes treated with 10^{-6} *M* dexamethasone followed by ^{35}S-methionine. Glucocorticoid treatment consistently induced a protein band at approximately 57 kDa.

similar molecular weights (but different isoelectric points). The 2-D gel autoradiograph in Figure 5 shows some protein inductions of approximate molecular weight 50 to 52 kDa and isolectric point ($_p$I) of approximately 5.0 to 5.2 which have been identified as consistent in three different experiments on two different cell lines. Two deinductions were also observed — at 56 to 58 kDa (pI region 8.3 to 8.5) — but these deinductions were not as consistent, as the inductions were not seen in one of the experiments.

C. ORGAN CULTURE STUDIES

The examination of irritant effects on specific protein synthesis in human skin organ culture was performed as a comparison for the cell culture system. Figure 6 shows the effect of these irritants on cultured keratinocytes. 10^{-6} *M* TPA caused induction of protein bands at molecular weights of approximately 64, 61, and 54 kDa and caused no major deinductions. There also appears to be a small relative induction of a band at 44 kDa which was observed in two different samples. For EPP, major inductions were noted at approximately 64, 61, 54, and 40 kDa. In addition, there was deinduction of a band at 44 kDa, a molecular weight similar to the small relative induction seen with TPA treatment.

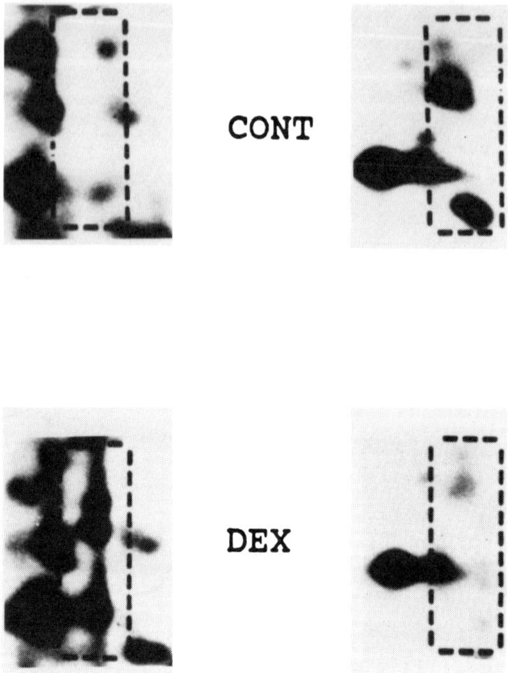

FIGURE 5. Autoradiograph of a 2-D gel of cultured keratinocytes treated with 10^{-6} M dexamethasone followed by ^{35}S-methionine labeling. The horizontal axis represents the isoelectric focusing axis (left side, acidic; right side, basic), and the vertical axis represents the axis in which proteins are separated by size in a 10% SDS-PAGE gel. The paired rectangles at the upper and lower left of the figure show a region of the gel of approximate molecular weight 50 to 52 kDa with a pI of approximately 5.0 to 5.2. The upper left rectangle is from a control dish and the lower right from a dexamethasone treated dish. The two spots shown induced by dexamethasone are not normally visualized in the 1-D gel electrophoresis due to overlying proteins of the same molecular weight, but of differing pI. Two deinductions of approximately 56 to 58 kDa, pI 8.3 to 8.5, are also shown on the right. These deinductions were not seen in every experiment.

D. EFFECTS OF IRRITANTS ON PGE_2

Radioimmunoassay for PGE_2 provided a useful means of quantifying the effects of these irritants on the cell and organ culture systems. The media were collected 16 to 24 h after exposure to the irritation/injury or after 5 d of exposure to 10^{-6} M dexamethasone. In cultured keratinocytes, 10^{-6} M TPA stimulated PGE_2 4.7 $\pm$ 0.7 fold, and 10^{-4} M EPP increased it 2.3 $\pm$ 0.7 fold. Maximal stimulation of PGE_2 occurred at 10^{-4} M EPP in this study and in other studies performed using this system. In some studies, however, maximal stimulation of PGE_2 by EPP occurred at 10^{-5} M; at 10^{-4} M there was either no stimulation or a slight depression in PGE_2 production, possibly due to a variable effect of toxicity to the cultured cells. PGE_2 was decreased by 10^{-4} and 3×10^{-4} H_2O_2 to 0.39 $\pm$ 0.09 and 0.32 $\pm$ 0.10, respectively, of the control. Treatment with 10^{-6} M dexamethasone also markedly decreased PGE_2 production to 0.07 $\pm$ 0.02 of control. It is interesting to note that in these experiments on TPA and H_2O_2, changes in PGE_2 always occurred at doses lower than those required to produce alterations in cellular morphology. For EPP, however,

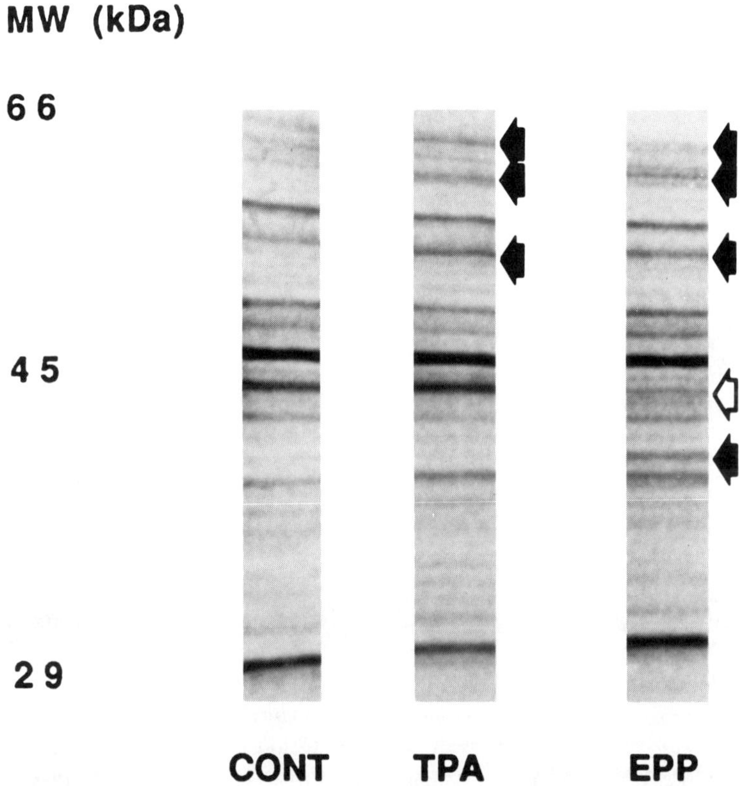

FIGURE 6. Effect of irritants on human skin in organ culture. The skin pieces were exposed to irritants followed by ^{35}S-methionine labeling; 10^{-6} *M* TPA caused inductions of protein bands at approximately 64, 61, and 54 kDa. In addition, a small relative induction was seen in a band at 44 kDa. EPP (10^{-4} *M*) caused inductions at approximately 64, 61, 54, and 40 kDa and deinduced a band at 44 kDa.

the changes in PGE_2 appeared at approximately the same doses which caused changes in morphology and protein synthesis.

The effects of irritants and dexamethasone on PGE_2 production in organ culture were similar to the effects observed in cultured keratinocytes; 10^{-6} *M* TPA stimulated PGE_2 5.3 ± 0.2 fold, and 10^{-4} *M* EPP stimulated PGE_2 2.5 ± 0.3 fold. Dexamethasone at 10^{-6} *M* decreased PGE_2 to 0.20 ± 0.10 of control. H_2O_2 has not yet been examined in the organ culture system.

IV. DISCUSSION

This work supports the view that isolated cell and organ culture models of human skin may be used to explore direct drug and hormone effects. The finding that the different irritants produce distinct effects in the protein gels is consistent with prior histological and physiological studies which have shown TPA and EPP to have different response patterns.[1,21,22]

The specific proteins altered by irritant exposure could play a role in (1) the injury response, (2) protection of the cells from initial or ongoing injury, or (3) repair of damage. Identification of the polypeptide inductions and deinductions through protein isolation and molecular biology approaches (e.g., subtraction-cloning cDNA library analysis) may help to elucidate the relevance of the individual keratinocyte responses and may provide new approaches to therapy.

It is of considerable interest that the molecular weight of the most prominent protein induced by TPA and H_2O_2 treatment of keratinocytes is approximately 44 kDa, the same as a possible induction seen with TPA in skin organ culture. In addition, this band is deinduced in the organ culture system by EPP treatment. The inductions and deinductions at 44 kDa must be analyzed using 2-D gels to fully appreciate the degree of change, since there is a potentially interfering band already present at approximately the same molecular weight in the controls in organ culture, but not keratinocyte cell culture.

Protein alterations produced by the irritant EPP appeared to be more dramatic in the tissue pieces than the keratinocyte cell culture system. Since the cultured keratinocytes do not respond with the protein inductions seen in the organ culture system, either another cell type is involved or something may be lost in the cell culture system which is preserved in organ culture.

The responses to irritants in the skin may involve necessary cell-cell interactions, appropriate spatial arrangements of the tissues, or a preservation of the differentiated characteristics of the cells in the organ *in situ* which is lost with serial propagation of the cells. The organ culture system contains several different kinds of cells in addition to keratinocytes, including fibroblasts, Langerhans' cells, and melanocytes. Based on histological studies of cross-sections of the dissected pieces of skin, the main cell type (i.e., over 90%) is clearly the keratinocyte, with fibroblasts being the second most common cell type in the dissected tissue pieces.[20] In our laboratory we are currently performing growth of keratinocytes on top of three-dimensional collagen networks in which fibroblasts are embedded; this approach may help to delineate the effect of environment on induction of proteins in the cultured cell system, which in some ways will reproduce the human skin environment.

Comparisons between the two irritants and H_2O_2 on PGE_2 levels also show interesting differences. While TPA and EPP treatment resulted in an increase in PGE_2 production, H_2O_2 produced a marked decrease in production of PGE_2. The decrease of PGE_2 production caused by H_2O_2 treatment is probably not a simple result of cell death, as H_2O_2-treated keratinocytes are clearly able to induce proteins in a dose-responsive manner after exposure to even higher doses of H_2O_2 than those necessary to decrease PGE_2 production. H_2O_2 may result in destruction of arachidonic acid and/or arachidonic acid processing enzymes in the outer cell membrane in a separate or related phenomenon to that of changes in protein synthesis. In our laboratory we have found marked differences in the PGE_2 responses of different cell types to H_2O_2 injury (e.g., H_2O_2 increased PGE_2 production in some cell types such as aortic endothelium and corneal endothelium, but decreased PGE_2 production in endothelium from the trabecular meshwork).

Thus, our ongoing studies of different irritants and anti-inflammatory agents in the skin are directed toward understanding individual differentiated cell responses in both cell and organ culture models. The information developed may help to classify the biochemical responses of the individual cells and to improve our understanding of different mechanisms of irritation. Evaluations of specific protein synthesis, changes in cellular second messengers, (e.g., see Chapter 20), different autacoids, and changes in the activities of specific inflammatory mediators may prove useful in this regard.

V. SUMMARY AND CONCLUSION

We have been defining the effects of different irritants and anti-inflammatory agents on the synthesis of specific proteins and production of PGE_2 in cultured human keratinocytes as a means of exploring the mechanism(s) of toxicity and therapeutic effects. TPA and EPP were compared because they are potent irritants which seem to have different mechanisms of action on the basis of animal studies. H_2O_2 also was investigated because oxidizing radicals have been implicated as one of the mechanisms involved in irritant and toxic effects.

The three agents produced different effects on protein synthesis and PGE_2 production. TPA treatment resulted in a major induction at 44 kDa, with a deinduction at 48 kDa. H_2O_2 also produced a major induction at 44 kDa and a smaller relative induction at approximately 49 kDa. EPP showed no major inductions, but deinduced proteins at 48 and 46 kDa. TPA and EPP stimulated PGE_2 production, whereas H_2O_2 treatment decreased it. In human skin organ culture, TPA produced the same major induction seen in the cell culture system, but EPP treatment resulted in different alterations in this system. Glucocorticoid treatment decreases PGE_2 production and shows protein inductions in the range of 50 to 58 kDa. Continued investigation of these and other biochemical markers of injury and glucocorticoid effects may help to better define cell-specific responses of human skin and also may help in the discovery of underlying subcellular mechanisms.

As presented in this brief review, we believe that *both* cell culture and organ culture models of the skin offer important new opportunities to examine inflammatory mediators and define drug effects in the skin. The preliminary findings we have obtained in these model systems suggest that different agents produce specific effects which appear relevant to *in vivo* toxicology. By examining individual eicosanoid responses in the different *in vitro* models, as well as different irritant effects on specific protein synthesis and inositol phosphate metabolism, we hope to contribute to the development of useful and relevant *in vitro* models of cutaneous toxicity. The initial results of our studies evaluating changes in specific protein synthesis, alterations in cellular morphology, and changes in PGE_2 secretion in keratinocytes appear encouraging. Continued development of the organ and cell culture systems could yield useful information concerning the biochemical responses of skin in normal and pathological states (especially with the use of more specific molecular probes) and may aid in the design of drugs with improved safety and reduced toxicity.

ACKNOWLEDGMENTS

We would like to thank Dr. Esther Patrick for expert advice concerning *in vivo* models of inflammation and application of the systems developed by her work in this area. We also appreciate the technical help of Holly Jones and Tracy Hydorn, who aided in the conduct of these studies.

Ernest Bloom is supported in part by a Colgate-Palmolive *In Vitro* Toxicology Fellowship.

REFERENCES

1. **Patrick, E., Maibach, H., and Burkhalter, A.,** Mechanisms of chemically induced skin irritation, *Toxicol. Appl. Pharmacol.*, 81, 476, 1985.
2. **Patrick, E., Burkhalter, A., and Maibach, H. I.,** Recent investigations of chemically induced skin irritation in laboratory mice, *J. Invest. Dermatol.*, 88 (Suppl. 3), 24s, 1987.
3. **Borenfreund, E. and Borrero, O.,** *In vitro* cytotoxicity assays. Potential alternatives to the Draize ocular allergy test, *Cell Biol. Toxicol.*, 1, 55, 1984.
4. **Stark, D. M., Shopsis, C., Borenfruend, E., and Bebich, H.,** Progress and problems in evaluating and validating alternative assays in toxicology, *Food Chem. Toxicol.*, 24, 449, 1986.
5. **Shopsis, C.,** *In vitro* ocular irritancy prediction: comparison of cell systems and assay conditions, *In Vitro Cell Dev. Biol.*, 23(3), 44A, 1987.
6. **Shopsis, C. and Eng, B.,** Rapid cytotoxicity testing using a semi automated protein determination on cultured cells, *Toxicol. Lett.*, 26, 1, 1985.
7. **DeLeo, V., Hanson, D., Scheide, S., Kong, B., and DeSalva, S.,** Primary cutaneous irritancy — an *in vitro* model, *Clin. Res.*, 36, 639a, 1988.

8. **DeLeo, V. A., Scheide, S., Horlich, H., Hanson, D., et al.,** The effect of 12-*O*-tetradecanoylphorbol-13-acetate (TPA) on phospholipid metabolism of human epidermal keratinocytes in culture, *Carcinogenesis,* 7, 937, 1986.
9. **Kvedar, J. and Levine, L.,** Modulation of arachidonic acid metabolism in a cultured newborn rat keratinocyte cell line, *Invest. Dermatol.,* 88, 124, 1987.
10. **Furstenberger, G., Richter, H., Fujenis, N. E., and Marks, F.,** Arachidonic acid and prostaglandin E_2 release and enhanced cell proliferation induced by the phorbol esther TPA in a murine epidermal cell line, *Cancer Lett.,* 11, 191, 1981.
11. **Cabral, F., Gottesman, M. M., and Yuspa, S. H.,** Induction of specific protein synthesis phorbol esters in mouse epidermal cell culture, *Cancer Res.,* 41, 2025, 1981.
12. **Kartasova, T., Cornelissen, B. J. C., Belt, P., and van de Putte, P.,** Effects of UV, 4-NQO and TPA on gene expression in cultured human epidermal keratinocyte, *Nucleic Acids Res.,* 15, 5945, 1987.
13. **Bloom, E., Maibach, H. I., Jones, H., Goldyne, M., and Polansky, J.,** *In vitro* studies of human keratinocytes and skin fibroblasts: TPA and dexamethasone responses, *Clin. Res.,* 36, 634A, 1988.
14. **Bloom, E., Maibach, H. I., Tammi, R., and Polansky, J.,** *In vitro* models of irritant effects on human skin, *Toxicologist,* 9, 6, 1988.
15. **Boyce, S. T. and Ham, R. G.,** Cultivation, frozen storage, and clonal growth of normal human epidermal keratinocytes in serum-free media, *J. Tissue Cult. Methods,* 9(2), 83, 1985.
16. **Tsao, M. C., Walthall, B. J., and Ham, R. G.,** Clonal growth of normal human epidermal keratinocytes in a defined medium, *J. Cell. Physiol.,* 110, 219, 1982.
17. **Boyce, S. T. and Ham, R. G.,** Calcium regulated differentiation of normal human epidermal keratinocytes in chemically defined clonal culture and serum-free serial culture, *J. Invest. Dermatol.,* 81(1), 33S, 1983.
18. **Tammi, R., Jansen, T., and Santti, R.,** Histiometric analysis of human skin in organ culture, *J. Invest. Dermatol.,* 73, 138, 1979.
19. **Tammi, R. and Jansen, T.,** Serum epidermal migration — skin factor — membrane ultrafiltration and physical stability characteristics, *Arch. Dermatol. Res.,* 267, 319, 1980.
20. **Tammi, R. and Santti, R.,** Ultrastructural study of hydrocortisone action in cultured human epidermis, *Br. J. Dermatol.,* 106, 65, 1982.
21. **Bloom, E., Maibach, H. I., Tammi, R., and Polansky, J. R.,** *In vitro* models for cutaneous effects of glucocorticoids using human skin organ and cell culture, *Models Dermatol.,* 4, 12, 1989.
22. **Raick, A. N. and Burdzy, K.,** Ultrastructural and biochemical changes induced in mouse epidermis by a hyperplastic agent, ethylphenylpropiolate, *Cancer Res.,* 33, 2221, 1973.
23. **Polansky, J. R. and Weinreb, R. N.,** Anti-inflammatory agents: steroids as anti-inflammatory agents, in *Handbook of Experimental Pharmacology,* Vol. 69, Sears, M. L., Ed., Springer-Verlag, Berlin, 1984, 459.
24. **Ivarie, R. D. and O'Farrell, R. H.,** The glucocorticoid domain: steroid-mediated changes in the rate of synthesis of rat hepatoma proteins, *Cell,* 13, 41, 1978.
25. **Bloom, E. and Polansky, J. R.,** unpublished data.

Section III. Experimental Investigations

Chapter 13

IRRITATION INSIGHTS: EPIDEMIOLOGY AND EXPERIMENTAL STATUS

Kaija Lammintausta and Howard I. Maibach

TABLE OF CONTENTS

I. EPIDEMIOLOGY

Data on the prevalence of irritant contact dermatitis are inadequate. The prevalence of hand dermatitis has been studied in European countries.[1-4] Hand dermatitis accounts for more than one half of all contact dermatitis in the general population, being highest in industrialized countries, where service occupations are important. On the basis of the available data, the prevalence of contact dermatitis it is impossible to estimate in the general population.

Skin diseases, more generally, have been studied systematically in certain occupations. Dermatitis occurs in 3 to 45% of workers, depending on the quality and extent of external exposure.[5-12] In a majority of occupational dermatitis an external etiology is important. Irritant dermatitis is a major problem in the occupations with the highest frequency of contact dermatitis.[9-11,13] Irritant contact dermatitis is most common when exposure to chemical and physical irritation and to wet work is greatest. Thus, epidemiological studies have often been directed to certain occupational risk groups. The relationship between irritant and allergic contact dermatitis varies between different countries even among same occupational groups due to differences in working practices.[11,14] On the other hand, the distinction between these diagnoses is often ill-defined. Irritation precedes and increases the risk for sensitization, or irritant dermatitis may complicate an allergic dermatitis.

In dermatology clinics of different countries, dermatitis accounts for 15 to 30% of the outpatient cases,[15-19] and the cause is exogenous in some 15 to 50%. Those patient populations are quite selected. Selection is even higher in specialized occupational skin disease clinics, where contact dermatitis accounts for 60 to 85% of the patients.[13,20] The proportion of the cases diagnosed as externally induced contact dermatitis shows wide variation between different clinics, reflecting the difficulty in deciding the relative importance between external and internal factors. The personal experience and opinions of dermatologists contribute in defining the final diagnosis, too. Multiple external factors have great influence on endogenous dermatitis and contact dermatitis. In the development and deterioration of an evident endogenous dermatitis, irritation is often important. Of contact dermatitis cases, slightly more than one half can be considered to be caused mainly by irritation factors. Even in those cases, the knowledge about constitutional and variable endogenous factors may be limited.

The occurrence of contact dermatitis in certain constitutional risk groups has been studied. In atopic subjects and nickel allergics, for example, irritant contact dermatitis appears to be increased.[9,10,21-24] Corresponding studies have not been done with subjects who have seborrheic skin or psoriasis. The associations between increased susceptibility and constitutional factors are inadequately known. Nickel sensitivity, for example, may be primary or secondary.

Studies on clinical patient materials in hospitals give epidemiologic information about long-lasting or chronic type irritant contact dermatitis because minimal cases often are treated by private practitioners. Simultaneously, the proportion of chronic endogenous dermatitis cases is overpresented. Field studies, in combination with these clinical studies, would increase information about risks.

II. RECENT EPIDEMIOLOGICAL FINDINGS AND ENDOGENOUS FACTORS

Comparisons between different age groups show that irritant contact dermatitis is common in young children in diapers[25-27] and decreases later in childhood.[28] The skin of a newborn is thin; thus, the capacity to resist external exposure may not be fully developed. There is also evidence that constitutional factors are important in diaper dermatitis. Atopic dermatitis, seborrheic dermatitis, and psoriasis are suggested diagnostic backgrounds.[26,27] An extended discussion on the biology of neonatal skin may be most important.

Irritant contact dermatitis is most common in hands; it frequently occurs at the age of 20 to 50.[10,21,22] Nickel-allergic subjects often develop hand dermatitis at that age, too,[21,22] whereas primary sensitization generally is seen at a younger age.[21,22,24] In adulthood atopic dermatitis the face and hands are commonly affected.[29] Those sites are also exposed to irritation the most, and environmental exposure is probably greatest at that age — somewhere between 15 and 50 years. Age-associated decreases occur in skin test reactivity to histamine; this is greatest in adulthood, decreasing after age 50.[30] Simultaneously, elderly subjects react more slowly and with a lower intensity to cutaneous irritants,[31] when compared with the reactivity of younger adults.

Elderly contact eczema patients are not common in the clinical patient populations. Field studies show that skin irritation problems are not uncommon among the elderly, although the diagnosis of irritant contact dermatitis is not often made by dermatologists for elderly subjects, who do not seek medical advice as frequently as the younger subjects.[32-34] While individual skin characteristics and the extent of exposure are principal factors in the development of irritant contact dermatitis, some age-related differences also exist. Adequate guidance would be important for all age groups. To be able to advise optimal skin care, we need data and results from experimental and epidemiologic studies. The reactivity of the skin at different ages and the extent and qualities of irritant contact dermatitis among the different age groups require further investigation.

Epidemiologic studies among different races reveal that different dermatologic problems are most important when the environmental exposure is different.[35-37] Thus, valid information about the associations between cutaneous irritability and race requires large-scale experimental and epidemiologic studies. The same is true for sex-related differences since the exposure is often different for both sexes. Sex- or race-related constitutional differences are still questionable.

III. PROSPECTIVE ASPECTS OF CLINICAL AND EPIDEMIOLOGIC STUDIES

In the future, epidemiologic studies should be done among the general population in different areas. Clinical and occupational risk groups need particular attention. Those studies should increase knowledge about the importance of the external factors in relation to structural and functional skin characteristics. It is known that the individual capacity to resist skin irritation is varied, the quality of reactions is multiple, and the etiology and pathomechanisms are multifactorial. In future studies, the subjects could be qualified according to constitutional characteristics with reference to the total quantity of reactivity in different pathogenetic aspects. The development of tolerance and "hardening" in the exposure to repeated irritation attacks will be one of the most interesting topics.

Subtyping of irritant dermatitis on the basis of the clinical appearance increases the information given to clinical dermatologists. Delayed skin irritation (for example, see References 38, 39, and 40) is infrequently reported, and dermatitis from repeated exposure to airborne irritants[41-43] may be inadequately diagnosed. Airborne irritant contact dermatitis is probably regarded as a constitutional dermatitis in a certain number of facial dermatitis cases, for example. Since the frequency of facial and hand dermatitis is high in all endogenous dermatitis patients,[42,43] associations to both constitutional and exogenous factors are suspected. The climate (temperature, wind, humidity, etc.) as a source of irritation needs particular attention.

IV. EXPERIMENTAL STUDIES ON IRRITANT CONTACT DERMATITIS

A. STRUCTURAL COMPONENTS

Stratum corneum lipids function as an important component of the epidermal water loss barrier.[44-49] The active and dynamic state of stratum corneum lipids[45] probably contributes to the stratum corneum water gradient.[47-50] Surfactants or solvents induce skin dryness. Later on, the increasing dryness becomes visible as chapping.[44,49] Of the stratum corneum lipids, membrane structures account for approximately 80%; 60 to 80% of them consist of neutral lipids and 15 to 35% of sphingolipids.[46] Of the sphingolipids, 80% are ceramides. Ceramides have polar and nonpolar ends, hydroxyl and amide groups. Sphingolipids or ceramides, functioning as amphipathic material to form lipid bilayers, may play an important role in the water-holding characteristics of skin.[48] In the proportions of individual skin lipids the interindividual variation is wide.[51-53]

Corneocytes are of great importance in protection against external irritation. Thus, the structural characteristics of these cells must, at least in part, determine the individual susceptibility for cutaneous irritation. Qualitative differences in corneocytes between dark-skinned subjects and Caucasoids were reported with increased corneocyte density in dark-skinned subjects.[54] Simultaneously, dark-skinned subjects seemed to react less to external irritation.[55] The importance of corneocyte density was not subsequently verified.[56] Instead, mean corneocyte area and the number of epidermal cells was found to correlate with individual irritability. Mean corneocyte area may be reduced[57] or normal[58] in atopic dermatitis patients regarded as susceptible to skin irritation due to constitutional factors.[59]

Stratum corneum thickness has been related to susceptibility to irritation.[60] Black skin needs fewer tape strippings to remove stratum corneum than does Caucasoid skin.[54] Stratum corneum was thicker in atopic dermatitis patients when compared with normal control subjects.[58] The number of cell layers may be a determining factor in cutaneous irritability. This hypothesis is supported by the observation that the number of cell layers decreased most in atopic dermatitis patients after soap and water washings.[61]

Dry skin is an ill-defined condition in relation to structural and functional aspects of skin.[62] Low hydration in the stratum corneum seems to be associated with high transepidermal water evaporation and reduced lipid content.[63] Atopic skin often appears dry. However, the documentation of the dryness and irritability of atopic skin is inadequate, although they are important criteria in the diagnosis of atopic dermatitis.[59] Decreased and normal water contents have been reported in atopic skin.[57,64-67] In "dryness" and/or irritability the quality of lipids may also be important. After soap and water washings the amount of sudanophilic material and surface lipids decreased most in atopic dermatitis patients.[61] Thus, the quality of skin lipids may be related to skin irritability. However, the clinical influence of soap and water washes may be beneficial even for atopic dermatitis patients.[68]

The histology of irritant contact dermatitis is not characteristic, but nonspecific alterations are seen in all contact dermatitis. Keratinocytes expressing major histocompatibility complex (MHC) II-related antigens are common in both irritant and allergic contact dermatitis.[69] Perivascular infiltrates of T-cells occur in allergic and irritant contact dermatitis, and the subtypes of T-cells are not characteristic.[70-73] Langerhans' cells (LC) are important in all contact dermatitis. Corresponding LC expression variation has been demonstrated in allergic and irritant contact dermatitis, but slight difference in the timing of the LC response may exist.[70]

B. SKIN FUNCTION AND PHYSIOLOGY

Modern bioengineering equipment has made it possible to evaluate the alterations in skin function and physiology related to irritant contact dermatitis.[74] Stratum corneum damage, vasodilatation, erythema, and edema are different aspects in the development of irritant dermatitis. Thus, multiple experimental equipment is employed to characterize irritant contact dermatitis. However, the clinical correlates of observations in individual experiments require elucidation.

Inflammation mediator interactions, association with individual functional cell types and their products, are multiple in skin inflammation.[75-78] In experimental animals, sodium lauryl sulfate (SLS) and alkyl dimethyl benzammonium chloride (ADBC) induced secretion of preformed mediators — histamine and lysozymal enzymes — from the cells.[79] The importance of several mediators is suggested: when the inhibitory response to cutaneous irritation showed wide variation or when the inhibition was induced with histamine antagonists, prostaglandin and kinin synthesis inhibitors and irritation had been induced by different irritants.[80] A peptidoleucotriene antagonist and an antagonist of platelet activating factor (PAF) were documented to be less effective in irritant contact dermatitis than in the allergic type; thus, cytotoxic effects may predominate in irritant contact dermatitis.[81,82] The lipoxygenase pathway was found to be enhanced in irritation, whereas it appeared inhibited in allergic contact dermatitis, thus giving the suggestion of cytotoxic mechanisms in irritant dermatitis.[81]

Studies on inflammation mediators also may explain the multiple individual differences in susceptibility to irritation. The variation in the releasibility of eicosanoids from leucocytes was shown in comparisons between healthy controls and atopic dermatitis patients. Spontaneous and stimulated release at prostaglandin E2, leukotriene B4, and leukotriene C4 were increased after immunological challenge.[83] The production of interleukin-1 from the monocytes of atopic dermatitis patients was decreased.[84] Associations with irritability can be discussed. The cell infiltrate in dermatitis does not show remarkable variation with relationship to the etiology and diagnosis. It is expected that the quality of the stimulation toward skin inflammation and dermatitis does not make a difference.

C. STUDIES RELATED TO ALTERATION IN IRRITATION REACTIVITY

An increase in the threshold for skin irritation after repeated irritant attacks is sometimes seen.[85-87] This desensitization or "hardening" is often related to an increase in skin thickness.[88] A corresponding decrease in reactivity is seen in allergic contact dermatitis after ultraviolet (UV) exposures.[89,90] Multiple alterations develop in mediator release when exposed to light.[91-95] These transitory phenomena, however, just may be a result of UV exposure. The releasability of mediators appears to have some importance, since the alteration in reactivity is specific for allergic or irritant reactions.

In clinical practice we observe that a decrease in the irritation threshold may be specifically related to the induction method, or it may be more nonspecific, without association with allergic or irritation mechanism. Simultaneously, the development of irritant dermatitis, the interindividual variation between subjects, and multiple individual alterations need much more investigation.

REFERENCES

1. **Agrup, G.,** Hand eczema and other dermatoses in South Sweden. Academic dissertation, *Acta Derm. Venereol.,* 49 (Suppl. 161), 1969.
2. **Lantinga, H., Nater, J. P., and Coenraads, P. J.,** Prevalence, incidence and course of eczema on the hand and forearm in a sample of the general population, *Contact Dermatitis,* 10, 135, 1984.
3. **Menné, T., Borgan, O., and Green, A.,** Nickel allergy and hand dermatitis in a stratified sample of Danish female population, *Acta Derm. Venereol.,* 62, 35, 1982.

4. **Kavli, G. and Forde, O. H.,** Hand dermatoses in Tromso, *Contact Dermatitis,* 10, 174, 1984.
5. **Högberg, M. and Wahlberg, J. E.,** Health screening for occupational dermatoses in house painters, *Contact Dermatitis,* 6, 100, 1980.
6. **Rystedt, I. and Fischer, T.,** Occupational skin diseases in shipyard workers, *Dermatosen,* 31, 87, 1983.
7. **Peltonen, L., Wickström, G., and Vaahtoranta, M.,** Occupational dermatoses in the food industry, *Dermatosen,* 33, 166, 1985.
8. **Mathias, C. G. T. and Morrison, J. H.,** Occupational skin diseases, United States: results from the bureau of labor statistics annual survey of occupational injuries and illnesses, 1973 through 1984, *Arch. Dermatol.,* 124, 1514, 1988.
9. **Nilsson, E.,** Individual and Environmental Risk Factors for Hand Eczema in Hospital Workers, Dissertation, University of Umeå, Umeå, Sweden, 1986.
10. **Lammintausta, K.,** Risk Factors for Hand Dermatitis in Wet Work, Dissertation, University of Turku, Turku, Finland, 1982.
11. **Cronin, E. and Kullavanijaya, P.,** Hand dermatitis in hairdressers, *Acta Derm. Venereol.,* 85 (Suppl.), 47, 1979.
12. **Fisher, T. and Rystedt, I.,** Hand eczema among hard metal workers, *Am. J. Ind. Med.,* 8, 381, 1985.
13. **Fregert, S.,** Occupational dermatitis in a 10-year material, *Contact Dermatitis,* 1, 96, 1975.
14. **Nethercott, J. R., MacPherson, M., Choi, C. K., and Nixon, P.,** Contact dermatitis in hairdressers, *Contact Dermatitis,* 14, 73, 1986.
15. **Bowker, N. C., Cross, K. W., Fairburn, E. A., and Wall, A. M.,** Sociological implications of an epidemiological study of eczema in the city of Birmingham, *Br. J. Dermatol.,* 95, 133, 1976.
16. **Johnson, M.-L. and Stern, D. S.,** Prevalence and ecology of skin disorders, in Fitzpatrick, T. B., Eisen, A. L., Wolff, K., Freedburg, J. H., and Austen, K. F., *Dermatology in General Medicine,* McGraw-Hill, New York, 1979, 3.
17. **Ratzer, M. A.,** The incidence of skin diseases in the West of Scotland, *Br. J. Dermatol.,* 81, 456, 1969.
18. **Mendenhall, R. C., Ramsay, D. L., Girard, R. A., DeFlorir, G. P., Weany, P. E., and Lloyd, J. S.,** A study of the practice of dermatology in the United States, *Arch. Dermatol.,* 114, 1456, 1978.
19. **Cronin, E.,** Clinical patterns of hand eczema in women, *Contact Dermatitis,* 13, 153, 1985.
20. **Wilkinson, D. S., Budden, M. G., and Hambly, E. M.,** A 10-year review of an industrial dermatology clinic, *Contact Dermatitis,* 6, 11, 1980.
21. **Peltonen, L.,** Nickel sensitivity in the general population, *Contact Dermatitis,* 5, 27, 1979.
22. **Gawkrodger, D. J., Vestey, J. P., Wong, W.-K., and Buxton, P. K.,** Contact clinic survey of nickel sensitive subjects, *Contact Dermatitis,* 14, 165, 1986.
23. **Rystedt, I.,** Factors influencing the occurrence of hand eczema in adults with a history of atopic dermatitis in childhood, *Contact Dermatitis,* 12, 247, 1985.
24. **Rystedt, I. and Fischer, T.,** Relationship between nickel and cobalt sensitization in hard metal workers, *Contact Dermatitis,* 9, 195, 1983.
25. **Seymour, J. L., Keswick, B. H., Hanifin, J. M., Jordan, W. P., and Milligan, M. C.,** Clinical effects of diaper types on the skin of normal infants and infants with atopic dermatitis, *J. Am. Acad. Dermatol.,* 17, 988, 1987.
26. **Jordan, W. E., Lawson, K. D., Berg, K. W., Franxman, J. J., and Marrer, A. M.,** Diaper dermatitis: frequency and security among a general infant population, *J. Am. Acad. Dermatol.,* 3, 198, 1986.
27. **Jordan, W. E. and Blaney, T. L.,** Factors influencing infant diaper dermatitis, in *Neonatal Skin,* Maibach, H. I. and Boisits, E. K., Eds., Marcel Dekker, New York, 1982, 205.
28. **Larsson, P. Å. and Liden, S.,** Prevalence of skin diseases among adolescents 12-16 years of age, *Acta Derm. Venereol.,* 60, 415, 1980.
29. **Rajka, G.,** The aetiology of atopic dermatitis, in *Atopic Dermatitis,* Rajka, G., Ed., W. B. Saunders, Philadelphia, 1975, 46.
30. **Skessa-Brociek, W., Manderscheid, J.-C., Michel, F.-B., and Bousquet, J.,** Skin test reactivity to histamine from infancy to old age, *Int. J. Allergy Clin. Immunol.,* 80, 711, 1987.
31. **Lejman, E., Stoudemayer, T., Grove, G., and Kligman, A. M.,** Age differences in poison ivy dermatitis, *Contact Dermatitis,* 11, 163, 1984.
32. **Grove, G. L.,** Biological aspects of aging skin, in *Skin Problems in the Elderly,* Fry, L., Ed., Churchill Livingstone, Edinburgh, 1985, 1.
33. **Droller, H.,** Dermatologic findings in a random sample of old persons, *Geriatrics,* 10 (55), 421, 1955.
34. **Beauregard, S. and Gilchrest, B. A.,** A survey of skin problems and skin care regimens in the elderly, *Arch. Dermatol.,* 123, 1638, 1987.
35. **Badame, J. A.,** Incidence of skin disease in rural Jamaica, *Int. J. Dermatol.,* 27, 109, 1988.
36. **Olumide, G.,** Hand dermatitis in Nigeria. I. Hand dermatitis in women, *Contact Dermatitis,* 17, 85, 1987.
37. **Goh, C. L. and Soh, S. D.,** Occupational dermatoses in Singapore, *Contact Dermatitis,* 11, 288, 1984.
38. **Malten, K. E., Arend, J., and Wigger, D. E.,** Delayed irritation: hexamediol diacrylate and butanediol diacrylate, *Contact Dermatitis,* 5, 178, 1979.

39. **Lovell, C. R., Rycroft, R. C. G., Williams, D. M. J., and Hanifin, J. W.,** Contact dermatitis from the irritancy (immediate and delayed) and allergenicity of hydroxyacrylate, *Contact Dermatitis,* 12, 117, 1985.
40. **Malten, K. E. and den Arend, J.,** Topical toxicity of various concentrations of DMSO recorded with impedance measurements and water vapour loss measurements. Recording of skin's adaptation to repeated DMSO irritation, *Contact Dermatitis,* 4, 80, 1978.
41. **Dooms-Goossens, E., Debuschere, K. M., Gerers, D. M., Dupré, K. M., Degreef, H. J., Loncke, J. P., and Snwuwaert, J. E.,** Contact dermatitis caused by airborne irritant, *J. Am. Acad. Dermatol.,* 15, 1, 1986.
42. **Crounse, R. G.,** Plant dermatitis due to Compositae (Asteraceae) family, *J. Am. Acad. Dermatol.,* 2, 417, 1980.
43. **Lachapelle, J. M.,** Occupational airborne irritant contact reaction to the dust of a food additive, *Contact Dermatitis,* 10, 250, 1984.
44. **Elias, P. M.,** Lipids and the epidermal permeability barrier, *Arch. Dermatol. Res.,* 270, 95, 1981.
45. **Elias, P.,** Epidermal lipids, barrier function and desquamation, *J. Invest. Dermatol.,* 80 (Suppl. 6), 44, 1983.
46. **Elias, P. M., Gopinethan, K. M., Grayson, S., and Brown, B. E.,** Membrane structural alterations in murine stratum corneum: relationship to the localization of polar lipids and phospholipases, *J. Invest. Dermatol.,* 91, 3, 1988.
47. **Grayson, S. and Elias, P. M.,** Isolation and lipid biochemical characterization of stratum corneum cell membrane complexes. Implications for the cutaneous permeability barrier, *J. Invest. Dermatol.,* 78, 128, 1982.
48. **Dowing, D. T., Stewart, M. E., Wertz, P. W., Colton, S. W., Abraham, W., and Strauss, J. G.,** Skin lipids: an update, *J. Invest. Dermatol.,* 88 (Suppl. 3), 2, 1987.
49. **Imokawa, G. and Hattori, M.,** A possible function of structural lipids in the water-holding properties of the stratum corneum, *J. Invest. Dermatol.,* 84, 282, 1985.
50. **Bowser, P. A. and White, R. J.,** Isolation barrier properties and lipid analyses of stratum compactum, a discrete region of stratum corneum, *Br. J. Dermatol.,* 112, 1, 1985.
51. **Green, S. C., Stewart, M. E., and Downin, D. T.,** Variation in sebum fatty acid composition among adult humans, *J. Invest. Dermatol.,* 83, 114, 1984.
52. **Naggaro-Porro, M., Passi, S., Boniforti, L., and Belsito, F.,** Effects of aging on fatty acids in skin surface lipids, *J. Invest. Dermatol.,* 73, 112, 1979.
53. **Lampe, M. A., Burlingane, A. L., Whitney, J., Williams, M. L., Brown, B. E., Roitman, E., and Elias, P.,** Human stratum corneum lipids, characterization and regional variations, *Lipid Res.,* 24, 120, 1983.
54. **Weigand, D. A., Haygood, C., Gaylor, J. R.,** Cell layers and density on Negro and Caucasian stratum corneum, *J. Invest. Dermatol.,* 62, 563, 1974.
55. **Weigand, D. A. and Gaylor, J. R.,** Irritant reaction in Negro and Caucasian skin, *South Med. J.,* 67, 548, 1974.
56. **Hamami, I. and Marks, R.,** Structural determinants of the response of the skin to chemical irritants, *Contact Dermatitis,* 18, 71, 1988.
57. **Al-Jaber, H. and Marks, H.,** Studies of the clinically uninvolved skin in patients with dermatitis, *Br. J. Dermatol.,* 111, 437, 1984.
58. **Finlay, A. Y., Nicholls, S., King, C. S., and Marks, R.,** The "dry" noneczematous skin associated with atopic eczema, *Br. J. Dermatol.,* 102, 249, 1980.
59. **Hanifin, J. M. and Rajka, G.,** Diagnostic features of atopic dermatitis, *Acta Derm. Venereol.,* 92 (Suppl.), 44, 1980.
60. **Frödin, T. and Andersson, C.,** Multiple parameter assessment of skin irritancy, *Contact Dermatitis,* 17, 92, 1987.
61. **White, M. L., Jenkinsen, D. M., Lloyd, D. H.,** The effect of washing on the thickness of the stratum corneum in normal and atopic individuals, *Br. J. Dermatol.,* 116, 525, 1987.
62. **Pierard, G. E.,** What does "dry skin" mean?, *Int. J. Dermatol.,* 26, 167, 1987.
63. **Thune, P., Nilsen, T., Hanstad, J. K., Gustavsen, T., and Dahl, H. L.,** The water barrier function of the askin in relation to the water content of stratum corneum, pH, and skin lipids, *Acta Derm. Venereol.,* 68, 277, 1988.
64. **Werner, Y., Lindberg, M., and Forslind, B.,** The water binding capacity of stratum corneum in dry non-eczematous skin of atopic eczema, *Acta Derm. Venereol.,* 62, 334, 1982.
65. **Werner, Y.,** The water content of the stratum corneum in patients with atopic dermatitis, *Acta Derm. Venereol.,* 66, 281, 1986.
66. **Werner, Y. and Lindberg, M.,** Transepidermal water loss in dry and clinically normal skin in patients with atopic dermatitis, *Acta Derm. Venereol.,* 65, 102, 1985.

67. **Gloor, M., Heyman, B., and Stuhlert, T.,** Infrared spectroscopic determination of water content of the horny layer in healthy subjects and in patient suffering from atopic dermatitis, *Arch. Dermatol. Res.*, 271, 429, 1981.
68. **Uehara, M. and Takada, K.,** Use of soap in the management of atopic dermatitis, *Clin. Exp. Dermatol.*, 10, 419, 1985.
69. **Gawkrodger, D. J., Carr, M. M., McVittie, E., Goy, K., and Hunter, A. A.,** Keratinocyte expression of MHC class III antigens in allergic sensitivity and challenge reactions and in irritant contact dermatitis, *J. Invest. Dermatol.*, 88, 11, 1987.
70. **Marks, J. G., Jr., Zaino, R. J., Bessler, M. F., and Williams, J. V.,** Changes in lymphocyte and Langerhans cell population in allergic and irritant contact dermatitis, *Int. J. Dermatol.*, 26, 354, 1987.
71. **Ferguson, J., Gibbs, J. H., Swanson Beck, J.,** Lymphocyte subsets and Langerhans cells in allergic and irritant patch test reactions: histometric studies, *Contact Dermatitis,* 13, 166, 1985.
72. **Scheynius, A., Fischer, T., and Forsum, U., and Klareskog, L.,** Phenotypic characterization *in situ* of inflammatory cells in allergic and irritant contact dermatitis in man, *Clin. Exp. Immunol.*, 55, 81, 1984.
73. **Christensen, O. B. and Wall, L. M.,** Long term effect on epidermal dendritic cells of four different types of exogenous inflammation, *Acta Derm. Venereol.*, 67, 305, 1987.
74. **Berardesca, E. and Maibach, H. I.,** Bioengineering and the patch test, *Contact Dermatitis,* 18, 3, 1988.
75. **Greaves, M. W. and Camp, R. D. R.,** Prostaglandines, leukotrienes, phospholipase, platelet activating factor and cytokines: an integrated approach to inflammation of human skin, *Arch. Dermatol. Res.*, 280 (Suppl.) 533, 1988.
76. **Fincham, N., Camp, R. D. R., and Leigh, I.,** Synthesis of arachidonate lipoxygenase products by epidermal cells, *J. Invest. Dermatol.*, 84, 447, 1985.
77. **Maurice, P. D. L., Barr, R. M., Kero, O., and Greaves, M. W.,** The effect of prostaglandin D2 on the cutaneous response to histamine in human subjects, *J. Invest. Dermatol.*, 87, 154, 1986.
78. **Archer, C. B., Cunningham, F. M., and Greaves, M. W.,** Actions of platelet activating factor (PAF) homologues and their combinations on neutrophil chemokinesis and cutaneous inflammatory responses in man, *J. Invest. Dermatol.*, 91, 82, 1988.
79. **Frosch, P. J. and Czarnetzki, B. M.,** Surfactants cause *in vitro* chemotaxis and chemokinesis of human neutrophils, *J. Invest. Dermatol.*, 80, 525, 1987.
80. **Patrick, E., Burkhalter, A., and Maibach, H. I.,** Recent investigations of mechanisms of chemically induced skin irritation in laboratory mice, *J. Invest. Dermatol.*, 88, 245, 1987.
81. **Csato, M. and Czarnetzki, B. M.,** Effect of BN 52021, a platelet activating factor antagonist, on experimental murine contact dermatitis, *Br. J. Dermatol.*, 118, 475, 1988.
82. **Rosenbach, T., Csato, M., and Czarnetzki, B. M.,** Studies on the role of leukotrienes in allergic contact dermatitis, *Br. J. Dermatol.*, 118, 475, 1988.
83. **Ruzicka, T. and Ring, J.,** Enhanced releasability of prostaglandin E2 and leukotrienes B4 and C4 from leukocytes of patients with atopic eczema, *Acta Derm. Venereol.*, 67, 469, 1987.
84. **Räsänen, L., Lehto, M., and Reunala, T.,** Decreased monocyte production of interleukin 1 and impaired lymphocyte proliferation in atopic dermatitis, *Arch. Dermatol. Res.*, 275, 215, 1987.
85. **Bjornberg, A., Lowhagen, G. B., and Tengberg, J. E.,** Relationship between intensities of skin test reaction to glass fibers and chemical irritants, *Contact Dermatitis,* 5, 171, 1979.
86. **Malten, K. E.,** Thoughts on irritant contact dermatitis, *Contact Dermatitis,* 7, 238, 1981.
87. **Lammintausta, K., Maibach, H. I., and Wilson, D.,** Human cutaneous irritation induced hyporeactivity, *Contact Dermatitis,* 17, 276, 1987.
88. **McOsker, D. E. and Beck, L. W.,** Characteristics of accomodated (hardened) skin, *J. Invest. Dermatol.*, 48, 372, 1967.
89. **Rosen, K., Mobacken, H., and Swanbeck, G.,** Chronic eczematous dermatitis of the hands: a comparison of PUVA and UVB treatment, *Acta Derm. Venereol.*, 67, 48, 1987.
90. **Thorvaldsen, J. and Volden, G.,** PUVA-induced diminution of contact allergic and irritant skin reactions, *Clin. Exp. Dermatol.*, 5, 43, 1980.
91. **Hansson, H.-A., Jansson, R., and Patruson, K.,** Transiently increased insulin-like growth factor. I. Immunoreactivity in UVB irradiated mouse skin, *J. Invest. Dermatol.*, 88, 328, 1988.
92. **De Leo, V., Scheide, S., Meshulan, J., Debra-Hanson, B.-S., and Cardullo, A.,** Ultraviolet radiation alters choline phospholipid metabolism in human keratinocytes, *J. Invest. Dermatol.*, 88, 303, 1988.
93. **Danno, K., Toda, K., and Horio, T.,** Ultraviolet B radiation suppresses mast cell degranulation induced by compound 48/80, *J. Invest. Dermatol.*, 87, 775, 1986.
94. **Ross, J. A., Howie, S. E. M., Norval, M., and Maingang, J.,** Two phenotypically distinct T cells are involved in ultraviolet-irradiated urocanic acid induced suppression of the efferent delayed type hypersensitivity response to Herpes simplex virus type 1 *in vivo, J. Invest. Dermatol.*, 89, 230, 1987.
95. **Harriot-Smith, T. G. and Halliday, W. J.,** Suppression of contact sensitivity by short-term ultraviolet irradiation. II. The role of urocanic acid, *Clin. Exp. Immunol.*, 72, 174, 1988.

Chapter 14

BIOENGINEERING APPROACHES: STATUS TO EXOGENOUS DERMATOSES

E. Berardesca and H. I. Maibach

TABLE OF CONTENTS

I. INTRODUCTION

Bioengineering tools provide assistance in investigating and monitoring exogenous dermatoses. Techniques to objectively quantify exogenous and endogenous dermatoses are developing rapidly. In spite of their cost, they allow one to obtain rapid, comparable, and reliable data on skin function.

Several bioengineering techniques have been applied to skin studies; few are widespread and commonly utilized. The more reliable and comparable ones documented in many publications are laser Doppler velocimetry to monitor blood flow, measurements of transepidermal water loss with evaporimetry, and measurements of skin thickness with ultrasound.

II. LASER DOPPLER VELOCIMETRY

Laser Doppler velocimetry (LDV) is based on the transmission of a helium-neon laser light through optical fibers to the skin. The incident radiation enters the skin and is scattered and reflected by nonmoving tissue components and by mobile red blood cells encountered as the radiation penetrates to a depth of 1 to 1.5 mm. A portion of the scattered and reflected incident radiation exits the skin and is collected by a second optical fiber that carries the light back to a photodetector, where it is converted to an electrical signal. Stationary skin tissue reflects and backscatters light at the same frequency as the incident source, while moving erythrocytes reflect the frequency-shifted radiation. The shift increases with increasing velocity. The LDV extracts the frequency-shifted signal and derives an output proportional to the flux of erythrocytes in the blood flow.[1] LDV is a reliable method to estimate cutaneous microcirculation. It has been used to assess percutaneous absorption of vasoactive molecules[2-4] and as a pharmacodynamic indicator of systematically induced vasodilatation.[5]

The first attempt to correlate LDV and skin irritant reactions was made by Nilsson et al.[6] in 1982. They applied to the volar forearm of healthy volunteers sodium lauryl sulfate (SLS) in concentrations ranging from 0.001 to 5% under occlusion for 24 h. The test sites were scored at three different times (26, 48, and 72 h) for a total number of approximately 950 recordings. They found an apparent relationship between visual scores and the recorded blood flow. In a mechanical flow model simulating the microvascular pattern of the skin, a linear relationship between LDV and the flow of erythrocytes was detected for low and moderate flow rates. For higher flow rates, photomultiple scattering in more than one moving erythrocyte and the increased light absorption due to higher erythrocyte volume fraction result in a nonlinearity in the flowmeter response. This nonlinearity causes a slight underestimation of the blood flow at high irritant concentration.

Subsequently, Wahlberg and Nilsson[7] evaluated the irritancy of propylene glycol in man. Repeated open exposures did not reveal an increase of skin blood flow, while single occlusive exposure caused both slight erythema and an increased blood flow. Staberg et al.[8] investigated 18 outpatients with eczematous skin not involving the back. They applied a standard patch test series of 31 substances and monitored the response by visual scoring and LDV after 48 h. A significant correlation between visual scoring and LDV values was found. In relation to the basal blood flow in the negative reaction sites, the blood flow increased fivefold in ?+ sites and tenfold in 1+ and 2+ sites. This finding confirms the previous report that LDV is useful in discriminating between negative and positive reactions, but fails to quantify strongly positive reactions. The possibility must be considered that maximum vasodilatation has been induced at 1+ levels and no further dilatation is possible despite the increased irritation. The same findings were reported in further studies.[9,10] Andersen and Staberg[9] found LDV useful in separating sensitized from nonsensitized animals after exposure to chlorocresol, while Blanken et al.[10] reported a good correlation between LDV and visible erythema, although they judged the naked eye to be more sensitive. Furthermore, inducing

TABLE 1
Differentiation of Allergic and Irritant Reactions with Laser Doppler Velocimetry (LDV), Evaporimetry (TEWL), and Measurements of Skin Thickness with Pulsed Ultrasound[23,26,33]

Technique	Irritation	Allergy
LDV	Increased within 2 d	Still increased after 1 week
TEWL	Early increase	Normal after 24 h; increased later
Skin thickness	Low superficial edema	Greater edema and thickness

irritation with SLS patch test exposure did not result in comparable concentration-dependent changes. Stronger vasodilatation did not necessarily imply an increase in LDV levels.

Recently, Willis et al.[11] further demonstrated the reliability of laser Doppler velocimetry for the detection of irritation. They assessed erythema in experimentally induced contact dermatitis, testing SLS, benzalkonium chloride, propylene glycol, dithranol, and other substances. Only 3 out of 100 tests gave a poor correlation between erythema and LDV response. Contrary to previous reports,[8,9] they found LDV capable of discriminating between the different visual gradings; sites with no visible erythema gave low blood flow values. Furthermore, skin staining induced by some irritants (e.g., dithranol) led to underestimation of erythema.

Staberg and Serup[12] went beyond the problem of noninvasively assessing irritation and attempted to differentiate, by means of LDV, between allergic and irritant reactions (Table 1). 12 subjects with known allergy to nickel were patch tested with nickel sulfate, nickel chloride, and SLS (1, 2, 5, and 10%). Although after 7 d the mean blood flow was higher in allergic reactions, no consistent differences were found.

III. TRANSEPIDERMAL WATER LOSS

Transepidermal water loss is the measure of the rate of water evaporation from the skin surface. The water diffusion through the stratum corneum is measurable at the skin surface as transepidermal water loss (TEWL) in grams per square meter per hour. *In vivo,* the concentration gradient is dependent on the relative humidity and the skin temperature.[13] As skin temperature increases, TEWL increases. The hydration of the stratum corneum, increasing the membrane diffusion constant, increases TEWL.[14] TEWL reflects the integrity of the water barrier function of the stratum corneum and is utilized to assess skin irritation.[15] *In vivo,* TEWL can be measured according to three different techniques:[16]

1. Closed chamber method: this consists of a capsule applied to the skin, collecting the vapor loss from the skin surface. The relative humidity inside the capsule is recorded with an electronic hygrosensor. The change in vapor loss concentration is initially fast and decreases proportionally as the humidity approaches 100%. The closed chamber method does not permit recording of continuous TEWL because skin evaporation ceases when the air inside the chamber is saturated.
2. Ventilated chamber method: this involves a chamber in which a gas of known water content passes through and is then applied on the skin. The water is picked up by the gas and measured through a hygrometer. This method allows the measurement of transepidermal water loss continuously, but if the carrier gas is too dry it artificially increases evaporation.
3. Open chamber method: this method utilizes a skin capsule open to the atmosphere. TEWL is calculated from the slope provided by two hygrosensors precisely oriented in the chamber. Air movement and humidity are the greatest drawbacks of this method when *in vivo* studies are performed.

In a correlation between TEWL and visual scoring irritation assessment[17] the skin of the volar forearm was exposed to 5% nickel sulfate, 5% SLS, 50% dimethyl sulfoxide (DMSO), and 5% phenol for 48 h. Of the 23 subjects studied, 14 exhibited positive skin reactions to nickel sulfate. The correlation between the erythema and increase in water vapor loss varied from substance to substance. The best correlation was found for 5% SLS, suggesting that cytotoxic damage of viable cells of epidermis parallels impairment of barrier function. This data confirms that reported by Malten and Den Arend.[18] Thus, TEWL measurements seem to correlate with impairment of the barrier function; however, besides a great individual and racial[19] variability, the elicited responses vary from irritant to irritant and do not allow comparison between different molecules.

Racial differences in TEWL and LDV responses have been demonstrated. Berardesca and Maibach[19,20] recently reported different skin responses to chemical irritation in races after exposure to two concentrations of SLS. The TEWL responses showed higher susceptibility of black skin to SLS and a different modulation of irritant reactions after exposure to different amounts of irritant. On the other hand, less erythema was detected with laser Doppler velocimetry in black than in white skin. Hispanics, with regard to water loss responses to SLS, showed the same pattern of reaction as blacks,[20] while erythematous reactions were strikingly similar to whites. Goh and Chia,[21] investigating irritant reactions to SLS in the Asian population, reported significantly higher TEWL values in irritated Chinese skin compared to Malay skin. Moreover, in this study the irritation index (defined as the difference of water loss between irritated and nonirritated skin) was not significantly different between the various ethnic groups, but was lower in males than females.

TEWL may detect and monitor nonvisible irritation[22] and discriminate between irritant and allergic reactions (Table 1).[23] In allergic reactions, initially the water barrier is not damaged if the application time of the test substance is short (i.e., 24 h), and TEWL increases after 48 h when the damage is secondary to the inflammatory infiltrate. Irritant reactions, on the contrary, show an earlier increase in TEWL values due to barrier damage.

IV. ULTRASOUND

Skin thickness is reliably evaluated by pulsed ultrasound measuring the total transit time of high-frequency pulsed ultrasound waves beamed into the skin and reflected off tissue interfaces. If tissue acoustical velocity is known, the measured time can be related to distances, thus transforming a transit time trace to a spatial representation of the internal structure of the skin.

The first attempt to correlate skin thickness and degree of skin irritancy was made by Wahlberg in 1983.[24] SLS in distilled water (1 and 5%) and nonanoic acid in 0.1 to 60% in propanolol was applied in rabbits, guinea pigs, and men. The guinea pig was more resistant than the rabbit. Increasing the concentration of the test substance caused an increase in fluid accumulation, establishing a dose-response relationship. Higher concentrations produced an earlier response. The study in man with nonanoic acid showed a maximal increase at the 60% concentration after 15 applications of the test substance. The development of edema was slow; a significant increase was observed on day 13. In a further work,[7] a comparison between LDV and skin fold thickness measurements for the assessment of skin irritancy was made after exposure to propylene glycol. While administration under occlusion caused an increase in skin blood flow, no change was detectable in skin thickness in man.

Serup et al.,[25] using a specially designed high-frequency pulsed ultrasound technique, reported differentiation of positive patch test reactions on the basis of skin thickness: doubtful reactions, <0.2 mm; + reactions, 0.3 to 0.7 mm; + + reactions, >0.8 mm; + + and + + + reactions could be differentiated by the absolute increase in the skin thickness, but less significantly.

Recently, Serup and Staberg[26] monitored the edema formation of a patch test with nickel sulfate, nickel chloride, and SLS in different concentrations in individuals with known nickel allergy. Edema was greater in allergic than in irritant reactions similar in strength according to the visual reading (Table 1). In allergic reactions edema was more transcutaneous, while in irritant reactions it was more superficial. Measurements of skin thickness seem reliable in differentiating the two types of reactions. Moreover, they suggested applying the patches on the forearm rather than the back, where subcutaneous reticulae may cause echoes which are difficult to distinguish from the echoes of the dermis-subcutaneous tissue interface.

Here we have reported the more commonly used techniques. These guarantee enough reliability and reproducibility. The development of other methods is in progress.

Prens et al.[27] reported a correlation between transcutaneous PO_2 pressure and the severity of patch test responses. Transcutaneous PO_2 is measurable with commercially available systems using a skin surface oxygen electrode. PO_2 decreases in relation to the severity of the skin response. The PO_2 on the skin surface depends on the capillary blood flow, PO_2 and O_2 conductivity related to epidermal water content, and thickness of the epidermis.[28] Measurements are made with an electrode kept at a temperature of 43°C to maximize blood flow in capillaries and avoid changes in recordings due to blood flow. In this way, surface PO_2 is influenced only by edema, spongiosis, and other local epidermal alterations. This technique, mainly useful in detecting edema and infiltration, quantifies strong skin responses.

Since skin irritation causes different patterns of reaction, the study of the skin surface by means of a polysulfide rubber replica may give important information.[29] Some irritants cause a papular reaction (hydrochloric acid, croton oil), while others elicit a nonpapular reaction (SLS). Video analyzing systems might be able to better quantify these responses.

Optical techniques were used by other groups. In 1986 Mendelow et al.,[30] using a hemalometer to assess patch tests, evaluated the hemoglobin index (proportional to the degree of erythema) by measuring the reflectance of the skin at 566 and 640 nm. The 566-nm light (green) is absorbed by blood; this is compared with the measurement taken in 640-nm (orange) light. The hemoglobin index is defined as the difference between the absorbances measured at these wavelengths. Negative reactions (clinical score 0) gave a hemoglobin index in the range of 3 to 13, while positive reactions (clinical score +3) gave a hemoglobin index falling in the range 120 to 159. The main disadvantages of this method could be the time required to do the measurement and the weight of the probe on the skin, which might produce blanching in the tested area.

Westerhof et al.[31] quantified UV-induced erythema on eight healthy volunteers after exposure to a solar simulator, reporting a linear relation between visual grading and chromameter readings. Similar findings are reported by Babulak et al.[32]

V. CONCLUSIONS

A "perfect" technique does not exist. Good evaluation of allergic and contact dermatitis may be accomplished correlating data obtained with different techniques, keeping in mind the limits of these techniques and the etiological factors that induced the dermatitis. Some exogenous dermatoses are edema-inducing and could be evaluated better with measurements of skin thickness; others mainly damage the water barrier, and still others develop only erythema with lack of edema. Laser Doppler velocimetry is useful in discriminating between doubtful reactions; ultrasound and PO_2 measurements help in quantifying stronger reactions.

Differentiation between allergic and irritant reactions is extremely interesting, and more work should be done in this area because bioengineering could help, not only in quantifying a phenomenon, but also in providing a diagnosis.

REFERENCES

1. **Nilsson, E. G., Tenland, T., and Oberg, P. A.,** Evaluation of a laser Doppler flowmeter for measurement of tissue blood flow, *IEEE Trans. Biomed. Eng.*, 27, 597, 1980.
2. **Guy, R. H., Wester, R. C., Tur, E., et al.,** Noninvasive assessment of the percutaneous absorption of methyl nicotinates in humans, *J. Pharm. Sci.*, 72, 1077, 1983.
3. **Tur, E., Guy, R. H., Tur, M., et al.,** Noninvasive assessment of local nicotinate pharmacodynamics by photopletysmography, *J. Invest. Dermatol.*, 80, 499, 1983.
4. **Wester, R. C., Maibach, H. I., Guy, R. H., et al.,** Minoxidil stimulates cutaneous blood flow in human balding scalps: pharmacodynamics measured by laser Doppler velocimetry and photopulse plethysmography, *J. Invest. Dermatol.*, 82, 515, 1984.
5. **Wester, R. C., Maibach, H. I., Guy, R. H., et al.,** Nitroglycerin pharmacodynamics monitored non-invasively by laser Doppler velocimetry, abstract presented at the 84th Annu. Meet. American Society for Clinical Pharmacology and Therapeutics, San Diego, March, 1983.
6. **Nilsson, G. E., Otto, U., and Wahlberg, J. E.,** Assessment of skin irritancy in man by laser Doppler flowmetry, *Contact Dermatitis,* 8, 401, 1982.
7. **Wahlberg, J. E. and Nilsson, G.,** Skin irritancy from propylene glycol, *Acta Derm. Venereol.*, 64, 286, 1984.
8. **Staberg, B., Klemp, P., and Serup, J.,** Patch test responses evaluated by cutaneous blood flow measurements, *Arch. Dermatol.*, 120, 741, 1984.
9. **Andersen, K. E. and Staberg, B.,** Quantitation of contact allergy in guinea pigs by measuring changes in skin blood flow and skin fold thickness, *Acta Derm. Venereol.*, 65, 37, 1985.
10. **Blanken, R., van der Valk, P. G. M., and Nater, J. P.,** Laser Doppler flowmetry in the investigation of irritant compounds on human skin, *Dermatosen,* 34, 5, 1986.
11. **Willis, C. M., Stephens, C. J. M., and Wilkinson, J. D.,** Assessment of erythema in irritant contact dermatitis. Comparison between visual scoring and laser Doppler flowmetry, *Contact Dermatitis,* 18, 138, 1988.
12. **Staberg, B. and Serup, J.,** Allergic and irritant reactions evaluated by laser Doppler flowmetry, *Contact Dermatitis,* 18, 40, 1988.
13. **Grice, K., Sattar, H., Sharrat, M., and Baker, H.,** Skin temperature and transepidermal water loss, *J. Invest. Derm.*, 57, 108, 1971.
14. **Dugard, P. H.,** Skin permeability theory in relation to measurements of percutaneous absorption in toxicology, in *Dermatotoxicology and Pharmacology,* Marzulli, F. N. and Maibach, H. I., Eds., John Wiley & Sons, New York, 1977, 525.
15. **Hassing, J. H., Nater, J. P., and Bleumink, E.,** Irritance of low concentrations of soap and synthetic detergents as measured by skin vapour loss, *Dermatologica,* 154, 314, 1982.
16. **Maibach, H. I., Bronaugh, R., Guy, R., Turr, E., Wilson, D., Jacques, S., and Chaing, D.,** Non-invasive techniques for determining skin function, in *Cutaneous Toxicity,* Drill, V. A. and Lazar, P., Eds., Raven Press, New York, 1984, 63.
17. **van Der Valk, P. G. M., Kruis-de Vries, M. H., Nater, J. P., Bleumink, E., and De Jong, M. C. J. M.,** Eczematous (irritant and allergic) reactions of the skin and barrier function as determined by water vapour loss, *Clin. Exp. Dermatol.*, 10, 185, 1985.
18. **Malten, K. E. and Den Arend, J.,** Topical toxicity of various concentrations of DMSO recorded with impedance measurements and vapour loss measurements, *Contact Dermatitis,* 4, 80, 1978.
19. **Berardesca, E. and Maibach, H. I.,** Racial differences in sodium lauryl sulphate induced cutaneous irritation:black and white, *Contact Dermatitis,* 18, 65, 1988.
20. **Berardesca, E. and Maibach, H. I.,** Sodium lauryl sulphate induced cutaneous irritation. Comparison of white and hispanic subjects, *Contact Dermatitis,* 19, 136, 1988.
21. **Goh, C. L. and Chia, S. E.,** Skin irritability to sodium lauryl sulphate — as measured by skin surface water vapour loss — by sex and race, *Clin. Exp. Dermatol.*, 13, 16, 1988.
22. **Berardesca, E. and Maibach, H. I.,** The effect of nonvisible irritation on the water holding capacity of the stratum corneum, in Abstr. 5th Eur. Symp. Contact Dermatitis, Heidelberg, West Germany, May 27 to 29, 1988.
23. **Serup, S. and Staberg, B.,** Differentiation of allergic and irritant reactions by transepidermal water loss, *Contact Dermatitis,* 16, 129, 1987.
24. **Wahlberg, J. E.,** Assessment of skin irritancy: measurements of skin fold thickness, *Contact Dermatitis,* 9, 21, 1983.
25. **Serup, J., Staberg, B., and Klemp, P.,** Quantification of cutaneous oedema in patch test reactions by measurements of skin thickness with high-frequency pulsed ultrasound, *Contact Dermatitis,* 10, 88, 1984.
26. **Serup, J. and Staberg, B.,** Ultrasound for assessment of allergic and irritant patch test reactions, *Contact Dermatitis,* 17, 80, 1987.

27. **Prens, E. P., van Joost, Th., and Steketee, J.,** Quantification of patch test reactions by transcutaneous PO_2 measurement, *Contact Dermatitis,* 16, 142, 1987.
28. **Lubbers, D. W.,** Theoretical basis of the transcutaneous blood gas measurement, *Crit. Care Med.*, 9, 721, 1981.
29. **Agner, T. and Serup, J.,** Skin reactions to irritants assessed by polysulfide rubber replica, *Contact Dermatitis,* 17, 205, 1987.
30. **Mendelow, A. Y., Forsyth, A., Feather, J. W., Baillie, A. J., and Florence, A. T.,** Skin reflectance measurements of patch test responses, *Contact Dermatitis,* 15, 73, 1986.
31. **Westerhof, W., van Hasselt, B. A. A. M., and Kammeijer, A.,** Quantification of UV-induced erythema with a portable computer controlled chromameter, *Photodermatology,* 3, 310, 1986.
32. **Babulak, S. W., Rhein, L. D., Scala, D. D., Simion, F. A., and Grove, G. Y.,** Quantitation of erythema in a soap chamber test using the Minolta Chroma (reflectance) meter: comparison of instrumental results with visual assessment, *J. Soc. Cosmet. Chem.*, 37, 475, 1986.
33. **Staberg, B. and Serup, J.,** Allergic and irritant skin reactions evaluated by laser Doppler flowmetry, *Contact Dermatitis,* 18, 40, 1988.

Chapter 15

EFFECTS OF DERMATOLOGICAL VEHICLES ON THE SKIN

Matti Hannuksela and Tuula Kinnunen

TABLE OF CONTENTS

I. INTRODUCTION

Tremendous development of dermatological formulations, both cosmetic and pharmacological, has taken place in the past 40 years. Most physicians think that the vehicle itself is inert. That is far from true. Vehicle components may increase or decrease transepidermal water loss and may alter the water balance of stratum corneum to a great degree.

Immediate contact reactions to various skin products are obviously much more common than is realized today. Transient itching, tingling, and redness from cosmetics and topical drugs are familiar to most users of these products. These symptoms may be milder forms of contact urticaria, which is caused by many chemical and physical agents.

Contact sensitivity to vehicle ingredients has been known for decades. This knowledge, however, has concerned only a few components, chiefly the preservatives contained in most skin products. In addition to preservatives and other antimicrobial agents, allergy to emulsifiers and propylene glycol (PG) has received some attention, but other components, such as oils, fats, and humectants, have been considered to be less interesting vehicle components.

In this chapter we briefly discuss the various components of dermatological vehicles and their effect on the water balance and barrier function of the skin. Immediate and delayed contact reactions are important in daily practice; therefore, they will be discussed in more detail.

II. COMPONENTS OF VEHICLES

Oil in water (o/w) emulsions contain at least water, one or more emulsifiers, and a preservative. Simple ointments may have only paraffins of various chain lengths. In most instances, however, cosmetic lotions, creams, and ointments and the vehicles of pharmaceutical products for the skin contain more components, often over ten different ingredients. Many substances may have two or more roles in the vehicles; for example, cetyl alcohol is an emulsifier but also a fat or thickener, and PG is sometimes added to the product to act as a humectant, solvent, penetrator, or antimicrobial agent. It is impossible to give a comprehensive list of the hundreds of ingredients used in cosmetics and the vehicles of topical drugs.

III. EFFECTS OF VEHICLES ON THE PERMEABILITY OF THE SKIN AND ON THE TRANSEPIDERMAL WATER LOSS

The stratum corneum is the skin's main barrier against foreign chemicals. It is like a wall, with the corneocytes acting as the bricks and the intercellular lipid bilayers as the mortar.

A. DIFFUSION THROUGH INTACT SKIN

Diffusion of various substances through the epidermis is a passive process that follows the laws of physics.[1-5] The diffusion rate depends on the solubility and polarity as well as on the size and shape of the substances. Water-soluble and polar molecules diffuse directly through the protein part of the stratum corneum (i.e., through the corneocytes) and lipid-soluble and nonpolar materials via the intercellular spaces.[6] The pilosebaceus units play a minor role in the diffusion process because their pores form only 0.1 to 1% of the total skin surface area.[7]

B. TRANSEPIDERMAL WATER LOSS (TEWL)

The water content of the intact keratin layer is 10 to 15%, but its weight can increase up to four to five times. Water evaporates from the stratum corneum. It is replaced by water diffusing from the deeper layers of the epidermis and dermis.

TABLE 1
Methods of Measuring the Hydration Status of the Keratin Layer

Method	Ref.
Electrical resistance	8—11
Impedance	9,10
Electrical conductance	4,10,12,13
Infrared spectroscopy	10,13

The principal factor in the water-binding capacity of the keratin layer is thought to be the "natural moisturizing factor", in which pyrrolidone carboxylic acid, inorganic salts, amino acids, and urea are the most important substances.

The hydration status of the horny layer can be measured with several methods which yield highly comparable results (Table 1).

TEWL can be measured in many ways. A device which measures the water vapor gradient near the skin surface, the Evaporimeter®,[7] is the most practical one for clinical studies. In ordinary environmental circumstances, the TEWL varies between 3 to 6 $g/m^2/h$.[7,15] The TEWL rate primarily reflects the condition of the barrier and, to a lesser extent, the hydration status of the horny layer.[16,17] The TEWL is clearly higher than normal in noneczematous and eczematous atopic skin;[18] in psoriatic lesions and erythrodermic skin it is even higher than in atopy.[16,19,20] On the other hand, in xerotic skin induced by a dry winter climate, the water content of the keratin layer has been found to be low, but the TEWL still remained in the normal range.[21]

C. THE EFFECT OF VEHICLES ON TEWL AND PENETRATION

1. Surfactants

The use of surfactants in skin cleansing lotions and creams may result in depletion of both the lipid material and the water-binding substances of the keratin layer, leading to impaired barrier function and increased penetration of both lipid- and water-soluble foreign materials into the skin.[22,23] Nonionic surface-active materials (e.g., common emulsifiers) have less of an effect on the barrier properties of the stratum corneum.[1] The alterations in the barrier function of the skin may be so small that they can only be demonstrated in special circumstances. Larmi et al.[24] pretreated healthy human back skin with 20% sorbitan sesquioleate in petrolatum three times in one day and found the immediate skin response to benzoic acid to be significantly increased.

2. Humectants

Humectants are substances which bind water readily. When applied to the skin they are thought to increase both the water content of the stratum corneum and also the TEWL.[3,25] It seems, however, that various humectants differ from each other in the capacity to affect TEWL.

Glycerol is a well-known humectant that hydrates the keratin layer, but it reduces the TEWL rate.[25] PG is also a humectant, but it is a good solvent for many substances, including topical corticosteroids, a penetrator, and a disinfectant. Although it enhances the penetration of many substances into the skin, it not unexpectedly also increases the TEWL.[19,26]

IV. ADVERSE REACTIONS

The adverse reactions to ingredients of vehicles include immediate and delayed irritant and allergic responses (Table 2). Immediate reactions are also known as contact urticaria or

TABLE 2
Contact Reactions to Vehicle Ingredients

Reaction	Remarks
Immediate	
Immunologic	IgE mediated
Nonimmunologic	Mediators unknown; prostaglandins may be responsible for at least some reactions; sorbic acid and benzoic acid are the best known substances in this group.
Delayed	
Irritant	Vehicle ingredients may act as coirritants and penetrants, but are seldom the only causes of irritant dermatitis.
Allergic	In unscented preparations emulsifiers, propylene glycol, and preservatives are the most common contact allergens; propylene glycol and emulsifiers may enhance the development of contact allergy to other ingredients.

as immediate contact reactions. The agents capable of producing immediate contact reactions, as well as the mechanisms underlying these reactions, are poorly known. Chemicals producing delayed irritant dermatitis are better known, but the factors influencing the skin response have attracted investigators' interest to a lesser extent. For several decades now, delayed contact allergy has been a far more important topic of basic and clinical studies than the other types of adverse reactions.

A. NONIMMUNOLOGIC CONTACT REACTIONS

Nonimmunologic immediate contact reactions (NIICRs) appear as redness or as wheal and flare reactions within minutes to 1 h and usually fade away within 1 or 2 h. The subjective symptoms include itching and tingling. Gollhausen and Kligman[27] proposed that contact urticaria be totally separated from other types of immediate contact reactions. This is, however, not always possible. In milder contact urticarial reactions, erythema is the only visible reaction.[28]

Epidemiological data on the prevalence and severity of NIICRs from dermatological vehicles is lacking. Burning and stinging sensations from various creams, ointments, lotions, and other products are known to be common, whereas real contact urticaria is very rare.

Many factors interfere in NIICRs; acetylsalicylic acid[28,29] and ultraviolet (UV) irradiation[30] are very potent inhibitors, and infrared irradiation[31] is a promoter of such reactions. Topical steroids and local anesthesia diminish the strength of NIICRs.[28,32] Pretreatment of the skin with 20% sorbitan sesquioleate increased the strength of NIICRs to benzoic acid (BA) incorporated in petrolatum, but 20% sorbitan sesquioleate in the vehicle diminished skin reactivity to BA.[33] This result indicates that an emulsifier may increase the penetration of foreign chemicals into the skin, but as a component of the vehicle itself an emulsifier can worsen the bioavailability of another ingredient of the vehicle.

Common dermatological vehicles contain many substances capable of producing NIICRs (Table 3). Among them, sorbic acid (SA)[27,28,32,34] and BA[27,32] have been studied most intensively.

1. Preservatives

SA causes nonallergic erythema and edema at as low a concentration as 0.05% in a mixture of isopropanol and water when applied without occlusion to the cheek or forehead.[28] Erythema and edema were also produced on the cheeks of 9 out of 20 subjects by creams and ointments containing SA as the preservative.[34]

BA is seldom used as a preservative in dermatological formulas. It is an active ingredient (6%, w/w) in Whitfield's ointment, which is still used for superficial dermatophytoses in many countries. The ointment does not cause any harm to the skin of the feet or palms, but

TABLE 3
Vehicle Ingredients Producing Immediate Contact Reactions

Preservatives
Sorbic acid
Benzoic acid
Chlorocresol
Formaldehyde
Benzyl alcohol
Emulsifiers
Sorbitan monolaurate
Polysorbate 60
Cetyl alcohol
Stearyl alcohol
Other
Sodium salt of pyrrolidone carboxylic acid
Fragrance materials
Balsam of Peru
Cinnamic alcohol and aldehyde
Eugenol
Geraniol
Anisyl alcohol
Benzyl alcohol

may produce intense erythema and edema when used on other skin areas. BA has also been mentioned in the lists of agents producing allergic immediate contact reactions,[35,36] but final proof of the allergic background is still lacking.

Freitas and Brandao[37] reported a case of contact urticaria to chlorocresol. Their patient, a 28-year-old woman, showed wheal and flare reactions followed by superficial necrosis in open and prick tests to 10%, but not to 1%, chlorocresol. The higher concentration elicited slight erythema in ten controls only occasionally, but not urticaria or necrosis. Another patient showing an immediate reaction to chlorocresol was reported by Goncalo et al.[38] This patient, a 35-year-old woman, showed an immediate reaction in a skin prick test with 1% chlorocresol in a mixture of alcohol and water, but she also had a delayed reaction in both the prick test and in the patch test. Controls reacted to 5%, but not to 1%, chlorocresol in the prick test reported by Goncalo et al.[38] Thus, it seems to us that the reaction to chlorocresol in the patient of Freitas and Brandao[37] may have been an immediate irritant reaction, but the patient reported by Goncalo et al.[38] obviously had both type I and type IV allergies to this preservative.

Formaldehyde is also a potential cause of contact urticaria.[39,40] However, the concentrations present in skin care products are so low that they are hardly likely to cause immediate skin reactions.

2. Emulsifiers

Cetyl and stearyl alcohols were the apparent cause of contact urticarial reaction to a canvas rubbing the left forearm skin of a female patient reported by Gaul.[41] Cetyl alcohol at 5% and stearyl alcohol at 3% in petrolatum produced contact urticaria in patch tests.

Sorbitan monolaurate[42] and polysorbate 60[43] have been found to be responsible for contact urticaria in one case each. Thus, it seems that contact urticaria to emulsifiers is very rare, and it is assumed, but not proven, that such reactions may be IgE-mediated allergic reactions.

3. Natural Moisturizing Factor

Sodium salt of pyrrolidone carboxylic acid (NaPCA), the chief component of the natural moisturizing factor, produced erythema on the upper back skin, but not on other skin sites, in 3 out of 13 healthy test subjects when tested at 12.5% in water.[44] The indication for this particular study was the occasional complaints of users of skin care products with either natural moisturizing factor or NaPCA.

4. Fragrances

In open testing, a majority of patients and controls developed contact urticaria to 7 out of 14 substances in the North American Contact Dermatitis Group perfume screening series.[45] These substances were tested at 1 to 5% in petrolatum. At concentrations present in perfumed dermatological preparations, the same substances do not apparently cause any NIICRs. Benzyl alcohol is used both as a preservative in unscented preparations and as an ingredient of perfume mixes. In higher concentrations it is a mild NIICR agent, but at the level (<0.5%) present in vehicles it hardly can elicit NIICRs.

B. CONTACT ALLERGY

Delayed contact allergy to vehicle ingredients has received much less attention than allergy to active ingredients of dermatological preparations. Even less attention has been paid to the enhancing effect of some vehicle ingredients (e.g., propylene glycol and emulsifiers) on the risk of sensitization to other vehicle components.

1. Contact Allergy to Vehicle Ingredients

Textbooks on contact allergy also refer to the many reports on contact allergy caused by vehicle ingredients. In particular, Nater and de Groot's[46] book and de Groot's[47] thesis should be mentioned in this connection. In Chapter 5 of this book, Bruze and his colleagues deal with one of the most fascinating new superallergens, the mixture of two isothiazolinones (Kathon®). In the following section we discuss contact allergy to emulsifiers, to humectants, and to a lesser extent to other vehicle components.

2. Emulsifier Allergy

Common emulsifiers, such as cetyl and stearyl alcohols and sorbitan esters, have been patch tested routinely in many dermatological departments. Allergy to these substance has been found in a number of patients with eczematous dermatitis (Table 4). The table includes substances used both as emulsifiers and as emollients and thickeners. The allergen in cetyl and stearyl alcohols must be an impurity because patch tests and repeated open application tests (ROATs) with chemically pure materials have had negative results.[56] Hayakawa and co-workers[57] found branched aliphatic C18 compounds to be the cause of lipstick dermatitis in an 18-year-old girl. Whether or not these compounds are present in commercial cetyl and stearyl alcohols is not known at present.

ROAT with 20% sorbitan sesquioleate was clearly positive in three out of four patients with a questionable or equivocal reaction to sorbitan sesquioleate in patch tests and in all five patients showing a + or + + patch test reaction.[56]

Contact allergy to other emulsifiers has not been investigated systemically. A considerable number of emulsifiers found to act as contact allergens in occasional cases are listed in Table 5. Cross allergy between sorbitan sesquioleate and the Spans® is common.[68] Oleamidopropyl dimethylamine is obviously an important contact allergen, at least in the Netherlands, and there seems to be cross allergy between it and a number of other dimethylamines.[47]

Wool alcohols are also used as emulsifiers. Contact allergy to them is usually encountered in 1 to 2% of eczema patients.[68,80-84] The prevalence of allergy to lanolin among the general population in Europe has been estimated to be as low as 9.7 per 1 million.[85]

3. Contact Allergy to Glycols

Of the polyalcohols, PG is the most important contact allergen. It was found to be a contact allergen in the early 1950s,[86] and this finding was verified about 20 years lat-

TABLE 4
Delayed Contact Allergy to Cetyl and Stearyl Alcohols and to Sorbitan Esters in Random Eczema Patients

	Tested	Positive		
	No.	No.	%	Ref.
Cetyl and stearyl alcohols	1644	4	0.2	48
	344	9	2.6	49
	1206	8	0.7	50
	330	7	2.1	51
	1011	27	2.7	52
	2064	81	3.9	53
	3095	22	0.7	56
	5202	56	1.1	54
	1374	11	0.8	55
Sorbitan sesquioleate	1206	5	0.5	50
	3095	15	0.5	55
	1374	10	0.7	55

er.[82,87-90] The exact number of eczema patients allergic to it has been a matter of controversy. While the number of patients considered to be allergic to PG has varied between 1.5 and 4.8% in Europe,[89,91-93] doubts have been voiced in the U.S. regarding this allergy.[94] In ROATs, all five patients showing a + or + + reaction to 1% PG in patch tests reacted to a cream containing 5% PG as a humectant.[56] In that particular study, 4 out of 12 patients with positive patch-test reactions to 30% PG, but not to lower concentrations, were ROAT positive to the cream with 5% PG. We recommend that eczema patients be tested routinely with 30% PG in water; in positive cases patch tests should be performed with a dilution series (1 or 2% and 10% PG in water), and ROAT should also be performed whenever possible. Those reacting to an isotonic (2%) or lower concentration of PG in patch tests should be considered to be allergic, but higher concentrations of PG also produce irritant patch test reactions; the higher the concentration, the more frequently this occurs.

Glycerol has been found to be a contact allergen in occasional cases only.[73,95]

Polyethylene glycols (PEGs) are also potent sensitizers. Braun[88] reported PEG allergy in 3 of 40 patients allergic to a topical antiseptic preparation containing PEG as the solvent. Out of 92 other patients with allergic contact dermatitis, 4 reacted positively to PEG 300.[88] PEG produced contact allergy in 1 of 200 subjects undergoing the Draize test with a bar soap.[96] That particular patient showed a positive response to 1% PEG 300, 600, 1000, 4000, and 6000 in subsequent patch tests, but a 7-d use test with 3% PEG 300 in petrolatum remained negative. ROAT was positive with PEG 300 and PEG 400 in one patient who was sensitized by PEG 300 in antibacterial ear drops.[56] A mixture of PEG 400 and PEG 3350 was the cause of contact allergy in a patient who used a mupirocin ointment for 4 weeks.[97] Both PEG 400 and PEG 3350 at 5% in petrolatum elicited a + + response in patch tests.

4. Contact Allergy to Preservatives

In most countries, parabens are usually tested in routine patch-test series. In addition to parabens, we have routinely tested chlorocresol and sorbic acid for many years. Allergy to these substances is extremely rare in Finland; for example, paraben allergy is seen in only 0.3% of patients with suspected allergic contact dermatitis.[68] The number of patients showing positive patch test reactions to parabens varies from one country to another. In Sweden the figure is 0.2 to 0.3%,[98] in Denmark 0.9%,[99] with about 2% in Germany[100] and as high as 4% in Saskatoon, Saskatchewan, Canada.[101]

Chloroacetamide is a potent sensitizer in cosmetics, paints, and industrial chemicals.[102]

TABLE 5
Contact Allergy to Emulsifiers Not Routinely Tested

Chemical	Ref.
Cocamide diethandamine	58
Cocamidopropyl betaine	58
Diethanolamine-dihydroxypalmityl phosphate + isopropylhydroxypalmityl ether	59
Diisopropanolamine	60
Emulgin RO/40	61
Glyceryl isostearate	57
Glyceryl stearate	62
Lauramide diethanolamine	58
Laureth-4	63
Methyl glucose sesquistearate	64
Octyl dodecanol	65
Oleamidopropyldimethylamine	58
PEG-32 stearate	58
Polyoxyethylene oxypropylene stearate	68
Polyoxyethylene sorbitol lanolin derivative	68
Sorbitan monolaurate	66,67
Sorbitan monooleate	50,69
Sorbitan monostearate	50
Stearamidoethyl diethylamine phosphate	70
Stearic acid	58
Triethanolamine (TEA)	68,72-76
TEA-coco-hydrolyzed animal protein	77
Trilaureth-4 phosphate	78
Tweens®	68,79

Several other preservatives have been found to cause contact allergy in occasional cases only[47,103] with the exception of the superallergen Kathon®, which is dealt with thoroughly in Chapter 5 of this book.

5. Antioxidants

Tosti and co-workers[104] reported 2 cases of contact allergy to butylated hydroxyanisole (BHA) and reviewed 14 further cases from the literature. In addition, the North American Contact Dermatitis Group found 3 BHA-sensitive patients among 713 patients with cosmetic dermatitis.[103] White et al.[105] found one patient allergic to butylated hydroxytoluene (BHT), five patients allergic to another antioxidant, tertiary butylhydroxyquinone, and seven allergic to BHA.

Propyl gallate is used in very low concentrations in preparations containing animal or plant oils and fats, either alone or together with other antioxidants. Contact allergy to it, as well as to other gallates, has been reported in case reports.[103,106,107]

6. Other Ingredients

Several other ingredients of dermatological vehicles have been reported to cause contact allergy (Table 6). Fragrances and masking agents are not discussed here. The table is far from complete, but it gives an idea of the types of chemicals which may be responsible for our patients' dermatitis. It is self-evident that oils and fats are not pure chemicals. This was nicely proved by Hayakawa et al.,[120] who found sesamolin and sesamin to be responsible for the contact allergic reactions to sesame oil.

V. ENHANCEMENT OF CONTACT SENSITIZATION

Various factors increase the risk of contact sensitization. Occlusive dressings, injuries in the barrier zone (e.g., leg ulcers and stasis dermatitis), and chemicals increasing the

TABLE 6
Contact Allergy to Fats and Oils in Dermatological Preparations.

Chemical	Ref.
Castor oil	108—111
Isopropyl myristate	112,113
Polyoxyethylated lanolin (Lanpol® 5)	114
Mineral oil	103,115
Myristyl alcohol	71,116
Oleyl alcohol	103,117
Olive oil	118
Sesame oil	119,120

TABLE 7
Contact Allergies to the Components of an Antiperspirant Lotion in 19 Patients with Armpit Dermatitis[122,123]

Substance(s)	No. positive
PB	7
TCC	1
PG	1
PB + TCC	2
PB + PG	6
PB + PG + TCC	2

Note: Trichlorocarbanilide (TCC) was tested at 2—5%, propantheline bromide (PB) at 5—10%, and propylene glycol (PG) at 1—10%.

penetration of chemicals into the skin greatly augment the sensitization rate. Electric current was the obvious aggravating factor in three cases of contact allergy to electrolyte gel; the allergen was PG in at least two of them.[121]

PG seems to be one of the most potent promoters of contact sensitization in dermatological formulas. In a number of cases, an antiperspirant containing 0.25% 3,4,4′-trichlorocarbanilide and 5% propantheline bromide in PG was found to cause contact allergy to one or more of its components in 2 to 16 weeks (Table 7).

PG also increased the rate of contact allergy in guinea pigs when open epicutaneous test and chamber test were used in the induction phase, but this enhancing effect was not seen in guinea pig maximization test.[124] In chromate-sensitive patients, potassium dichromate gave positive results more readily when patch tested in a mixture of PG and water than when tested alone in water.[125] The mixture was also superior to petrolatum as a test vehicle.

Topical zinc sulfate augmented cutaneous reactivity to *Candida* extract,[126] but little is known about its role in delayed contact allergy.

REFERENCES

1. **Scheuplein, R. J. and Blank, I. H.,** Permeability of the skin, *Physiol. Rev.*, 51, 702, 1971.
2. **Barry, B. W.,** *Dermatological Formulations: Percutaneous Absorption*, Marcel Dekker, New York, 1983, 160.
3. **Idson, B.,** Water and the skin, *J. Soc. Cosmet. Chem.*, 24, 197, 1973.
4. **Cooper, E. R. and Berner, B.,** *Methods in Skin Research*, John Wiley & Sons, New York, 1985, 407.
5. **Schaefer, H. and Jamoulle, J. C.,** Skin pharmacokinetics, *Int. J. Dermatol.*, 27, 351, 1988.
6. **Albery, W. J. and Hadgraft, J.,** Percutaneous absorption: *in vivo* experiments, *J. Pharm. Pharmacol.*, 31, 140, 1979.
7. **Idson, B.,** Percutaneous absorption, *J. Pharm. Sci.*, 64, 901, 1975.
8. **Tagami, H., Ohi, M., Iwatsuki, K., Kanamaru, Y., Yamada, M., and Ichijo, B.,** Evaluation of the skin surface hydration *in vivo* by electrical measurement, *J. Invest. Dermatol.*, 75, 500, 1980.
9. **Werner, Y., Lindberg, M., and Forslind, B.,** The water-binding capacity of stratum corneum in dry non-eczematous skin of atopic eczema, *Acta Derm. Venereol.*, 62, 334, 1982.
10. **Triebskorn, A., Gloor, M., and Greiner, F.,** Comparative investigations on the water content of the stratum corneum using different methods of measurement, *Dermatologica*, 167, 64, 1983.
11. **Blank, I. H., Moloney, J., Emslie, A. G., Simon, I., and Apt, C.,** The diffusion of water across the stratum corneum as a function of its water content, *J. Invest. Dermatol.*, 82, 188, 1984.
12. **Blichmann, C. W. and Serup, J.,** Assessment of skin moisture, *Acta Derm. Venereol.*, 68, 284, 1988.
13. **Frödin, T., Helander, P., Molin, L., and Skogh, M.,** Hydration of human stratum corneum studied *in vivo* by optothermal infrared spectometry, electrical capacitance measurement, and evaporimetry, *Acta Derm. Venereol.*, 68, 461, 1988.
14. **Nilsson, G. E.,** Measurement of water exchange through skin, *Med. Biol. Eng. Comput.*, 15, 209, 1977.
15. **Leveque, J. L., Garson, J. C., and de Rigal, J.,** Transepidermal water loss from dry and normal skin, *J. Soc. Cosmet. Chem.*, 30, 333, 1979.
16. **Tagami, H. and Yoshikuni, K.,** Interrelationship between water-barrier and reservoir functions of pathologic stratum corneum, *Arch. Dermatol.*, 121, 642, 1985.
17. **Thune, P., Nilsen, T., Hanstad, I. K., Gustavsen, T., and Lövig Dahl, H.,** The water barrier function of the skin in relation to the water content of stratum corneum, pH and skin lipids, *Acta Derm. Venereol.*, 68, 277, 1988.
18. **Werner, Y. and Lindberg, M.,** Transepidermal water loss in dry and clinically normal skin in patients with atopic dermatitis, *Acta Derm. Venereol.*, 65, 102, 1985.
19. **Baker, H.,** Experimental studies on the influence of vehicles on percutaneous absorption, *J. Soc. Cosmet. Chem.*, 20, 239, 1969.
20. **Blichmann, C. and Serup, J.,** Hydration studies on scaly hand eczema, *Contact Dermatitis*, 16, 155, 1987.
21. **Leveque, J. L., Grove, G., de Rigal, J., Corcuff, P., Kligman, A. M., and Saint Leger, D.,** Biophysical characterization of dry facial skin, *J. Soc. Cosmet. Chem.*, 82, 171, 1987.
22. **Zesch, A.,** Short and long-term risks of topical drugs, *Br. J. Dermatol.*, 115 (Suppl. 31), 63, 1986.
23. **Scheuplein, R. J. and Ross, L.,** Effects of surfactants and solvents on the permeability of epidermis, *J. Soc. Cosmet. Chem.*, 21, 853, 1970.
24. **Larmi, E., Lahti, A., and Hannuksela, M.,** Immediate contact reactions to benzoic acid and sodium salt of pyrrolidone carboxylic acid. Comparison of various skin sites, *Contact Dermatitis*, 20, 38, 1989.
25. **Batt, M. D. and Fairhurst, E.,** Hydration of the stratum corneum, *Int. J. Cosmet. Sci.*, 8, 253, 1986.
26. **Wehr, R. F. and Krochmal, L.,** Considerations in selecting a moisturizer, *Cutis*, 39, 512, 1987.
27. **Gollhausen, R. and Kligman, A. M.,** Human assay for identifying substances which induce non-allergic contact urticaria: the NICU-test, *Contact Dermatitis*, 13, 98, 1985.
28. **Soschin, D. and Leyden, J. J.,** Sorbic acid-induced erythema and edema, *J. Am. Acad. Dermatol.*, 14, 234, 1986.
29. **Lahti, A., Väänänen, A., Kokkonen, E.-L., and Hannuksela, M.,** Acetylsalicylic acid inhibits non-immunologic contact urticaria, *Contact Dermatitis*, 16, 133, 1987.
30. **Larmi, E., Lahti, A., and Hannuksela, M.,** Ultraviolet light inhibits non-immunologic immediate contact reactions to benzoic acid, *Arch. Dermatol. Res.*, 280, 420, 1988.
31. **Larmi, E., Lahti, A., and Hannuksela, M.,** Effects of infra-red and nedymium yttrium aluminum garnet laser irradiation on non-immunologic immediate contact reactions to benzoic acid and methyl nicotinate, *Dermatosen*, 37, 210, 1989.
32. **Lahti, A.,** Non-immunologic contact urticaria, *Acta Derm. Venereol.*, 60 (Suppl. 91), 1, 1980.
33. **Larmi, E., Lahti, A., and Hannuksela, M.,** Effects of sorbitan sesquioleate on non-immunologic immediate contact reactions to benzoic acid, *Contact Dermatitis*, 19, 368, 1988.
34. **Fryklöf, L.-E.,** A note on the irritant properties of sorbic acid in ointments and creams, *J. Pharm. Pharmacol.*, 10, 719, 1958.

35. **von Krogh, G. and Maibach, H. I.**, The contact urticaria syndrome — 1982, *Semin. Dermatol.*, 1, 59, 1982.
36. **Lahti, A. and Maibach, H. I.**, Immediate contact reactions, *Exogenous Dermatoses*, in press, 1989.
37. **Freitas, J. P. and Brandao, F. M.**, Contact urticaria to chlorocresol, *Contact Dermatitis*, 15, 252, 1986.
38. **Goncalo, M., Goncalo, S., and Moreno, A.**, Immediate and delayed sensitivity to chlorocresol, *Contact Dermatitis*, 17, 46, 1987.
39. **Lindskov, R.**, Contact urticaria to formaldehyde, *Contact Dermatitis*, 8, 333, 1982.
40. **Andersen, K. E. and Maibach, H. I.**, Multiple application delayed onset contact urticaria: possible relation to certain unusual formalin and textile reactions?, *Contact Dermatitis*, 10, 227, 1984.
41. **Gaul, L. E.**, Dermatitis from cetyl and stearyl alcohols, *Arch. Dermatol.*, 99, 593, 1969.
42. **Boyle, J. and Kennedy, C. T. C.**, Contact urticaria and dermatitis to Alphaderm®, *Contact Dermatitis*, 10, 178, 1984.
43. **Maibach, H. and Conant, M.**, Contact urticaria to a corticosteroid cream: polysorbate 60, *Contact Dermatitis*, 3, 350, 1977.
44. **Larmi, E., Lahti, A., and Hannuksela, M.**, Immediate contact reactions to benzoic acid and sodium salt of pyrrolidone carboxylic acid. Comparison of various skin sites, *Contact Dermatitis*, 20, 38, 1989.
45. **Emmons, W. W. and Marks, J. G., Jr.**, Immediate and delayed reactions to cosmetic ingredients, *Contact Dermatitis*, 13, 258, 1985.
46. **Nater, J. P. and de Groot, A. C.**, *Unwanted Effects of Cosmetics and Drugs Used in Dermatology*, 2nd ed., Elsevier, Amsterdam, 1985.
47. **de Groot, A. C.**, Adverse Reactions to Cosmetics, Ph.D. thesis, State University of Groningen, Groningen, The Netherlands, 1988.
48. **Hjorth, N. and Trolle-Lassen, C.**, Skin reactions to preservatives in creams with special regards to paraben esters and sorbic acid, *Am. Perfum.*, 77, 146, 1962.
49. **Braun, W.**, Iatrogene Sensibilisierung und Beinleiden, *Med. Klin. Munich*, 65, 506, 1970.
50. **Hannuksela, M., Kousa, M., and Pirilä, V.**, Contact sensitivity to emulsifiers, *Contact Dermatitis*, 2, 201, 1976.
51. **Blondeel, A. and Achten, G.**, Contact allergy in 330 dermatological patients, *Contact Dermatitis*, 4, 270, 1978.
52. **Keilig, W.**, Ein Beitrag zur Kontaktallergie gegen Lanette N, Inaugural Dissertation, University of Munich, Munich, West Germany, 1980.
53. **Keilig, W.**, Kontaktallergie auf Cetylstearylalkohol (Lanette O) als therapeutisches Problem bei Stauungsdermatitis und Ulcus cruris, *Dermatosen*, 31, 50, 1983.
54. **Broeckx, W., Blondeel, A., Dooms-Goossens, A., and Achten, G.**, Cosmetic intolerance, *Contact Dermatitis*, 16, 189, 1987.
55. **Hannuksela, M.**, Skin contact allergy to emulsifiers, *Int. J. Cosmet. Sci.*, 10, 9, 1988.
56. **Hannuksela, M. and Salo, H.**, The repeated open application test (ROAT), *Contact Dermatitis*, 14, 221, 1986.
57. **Hayakawa, R., Matsunaga, K., Suzuki, M., Arima, Y., and Ohkido, Y.**, Lipstick dermatitis due to C18 aliphatic compounds, *Contact Dermatitis*, 16, 215, 1987.
58. **de Groot, A. C., Bruynzeel, D. P., Bos, J. D., van der Meren H. L. M., van Joost, T., Jagtman, B. A., and Weyland, W.**, The allergens in cosmetics, *Arch. Dermatol.*, 124, 1525, 1988.
59. **Dooms-Goossens, A., Debusschere, K., Gladys, K., and Degreef, H.**, Contact allergy to an emulsifier in a cosmetic lotion, *Contact Dermatitis*, 18, 249, 1988.
60. **Cronin, E.**, *Contact Dermatitis*, Churchill Livingstone, Edinburgh, 1980, 149.
61. **Jones, S. K. and Kennedy, C. T. C.**, Contact dermatitis from Emulgin RO/40, an emulsifier in Hioxyl cream, *Contact Dermatitis*, 18, 108, 1988.
62. **Millard, L. F. and Barrett, P. L.**, Contact allergy from Mexenone masquerading as an exacerbation of light sensitivity, *Contact Dermatitis*, 6, 222, 1980.
63. **Svensson, Å**, Allergic contact dermatitis to laureth-4, *Contact Dermatitis*, 18, 113, 1988.
64. **Dooms-Goossens, A., Vandekerckhove, M., Verschave, H., and Degreef, H.**, Cosmetic dermatitis due to methyl glucose sesquisterarate, *Contact Dermatitis*, 10, 312, 1984.
65. **Tucker, W. F. G.**, Contact dermatitis to Eutanol G, *Contact Dermatitis*, 9, 88, 1983.
66. **Finn, O. A. and Forsyth, A.**, Contact dermatitis due to sorbitan monolaurate, *Contact Dermatitis*, 1, 318, 1975.
67. **Lawrence, C. M. and Smith, A. G.**, Ampliative medicament allergy: concomitant sensitivity to multiple medicaments including yellow soft paraffin, white soft paraffin, gentian violet and Span 20, *Contact Dermatitis*, 8, 240, 1982.
68. **Hannuksela, M., Kousa, M., and Pirilä, V.**, Allergy to ingredients of vehicles, *Contact Dermatitis*, 2, 105, 1976.
69. **Austad, J.**, Allergic contact dermatitis to sorbitan monooleate (Span 80), *Contact Dermatitis*, 8, 426, 1982.

70. **Taylor, J. S., Jordan, W. P., and Maibach, H. I.,** Allergic contact dermatitis from stearamidoethyl diethylamine phosphate: a cosmetic emulsifier, *Contact Dermatitis,* 10, 74, 1984.
71. **de Groot, A. C., Bruynzeel, D. P., van Joost, T., and Weyland, J. W.,** Cosmetic allergy from myristyl alcohol, *Contact Dermatitis,* 19, 76, 1988.
72. **Curtis, G. and Netherton, E. W.,** Cutaneous hypersensitivity to triethanolamine, *Arch. Dermatol. Syphilol.,* 41, 729, 1940.
73. **Thyresson, N., Lodin, A., and Nilzen, A.,** Eczema of the hands due to triethanolamine in cosmetic handlotions for housewives, *Acta. Derm. Venereol.,* 36, 355, 1956.
74. **Cronin, E.,** *Contact Dermatitis,* Churchill Livingstone, Edinburgh, 1980, 100.
75. **Edwards, E. K., Jr. and Edwards, E. K.,** Allergic reaction to triethanolamine stearate in a sunscreen, *Cutis,* 31, 195, 1983.
76. **Jones, S. K. and Kennedy, C. T. C.,** Contact dermatitis from triethanolamine in E45 cream, *Contact Dermatitis,* 19, 230, 1988.
77. **Emmett, E. A. and Wright, R. C.,** Allergic contact dermatitis from TEA-coco hydrolyzed protein, *Arch. Dermatol.,* 112, 1008, 1976.
78. **Neill, S. M. and Du Vivier, A.,** Contact dermatitis to trilaureth phosphate, *Contact Dermatitis,* 11, 130, 1984.
79. **Ebner, H. and Lindemayr, H.,** Ulcus cruris and allergisches Kontakt-ekzem, *Wien. Klin. Wochenschr.,* 89, 184, 1977.
80. **Hjorth, N. and Trolle-Lassen, C.,** Skin reactions to ointment bases, *Trans. St. John's Hosp. Dermatol. Soc.,* 49, 127, 1963.
81. **Bandmann, H.-J., Calnan, C. D., Cronin, E., Fregert, S., Hjorth, N., Magnusson, B., Maibach, H., Malten, K., Meneghini, C., Pirilä, V., and Wilkinson, D.,** Dermatitis from applied medicaments, *Arch. Dermatol.,* 106, 335, 1972.
82. **Fisher, A. A., Pacher, F., and Kanof, N. B.,** Allergic contact dermatitis due to ingredients of vehicles, *Arch. Dermatol.,* 104, 286, 1971.
83. **Epstein, E.,** The detection of lanolin allergy, *Arch. Dermatol.,* 106, 676, 1972.
84. **Sugai, T. and Higashi, J.,** Hypersensitivity to hydrogenated lanolin, *Contact Dermatitis,* 1, 146, 1975.
85. **Clark, E. W.,** Estimation of the general incidence of specific lanolin allergy, *J. Soc. Cosmet. Chem.,* 26, 323, 1975.
86. **Warshaw, T. G. and Herrmann, F.,** Studies of skin reactions to propylene glycol, *J. Invest. Dermatol.,* 19, 423, 1952.
87. **Agrup, G.,** Hand eczema, *Acta Derm. Venereol.,* 49 (Suppl. 61), 1969.
88. **Braun, W.,** Kontaktallergien gegen Polyaethylenglycol, Zentralbl. Haut- Geschlechtskrankheiten, 44, 385, 1969.
89. **Hannuksela, M., Pirilä, V., and Salo, O. P.,** Skin reactions to propylene glycol, *Contact Dermatitis,* 1, 112, 1975.
90. **Hannuksela, M. and Förström, L.,** Reactions to peroral propylene glycol, *Contact Dermatitis,* 4, 41, 1978.
91. **Blondeel, A., Oleffe, J., and Achten, G.,** Contact topical allergy in 330 dermatological patients, *Contact Dermatitis,* 4, 270, 1978.
92. **Oleffe, J. A., Blondeel, A., and Achten, G.,** Allergy to chlorocresol and propylene glycol in a steroid cream, *Contact Dermatitis,* 5, 53, 1979.
93. **Angelini, G. and Meneghini, C. L.,** Contact allergy from propylene glycol, *Contact Dermatitis,* 7, 197, 1981.
94. **Trancik, R. J. and Maibach, H. I.,** Propylene glycol: irritation or sensitization?, *Contact Dermatitis,* 8, 185, 1982.
95. **Hannuksela, M. and Förström, L.,** Contact hypersensitivity to glycerol, *Contact Dermatitis,* 2, 291, 1976.
96. **Maibach, H. I.,** Polyethyleneglycol: allergic contact dermatitis potential, *Contact Dermatitis,* 1, 247, 1975.
97. **Daly, B. M.,** Bactroban allergy due to polyethylene glycol, *Contact Dermatitis,* 17, 48, 1987.
98. **Edman, B. and Möller, H.,** Trends and forecasts for standard allergens in a 12-year patch test material, *Contact Dermatitis,* 8, 95, 1982.
99. **Menné, T. and Hjorth, N.,** Routine patch testing with paraben esters, *Contact Dermatitis,* 19, 189, 1988.
100. **Gollhausen, R., Enders, F., Przybilla, B., Burg, G., and Ring, J.,** Trends in allergic contact sensitization, *Contact Dermatitis,* 18, 147, 1988.
101. **Hogan, D. J., Hill, M., and Lane, P. R.,** Results of routine patch testing of 542 patients in Saskatoon, Canada, *Contact Dermatitis,* 19, 120, 1988.
102. **Jones, S. K. and Kennedy, T. C.,** Cloracetamide as an allergen in the paint industry, *Contact Dermatitis,* 18, 304, 1988.
103. **Adams, R. M. and Maibach, H. I.,** A five-year study of cosmetic reactions, *J. Am. Acad. Dermatol.,* 13, 1062, 1985.

104. **Tosti, A., Bardazzi, F., and Russo, R.**, Contact dermatitis from butylated hydroxyanisole, *Contact Dermatitis,* 17, 257, 1987.
105. **White, I. R., Lovell, C. R., and Cronin, E.**, Antioxidants in cosmetics, *Contact Dermatitis,* 11, 265, 1984.
106. **Bojs, G., Nicklasson, B., and Svensson, Å.**, Allergic contact dermatitis to propyl gallate, *Contact Dermatitis,* 17, 294, 1987.
107. **Heine, A.**, Contact dermatitis from propyl gallate, *Contact Dermatitis,* 18, 313, 1988.
108. **Brändle, I., Boujnah-Khouadja, A., and Foussereau, J.**, Allergy to castor oil, *Contact Dermatitis,* 9, 424, 1983.
109. **Sai, S.**, Lipstick dermatitis caused by castor oil, *Contact Dermatitis,* 9, 75, 1983.
110. **Sai, S.**, Lipstick dermatitis caused by ricinoleic acid, *Contact Dermatitis,* 9, 524, 1983.
111. **Andersen, K. E. and Nielsen, R.**, Lipstick dermatitis related to castor oil, *Contact Dermatitis,* 11, 253, 1984.
112. **Calnan, C. D.**, Contact dermatitis from drugs, symposium on drug sensitization, *Proc. R. Soc. Med.,* 55, 39, 1962.
113. **Fisher, A. A.**, Allergic reaction to feminine hygiene sprays, *Arch. Dermatol.,* 108, 801, 1973.
114. **Rademaker, M., Kirby, J. D., and White, I. R.**, Contact cheilitis to shellac, Lanpol 5 and colophony, *Contact Dermatitis,* 15, 3, 1986.
115. **Cronin, E.**, *Contact Dermatitis,* Churchill Livingstone, Edinburgh, 1980, 102.
116. **Pecegueiro, M., Brandao, M., and Concalo, S.**, Contact dermatitis to Hirudoid® cream, *Contact Dermatitis,* 17, 290, 1987.
117. **Calnan, C. D. and Sarkany, I.**, Studies in contact dermatitis. XII. Sensitivity to oleyl alcohol, *Trans. St. John's Hosp. Dermatol. Soc.,* 44, 47, 1960.
118. **van Joost, Th., Sillevis Smitt, J. H., and van Ketel, W. G.**, Sensitization to olive oil (olea europeae), *Contact Dermatitis,* 7, 309, 1981.
119. **van Dijk, E., Neering, H., and Vitányi, B. E. J.**, Contact hyper-sensitivity to sesame oil in patients with leg ulcers and eczema, *Acta Derm. Venereol.,* 53, 133, 1973.
120. **Hayakawa, R., Matsunaga, K., Suzuki, M., Hosokawa, K., Arima, Y., Shin, C. S., and Yoshida, M.**, Is sesamol present in sesame oil?, *Contact Dermatitis,* 17, 133, 1987.
121. **Castelain, P.-Y. and Chabeau, G.**, Contact dermatitis after trans-cutaneous electric analgesia, *Contact Dermatitis,* 15, 32, 1986.
122. **Hannuksela, M.**, Allergy to propantheline in an antiperspirant (Ercoril® lotion), *Contact Dermatitis,* 1, 244, 1975.
123. **Ågren-Jonsson, S. and Magnusson, B.**, Sensitization to propantheline bromide, trichlorocarbanilide and propylene glycol in an anti-perspirant, *Contact Dermatitis,* 2, 79, 1976.
124. **Kero, M. and Hannuksela, M.**, Guinea pig maximization test, open epicutaneous test and chamber test in induction of delayed contact hypersensitivity, *Contact Dermatitis,* 6, 341, 1980.
125. **Hannuksela, M.**, Propylene glycol promotes allergic patch test reactions, *Boll. Dermatol. Allergol. Prof.,* 2, 40, 1987.
126. **Lin, R. Y., Busher, J., Bogden, G. J. D., and Schwartz, R. A.**, Topical zinc sulfate augmentation of human delayed type skin test response, *Acta Derm. Venereol.,* 65, 190, 1985.

Chapter 16

VULVAR PHYSIOLOGY

Howard A. Oriba, Peter Elsner, and Howard I. Maibach

TABLE OF CONTENTS

I. PERCUTANEOUS ABSORPTION

In recent years there has been a growing interest in the development of transdermal drug delivery systems as an alternative to parenteral/oral drug therapy. Topical administration avoids the first pass metabolism and may provide a constant, controlled delivery of drugs. In addition, patient compliance may be enhanced, especially in the elderly. However, the flux of a drug through skin is generally limited by the stratum corneum,[1] rather than the transdermal drug delivery unit itself. Moreover, there is a differential regional penetration of various topically applied chemicals, differences in absorption being as great as 100-fold.[2] This observation continues to create much interest, and its nature remains an elusive entity. The first applied transdermal, testosterone, utilizes the scrotum to provide enhanced penetration.[3]

Using hydrocortisone as a model compound, its absorption was found to be 1% absorbed through forearm skin, 3% through the scalp, 6% through the forehead, 13% through the jaw, and 42% through scrotal skin.[4] Early work in vulvar percutaneous absorption of ^{14}C-hydrocortisone showed the vulva absorbed 7.7 ± 2.5% of the applied dose and that the forearm absorbed 1.37 ± 1.9%.[5] This enhanced percutaneous penetration of vulvar skin has probable relevance to various therapeutic agents that may be used in this region. Using a mass balance technique, the percutaneous absorption of testosterone and hydrocortisone was investigated[6] (Table 1). The vulvar flux to testosterone and hydrocortisone was significantly higher in comparison to volar forearm. Pre- and postmenopausal differences in absorption were noted at the vulva for hydrocortisone, but not for testosterone.

II. TRANSEPIDERMAL WATER LOSS AS AN INDEX OF STRATUM CORNEUM

A. FUNCTION

Vertebrates have evolved a highly specialized, stratified structure, the stratum corneum, to provide an interface to retain water. In the absence of sweating, water diffuses from the viable dermal tissues through the stratum corneum to the external environment as a result of the gradient that exists. The rate at which water diffuses across the stratum corneum is called the transepidermal water loss, or TEWL. Since TEWL is dependent upon an intact stratum corneum, it is also an *in vivo* index of the efficiency and integrity of the stratum corneum. Damage to the barrier by skin diseases or by physical or chemical trauma results in an increased TEWL from baseline values. TEWL is dependent on numerous factors, including skin temperature, relative humidity of the air, and cutaneous blood flow.[7]

Early work on TEWL of vulvar skin (Figure 1) revealed the mean TEWL to be 1.45×10^3 μg water per square centimeter per hour, while the forearm was 7.7×10^2 μg water per square centimeter per hour ($p < 0.0001$).[8] Thus, the vulva appears to be far more permeable to water than the forearm. To eliminate possible eccrine sweat contamination, subjects received subcutaneous injections of atropine in both forearm and vulva test sites. The average mean forearm TEWL was 8.7×10^2 μg/cm²/h, and the vulvar TEWL was 1.4×10^3 μg/cm²/h.[8] It should be noted that periodic bursts of markedly increased water loss were noted in a number of vulvar test sites, but none in forearm sites. It is unclear whether this increased emotional sweating in the vulva is simply a function of an increased number of eccrine sweat glands in the vulva or is an intrinsic property of vulvar tissue. Since the feeling of wetness in this area is a common complaint, we are undertaking further studies to better understand the mechanism involved and practical ramifications.

B. STRATUM CORNEUM WATER-HOLDING CAPACITY

Stratum corneum water content is a function of at least two properties: (1) hygroscopicity

TABLE 1
Percutaneous Absorption of Hydrocortisone and Testosterone (Pre- vs. Postmenopausal)

Compound	Site	Pre vs. post	% Absorbed
Hydrocortisone	Arm	Pre	2.8 ± 2.4
		Post	1.6 ± 1.2
	Vulva	Pre	8.1 ± 4.1
		Post	4.4 ± 2.8
Testosterone	Arm	Pre	20.2 ± 8.1
		Post	14.7 ± 4.2
	Vulva	Pre	25.2 ± 6.8
		Post	24.3 ± 5.4

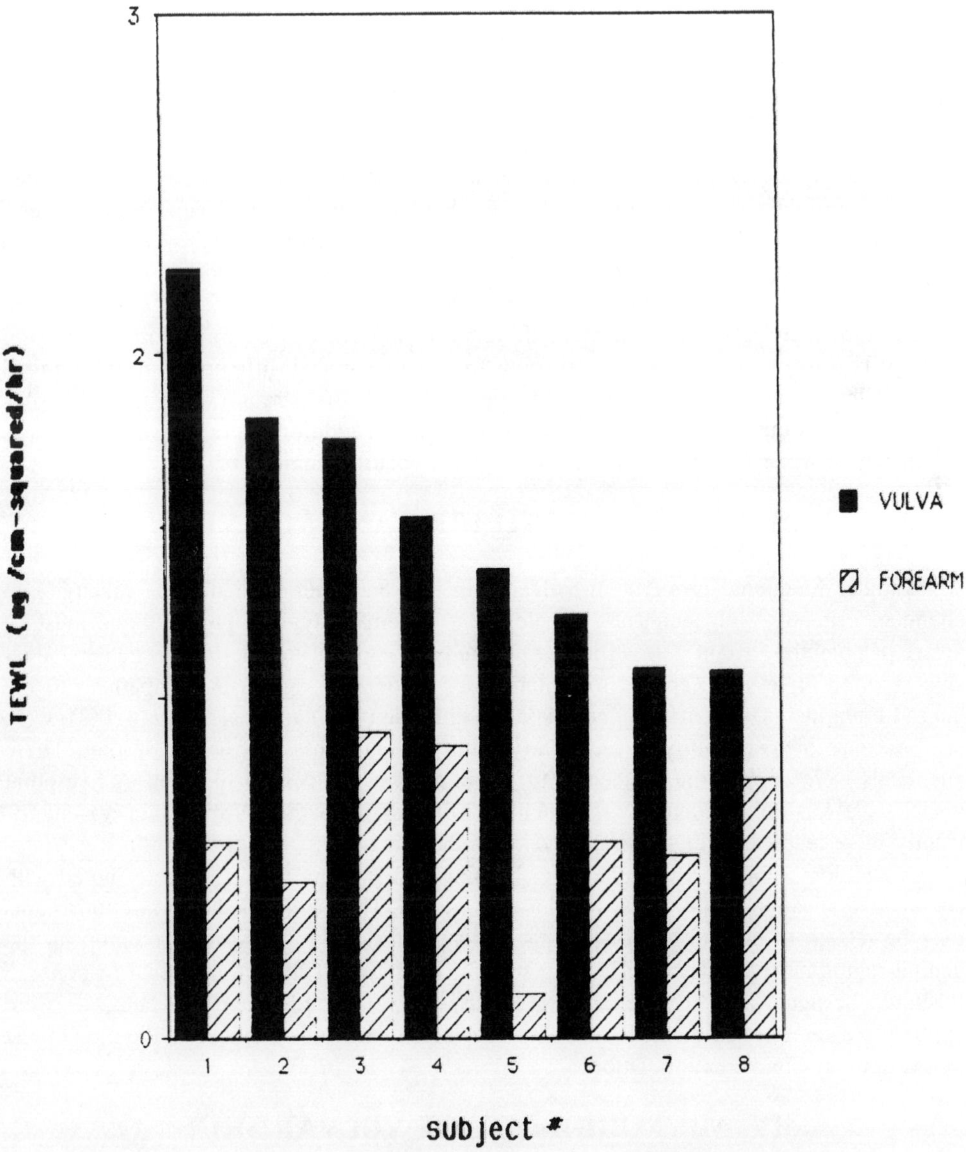

FIGURE 1. Transepidermal water loss from vulva and forearm.

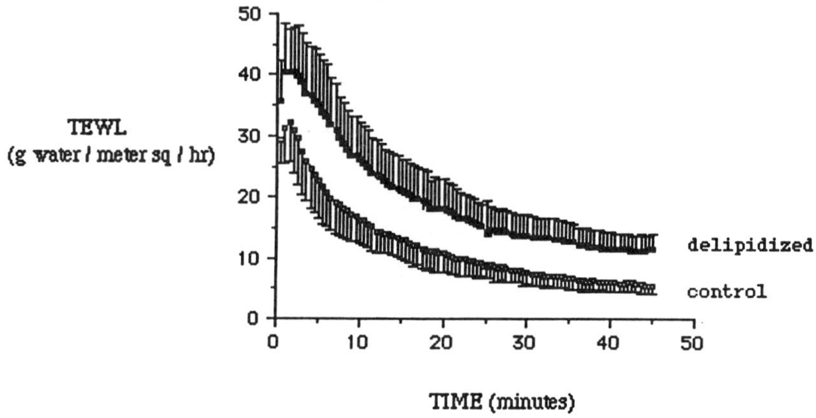

FIGURE 2. Typical transepidermal water loss decay curves fitted to biexponential equation, $y = Ae - {}^{B}1^{t} + Be - {}^{B}2^{t}$, where A and B are coefficients, B_1 and B_2 are decay constants of initial and terminal phase of decay curve, respectively, and t is time in minutes.

(i.e., the ability to take up water) and (2) the ability to retain water. Recently, we examined the behavior of vulvar stratum corneum to water loading using transepidermal decay curves.[9] Serial TEWL measurements were made of vulvar and forearm skin which were loaded with water by occluding with plastic for 4 h. These transepidermal decay curves were analyzed by curve-fitting analysis and fitted to a biexponential equation (Figure 2). In comparison to the forearm, there is no fundamental difference in the kinetics of water loss after water loading; however, upon delipidization, both sites demonstrated differences in the terminal phase of the decay curve. This probably reflects the role that stratum corneum lipids play in the water-holding capacity of the stratum corneum.[10] In addition, there were no differences in kinetics of water loss between pre- and postmenopausal women.

III. PROCLIVITY TO IRRITANT ACTIVITY

Another functional property of vulvar skin that has been studied is its reactivity to chemical irritants. With increasing development of female hygiene products one must be aware of potential irritant dermatitides. One study has sought to determine whether vulvar skin is more or less susceptible to irritation from chemical agents in comparison to the ventral forearm.[11] Two irritants, benzalkonium chloride (17%) and maleic acid (20%), were used because of their ability to induce erythema without gross destruction of skin. These chemicals were applied unoccluded at the labia majora and ventral forearm. A five-point visual scale was used to quantify the intensity of the skin reaction. Vulvar skin was significantly more reactive to these compounds than the forearm.

In addition, subjective complaints of burning and stinging were frequently noted with chemical application to the vulva, but not the forearm. This heightened response to irritants may be related to the innervation of this region of the body. The nerve receptors in the genital region have a lower threshold to stimuli, reflective of its role in sexual response.[12] It should be noted that this increased reactivity only occurred when the chemicals were applied to a carefully delineated area. This specialized skin hides erythema; thus, one must be cautious in clinical extrapolation.

IV. MICROBIOLOGY OF VULVAR SKIN

The microbial ecology of vulvar skin is distinct. Numerous reports have characterized vaginal flora, but only one study has examined normal vulvar skin flora.[13] Counts per square

TABLE 2
Bacterial Counts (Organisms/cm²) Between Vulva and Forearm Skin[13]

Organism	Vulva	Forearm
Staphylococcus aureus	4.1×10^4	1.4×10
Coagulase-negative staphylococci	5.7×10^5	1.8×10^2
Micrococci	5.1×10^5	2.9×10^2
Streptococci	3.7×10^2	0.5×10
Lipophilic diphtheroids	7.9×10^5	1.1×10^2
Nonlipophilic diptheroids	4.6×10^5	1.1×10
Lactobacillus	4.6×10^5	1.0×10
Bacillus	0	1.2×10
Gram-negative rods	1.8×10^3	0.1×10
Yeasts	8.2×10	0.8×10

centimeter of area were higher on the vulva than on the forearm for *Staphylococcus aureus,* coagulase-negative staphylococci, micrococci, streptococci, lipophilic diphtheroids, lactobacilli, gram-negative rods, and yeasts (see Table 2). Only *Bacillus* was noted to be higher in both incidence and microbial count on the forearm. Lipophilic diphtheroids were the predominant bacteria in the vulva, and yeasts were the least numerous.

The highest incidence of *S. aureus* was noted on the vulva (67%), followed by the perineum (perianal) and nose (30% each), and then the forearm (11%). The staphylococcal carriage for nose and perianal skin was comparable to previous reports.[14]

Several factors may account for these quantitative and qualitative differences. The vulva is juxtaposed between an intertriginous area and vaginal/rectal orifices. The greater density of organisms of the vulva is consistent with other partially occluded anatomical sites, such as the axilla and toe webs, where normal flora are denser.[15] The high incidence of *S. aureus* on the vulva is interesting. *S. aureus* organisms have a selective adherence to vulvar keratinocytes.[16] Since the initial event in many infectious processes is the binding of an organism to the host epithelial cell, the vulvar skin may play an important role in the pathogenesis of *S. aureus*-related infections. The dispersion of *S. aureus* from vulvar skin may be important in the epidemiology of cross infection between individuals, as well as foci of chronic staphylococcal infections. Furthermore, determination of staphylococcal carrier status should include culture of vulvar skin, in addition to the perineum and nares. The clinical implications of minimizing patient microbial disease and preventing nosocomial infections should also be considered.

V. SKIN BLOOD FLOW AND RESPONSE TO HISTAMINE

Various techniques have been used to measure skin blood flow, including ^{133}Xe washout after subcutaneous injection of the isotope.[17] An improved method using the LDV 5000 (MedPacific Corp., Seattle, WA) has the advantages of being noninvasive and capable of continuous measurement. It utilizes a low-power helium-neon laser as a light source. A probe is placed on the skin, and light striking moving red blood cells is reflected with a shift in frequency (the Doppler effect).

The mean Doppler frequency is predicted by theory to be linearly related to blood flow.[18] The mean blood flow through labia majora skin was twice that of forearm skin ($p < 0.003$).[19] Vulvar skin also showed a significant increase in blood flow after application of histamine ($p < 0.04$), while the forearm had no significant increase after this dose of histamine was applied (see Table 3). The increased permeability of vulvar skin to histamine might have been expected in light of the previous finding of enhanced penetration to hydrocortisone.

TABLE 3
Blood Flow in Forearm and Vulvar Skin[19]

Subject	Forearm	Vulva
1	40 (100)	80 (400)
2	40 (70)	142 (250)
3	75 (80)	95 (400)
4	45 (45)	160 (140)
5	40 (40)	115 (130)
6	70 (70)	103 (85)
7	70 (75)	82 (137)
8	100 (92)	90 (80)
9	80 (300)	191 (450)
10	55 (60)	173 (207)
Average	61 ± 21* (93 ± 75)	123 ± 40* (227 ± 140)

Note: *p <0.003; values in parenthesis are after histamine injection. In arbitrary units (a.u.).

Increased baseline blood flow of the vulva was not unexpected, since TEWL has been shown to be partially dependent on blood flow. Tur et al[20] extensively summarized our knowledge of regional variation in blood flow; it is of interest that the face and vulva share the properties of increased penetration of chemicals and increased blood flow.

VI. HISTOLOGY

Harper and McNicol[21] studied the age-dependent variations in normal vulvar skin at the junction between the upper one third and lower two thirds of the labium majus. Five specimens were obtained for each 5-year age group between fetal age and senescence (80 years). Looking at epidermal patterns, they found a variation with age. Epidermal ridges are absent until after puberty; then they increase in depth and are regular during childbearing age. After menopause, a gradual thinning to childhood levels can be noted. The keratin layer shows gradual thickening throughout life, but changes are not prominent . No infiltrates are present in the dermis until 20 years of age; there is then a gradual appearance of an infiltrate, comprised of a dense layer of lymphocytes, from 30 to 50 years of age. Plasma cells are mainly seen in the age group above 70 years. An increase in the diameter and an increased density of the elastic fiber network in the dermis were noted with advancing age.

Jones[22] studied sections of labia majora and labia minora at defined positions of the vulva. No significant differences between the various positions of the vulva were found. Significant age-dependent increases were also noted for keratin, epithelium thickness, dermal thickness, number of rete pegs, and parakeratosis; however, the changes in the dermal infiltrate were not observed as in previous studies. On the lateral aspects of labia minora, stratum corneum is always present.[22] For the medial side of labia minora, this was only true in 62% of the cases.

Stratum corneum replacement rate is an indirect measurement of metabolic activity of the integumentary epithelia. The rates of replacement depend on a number of factors, particularly the region of the body,[23] stratum corneum thickness, and duplication rate of the germinative cell population. The effects of the disease state and drugs on the vulvar stratum corneum replacement rate have not been thoroughly studied. Age-dependent differences have been observed in stratum corneum turnover of labia majora using the dansyl chloride method. Postmenopausal women had slower turnover rates, averaging about 14.8 ± 1.7 d compared to 9.8 ± 1.6 d for premenopausal women.[24]

VII. CYTOLOGY

Compared to cervical cytology, little work has been done on vulvar cytology until recently. Nauth et al.[25] compiled most of the present knowledge on the cytology of the normal vulva. Smears from normal vulvar skin revealed a high correlation between age and presence of nucleated keratinized squames (parakeratotic cells), which occurred most often (>50%) in the third and rarely in the eighth decade. The occurrence of parakeratotic cells is hormone dependent. An increase of parakeratosis occurs around ovulation,[26] and the local application of testosterone on vulvar skin reduces vulvar parakeratosis.[27]

VIII. CONCLUSION

The vulvar skin is a form of specialized skin possessing unique morphological and functional properties. Vulvar dermatological diseases can show a different morphology and occasionally manifest solely at this site. The vulva tends to be more permeable than other sites and have increased proclivity to irritant activity, unique microbial ecology, and increased blood flow. One must be cognizant of these differences of this specialized skin, especially when administering therapy. Hydrocortisone and testosterone, and probably plethora of other drugs, have increased absorption at this site. The nature of this increased flux to drugs remains unanswered, but hopefully will be elucidated as we learn more about vulvar skin physiology. There are age-dependent differences at both the anatomical and functional levels of vulvar skin. The vulva may serve as a model of aging, especially to evaluate the effects of sex hormone cessation. Much more work is needed for this underexplored, specialized skin.

REFERENCES

1. **Scheuplin, R. J. and Blank, I. H.,** Permeability of the skin, *Physiol. Rev.*, 51, 702, 1971.
2. **Maibach, H. I., Feldman, R. J., Milby, T. H., et al.,** Regional variation in percutaneous penetration in man, *Arch. Environ. Health*, 23, 208, 1967.
3. **Place, V.,** personal communication.
4. **Feldman, R. J. and Maibach, H. I.,** Regional variation in percutaneous penetration of ^{14}C-cortisone in man, *J. Invest. Dermatol.*, 48, 181, 1967.
5. **Britz, M. B. and Maibach, H. I.,.** Human percutaneous penetration of hydrocortisone: the vulva, *Arch. Dermatol. Res.*, 267, 313, 1980.
6. **Oriba, H. A. and Maibach, H. I.,** Age-related differences in percutaneous absorption at the vulva, in preparation.
7. **Hattingh, J.,** The influence of skin temperature, environmental temperature, and relative humidity on transepidermal water loss, *Acta Derm. Venereol.*, 48, 181, 1967.
8. **Britz, M. B. and Maibach, H. I.,** Human labia major skin: transepidermal water loss *in vivo, Acta Derm. Venereol.*, Suppl., 85, 23, 1979.
9. **Oriba, H. and Maibach, H. I.,** Vulvar transepidermal water loss decay curves: effect of site, age, and delipidization, submitted.
10. **Tagami, H. and Yoshikuni, K.,** Interrelationship between water-barrier and reservoir functions of pathologic stratum corneum, *Arch. Dermatol.*, 121, 642, 1985.
11. **Britz, M. B. and Maibach, H. I.,** Human cutaneous vulvar reactivity to irritants, *Contact Dermatitis*, 5, 375, 1979.
12. **Erickson, K. L. and Montagna, W.,** New observations on the anatomical features of the female genitalia, *J. Am. Med. Womens Assoc.*, 27, 573, 1972.
13. **Aly, R., Britz, M. B., and Maibach, H. I.,** Quantitative microbiology of human vulva, *Br. J. Dermatol.*, 101, 445, 1979.
14. **Noble, W. C. and Somerville, D. A.,** *Microbiology of Human Skin,* W. B. Saunders, Philadelphia, 1974.

15. **Aly, R. and Maibach, H. I.,** Aerobic microbial flora of intertrigenous skin, *Appl. Environ. Microbiol.,* 33, 97, 1977.
16. **Bibel, D. J., Aly, R., Lahti, I., Shinefield, H. R., and Maibach, H. I.,** Microbial adherence to vulvar epithelial cells, *J. Med. Microbiol.,* 23, 75, 1987.
17. **Henricksen, O.,** Local sympathetic reflex mechanisms in regulation of blood flow in human subcutaneous adipose tissue, *Acta Physiol. Scand. Suppl.,* 450, 1, 1977.
18. **Bonner, R. F. and Clem, T. R.,** Laser-Doppler continuous real-time monitor of pulsatile and mean blood flow in tissue microcirculation, in *Scattering Techniques Applied to Supramolecular and No-Equilibrium Systems,* Chen, S. et al., Eds., Plenum Press, New York, 1981, 685.
19. **Britz, M. and Maibach, H. I.,** Normal vulvar skin. A model for specialized skin, in *Models in Dermatology,* Vol. 1, Maibach H. I. and Lowe, N. J., Eds., S. Karger, Basel, 1985, 83.
20. **Tur, E., Tur, M., Maibach, H. I., and Guy, R.,** Basal perfusion of the cutaneous microcirculation: measurements as a function of anatomic position, *J. Invest. Dermatol.,* 81, 442, 1983.
21. **Harper, W. F. and McNicol, E. M.,** A histological study of normal vulvar skin from infancy to old age, *Br. J. Dermatol.,* 96, 249, 1977.
22. **Jones, I. S.,** A histological assessment of normal vulvar skin, *Clin. Exp. Dermatol.,* 8, 513, 1983.
23. **Jansen, L. H., Hojyo-Tomoko, M. T., and Kligman, A. M.,** Improved fluorescence staining technique for estimating turnover of the human stratum corneum, *Br. J. Dermatol.,* 90, 9, 1974.
24. **Oriba, H. A. and Maibach, H. I.,** unpublished observations.
25. **Nauth, H. F. and Boeger, A.,** New aspects of vulvar cytology, *Acta Cytol.,* 26, 1, 1982.
26. **Nauth, H. F. and Haas, M.,** Zytologische und histologische Untersuchungen zur Hormonabhaengigkeit der Haut des aeusseren weiblichen Genitales, *Geburtshilfe Frauenheilkd.,* 44, 451, 1984.
27. **Nauth, H. F.,** Zur lokalen Testosterontherapie des Lichen sclerosus et atrophicus der Vulva, *Geburtshilfe Frauenheilkd.,* 42, 476, 1982.

Chapter 17

PERCUTANEOUS ABSORPTION IN DISEASED SKIN: RELATIONSHIP TO THE EXOGENOUS DERMATOSES

Kee Chan Moon and Howard I. Maibach

TABLE OF CONTENTS

I. INTRODUCTION

Skin has many important functions, e.g., protection of the host from an external hazardous environment, prevention of water loss, and regulation of body temperature. The most important protective barriers to external chemicals or internal body water loss are believed to reside in the outermost layer of the skin, i.e., the stratum corneum.

While studies about skin absorption or transepidermal water loss (TEWL) using intact skin are abundant, little is known about the permeability of abnormal or diseased skin. In the real world, topical products are formulated to treat diseased skin, and it is most probable that pharmacokinetics and pharmacodynamics would be different in such altered skin. Therefore, there is a need for a well-characterized model for studying the physicochemical properties governing the rate of absorption of chemicals through abnormal skin.

This review is based on the sporadic data collected from the literature. As study models, some investigators have used artificially induced altered or damaged skin, while others observed penetration from clinically diseased animal or human skin. Table 1 shows categories of damaged or diseased skin models used in penetration studies. Literature data are summarized in Table 2.

II. TAPE STRIPPING, ABRASION, OR RUBBING

It is generally believed that a major barrier to water or chemicals resides in the stratum corneum. Consequently, damage affecting this layer should result in alteration of percutaneous absorption. Among the various modalities which disintegrate the physicochemical properties of the stratum corneum, removal of the layer using cellophane tape has been popular, mainly due to its simplicity, and many have studied the tape-stripping effect on the percutaneous absorption. Feldmann and Maibach[1] studied the effect of stripping and occlusion on penetration of hydrocortisone in human skin. Percentage of applied dose recovered in the urine over a period of 10 d were analyzed. Stripping the skin doubled this amount and significantly altered the absorption rate curve. Stripping and occlusion increased 32 times that of unmodified skin.

Bronaugh and Stewart[6] compared the *in vitro* permeation of compounds through skin damaged by several different methods, e.g., tape stripping and abrasion line with a hypodermic needle. Water permeation through tape-stripped rat skin was much higher (80.6%) than through intact skin (4.1%) in terms of dose absorbed. With human skin, the values were 56.9% vs. 2.7%, respectively. Needle abrasion produced an intermediate effect. Scott and Dugard,[4] in an *in vitro* rat model, used a scalpel blade and sandpaper for superficial skin alteration and also included suction blisters. These alterations increased water permeation. However, Sinha and Shaw[9] found no significant differences in the *in vivo* permeation of diflorasone through normal and scalpel-rubbed abrased guinea pig or monkey skin.

McMaster et al.[10] attempted to ascertain whether simple rubbing enhanced *in vivo* absorption. They applied cortisone in hydrophilic ointment to a retroauricular skin area in the guinea pig and on the abdomen in the rhesus monkey. However, rubbing with 100 strokes using a metal spatula caused no significant absorption increase.

III. FREEZING OR BRANDING

Recent evidence suggests that freezing skin for *in vitro* absorption studies alters its permeability. Kemppainen et al.[12] studied the absorption of T-2 toxin in fresh and frozen human and monkey skin *in vitro*. For skin of both species, the permeability rate increased twofold when stored at −60°C for no longer than 10 d. Similar results were obtained by Hawkins and Reifenrath.[11] The penetration of *N,N*-diethyl-*m*-toluamide was determined using

TABLE 1
Study Models of Altered or Diseased Skins

Mechanically injured skin models
Tape stripping
Abrasion
Rubbing
Thermally injured skin models
Freezing
Heating or branding
Chemically injured skin
Delipidization
Protein denaturation
Irritation
Allergic contact dermatitis
Ultraviolet irradiation
Clinical diseases
Animal disease model
Human skin diseases

pig skins after a freezing period of 1 to 6 weeks at −80°C. Absorption rate increased linearly with time of storage. Berenson and Burch[13] believed that the rate of water transfer through isolated skin was not altered significantly after several postmortal days even after being frozen for a period of at least 1 month and then thawed, provided deterioration of the corneum did not occur. It is probable that permeant or chemicals which are highly altered by skin metabolism may be much more subject to freezing because freezing and thawing can destroy metabolic activity of the tissue.

On the other hand, heating or branding the skin consistently increased permeation. Moderate to severe burns produce bulla, which results in denuded skin with total destruction of barrier function. For the management of the burn patient, it is essential to restore barrier function and to prevent further tissue fluid loss from inside the body or secondary infections or toxic agent from outside the body as well. Boosalis et al.[16] found markedly increased serum concentrations and urinary excretion of silver in burn patients treated with topical silver sulfadiazine. Experimentally, heating or branding has no significant effect on permeation until burn temperature reaches 70°C.[15,33] At higher temperatures an exaggerated increase in permeation rate was noted. Permeability of methanol was altered over 50 times the control rate by 150°C scalding.

IV. DELIPIDIZATION

The barrier function of the stratum corneum depends on its mosaic of lipid and protein conformation.[34] The barrier capacity of the skin should be changed by its delipidization process through extraction of the fatty materials. Organic solvents are commonly used in the extraction of lipids or in degreasing in modern industry. Such solvents range from water-miscible methanol, ethanol, and acetone to the water-immiscible aliphatic hydrocarbon hexane and various chlorohydrocarbons such as chloroform and carbon tetrachloride. Mixtures of these solvents efficiently extracted skin lipids. Several studies have dealt with skin absorption following its delipidization. The *in vitro* water permeation experiments with excised skin have shown increased rates.[13,35-38] These effects are quantitatively different with different solvents. Acetone alone was found less effective in increasing water permeation, but when mixed with diethyl ether it was much more effective.[13-35] Scheuplein and Ross[38] graded the effectiveness of delipidizing solvent on water permeation as follows: ethanol < acetone < diethyl ether < chloroform < chloroform-methanol mixture. Butanol and butyric

TABLE 2
Summary of Literature in Absorption Studies in Diseased Skin

Disease model	*In vitro/ in vivo*	Skin	Method	Penetrant	Result	Ref.
ST	*In vivo*	Human	Urine	^{14}C HC	×2	1
ST	*In vitro*	Pig mucosa	Diff cell, K_p	$^{3}H_2O$,HRPO	×2—8	2
ST	*In vivo*	Guinea pig	(×5—15), glass vessel	SA, carbinoxamine	×10, ×3	3
ST	*In vitro*	Rat	(×15), diff cell, K_p	$^{3}H_2O$	†	4
ST	*In vivo*	Rat	(×15), chamber, TEWL	H_2O	†	4
ST	*In vivo*	Rat	(×15), urine	^{3}H-mann, ^{14}C-OB	×100, ×2	5
ST	*In vitro*	Rat	(×12), diff cell	$^{3}H_2O$	×20	6
ST	*In vitro*	Human	(×12), diff cell	$^{3}H_2O$	×20	6
ST	*In vitro*	Rat	(×12), diff cell	^{3}H-Cort, ^{14}C-NA	×2.5, ×11	6
ST	*In vivo*	Human	Urine	Pesticides	×3—7	7
ST	*In vivo*	Guinea pig	(×30), glass plate	^{14}C-NA,MN,BN	†	8
ST	*In vitro*	Guinea pig	(×30), diff cell	^{14}C-NA,MN,BN	†	8
AB	*In vivo*	Monkey, rat	Rub, serum, and excreta	^{3}H-diflorasone	—	9
AB	*In vivo*	Rat	Needle, urine	^{3}H-Cort, ^{14}C-NA	×3, ×7	6
AB	*In vivo*	Guinea pig, monkey	Stroke, urine	^{14}C-Cort		10
FR	*In vitro*	Pig	−80°, diff cell	*m*-Deet	1.1—1.4	11
FR	*In vitro*	Human, monkey	−60°, diff cell	T-2 toxin	×2	12
FR	*In vitro*	Human	Water transfer	H_2O	—	13
BU	*In vitro*	Hairless mice	100—150°C, diff cell	^{14}C-Ph, ^{3}H-MtOH	×7, ×50	14
BU	*In vitro*	Hairless mice	60°C, diff cell	^{14}C-*n*-alkanols	×1.5—4	15
BU	*In vivo*	Human	Burn, serum, and urine	Silver sulfadiazine	†	16
UV	*In vivo*	Rat	UVA + UVB	^{14}C-NA	×7.5	6
UV	*In vitro*	Rat	UVA + UVB	^{14}C-NA	×2.5	6
UV	*In vitro*	Hairless mice	UVB, biopsy	^{14}C-HC	†	17
UV	*In vitro*	Hairless mice	UVC, diff cell	^{14}C-HC	†	18
UV	*In vitro*	Human	UVA, water transfer	H_2O	—	13
CH	*In vivo*	Guinea pig	pH, glass vessel	SA, carbinoxamine	v	3
CH	*In vitro*	Human	Phenol, diff cell	Phenols	†	19
CH	*In vitro*	Hairless mice	Vit A, acetic acid, diff cell	^{14}C-HC	†	18
CH	*In vitro*	Human	NaOH, KCN, water transfer	H_2O	—	13
CH	*In vivo*	Human	NaOH (pH12), Na_3PO_4, water loss	H_2O	×3.5	20
CH	*In vitro*	Human	Acetone, ether, alcohol, diff cell, water loss	H_2O	†	13
CH	*In vivo*	Guinea pig	DNCB irr & allergy, glass chamber	^{14}C-NA,MN,BN	†	8
CL	*In vivo*	Human	Atopic, ^{125}I-Cort, RIA, serum	HC	†	21
CL	*In vivo*	Human	Psoriasis, urine	^{14}C-HC, anthralin	—	22
CL	*In vivo*	Human	Psoriasis, urine	^{14}C-HC	—	23
CL	*In vivo*	Human	Psoriasis, flame photometry, TEWL	H_2O	×2—5.6	24

TABLE 2 (continued)
Summary of Literature in Absorption Studies in Diseased Skin

Disease model	*In vitro/ in vivo*	Skin	Method	Penetrant	Result	Ref.
CL	*In vitro*	Human	Psoriasis, diff cell, HPLC	Anthralins	†	25
CL	*In vivo*	Monkey	Steroid, urine	^{14}C-HC	†	26
CL	*In vitro*	Monkey	Eczema, diff cell	^{3}H-HC, TA	×2	27
CL	*In vivo*	Human	Atopic/seborrheic, ^{125}I-RIA, serum	HC	†	28
CL	*In vivo*	Human	Atopic/seborrheic, ^{125}I-RIA, serum	HC	†	29
CL	*In vitro*	Hairless mice	EFAD, humidity chamber	^{14}C-HC	×2	30
CL	*In vivo*	Hairless rat	EFAD, tissue and blood, urine	^{14}C-HC	†	31
CL	*In vivo*	Human	Mycosis fungoides, urine	^{14}C-carmustine	†	32

Note: Abbreviations are as follows: Disease model — ST: stripping, AB: abrasion, FR: freezing, BU: burn, UV: ultraviolet, CH: chemical and delipidization, CL: clinical disease; Method — diff cell: diffusion cell, urine: urine excretion of drug, K_p: permeation coefficient, TEWL: transepidermal water loss, DNCB: dinitrochlorobenzene, RIA: radioimmunoassay, HPLC: high performance liquid chromatography, UVA: ultraviolet-A, UVB: ultraviolet-B, EFAD: essential fatty acid deficiency; Penetrant — HC: hydrocortisone, HRPO: horseradish peroxidase, SA: salicylic acid, mann: mannitol, OB: octyl benzoate, Cort: cortisone, NA: nicotinic acid, MN: methyl nicotinate, BN: butyl nicotinate, m-Deet: N,N′-diethyl-*m*-toluamide, Ph: phenol, MtOH: methanol, TA: triamcinolone acetonide, H_2O: water; Result — ×: ratio of increase, †: increase, —: no change, v: variable.

acid penetration through the octanol-treated skin nearly quadrupled. *In vivo* studies are rare, and results are variable depending on the animal species and solvents used. Bucks et al.[39] found no difference in hydrocortisone topical absorption through human skin treated by trichlorethane, but a recent hairless guinea pig study showed a fivefold increase in hydrocortisone after chloroform/methanol (2:1) treatment.[46]

V. CHEMICAL IRRITATION

Most chemicals and drugs are capable of derangement of the skin to some extent when applied topically, although the nature and concentration of the chemicals may be more important factors. We have substantial evidence for the profound effect of organic solvent on skin permeation by delipidization. Since one of the basic materials of the cornified layer of epidermis consists of keratinous protein, it is conceivable that alteration of quantity and/or physicochemical state of that substance in the epidermis could alter permeation properties. Some work has been done with keratin, but results are not consistent. Berensen and Burch[13] treated with alkali (0.1 *M* NaOH up to 24 h) following extraction of the lipids from excised human skin and found no further change in the rate of water diffusion. They also observed no significant variation following addition of 0.001 *M* potassium cyanide.

In an *in vivo* human experiment, Malten and Thiele[20] studied the effect of sequential skin exposures to alkali and phosphate solutions on water loss by skin. The NaOH (pH 12) and Na_3PO_4 (pH 12) caused an increase in water loss of about 350%. They observed marked inflammation and vesicles. They presumed the effect to be due to a conformational collapse of the protein of this layer and a consequent increase in porosity by damaging the ductal epithelium of the sweat glands.

Under certain conditions phenol damages skin, leading to increased penetration rates.

Phenolic compounds appear to interact with the skin by protein denaturation. Roberts et al.[19] examined the influence of concentration and structure of various phenolic compounds on the permeability of human skin. The threshold concentration of phenol for damage was 1.5% w/v. For higher concentrations, the steady state K_p increased as a result of damage to the stratum corneum.

Solomon and Lowe[18] produced an abnormal epidermis by three different methods: ultraviolet (UV) light irradiation, topical vitamin A acid, and topical acetic acid. Topical application of retinoic acid to the skin of hairless mice produces an acanthotic epidermis with an increased mitotic index,[40] and topical application of acetic acid promotes epidermal DNA synthesis and hyperplasia.[41] *In vitro* absorption of hydrocortisone is significantly increased through those types of abnormal epidermis. These observations may be attributed to both epidermal proliferation and altered keratinization. There is evidence that results with topical application of retinoic acid also may reflect abnormalities in the phospholipid structure of the cell membranes.[42]

Osamura et al.[8] compared skin penetration of nicotinic acid compounds through several skin conditions in the guinea pig. They produced allergic contact dermatitis by sensitization with 2,4-dinitro-1-chlorobenzene (DNCB) in ethanol. The release of nicotinic acid and methylnicotinate from water and olive oil was greatest in the tape-stripped skin, followed by irritated, allergic, and normal skin condition.

A recent *in vivo* study[46] done in the hairless guinea pigs showed more than a twofold increase of hydrocortisone and benzoic acid through the sodium lauryl sulfate-irritated skin.

VI. ULTRAVIOLET IRRADIATION

UV irradiation causes various effects on the skin. In hairless mouse skin, UV light produced increased mitoses, increased DNA synthesis, and epidermal hyperplasia.[43] UV light also can alter cutaneous lipids, leading to a decrease in phospholipids which are essential as membrane components.[44]

Bronaugh and Stewart[6] observed the effect of UV irradiation on skin absorption. The backs of the rats were irradiated with a UV lamp, which produced mild and intense erythema with 1.5 and 6.0 min irradiation, respectively. Percutaneous penetration through the mild erythema resulted in a 3.2-fold increase in nicotine absorption. The severe erythema caused by 6 min irradiation resulted in 51% absorption of the applied dose.

Lamaud and Schalla[17] measured hydrocortisone absorption into the skin after ultraviolet-B (UV-B) irradiation in the hairless rat. Compared with normal skin, UV irradiation increased hydrocortisone concentrations in all skin layers examined by biopsy.

Solomon and Lowe[18] did an *in vitro* study after ultraviolet-C (UV-C) irradiation in mice and observed increased absorption of 0.1% hydrocortisone. However, Berensen and Burch[13] found no difference in the rate of water diffusion.

UV irradiation offered the possibility of inducing a defective skin-barrier function and an inflammatory reaction without mechanical perturbation of the skin. However, elevated skin temperature or blood flow may have contributed to the increased drug penetration in these experiments.

VII. CLINICAL DISEASE

A. ESSENTIAL FATTY ACID-DEFICIENT ANIMAL

Essential fatty acid deficiency (EFAD) results in epidermal hyperproliferation with acanthosis and hyperkeratosis. EFAD mice have been used as a model of experimental epidermal hyperproliferation. Solomon et al.[30] used EFAD hairless mice to study the percutaneous absorption of a 0.1% solution of 4-^{14}C-hydrocortisone. They showed significantly increased (two- to threefold) *in vitro* absorption through EFAD compared with normal hairless mouse skin.

Lambrey et al.[31] produced an EFAD rat model and studied hydrocortisone absorption. On the first day, 19% of the total applied radioactivity was recovered in the urine of the EFAD rats, while 1% was recovered from the controls.

B. PSORIASIS

Psoriasis is a proliferative and inflammatory skin disorder manifested by thick scaling patches. Histologically, it shows hyperkeratosis, parakeratosis, absence of granular layer, and prominent vasculatures in the dermal papillae. In many skin diseases, especially psoriasis, the integrity of the stratum corneum is either partially or completely damaged, with a corresponding loss of its barrier function. Several studies have dealt with absorption in the psoriatic skin.

Grice et al.[24] studied TEWL and its sodium and potassium content in normal and psoriatic skin. The TEWL range was from 2.0 to 5.6 $\mu g/cm^2/h$ in 6 psoriatic patients, while the mean TEWL in 15 normal subjects was 0.3 ± 0.1 $\mu g/cm^2/h$. The result showed a preservation of the electrolyte barrier in active psoriasis despite serious impairment of the water barrier. Wang et al.[25] investigated the release and permeation of anthralin products with use of an *in vitro* Franz diffusion cell unit. Involved psoriatic skin was much more permeable to anthralin than was uninvolved psoriatic skin.

Wester et al.,[22,23] however, reported no difference in the percutaneous absorption of hydrocortisone or dithranol between the psoriatic patients and normal subjects. They presumed that the vasoconstriction activity of the drug might have counteracted the disrupted barrier properties.

C. ATOPIC DERMATITIS

Turpeinen et al.[21] studied percutaneous absorption of hydrocortisone in 18 young adults after the acute phase of atopic dermatitis by measuring serum cortisol level using a [^{125}I]cortisol radioimmunoassay kit. Endogenous secretion of cortisol was suppressed with dexamethasone. Patients with widespread and high-grade dermatitis showed a median serum concentration of 314 nmol/l, compared to 114 nmol/l for the widespread but low-grade dermatitis and 10 nmol/l in the remission group. Turpeinen et al.[28,29] observed increased absorption of topical 1% hydrocortisone in children with severe dermatitis.

D. EXFOLIATIVE DERMATITIS

In cases of exfoliative dermatitis, a significant amount of sodium is absorbed from baths of corn starch in a 0.275 *M* solution of sodium bicarbonate.[45] Penetration of hydrocortisone is greater than 50% in this situation.[47]

E. ECZEMA

Using monkey skin *in vitro,* Bronaugh et al.[27] found that percutaneous absorption of hydrocortisone through eczematous skin was twice as high as through normal skin.

F. MYCOSIS FUNGOIDES

Zackheim et al.[32] measured *in vivo* percutaneous absorption of 1,3-bis(2-chloroethyl)-1-nitrosourea (BCNU) (Carmustine) in five patients with mycosis fungoides. Absorption through diseased skin was greater than through uninvolved skin. Increased penetration was explained as an impaired stratum corneum barrier, as represented by parakeratosis, inter- and intracellular edema, vascular dilatation, or other changes often found in lesions of mycosis fungoides.

VIII. SUMMARY

Due to the relative paucity of studies about percutaneous absorption through the altered or diseased skin, it is difficult to obtain consistent results and facts about pharmacokinetics and pharmacodynamics of topically applied drugs in such conditions. However, the general view is that in most injured or diseased skin the balance is tipped to decreased or reduced barrier state, that absorption or penetration of most chemicals or drugs is increased. Industrial workers who are dealing with chemical solvents should pay attention because toxic agents may easily penetrate through solvent-damaged skin.

As most therapeutic topical medications are applied on the "abnormal" skin, knowledge of fundamental pharmacokinetics and pharmacodynamics of chemicals through such skin is required of both physicians and patients.

Much work remains to be done with chemicals of varying properties under many dosing circumstances before we will have a clear view of the quantitative effect of damaged skin on percutaneous penetration.

REFERENCES

1. **Feldmann, R. J. and Maibach, H. I.,** Penetration of ^{14}C-hydrocortisone through normal skin: the effect of stripping and occlusion, *Arch. Dermatol.,* 91, 661, 1965.
2. **Squier, C. A. and Hall, B. K.,** *In vitro* permeability of porcine oral mucosa after epithelial separation, stripping and hydration, *Arch. Oral Biol.,* 30(6), 485, 1985.
3. **Washitake, M., Yajima, T., Anmo, T., Arita T., and Hori, R.,** Studies on percutaneous absorption of drugs. III. Percutaneous absorption of drugs through damaged skin, *Chem. Pharm. Bull.,* 21, 2444, 1973.
4. **Scott, R. C., Dugard, P. H., and Doss, A. W.,** Permeability of abnormal rat skin, *J. Invest. Dermatol.,* 86(2), 201, 1986.
5. **Scott, R. C. and Dugart, P. H.,** A model for quantifying absorption through abnormal skin, *J. Invest. Dermatol.,* 86(2), 208, 1986.
6. **Bronaugh, R. L. and Stewart, R. T.,** Methods for *in vitro* percutaneous absorption studies. V. Permeation through damaged skin, *J. Pharm. Sci.,* 74(10), 1062, 1985.
7. **Wester, R. C. and Maibach, H. I.,** Dermatopharmacokinetics: a dead membrane or a complex multifunctional viable process?, in *Progress in Drug Metabolism,* Bridges, J. W. and Chasseaud, L. F., Eds., Taylor & Francis, London, 1986, 95.
8. **Osamura, H., Jimbo, Y., and Ishihara, M.,** Skin permeation of nicotinic acid, methyl nicotinate and butyl nicotinate in the guinea pig — comparison of *in vivo* and excised skin and the effects of four dermatologic conditions, *J. Dermatol.,* 11, 471, 1984.
9. **Sinha, A. L. W., Shaw, S. R., and Weber, D. J.,** Percutaneous absorption and excretion of tritium-labeled diflorasone diacetate, a new corticosteroid in the rat, monkey and man, *J. Invest. Dermatol.,* 71, 372, 1978.
10. **McMaster, J., Maibach, H. I., Wester, R. C., and Bucks, D. A. W.,** Does rubbing enhance *in vivo* dermal absorption?, in *Percutaneous Absorption,* Bronaugh, R. I. and Maibach, H. I., Eds., Marcel Dekker, New York, 1985, 359.
11. **Hawkins, G. S., Jr. and Reifenrath, W. G.,** Development of an *in vitro* model for determining the fate of chemicals applied to skin, *Fundam. Appl. Toxicol.,* 4, 133, 1984.
12. **Kemppainen, B. W., Riley, R. T., Pace, J. G., and Hoerr, F. J.,** The effects of skin storage conditions and concentration of applied dose on [3H]T-2 toxin penetration through excised human and monkey skin, *Food Chem. Toxicol.,* 24, 221, 1986.
13. **Berenson, G. S. and Burch, G. E.,** Studies of diffusion of water through dead human skin: the effect of different environmental states and of chemical alterations of the epidermis, *Am. J. Trop. Med. Hyg.,* 31, 842, 1951.
14. **Behl, C. R., Linn, E. E., Flynn, G. L., Ho, N. F. H., Higuchi, W. I., and Pierson, C. L.,** Permeation of skin and eschar by antiseptics. II. Influence of controlled burns on the permeation of phenol, *J. Pharm. Sci.,* 72(4), 397, 1983.

15. **Behl, C. R., Flynn, G. L., Barrett, M., Walters, K. A., Lim, E. E., Mohamed, Z., Kurihara, T., Ho, N. F. H., and Higuchi, W. I.,** Permeability of the thermally damaged skin. II. Immediate influences of branding at 60°C on hairless mouse skin permeability, *Burns,* 7(6), 389, 1981.
16. **Boosalis, M. G., McCail, J. T., Ahrenholz, D. H., Solem, L. D., and McClain, C. T.,** Serum and urinary silver levels in thermal injury patients, *Surgery,* 101(1), 40, 1987.
17. **Lamaud, E. and Schalla, W.,** Influence of UV irradiation on penetration of hydrocortisone. *In vivo* study in hairless rat skin, *Br. J. Dermatol.,* 111 (Suppl. 27), 152, 1984.
18. **Solomon, A. E. and Lowe, N. J.,** Percutaneous absorption in experimental epidermal disease, *Br. J. Dermatol.,* 100, 717, 1979.
19. **Roberts, M. S., Anderson, R. A., and Swarbrick, J.,** Permeability of human epidermis to phenolic compounds, *J. Pharm. Pharmacol.,* 29, 677, 1977.
20. **Malten, K. E. and Thiele, F. A. J.,** Evaluation of skin damage. II. Water loss and carbon dioxide release measurements related to skin resistance measurements, *Br. J. Dermatol.,* 89, 565, 1973.
21. **Turpeinen, M., Mashkilleyson, N., Björksten, F., and Salo, O. P.,** Percutaneous absorption of hydrocortisone during exacerbation and remission of atopic dermatitis in adults, *Acta Derm. Venereol.,* 68, 331, 1988.
22. **Wester, R., Mobayen, M., Ryatt, K., Bucks, D., and Maibach, H. I.,** *In vivo* percutaneous absorption of dithranol in psoriatic and normal volunteers, in Psoriasis: Proc. 4th Int. Symp., Farger, E. M., Nall, L., Morhenn, V., and Jacobs, P. H., Eds., Elsevier, New York, 1987, 429.
23. **Wester, R. C., Bucks, D. A. W., and Maibach, H. I.,** *In vivo* percutaneous absorption of hydrocortisone in psoriatic patients and normal volunteers, *J. Am. Acad. Dermatol.,* 8, 645, 1983.
24. **Grice, K. A., Sattar, H., and Baker, H.,** The cutaneous barrier to salts and water in psoriasis and in normal skin, *Br. J. Dermatol.,* 88, 459, 1973.
25. **Wang, J. C., Patel, B. G., Ehmann, C. W., and Lowe, N.,** The release and percutaneous permeation of anthralin products using clinically involved and uninvolved psoriatic skin, *J. Am. Acad. Dermatol.,* 16(4), 812, 1987.
26. **Wester, R. C., Noonan, P. K., and Maibach, H. I.,** Percutaneous absorption of hydrocortisone increases with long-term administration. (*in vivo* studies in the rhesus monkey), *Arch. Dermatol.,* 116, 186, 1980.
27. **Bronaugh, R. L., Weingarten, D. P., and Lowe, N. J.,** Differential rates of percutaneous absorption through the eczematous and normal skin of a monkey, *J. Invest. Dermatol.,* 87(4), 451, 1986.
28. **Turpeinen, M.,** Influence of age and severity of dermatitis on the percutaneous absorption of hydrocortisone in children, *Br. J. Dermatol.,* 118, 517, 1988.
29. **Turpeinen, M., Salo, O. P., and Leisti, S.,** Effects of percutaneous absorption of hydrocortisone on adrenocortical responsiveness in infants with severe skin disease, *Br. J. Dermatol.,* 116(4), 475, 1986.
30. **Solomon, A. E. and Lowe, N. J.,** Percutaneous absorption in experimental epidermal proliferation, *Arch. Dermatol.,* 114, 1029, 1978.
31. **Lambrey, B., Schalla, W., Kail, N., Lahmy, J. P., and Schaefer, H.,** Influence of an essential fatty acid deficient diet on absorption of topical hydrocortisone in the rat, *Br. J. Dermatol.,* 117(5), 607, 1987.
32. **Zackheim, H. S., Feldmann, R. J., and Maibach, H. I.,** Percutaneous absorption of 1,3-bis(2-chloroethyl)-1-nitrosourea (BCNU, Carmustine) in mycosis fungoides, *Br. J. Dermatol.,* 97, 65, 1977.
33. **Behl, C. R., Linn, E. E., Flynn, G. L., Pierson, C. L., Higuchi, W. I., and Ho, N. F. H.,** Permeation of skin and eschar by antiseptics. I. Base line studies with phenol, *J. Pharm. Sci.,* 72(4), 391, 1983.
34. **Fukuyama, K., Inoue, N., Suzuki, H., and Epstein, W. L.,** Keratinization, *Int. J. Dermatol.,* 15, 473, 1976.
35. **Mali, J. W. H.,** The transport of water through the human epidermis, *J. Invest. Dermatol.,* 37, 451, 1956.
36. **Onken, H. D. and Moyer, C. A.,** The water barrier in human epidermis, *Arch. Dermatol.,* 87, 584, 1963.
37. **Matoltsy, A. G., Downes, A. M., and Sweeney, T. M.,** Studies of the epidermal water barrier. II. Investigation of the chemical nature of the water barrier, *J. Invest. Dermatol.,* 50(1), 19, 1968.
38. **Scheuplein, R. J. and Ross, L.,** Effects of surfactants and solvents on the permeability of epidermis, *J. Soc. Cosmet. Chem.,* 21, 853, 1970.
39. **Bucks, D. A. W., Maibach, H. I., Menczel, E., and Wester, R. C.,** Percutaneous penetration of hydrocortisone in humans following skin delipidization by 1:1:1 trichlorethane, *Arch. Dermatol. Res.,* 275, 242, 1983.
40. **Zil, J. S.,** Vitamin A acid effects on epidermal mitotic activity, thickness and cellurarity in the hairless mouse, *J. Invest. Dermatol.,* 59(3), 228, 1972.
41. **Slaga, T. J., Bowden, G. T., and Boutwell, R. K.,** Acetic acid, a potent stimulator of mouse epidermal macromolecular synthesis and hyperplasia but with weak tumor-promoting ability, *J. Natl. Cancer Inst.,* 55(4), 983, 1975.
42. **Aso, K., Rabinowitz, I., and Farber, E. M.,** The role of prostaglandin E, cyclic AMP and cyclic GMP in the proliferation of guinea-pig ear skin by topical application of vitamin A acid, *J. Invest. Dermatol.,* 67, 231, 1976.

43. **Du Vivier, A. and Stoughton, R. B.,** An animal model for screening topical and systemic drugs for potential use in the treatment of psoriasis, *J. Invest. Dermatol.,* 65, 235, 1975.
44. **Johnson, B. E. and Daniels, F., Jr.,** Lysosomes and the reactions of skin to ultraviolet light radiation, *J. Invest. Dermatol.,* 53(2), 85, 1969.
45. **Elliott, J. A., Jr. and Odel, H. M.,** Percutaneous absorption of sodium in cases of exfoliative dermatitis, *J. Invest. Dermatol.,* 15, 389, 1950.
46. **Moon, K. C., Wester, R. C., and Maibach, H. I.,** Diseased skin models in the hairless guinea pig: *in vivo* percutaneous absorption, *Dermatologica,* 180, 8, 1990.
47. **Maibach, H. I.,** unpublished data.

Chapter 18

TRANSDERMAL DRUG DELIVERY SYSTEMS: ADVERSE REACTIONS — DERMATOLOGIC OVERVIEW

Daniel J. Hogan and Howard I. Maibach

TABLE OF CONTENTS

I. INTRODUCTION

Research over the last decade has demonstrated that drug delivery through the skin is feasible for many simple potent drug molecules (<1000 Da) via transdermal drug delivery systems.[1] Currently, six drugs are marketed in these systems (estradiol, nitroglycerin, clonidine, scopolamine, fentanyl, and isosorbide dinitrate), and research on the transdermal delivery of other drugs is continuing.[2,3] These therapeutic systems are custom made for each drug, as drugs vary in their potency, half-life, clearance, percutaneous absorption, cost, etc.[2] These devices are discs that generally consist of a backing membrane (impermeable to the drug), a reservoir containing the drug, and a rate-controlling membrane attached to the skin with an adhesive. The clonidine and scopolamine dosage forms are multilaminates; the estradiol and nitroglycerin dosage forms use a semisolid or liquid drug reservoir.[2] The simplest systems are essentially adhesive tape with incorporated drug.

II. ADVANTAGES OF TRANSDERMAL DRUG DELIVERY SYSTEMS

Transdermal drug delivery systems allow the continuous and sustained administration of drugs through the skin. Gastrointestinal and hepatic first-pass metabolism are avoided.[1] Gastric discomfort is less than that experienced with oral administration for many medications.[2] A more even and predictable serum drug level may be achieved.[1] Steady-state plasma concentrations of a drug are achieved without the high peak levels associated with oral therapy.[1] The avoidance of high peak levels may help minimize the side effects of certain drugs. The frequency of drug administration may be reduced with transdermal drug delivery. Multiday continuous drug delivery is possible even for drugs with a short half-life.[2]

III. DISADVANTAGES OF TRANSDERMAL DRUG DELIVERY SYSTEMS

Relatively low plasma concentrations of a drug are achieved using transdermal drug delivery. The skin is an excellent barrier for some compounds, and only a low percentage of many topically applied drugs will be absorbed percutaneously.[1] The expense of transdermal drug delivery is only justified for drugs for which conventional administration has severe disadvantages.[2]

Transdermal drug delivery should improve patient compliance, but dermatologic side effects may interfere with this novel drug delivery system. Table 1 summarizes the incidence of dermatologic reactions in patients using the devices most commonly employed to date. We shall review these unwanted side effects and suggest measures to minimize the interference of dermatologic side effects with treatment.

IV. GENERAL EFFECTS OF TRANSDERMAL DRUG DELIVERY SYSTEMS ON THE SKIN

A. NONIMMUNOLOGIC

1. Effect of Occlusion

Transdermal drug delivery systems are designed for application to the skin for 1 to 7 d.[3] Discs should not be reapplied to the same site. Long-term skin occlusion inevitably causes sweat and water vapor to accumulate on the skin's surface. Evaporation of this water is inhibited. Sweat ducts also may become occluded by this sustained occlusion and heat. Sweat may break through sweat duct walls and induce a pruritic intraepidermal inflammatory reaction (miliaria rubra). Such reactions are limited to the application site. These common

TABLE 1
Contact Dermatitis in a Series of Patients Using Transdermal Drug Delivery Systems

Drug	No. patients treated	% contact dermatitis irritant/allergic	Positive oral challenge	Year	Ref.
Clonidine	35	11/31	0/4	1985	23
	29	?/38	ND	1983	24
	20	?/50	0/2	1988	22
	101	25/5	ND	1988	21
	673	?/19	1/217	1987	27
	3079	?/12.5	1/29		
Estradiol	448	24/?	ND	1987	37
Nitroglycerin	30	10/0	ND		46
Scopolamine	11	27/0	ND	1983	14
	164	0/10	ND	1989	48

Note: ND = not determined.

reactions resolve within 1 d. The experimental incorporation of hydrogels that absorb water helped minimize cutaneous irritation from a transdermal drug delivery system in one study.[4]

Occlusion creates an ideal climate for overgrowth of bacteria and yeast. Pretreatment with chlorhexidine helped minimize bacterial overgrowth under discs that contained gels that did not absorb water.[4]

2. Irritant Contact Dermatitis

Irritant contact dermatitis may occur with any transdermal drug delivery system. Individuals vary in their susceptibility to irritant contact dermatitis; those with atopic dermatitis or other active dermatitis are most susceptible. The greater the period of occlusion, the greater the risk of irritant contact dermatitis. Repetitive application of therapeutic patches to a site increases the risk of irritant contact dermatitis.

3. Transient Erythema

The removal of a therapeutic system affixed to the skin with a pressure-sensitive adhesive may cause transient erythema.[5]

4. Burns from Microwave Ovens

A report describes a man who received a second-degree burn on his chest, at the site of a nitroglycerin disc, from a microwave oven that had a leak.[6] It was assumed that the metallic element of the disc was heated by the microwave radiation. A recent study found it unlikely that exposure of the most common transdermal delivery systems to microwave diathermy fields or low level leakage radiation from microwave ovens would cause burns in users.[6a]

B. INADVERTENT TRANSFER OF MEDICATION

1. Transfer of Medication to Another Site

Several reports describe anisocoria due to scopolamine therapeutic systems. Scopolamine may be transferred to the eye from the medication disc via the patient's fingers.[7,8] Patients who insert and remove contact lens should be particularly careful when handling scopolamine discs.[8]

2. Person-to-Person Transfer

Occasionally, drug delivery systems may become dislodged inadvertently. One report describes the inadvertent transfer of a clonidine disc from a parent to a child.[9] The child developed systemic symptoms. Loss of adhesion of a transdermal system for testosterone to scrotal skin occurred frequently during the initiation of this treatment in a clinical trial.[10]

C. IMMUNOLOGIC

Allergic reactions to drug delivery systems are usually caused by a hypersensitivity in some patients to one or more compounds in the adhesive layer or to the drug itself. Cutaneous irritation from prolonged occlusion may facilitate sensitization to constituents of a disc.[11] The longer the period of occlusion the greater the cutaneous irritation induced. These conditions simulate the "maximization procedure" devised by Kligman[12] to predict the ability of a chemical to provoke allergic contact dermatitis.

V. SCOPOLAMINE

Scopolamine, the first drug approved for transdermal drug delivery, is for the prevention and treatment of motion sickness.[2,3,13,14] In contrast to the other drugs marketed in transdermal drug delivery systems, the discs are applied to the postauricular area rather than to the trunk.[13] The absorption of topically applied medications is greater through the skin of the postauricular area than through the skin of the trunk.[2,15]

A. ALLERGIC CONTACT DERMATITIS

There are 19 published cases of allergic contact dermatitis to scopolamine drug delivery systems.[16,17,48] These reactions precluded the further use of these discs by these patients and occurred in 10% of Israeli sailors after 1.5 to 15 months of use.[48] Cross-reactions to atropine have not been reported in patients with allergic contact dermatitis to scopolamine. Allergic contact dermatitis to scopolamine cream and allergic keratoconjunctivitis to scopolamine have also been reported.[16] Patients suspected of allergic contact dermatitis to scopolamine should be patch tested to 1% scopolamine in petrolatum or water.[5]

B. IRRITANT CONTACT DERMATITIS

Skin irritation may occur with both the placebo and active scopolamine patches. The incidence of skin irritation increased with the length of occlusion and was maximal at the longest period of occlusion studied (72 h).[14]

C. DELIRIUM AND OTHER DISTAL EFFECTS FROM TRANSDERMAL SCOPOLAMINE

Transdermal scopolamine is not recommended for use in children.[13] Delirium has been attributed to transdermal scopolamine in children and the elderly. Patients may not remember that these patches are medications.[18] The presence of a patch may only be discovered on physical examination. Transdermal scopolamine has been reported to precipitate glaucoma;[19,20] therefore, this device should not be used by those predisposed to this disease.

VI. CLONIDINE

Transdermal drug delivery systems for clonidine are available in the U.S. for the treatment of hypertension.[21] They are more expensive than oral clonidine and considerably more expensive than thiazide diuretics. The most common reason for discontinuing treatment with clonidine discs is local, severe skin reactions. Numerous reports document allergic contact dermatitis to transdermal clonidine;[22-29] the incidence of sensitization is up to 50%.[22] These systems are designed to be applied for 1 week, several days longer than the application period for transdermal drug delivery systems for other drugs.[3] It has been suggested, but not documented, that shortening the period between changes of the patches from 7 to 3 d may help decrease the incidence of sensitization, which is higher in women and lower in blacks.[21,25]

Rebound hypertension has been reported in elderly hypertensive patients who discontinue transdermal clonidine. One report suggested that allergic contact dermatitis to clonidine impaired the percutaneous absorption of transdermal clonidine, which led to the development of rebound hypertension in one patient.[29] Most patients with allergic contact dermatitis to transdermal clonidine developed symptoms within the first 26 weeks of treatment. An occasional patient with allergic contact dermatitis to transdermal clonidine has developed a generalized eruption.[27] Most patients with allergic contact dermatitis to clonidine who were challenged with oral clonidine experienced no flare of dermatitis.[22,26,27] The high incidence of allergic contact dermatitis to transdermal clonidine may limit its usefulness.[22,25,30] Patients suspected of having developed allergic contact dermatitis to clonidine should be patch tested to 9% clonidine in petrolatum.[27] One case of allergic contact dermatitis to polyisobutylene in the clonidine disc has been reported.[23,24]

VII. NITROGLYCERIN

The half-life of nitroglycerin (only 3 min) stimulated the development of its transdermal drug delivery system.[2] Erythema under a nitroglycerin (glyceryl trinitrate) disc is common, since this drug is a vasodilator.[5] These are case reports of allergic contact dermatitis to nitroglycerin in ointments and in transdermal drug delivery systems used to treat angina.[31-35] It is of interest that one of the first cases developed allergic contact dermatitis to nitroglycerin only after he switched from nitroglycerin ointment to a nitroglycerin patch.[31] Repetitive application of nitroglycerin patches to the same site has been emphasized as a risk factor for the development of sensitization to nitroglycerin.[32] Multifocal areas of allergic contact dermatitis to nitroglycerin in a transdermal drug delivery system with extensive postinflammatory hypermelanosis has also been reported.[33] Nitroglycerin should be patch tested at a concentration of 0.2 mg of nitroglycerin per milliliter of water.[31] One report found that patients allergic to nitroglycerin could continue to use nitroglycerin ointment if, on removal of nitroglycerin ointment, they applied a potent topical corticosteroid cream (0.05% fluocinonide) to the site of application.[35] Patients with a history of allergic contact dermatitis to nitroglycerin usually do not develop dermatitis if treated with oral glyceryl trinitrate.[31,33]

VIII. ESTRADIOL

Concern for the hepatotoxicity of oral estradiol was one of the factors that stimulated the development of estradiol in a transdermal drug delivery system.[2] The most common adverse reaction to estradiol in a transdermal drug delivery system during clinical trials was erythema and cutaneous irritation at the application sites. This occurred in 17% of the women treated and caused 2% to discontinue treatment.[36] Ethanol is present in the formulation as an enhancer of percutaneous penetration of estradiol.[3] No experimental sensitization or phototoxic reactions to estradiol in a transdermal drug delivery system were obtained in a small number of volunteers.[37] Skin irritation to transdermal estradiol is said to be more common in warm or humid climates, which accentuate the effects of cutaneous occlusion.[38] Allergic contact dermatitis has been attributed to the adhesive for this device and to hydroxypropyl cellulose, which is the major component of the drug reservoir of the transdermal estradiol patch.[39,40] It appears that there are other potential allergens in the drug reservoir.[39] Allergic contact dermatitis to the estradiol in these patches has not been reported.

IX. THE FUTURE

Transdermal drug delivery systems for coumarin, isoproterenol HCl, indomethacin, nicotine, testosterone, PHNO (a dopamine agonist), timolol, and antihistamines may be available in the near future.[2,10,41-45] Unlike other transdermal drug delivery systems, the system for testosterone is applied to scrotal skin, which has the highest rate of percutaneous absorption for steroids. In contrast to the therapeutic system which is designed to deliver micrograms of estradiol, the transdermal system for testosterone is designed to deliver milligrams of testosterone. Transient local pruritus has been reported in a few of the men treated with the placebo delivery system for testosterone.[10]

Improved predictive tests for the potential development of allergic contact dermatitis to drugs in these systems are needed. Standard predictive tests in guinea pigs and conventional clinical trials had indicated that clonidine was not a contact allergen, but many users of clonidine in transdermal drug delivery systems developed allergic contact dermatitis to clonidine.[41] Better definition of the relationship between length of occlusion and risk of sensitization is indicated.

The incorporation of agents in transdermal drug delivery systems to enhance percutaneous penetration of drugs may increase the range of drugs that can be delivered by these systems, but dermatologic reactions to these enhancers of percutaneous absorption could limit their usefulness. The role of topical steroids in the suppression of dermatitis to transdermal drug delivery systems needs to be better defined, especially for patients who require continued treatment with a therapeutic device. Improved patient education to minimize reapplication of these devices to the same site would help reduce the risk of sensitization. These therapeutic systems will remain of interest to dermatologists and nondermatologists alike.

REFERENCES

1. **Guy, R. H., Hadgraft, J., and Bucks, D. A. W.,** Transdermal drug delivery and cutaneous metabolism, *Xenobiotica,* 17, 325, 1987.
2. **Shaw, J. E., Cramer, M. P., and Gale, R.,** Rate-controlled transdermal therapy utilizing polymeric membranes, in *Transdermal Delivery of Drugs,* Vol. 1, CRC Press, Boca Raton, FL, 1987, 102.
3. **Brown, L. and Langer, R.,** Transdermal delivery of drugs, *Annu. Rev. Med.,* 39, 221, 1988.
4. **Hurkmans, M., Bodde, H. E., van Driel, L. M. J., van Doorne, H., and Junginger, H. E.,** Skin irritation caused by transdermal drug delivery systems during long-term (5 days) application, *Br. J. Dermatol.,* 112, 461, 1985.
5. **Fisher, A. A.,** Dermatitis due to transdermal therapeutic systems, *Cutis,* 34, 526, 1984.
6. **Murray, K. B.,** Hazards of microwave ovens to transdermal delivery system, *N. Engl. J. Med.,* 310, 721, 1984.

6a. **Mosley, H., Johnston, S., and Allen, A.,** The influence of microwave radiation on transdermal delivery systems, *Br. J. Dermatol.,* 122, 361, 1990.

7. **McCrary, J. A., III and Webb, N. R.,** Anisocoria from scopolamine patches, *J. Am. Med. Assoc.,* 243, 353, 1982.
8. **Carlston, J. A.,** Unilateral dilated pupil from scopolamine disk, *J. Am. Med. Assoc.,* 248, 31, 1982.
9. **Reed, M. T. and Hamburg, E. L.,** Person-to-person transfer of transdermal drug-delivery systems: a case report, *N. Engl. J. Med.,* 314, 1120, 1986.
10. **Bals-Pratsch, M., Yoon, Y. D., Knuth, U., and Nieschlag, E.,** Transdermal testosterone substitution therapy for male hypogonadism, *Lancet,* 2, 943, 1986.
11. **Neiboer, C., Bruynzeel, D. P., and Boorsma, D. M.,** The effect of the skin with transdermal therapeutic system on Langerhans' cells and the induction of skin irritation, *Arch. Dermatol.,* 123, 1499, 1987.
12. **Kligman, A. M.,** The identification of contact allergens by human assay. III. The maximization test: a procedure for screening and rating contact sensitizers, *J. Invest. Dermatol.,* 47, 393, 1966.
13. Transderm-V, CIBA-GEIGY, in *Compendium of Pharmaceuticals and Specialities,* 24th ed., Krogh, C. M. E., Ed., Canadian Pharmaceutical Association, Ottawa, Ontario, 1989, 1059.

14. **Homick, J. L., Kohl, R. L., Reschke, M. F., Degioanni, J., and Cintron-Trevino, N.,** Transdermal scopolamine in the prevention of motion sickness: evaluation of the time course of efficacy, *Aviat. Space Environ. Med.*, 54, 994, 1983.
15. **Scheuplein, R. J. and Fronaugh, R. L.,** Percutaneous asorption, in *Biochemistry and Physiology of the Skin,* Goldsmith, L. A., Ed., Oxford University Press, New York, 1983, 1255.
16. **Trozak, D. J.,** Delayed hypersensitivity to scopolamine delivered by a transdermal device, *J. Am. Acad. Dermatol.,* 13, 247, 1985.
17. **van der Willigen, A. H., Oranje, A. P., Stolz, E., and van Joost, T.,** Delayed hypersensitivity to scopolamine in transdermal therapeutic systems, *J. Am. Acad. Dermatol.,* 18, 146, 1988.
18. **MacEwan, G., William, R., Noone, R. A., and Joseph, A.,** Psychosis due to transdermally administered scopolamine, *Can. Med. Assoc. J.,* 133, 4331, 1985.
19. **Hamill, M. B., Suelflow, J. A., and Smith, J. A.,** Transdermal scopolamine delivery system (TRANS-DERM-V) and acute angle-closure glaucoma, *Ann. Ophthalmol.,* 15, 1011, 1983.
20. **Fraunfelder, F. T.,** Transdermal scopolamine precipitating narrow-angle glaucoma, Letter to the Editor, *N. Engl. J. Med.,* 307, 1079, 1982.
21. **Catapress-TTS,** in Physicians' Desk Reference, 42nd ed., Huf, B. B., Ed., Medical Economics Company, Oradell, NJ, 1988, 719.
22. **Horning, J. R., Zawada, E. T., Jr., Simmons, J. L., Williams, L., and McNulty, R.,** Efficacy and safety of two-year therapy with transdermal clonidine for essential hypertension, *Chest,* 93, 941, 1988.
23. **Groth, H., Vetter, H., Kneusel, J., and Better, W.,** Allergic skin reactions to transdermal clonidine, *Lancet,* 1, 850, 1983.
24. **Groth, H., Vetter, H., Knuesel, J., Baumgart, P., and Vetter, W.,** Transdermal clonidine in essential hypertension, in *Problems During Long-Term Treatment. Low Dose Oral and Transdermal Therapy of Hypertension,* Weber, M. A., Drayer, J. I. M., and Kolloch, R., Eds., Steinkopff Verlag, Darmstadt, West Germany, 1985, 60.
25. Discussion, in Problems during Long-Term Treatment. Low Dose Oral and Transdermal Therapy of Hypertension, Weber, M. A., Drayer, J. I. M., and Kolloch, R., Eds., Steinkopff Verlag, Darmstadt, West Germany, 1985, 66.
26. **Grattan, C. E. H. and Kennedy, C. T. C.,** Allergic contact dermatitis to transdermal clonidine, *Contact Dermatitis,* 19, 225, 1985.
27. **Maibach, H.,** Oral substitution in patients sensitized by transdermal clonidine treatment, *Contact Dermatitis,* 16, 1, 1987.
28. **White, T. M. and Guidry, J. R.,** Rebound hypertension associated with transdermal clonidine and contact dermatitis, *West. J. Med.,* 145, 104, 1986.
29. **Metz, S., Klein, C., and Morton, N.,** Rebound hypertension after discontinuation of transdermal clonidine therapy, *Am. J. Med.,* 82, 17, 1987.
30. **Mougeolle, J. M.,** Allergies medicamenteuses: cas particulier des systemes therapeutiques transdermiques, *Nouv. Dermat.,* 7 (Suppl. 4), 511, 1988.
31. **Rosenfeld, A. S. and White, W. B.,** Allergic contact dermatitis secondary to transdermal nitroglycerin, *Am. Heart J.,* 108, 1061, 1984.
32. **Fischer, R. G. and Tyler, M.,** Severe contact dermatitis due to nitroglycerin patches, *South. Med. J.,* 78, 1523, 1985.
33. **Harari, Z., Sommer, I., and Boleslaw, K.,** Multifocal contact dermatitis to nitroderm TTS 5 with extensive postinflammatory hypermelanosis, *Dermatologica,* 174, 249, 1987.
34. **Topaz, O. and Abraham, D.,** Severe allergic contact dermatitis secondary to nitroglycerin in a transdermal therapeutic system, *Ann. Allergy,* 59, 365, 1987.
35. **Hendricks, A. A. and Dec, W.,** Contact dermatitis due to nitroglycerin ointment, *Arch. Dermatol.,* 115, 853, 1979.
36. Estraderm, CIBA, in *Compendium of Pharmaceutical and Specialities,* 24th ed., Krogh, C. M. E., Ed., Canadian Pharmaceutical Association, Ottawa, Ontario, 1989, 379.
37. **Utian, W. H.,** Transdermal estradiol overall safety profile, *Am. J. Obstet. Gynecol.,* 156, 1336, 1987.
38. **Judd, H., Ravnikar, V., Utain, W. H., and Lawrence, W.,** Transdermal estradiol overall safety profile, panel discussion III, *Am. J. Obstet. Gynecol.,* 156, 1338, 1987.
39. **McBurney, E. I., Noel, S. B., and Collins, J. H.,** Contact dermatitis to transdermal estradiol system, *J. Am. Acad. Dermatol.,* 20, 503, 1989.
40. **Schwartz, B. K. and Clendenning, W. E.,** Allergic contact dermatitis from hydroxypropyl cellulose in a transdermal estradiol patch, *Contact Dermatitis,* 18, 106, 1988.
41. **Shaw, J. E., Prevo, M. E., and Amkraut, A. A.,** Testing of controlled-release transdermal dosage forms, *Arch. Dermatol.,* 123, 1548, 1987.
42. **Ritschel, W. A. and Barkhaus, J. K.,** Use of sorption promoters to increase systemic absorption of coumarin from transdermal drug delivery systems, *Arzneim. Forsch.,* 38, 1774, 1988.

43. **Patel, R. A. and Vasavada, R. C.,** Transdermal delivery of isoproterenol HCL: an investigation of stability, solubility, partition coefficient, and vehicle effects, *Pharm. Res.*, 5, 116, 1988.
44. **Chien, Y. W., Xu, H. L., Chians, C. C., and Huang, Y. C.,** Transdermal controlled administration of indomethacin. I. Enhancement of skin permeability, *Pharm. Res.*, 5, 103, 1988.
45. **Lang, A. E.,** Additional comments on Parkinson's disease, *Can. Med. Assoc. J.*, 140, 513, 1989.
46. **Kapoor, A. S., Dang, N. S., and Reynolds, R. D.,** Sustained effects of transdermal nitroglycerin in patients with angina pectoris, *Clin. Pharmacol. Ther.*, 7, 674, 1985.
47. **Hogan, D. J. and Maibach, H. I.,** Adverse reactions to transdermal drug delivery systems, *J. Am. Acad. Dermatol.*, 22, 811, 1990.
48. **Gordon, D. R., Shupak, A., Doweck, I., and Spitzer, O.,** Allergic contact dermatitis caused by transdermal hyoscine, *Br. Med. J.*, 298, 1220, 1989.

Chapter 19

SUSCEPTIBILITY TO SODIUM LAURYL SULFATE-INDUCED IRRITANT DERMATITIS: RELATION TO BIOPHYSICAL SKIN PROPERTIES

Klaus-P. Wilhelm and Howard I. Maibach

TABLE OF CONTENTS

I. INTRODUCTION

Contact dermatitis accounts for the major source of disability in occupational dermatology and to the citizen at large.[1,2] Although at least one half of these cases are considered to be of an irritant type, relatively little is known about its pathophysiology.[1-3] By definition, acute (primary) irritant dermatitis, in contrast to allergic contact dermatitis, is elicited at first exposure in most of the exposed population.[4] However, this is only true for strong irritants at high concentrations. Irritant thresholds and dose responses vary considerably among individuals when tested with milder irritants.[5-8] Relatively little is known about endogenous factors that contribute to this variability. Since irritation changes skin physiology in many ways, we evaluated whether certain biophysical parameters of apparently healthy skin were related to the vulnerability of skin to the model irritant sodium lauryl sulfate (SLS). We measured skin surface pH, pH in deeper layers of stratum corneum (SC), sebum content, baseline transepidermal water loss (TEWL), stratum corneum turnover time (SCTT), and water-holding capacity of SC (TEWL decay, post-24-h occlusion) in a study population of young, healthy Caucasian female volunteers. These parameters were compared with the intensity of experimentally induced irritant dermatitis by simple linear correlation and factor analysis.

II. EXPERIMENTAL METHODS

A. STUDY POPULATION

Ten healthy Caucasian women between 29 and 42 years of age (31.1 ± 1.3) were the volunteer subjects after having given their written informed consent. The study had been approved by the University of California-San Francisco Committee for Human Research.

B. APPLICATIONS

Sodium lauryl sulfate (0.1 ml of a 1% solution; Sigma, St. Louis, MO) in distilled water was applied to premarked test sites on the midvolar forearm and on the upper back for 24 h using occlusive plastic chambers (Hilltop Laboratory, Cincinnati, OH). Water served as vehicle control. The patches were fixed with nonocclusive paper tape (Scanpore, Norgeplaster, Oslo, Norway).

C. MEASUREMENTS

TEWL was measured with an Evaporimeter (EP1, Servo Med, Stockholm, Sweden) 24 h after patch removal. This instrument uses the method of vapor pressure gradient calculation described in detail by Nilsson.[9] The probe was held in place for each measurement until a stable TEWL value had been established (approximately 30 s). Skin surface pH was measured at three different sites on the volar forearm with a flat surface electrode for pH measurements (Model 91-35, Orion Research, Boston, MA) which was connected to a pH meter (Model 130, Corning, Medfield, MA). Measurements were repeated after removal of superficial SC layers by five tape strips (Scotch® 600, 3M, St. Paul, MN). SCTT was determined using the dansyl chloride staining method.[10] Briefly, stratum corneum proteins are stained with the fluorescent dansyl chloride and are subsequently examined daily with ultraviolet illumination. SCTT is defined as the time (in days) from staining until disappearance of fluorescence.

SC water-holding capacity was evaluated by occluding midvolar forearm skin for 24 h with a polypropylene plastic chamber lacking a cotton pad. After removal, TEWL was recorded in 1-min intervals for 45 min.

Since forearm sebum levels were often below the sensitivity of the instrument, the upper back (lower one third of the scapula) was chosen for this parameter. Measurements were performed in six replicas adjacent to the SLS patch with a Sebumeter (SM 810 PC, Courage & Khazaka, Cologne, West Germany). The method used by this appliance is based on photometry of a special plastic film which becomes transparent by lipids.[11] Before measurement, all volunteers had been resting at room temperature for at least 15 min. All measurements were performed under constant ambient conditions (20 ± 2°C; 45 to 65% relative humidity).

D. STATISTICS

Differences in TEWL were tested for their significance with the Student's *t*-test for paired samples.[12] The area under the curve (AUC) for TEWL post-occlusion was calculated by integrating the computer-derived curves of best fit. The relationship between skin baseline parameters and SLS-induced TEWL increase was studied by simple linear correlation[12] and by factor analysis.[13] Factors were extracted by their principal components and transformed by the Orthotran/Varimax procedure. The number of factors to be extracted was determined by the 75% variance rule.[13] All statistical computations were performed with a statistics program (StatView512$^+$, BrainPower, Inc., Calabasas, CA) on a personal computer (Macintosh SE, Apple Computer, Cupertino, CA).

III. RESULTS

pH values are means of three replicas per individual. Skin surface values (5.5 ± 0.1; means ± SEM) decreased to 5.1 ± 0.1 after removal of superficial SC layers by five tape strips. The difference between surface pH and pH after five tape strips tended to be greater with higher surface values (Figure 1). SCTT, as measured by fluorescence of forearm skin stained with dansyl chloride, was 19.3 ± 0.8 d (data not shown).

A TEWL of 51.8 $g/m^2/h$, 1 min after removal of the occlusive chamber, was the result of 24 h occlusion. Pretreatment values were reached approximately 45 to 60 min post-occlusion. A typical curve for TEWL decay over time is shown in Figure 2. The TEWL decay parameters, first minute value, and AUC are summarized in Table 1.

Sebum levels on the volar forearm were often below the sensitivity of the employed method. Therefore, sebum content was measured at six sites on the upper back adjacent to the SLS patch application (Table 2). The correlation coefficients between TEWL after SLS treatment and pretreatment baseline parameter are shown in Table 3. Factor analysis showed a high load of skin surface pH, pH after five tape strips, and TEWL of SLS-treated skin on factor 1 which might be interpreted as an individual's irritability factor (Table 4). Baseline TEWL, first minute TEWL, and the AUC TEWL post-occlusion loaded high on factor 2, which might correlate with SC hydration and SC water-binding capacity. SCTT and sebum levels were not included in the factor analysis since they were measured at different anatomic sites or in different individuals, respectively.

Of all measured parameters, only skin surface pH ($r = 0.66$; $p \leq 0.02$) and skin pH after five tape strips ($r = 0.81$; $p \leq 0.005$) correlated significantly with TEWL after SLS exposure (Figure 3).

IV. DISCUSSION

Skin irritation in man exhibits great variability, ranging from no reaction in some individuals to severe blistering or corrosive reactions in others.[5-8,14] Relatively little is known about individual factors that contribute to this variability. Atopics are generally considered to have a higher risk of developing irritant dermatitis later in life.[15] This greater susceptibility has also been demonstrated experimentally.[16] Other risk factors for developing irritant dermatitis derived from retrospective epidemiologic studies are female sex, young age, and fair

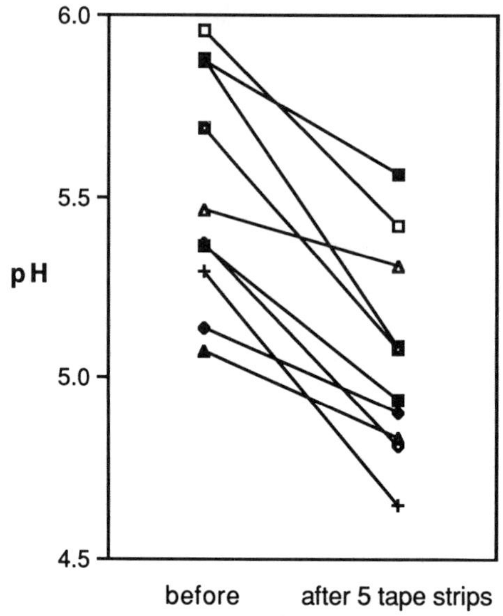

FIGURE 1. Skin surface pH before and after removal of superficial SC layers by five tape strips (data represent means of three measurements, $n = 10$ volunteers). After removal of superficial stratum corneum by five tape strips, skin pH decreased 0.5 units on average. The decrease tended to be higher for individuals with high baseline values.

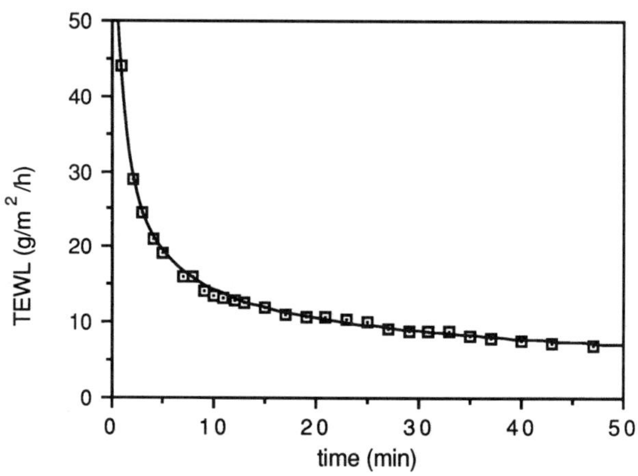

FIGURE 2. Typical TEWL deconvolution curve after 24 h occlusion of volar forearm skin (32-year-old female). This figure illustrates the exponential TEWL deconvolution profile. After 45 min, baseline TEWL is almost reached.

skin (reviewed in References 3 and 17). However, retrospective data inherently include a population bias and lack controlled exposure to the agent of question; i.e., it is unknown whether it is the female sex characteristic or the work or hobby activity that is responsible for the higher risk for females than males. Therefore, it is not surprising that these risk factors are often questioned and are especially difficult to reproduce in a controlled experimental approach.[18-21] Lammintausta and colleagues[28] induced irritant dermatitis by two

TABLE 1
Skin Baseline Parameter and TEWL after Sodium Lauryl Sulfate (SLS)

Person	TEWL			pH		TEWL post-occlusion	
	Pre	Control	1% SLS	Surf	Strip	1 min	AUC
1	3.9	4.6	12.1	5.88	5.09	42.5	9.5
2	4.9	5.4	6.5	5.14	4.90	40.5	7.9
3	3.5	3.7	13.7	5.69	5.08	62.4	13.1
4	3.8	3.5	8.4	5.37	4.81	62.2	10.5
5	5.5	7.8	9.8	5.36	4.94	41.0	14.1
6	4.7	4.3	17.4	5.96	5.42	43.3	10.6
7	4.2	5.4	10.4	5.07	4.83	50.4	14.9
8	2.9	3.8	16.4	5.46	5.31	50.3	11.4
9	5.5	6.0	12.4	5.87	5.56	76.7	17.5
10	2.8	2.5	3.6	5.29	4.65	48.7	10.8
Mean	4.2	4.7	11.1*	5.51	5.06*	51.8	12.0
SEM	0.3	0.5	1.4	0.10	0.09	3.7	0.9

Note: TEWL ($g/m^2/h$) of volar forearm skin pretreatment (pre) (mean of four replicas), 24 h after 24-h application of 1% SLS or water (control); pH values at skin surface (surf) and after five tape strips (strip) are means of three replicas per individual. TEWL after 24 h occlusion: 1 min = first minute TEWL value, AUC = area under the curve; representing water loss in the first hour (g/m^2). * Significantly different from control: $p \leq 0.05$ (paired *t*-test).

TABLE 2
Sebum Content and TEWL on the Upper Back

Subject	TEWL[a]			
	Pre	Control	1% SLS	Sebum[b]
1	3.5	4.4	29.1	14.0
2	3.5	3.2	52.3	10.2
3	3.2	4.7	45.1	6.8
4	1.3	1.4	6.0	12.7
5	2.5	2.5	46.5	12.5
6	3.9	3.7	37.5	9.7
7	4.4	5.6	44.3	11.2
8	3.1	3.6	41.3	16.0
9	3.9	4.5	38.5	57.3
10	4.3	4.2	43.0	31.7
Mean ± SEM	3.4 ± 0.3	3.7 ± 0.3	38 ± 4.1*	18.2 ± 4.8

Note: TEWL measured 24 h after 24-h sodium lauryl sulfate application was increased tenfold on the upper back; pre = pretreatment; Note the high interindividual variability factor 100) in casual sebum content. *Significantly different from control; $p \leq 0.05$ (paired *t*-test).

[a] TEWL ($g/m^2/h$).
[b] $\mu g/cm^2$ (mean of six measurements).

TABLE 3
Correlation Matrix

	TEWL 1	TEWL 2	pH 1	pH 2	1st min TEWL	TEWL AUC	SCTT	Sebum
TEWL 1	1							
TEWL 2	0.067	1						
pH 1	0.154	0.658*	1					
pH 2	0.348	0.814**	0.769**	1				
1st min TEWL	0.035	0.113	0.273	0.356	1			
TEWL AUC	0.375	0.207	0.128	0.353	0.633*	1		
SCTT	0.149	0.282	n.d.	n.d.	n.d.	n.d.	1	
Sebum	0.287	0.004	n.d.	n.d.	n.d.	n.d.	n.d.	1

Note: TEWL 1 — baseline TEWL (mean of four replicas; forearm except for correlation with sebum [upper back]); TEWL 2 — TEWL 24 h after 24-h SLS exposure (forearm except for comparison with sebum [upper back]); pH 1 — pH on skin surface; pH 2 — pH after five tape strips; 1st min TEWL (TEWL 1 min after 24-h occlusion); TEWL AUC (area under this curve for TEWL after 24-h occulusion). SCTT — stratum corneum turnover (dansyl chloride method); sebum content (upper back; mean of six replicas). Level of significance: $*p \leq 0.05$; $**p \leq 0.005$; not significant unless otherwise indicated; n.d. = not determined.

TABLE 4
Factor Analysis

Orthogonal Transformation Solution-Varimax

	Factor 1	Factor 2
Basal TEWL	0.142	0.448
pH	0.879	0.118
pH post-strip	0.898	0.342
1 min POST	0.116	0.798
1 h water 1 . . .	0.077	0.918
1% SLS d3	0.909	0.050

Note: Factor analysis of skin baseline parameters and TEWL of SLS-irritated skin. Factors were extracted by their principal components and transformed by the Orthotran/Varimax Procedure.[13] The number of factors to be extracted was determined by the 75% variance rule.[13] Factor analysis showed a high load of skin surface pH, pH after five tape strips, and TEWL of SLS-treated skin on factor 1, which might be interpreted as an individual's irritability factor. Baseline TEWL, first minute TEWL, and the AUC TEWL post-occlusion loaded high on factor 2, which might correlate with SC hydration and SC water-binding capacity.

approaches: single patch application or repeated open application. The responses were evaluated visually by TEWL, skin blood flow, and SC water content measurements, showing no sex-related differences.[22] Their results and the study by Goh and Chia,[20] who observed lower baseline TEWL in males than in females but no sex-related differences in SLS-irritated skin, supported an earlier study by Björnberg.[18] He tested 21 pairs of men and women with

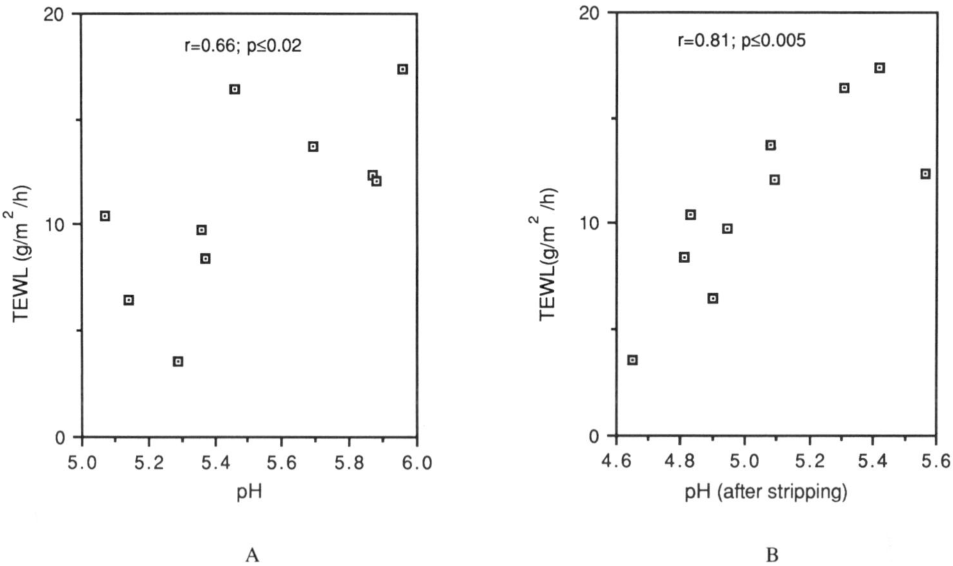

FIGURE 3. pH of untreated volar forearm skin (A) and pH after removal of superficial SC layers by five tape strips (B) significantly correlated with SLS-induced TEWL increase (n = 10 volunteers). (From Wilhelm, K. P. and Maibach, H. I., Susceptibility to SLS-induced irritant dermatitis: relation to skin pH, TEWL, sebum concentration, and stratum corneum turnover time, *J. Am. Acad. Dermatol.*, accepted for publication, 1990. With permission.)

hand eczema, matched for age, and 21 pairs of noneczematous subjects with an array of 13 primary irritants at various concentrations.[18] With a few exceptions, no significant differences between the sexes were demonstrable.

Even less is known about endogenous factors (other than atopic traits) that might contribute to the great variability in irritant reaction in a comparatively homogeneous study population, e.g., young, healthy white women.[7] Irritation changes several measurable biophysical skin properties. SLS, a well-studied model irritant, increases TEWL, skin blood flow, SCTT, skin pH, water-holding capacity of SC, and, initially, SC water content.[7,19,23,24] Our objective has been to determine whether baselines of those parameters changed by an irritant correlated themselves to the severity of the reaction induced by that irritant. No significant correlation between TEWL before and after SLS patch application was seen in the present study. This is in confirmation with earlier studies from this and other laboratories,[20,27,28] but in contrast with findings by Murahata and collaborators[25] and Tupker and co-workers.[26] Both groups found a significant correlation between baseline TEWL and TEWL after SLS application. However, in their studies, SLS was applied cumulatively for 5 d, several hours per day. This difference in experimental design might explain the controversial observations. Additionally, we measured TEWL 1 d after SLS patch removal to avoid artifacts caused by the increased hydration, a measure not followed by the other groups.

The role of skin pH in skin function has been discussed controversially since it was first measured by Schade and Marchionini.[30] The term "acid mantle", which was coined by them namely because of false (too low) values (pH 1.8 to 5.0), suggested that an acid skin surface is a better barrier against microbial influences. Later, Locher[31] suggested that higher skin pH was due to swollen keratin and reflected a less effective barrier against external influences. More recently, high skin pH was correlated to higher TEWL values and reduced surface lipid concentrations.[32] However, no influence of the pH of different soap or detergent solutions on their irritant potential was noted in the range of pH 4.0 to 10.5.[5,29]

In the present study, a significant correlation was observed between skin surface pH and, even more strongly, of skin pH after five tape strips with the severity of SLS-induced dermatitis. The fact that pH after five tape strips correlated more strongly with irritation suggests that not only the surface film (composed of triglycerides, phospholipids, esterified cholesterols, and other materials released by holocrine sebaceous glands as well as salts and water released by eccrine sweat glands which normally cover the surface and are commonly referred to as the "acid mantle"), but the chemical composition of SC itself, roughly characterized by its pH, determines its barrier capabilities. The observed correlation between skin pH and SLS-induced TEWL increase might be explained by lower lipid concentration[13] and/or more hydrated SC.[14] Both lower lipid concentration[13] and more hydrated SC[14] have been related to higher skin pH, and both would weaken the skin barrier against the hydrophilic SLS.

TEWL relaxation after 24 h occlusion was a more sensitive parameter than baseline TEWL in demonstrating (1) differences between young and old persons[33] and (2) subclinical changes induced by SLS.[23] In our experiments neither parameter of this functional assay (first minute AUC) provided additional information for characterizing sensitive individuals.

Factor analysis is a procedure commonly used in psychology and social sciences to elucidate which parameter(s) best reflect certain multifactorial determined variables, e.g., the wealth of a nation. In this investigation factor analysis substantiated our observations by demonstrating a high load of skin surface pH, pH after five strips, and TEWL of SLS-irritated skin on factor 1. This factor might be discussed as an individual's irritability factor. Baseline TEWL and TEWL post-occlusion did load very low on factor 1, but loaded high on factor 2, which might be defined as the SC water-binding capacity factor. Since SLS-induced dermatitis is characterized by rough, dry skin, we compared the sebum content in an area with usually high sebum levels with the susceptibility to irritant dermatitis in the same area (upper back). The reproducibility of sebum measurements with six replicas was satisfactory. However, a great interindividual variability (factor 10) was observed. Despite these marked interindividual differences, sebum content did not correlate to the magnitude of SLS-induced dermatitis.

In conclusion, of the parameters under investigation, only skin pH was significantly correlated to an individual's susceptibility to SLS-induced dermatitis. Since pH reflects chemical composition of SC, further research on SC composition may help explain proclivity to SLS irritation.

V. SUMMARY

Considerable interindividual differences in susceptibility to irritant dermatitis are seen in experimental skin toxicology as well as in clinical practice. This study questions whether certain baseline values of biophysical skin parameters influenced by irritation relate to differences in the susceptibility to irritant dermatitis.

The following parameters of apparently healthy skin were measured: surface pH, pH in deeper layers of stratum corneum (SC) (after removal of superficial SC by five tape strips), sebum content, stratum corneum turnover time (SCTT) baseline transepidermal water loss (TEWL), and SC water-holding capacity determined by TEWL decay, post 24 h occlusion. Subsequently, the severity of irritant dermatitis induced by 24 h application of the model irritant sodium lauryl sulfate (SLS) quantified by TEWL was compared with the parameters of apparently healthy skin.

Of the parameters under investigation only skin surface pH ($r = 0.66$; $p \leq 0.02$) and, even more strongly, pH in deeper layers of the SC ($r = 0.81$; $p \leq 0.005$) correlated with the severity of SLS-induced dermatitis as quantified by TEWL. Evaluation of the data by factor analysis showed a high load of surface pH, pH after five strips, and TEWL of SLS-treated skin on factor 1 (which may be defined as an irritability factor), whereas baseline

TEWL and TEWL post-occlusion loaded high on factor 2 (SC hydration factor). The finding that in healthy volunteers the pH in deeper layers of SC more closely correlated to the severity of irritant dermatitis than the surface pH suggests that the chemical composition of SC roughly characterized by pH is of greater importance for the skin barrier function than skin surface hydrogen ion concentration.

ACKNOWLEDGMENTS

KPW is the recipient of a postdoctoral fellowship from the Deutsche Forschungsgemeinschaft (DFG, Bonn, West Germany), which is gratefully acknowledged.

REFERENCES

1. **Andersen, K. E., Benezra, C., Burrows, D., et al.,** Contact Dermatitis. A review, *Contact Dermatitis,* 16, 55, 1987.
2. **Mathias, C. G. T. and Morrison, J. H.,** Occupational skin disease, United States. Results from the bureau of labor statistics annual survey of occupational injuries and illnesses, 1973 through 1984, *Arch. Dermatol.,* 124, 1519, 1988.
3. **Wilhelm, K. P. and Maibach, H. I.,** Susceptibility to SLS-induced irritant dermatitis: relation to skin pH, TEWL, sebum concentration, and stratum corneum turnover time, *J. Am. Acad. Dermatol.,* accepted for publication, 1990.
4. **Wilkinson, J. D. and Rycroft, R. J. G.,** Contact dermatitis, in *Textbook of Dermatology,* Vol. 1, 4th ed., Rook, A., Wilkinson, D. S., Ebling, F. J. G., Champion, R. H., and Burton, J. L., Eds., Blackwell Scientific, Oxford, 1986, 435.
5. **Frosch, P. J.,** Irritancy of soaps and detergent bars, in *Principles of Cosmetics for the Dermatologist,* Frost, P. and Horwitz, S. N., Eds., C. V. Mosby, St. Louis, 1982, 5.
6. **Frosch, P. J. and Wissing, C.,** Cutaneous sensitivity to ultraviolet light and chemical irritants, *Arch. Dermatol. Res.,* 272, 269, 1982.
7. **Wilhelm, K. P., Surber, C., and Maibach, H. I.,** Quantification of sodium lauryl sulfate irritant dermatitis in man: comparison of four techniques: skin color reflectance, transepidermal water loss, laser Doppler flow measurement and visual scores, *Arch. Dermatol. Res.,* 281, 293, 1989.
8. **Marshall, E. K., Jr., Lynch, V., and Smith, H. W.,** On dichlorethylsulphide (mustard gas). II. Variations in susceptibility of the skin to dichlorethylsulphide, *J. Pharmacol. Exp. Ther.,* 12, 291, 1919.
9. **Nilsson, G. E.,** Measurement of water exchange through skin, *Med. Biol. Eng. Comput.,* 15, 209, 1977.
10. **Jansen, L. H., Hojyo-Tomoko, M. T., and Kligman, A. M.,** Improved fluorescence staining technique for estimating turnover of the human stratum corneum, *Br. J. Dermatol.,* 90, 9, 1974.
11. **Dikstein, S., Hartzshtark, A., Bercovici, R., et al.,** Distribution of sebum measurement in normal adult woman (abstract), 4th Int. Symp. Bioengineering and the Skin, Besancon, France, September 1983.
12. **Zar, J. H.,** *Biostatistical Analysis,* Prentice-Hall, Englewood Cliffs, NJ, 1974, 121.
13. **Gorsuch, R. T.,** *Factor Analysis,* Lawrence Erlbaum Associates, Hilldale, NJ, 1983.
14. **Frosch, P. J. and Kligman, A. M.,** Rapid blister formation in human skin with ammonium hydroxide, *Br. J. Dermatol.,* 96, 461, 1977.
15. **Rystedt, I.,** Hand eczema in patients with history of atopic manifestations in childhood, *Acta Derm. Venereol.,* 65, 305, 1985.
16. **Gollhausen, R., Göttsberger, K., Winter, H., et al.,** Skin sensitivity in patients with atopic eczema and normals. Evaluation by visual assessment, evaporimeter and laser Doppler flowmeter, *J. Invest. Dermatol.,* 90 (Abstr.), 241, 1988.
17. **Lammintausta, K. and Maibach, H. I.,** Exogenous and endogenous factors in skin irritation, *Int. J. Dermatol.,* 27, 213, 1988.
18. **Björnberg, A.,** Skin reactions to primary irritants in men and women, *Acta Derm. Venereol.,* 55, 191, 1975.
19. **Elsner, P., Wilhelm, D., and Maibach, H. I.,** Irritation on vulvar and forearm skin, submitted.
20. **Goh, C. L. and Chia, S. E.,** Skin irritability to sodium lauryl sulphate — as measured by skin water vapour loss — by sex and race, *Clin. Exp. Dermatol.,* 13, 16, 1988.

21. **Berardesca, E. and Maibach, H. I.,** Racial differences in sodium lauryl sulphate induced cutaneous irritation: black and white, *Contact Dermatitis,* 18, 65, 1988.
22. **Holst, R. and Möller, H.,** One hundred twin pairs patch tested with primary irritants, *Br. J. Dermatol.,* 93, 145, 1975.
23. **Berardesca, E. and Maibach, H. I.,** Effect of nonvisible damage on the water-holding capacity of the stratum corneum, utilizing the plastic occlusion stress test (POST), in *Current Topics in Contact Dermatitis,* Frosch, P. J., Dooms-Goosens, A., Lachapelle, J.-M., Rycroft, R. J. G., and Sheper, R. J., Eds., Springer-Verlag, Heidelberg, 1989, 554.
24. **Wilhelm, K. P., Saunders, J. C., and Maibach, H. I.,** Increased stratum corneum turnover induced by subclinical irritant dermatitis, *Br. J. Dermatol.,* 1990, in press.
25. **Murahata, R. I., Crowe, D. M., and Roheim, J. R.,** The use of transepidermal water loss to measure and predict the irritation response to surfactants, *Int. J. Cosmet. Sci.,* 8, 225, 1986.
26. **Tupker, R. A., Coenraads, P. J., Pinnagoda, J., et al.,** Baseline transepidermal water loss (TEWL) as a prediction of susceptibility to sodium lauryl sulphate, *Contact Dermatitis,* 20, 265, 1989.
27. **Freeman, S. and Maibach, H. I.,** Study of irritant contact dermatitis produced by repeat patch testing with sodium lauryl sulphate (SLS) and assessed by visual methods, transepidermal water loss and laser doppler, *J. Am. Acad. Dermatol.,* 19, 496, 1988.
28. **Lammintausta, K., Maibach, H. I., and Wilson, D.,** Susceptibility to cumulative and acute irritant dermatitis. An experimental approach in human volunteers, *Contact Dermatitis,* 19, 84, 1988.
29. **Murahata, R. I., Toton-Quinn, R., and Finkey, M. B.,** Effect of pH on the production of irritation in a chamber irritation test, *Am. Acad. Dermatol.,* 18, 62, 1988.
30. **Schade, H. and Marchionini, A.,** Der Säuremantel der Haut (nach Gaskettenmessungen), *Klin. Wochenschr.,* 7, 12, 1928.
31. **Locher, G.,** Permeabilitätsprüfung der Haut Ekzemkranker und Hautgesunder für den neuen Indikator Nitrazingelb ''Geigy'', Modifzierung der Alkaliresistenzprobe, pH-Verlauf in der Tiefe des stratum corneum, *Dermatologica,* 124, 159, 1962.
32. **Thune, P., Nilsen, T., Hanstad, I. K., et al.,** The water barrier function of the skin in relation to the water content of stratum corneum, pH and skin lipids. The effect of alkaline soap and syndet on dry skin in elderly, non-atopic patients, *Acta Derm. Venereol.,* 68, 277, 1988.
33. **Roskos, K. V.,** Assessment of skin barrier function using transepidermal water loss — effect of age, *Pharm. Res.,* 6, 949, 1989.

Chapter 20

REGULATION OF PROSTAGLANDIN E_2 AND INOSITOL PHOSPHATE METABOLISM IN CULTURED HUMAN KERATINOCYTES

Richard B. Crook, WeiDong Huang, Jon R. Polansky, and Ernest Bloom

TABLE OF CONTENTS

I. INTRODUCTION

Studies of the mechanisms by which irritants and inflammatory mediators produce alterations in the skin are currently investigating actions on individual cell types and specific biochemical pathways. Since responses to inflammatory mediators often develop over several hours, the time courses for the individual cellular and biochemical changes may play an important role. The current chapter considers both initial and delayed effects of a variety of agents on human keratinocytes which may provide information regarding exogenous dermatoses, using changes in prostaglandin E_2 (PGE_2) and inositol phosphate (IP) metabolism as biochemical measures of cellular responses.

The ability to propagate keratinocytes in serial cell culture has made it possible to investigate a number of potentially important responses of this unique cell type.[1-5] Although the mechanisms by which the keratinocyte responds to different stimuli remain to be defined, it is likely that second messenger signals produced after exposure of the cells to irritants and drugs play a role in the mediation of inflammatory and wound healing responses. Therefore, understanding of these signals may aid development of therapeutic protocols.

Over the past few years we have been investigating the release of regulators from the eicosanoid pathway (prostaglandins [PGs] and other arachidonic acid metabolites; see References 6 to 9 and Chapter 12) and more recently have been examining the stimulation of IP formation, a key step in the inositol phosphate/diacylglycerol (IP/DAG) signaling pathway. PGs and other products of arachidonic acid metabolism are believed to be potent local regulators, or autacoids, in the skin and other tissues. The IP/DAG pathway has been described in numerous tissues to mediate such functions as secretion,[10] ion transport,[11] and mitosis.[12] Both PGs and the IP/DAG pathway have been shown to respond to inflammatory mediators and may be important markers of the inflammatory response. Interest in the IP/DAG pathway in the skin is also evidenced by the reports of Talwar et al.[13,14] and Fisher et al.,[15] who demonstrated that a variety of inflammatory mediators stimulate the IP/DAG pathway in keratinocytes. Stimulation of PG synthesis and stimulation of IP/DAG metabolism can occur through separate pathways, but are related to one another through a common stimulation of arachidonic acid.[16] PGs also can be produced by activation of cell membrane phospholipases and by cell injury.

Since effects on PGE_2 and IPs may provide separate sources of evidence for the effects of irritants and inflammatory mediators, experiments from the two pathways in cultured human keratinocytes are reported.

II. MATERIALS AND METHODS

A. CHEMICALS

Histamine, bradykinin, [D-Phe^7]bradykinin, arginine-vasopressin, norepinephrine, angiotensin II, α-melanocyte stimulating hormone, prostaglandin $F_{2\alpha}$, PGE_2, substance P, carboxypeptidase A, trypsin, papain, plasmin, hirudin, leupeptin, soybean trypsin inhibitor, diphenhydramine, tetradecahoylphorbolacetate (TPA), H_2O_2, and calcium ionophore A23187 were from Sigma Chemical Company (St. Louis, MO). Ethylphenyl propiolate (EPP) was from Aldrich Chemical Company (Milwaukee, WI). Metiamide was a gift from Ms. Julie Christie of Smith, Kline and French Laboratories (King of Prussia, PA). Thrombin and diisofluorophosphate treated thrombin (DIP-thrombin) were kind gifts from Dr. Marc Shuman (Cancer Research Institute, University of California-San Francisco. [^{3}H]-myoinositol (10 to 20 Ci/mmol) was from American Radiolabelled Chemicals (St. Louis, MO). AG 1X8 ion exchange resin was from BioRad (Richmond CA). All other chemicals were reagent grade from Fisher Scientific (Pittsburgh, PA).

B. CELL ISOLATION AND CULTURE

Human keratinocytes were propagated from normal skin specimens obtained from surgeries performed for reasons unrelated to skin diseases. The skin specimens were washed in calcium-magnesium-free phosphate buffered saline (PBS-CMF) and dissected into 4-mm-wide strips. The strips were then incubated in 625 U/ml Type I collagenase (CLS I; Worthington Incorporated, Freehold, NJ) overnight. After collagenization, the tissue pieces were washed in PBS-CMF and the epidermis removed as an intact sheet using sterilized forceps. The epidermal sheets were then briefly exposed to a trypsin solution (STV — 0.025% trypsin, 0.01% EDTA, and 0.15 *M* NaCl in 50 m*M* phosphate buffer, pH 7.4) and gently dounced until the solution turned cloudy, indicating release of cells. The STV was then neutralized with calcium and soybean trypsin inhibitor, and the mixture was centrifuged at 200 × g for 1 min to pellet the cells. The supernatant was discarded and the pellet raised in medium and plated using either tissue culture dishes or flasks at a density of approximately 50,000 cells per square centimeter. The primary cultures of keratinocytes were considered ready for trypsinization and subculture or freezing when they were approximately one half confluent. Cultures were cryopreserved by standard methods in liquid nitrogen and recovered as needed for experiments. The culture medium for the cells was keratinocyte growth medium (KGM; Clonetics Corporation, San Diego, CA). All experiments were performed on second and third passage cells.

C. PGE_2 ASSAYS

Media samples were collected at the time periods indicated and PGE_2 levels determined by radioimmunoassay (as described previously; see Reference 6). Briefly, PGE_2 was measured using rabbit anti-prostaglandin E_2 antibody (Sigma Chemical, St. Louis, MO), with a lower limit of sensitivity of approximately 5 pg/ml. Media samples were diluted appropriately to give prostaglandin concentrations in the range of the standard curve (5 to 250 pg/ml). The intra-assay coefficients of variation were <10%. Cross-reactivities of the antibody with other prostaglandins were <1%.

D. INOSITOL PHOSPHATE MEASUREMENT

Cultured keratinocytes were labeled with 15 μCi/ml [^{3}H]-myoinositol for 20 to 26 h at 37°C. The cells were then changed to KGM medium containing 10 m*M* $LiCl_2$[17] and incubated 5 min at 37°C. Bioactive compounds were added in 100× concentrated solutions and the cells incubated at 37°C for different periods of time. To terminate the incubations, the medium was aspirated and the cell monolayers were lysed with 100% methanol and scraped off the surface with a rubber policeman, and the inositol phosphates extracted as described by Schacht.[18] Ion exchange chromatography by the method of Berridge[19] was carried out as described previously.[20] Total inositol phosphates (inositol monophosphate [$InsP_1$], inositol bisphosphate [$InsP_2$], and inositol triphosphate [$InsP_3$] were eluted from 1.2-ml AG-1X8 columns with 8 ml of 1 *M* $HCOONH_4$/0.1 *M* HCOOH. $InsP_1$, $InsP_2$, and $InsP_3$ were eluted individually as described.[20] Sample (1 ml) was added to 7 ml of Scintiverse II (Fisher Scientific) and counted in a Packard scintillation counter with 35% counting efficiency. Elution characteristics of $InsP_1$, $InsP_2$, and $InsP_3$ were determined using tritiated standards (Amersham, Arlington Heights, IL).

III. RESULTS

To explore the role of irritants and specific mediators on keratinocytes (Figure 1), it was found that (1) release of PGE_2 in the cell culture media was an effective measure of the effects of these agents over time and (2) stimulation of the IP/DAG pathway was useful for evaluating direct mechanisms and inhibitor effects.

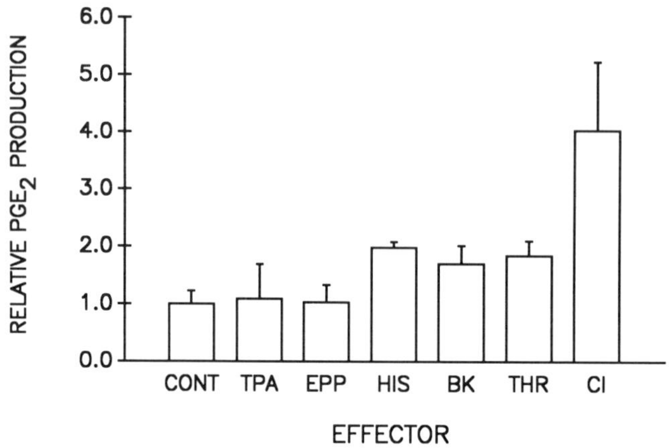

FIGURE 1. Survey of the ability of various effectors to stimulate PGE_2 secretion in human keratinocytes. Media were sampled 30 min after exposure to irritants as described in Section II. Data represent averages of duplicate samples whose values are indicated by error bars. CONT = control, TPA = 1 μ*M* tetradecanoyl phorbol acetate, EPP = 100 μ*M* ethylphenyl propiolate, HIS = 100 μ*M* histamine, BKN = 1 μ*M* bradykinin, THR = 3 U/ml thrombin, CI = 10 μ*M* calcium ionophore A23187.

A. PROSTAGLANDIN (PG)E_2 RESPONSES TO IRRITANTS AND INFLAMMATORY MEDIATORS

Figure 1 shows the production of PGE_2 0.5 h after addition of concentrations of TPA (1 μ*M*), EPP (100 μ*M*), histamine (100 μ*M*), bradykinin (1 μ*M*), thrombin (3 U/ml), and calcium ionophore (10 μ*M*), which were known to be maximal stimulating concentrations in other cell types examined. Only small stimulations (10 to 25%) were observed with TPA and EPP, while histamine, bradykinin, and thrombin all showed 1.8- to 2-fold stimulations. The greatest effect was seen with calcium ionophore (CI), which produced a fourfold stimulation.

Since a 1- to 2-h delay in PG stimulation after treatment with TPA and EPP in the mouse ear model was observed, it was next examined whether cultured human keratinocytes would show a delay in their PGE_2 response to these irritants. Interestingly, the keratinocytes showed a time delay consistent with the *in vivo* mouse ear data (see Figures 2A and B). The time-course effects of other irritants, mediators, and potentially toxic agents are compared with the TPA and EPP data in Table 1. H_2O_2 (100 μ*M*) injury to the keratinocyte resulted in a substantial decrease in (and later recovery of) PGE_2 production; higher H_2O_2 concentrations (>100 μ*M*) also produced an initial drop in PGE_2 with no recovery due to cell death. By contrast, calcium ionophore (0.1 μ*M*) treatment resulted in a very large initial stimulation of PGE_2 production followed by cell death after 2 h, with minimal PGE_2 production after this period. Histamine, thrombin, and fetal calf serum (FCS) showed progressive increases with time relative to control (similar to TPA), while bradykinin did not increase PGE_2 production after its initial stimulatory effect. It is possible that the agents which produce delayed progressive effects may do so by stimulating multiple pathways. The reasons for the different effects of bradykinin compared to histamine and thrombin in the later time periods are not known, but may involve effects on separate pathways or toxic effects.

B. STIMULATION OF INOSITOL PHOSPHATE METABOLISM AND EFFECTS OF SPECIFIC ANTAGONISTS

Based on the PGE_2 data and on earlier pilot studies, it appeared that histamine, bradykinin, and thrombin may stimulate on the IP/DAG pathway; this pathway may serve as

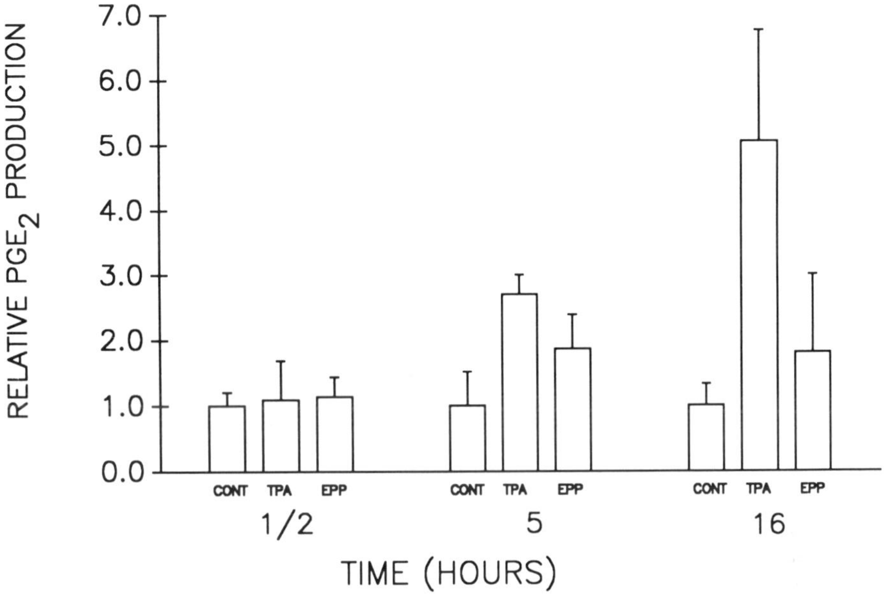

A

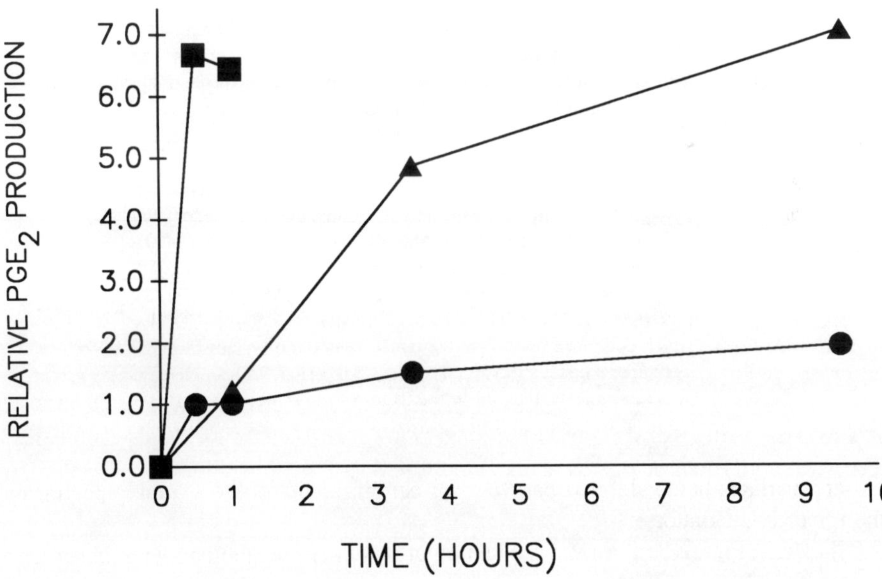

B

FIGURE 2. Stimulation of PGE_2 by irritants as a function of time. (A) Cultured keratinocytes were exposed to either 1 μM TPA or 100 μM EPP, and media samplings were taken at 0.5, 1, and 2 h. (B) Time course of PGE_2 production after exposure to 10 μM calcium ionophore A23187 (■) compared to TPA (▲) and untreated control dishes (●). 10 μM A23187 resulted in cell death (see text).

TABLE 1
Irritant and Stimulator Effects on Prostaglandin E_2 Production in Cultured Human Keratinocytes: Time Course of the Responses

	Early (30 min)	Delayed (2—5 h)	Prolonged (16—24 h)
TPA	0	+1	+3
EPP	0	+1	+1
H_2O_2	−1	−½	0
CI	+4	—	—
HIS	+1	+3	+3
BK	+1	+1	+1
THR	+1	+4	+3
FCS	0	+4	+4
CONT	0	0	0

Note: TPA (1 μ*M*), EPP (100 μ*M*), H_2O_2 (100 μ*M*), calcium ionophore A23187 (CI, 10 μ*M*), histamine (HIS, 100 μ*M*), bradykinin (BK, 1 μ*M*), thrombin (THR, 3 U/ml), and fetal calf serum (FCS, 10%) were added to second-passage human foreskin keratinocytes, with PGE_2 levels determined within the time period evaluated in our experimental studies. The results of responses to the various agents were graded on a relative scale (+1 = 0.5- to 2-fold increase, +2 = 2- to 3-fold increase, +3 = 3- to 6-fold increase, +4 = 6- to 15-fold increase; negative numbers represent a similar scale for full decrease relative to control). Concentrations of H_2O_2 or EPP greater than those shown were uniformly toxic to the cells and showed cell death. CI treatment of the cells produced cell death after the first 30-min period, and there was essentially no further production of PGE_2 by the cells with incubations >2 h with this agent.[15] Platelet activating factor (PAF) at 100 n*M*, a concentration 100 times that required for maximum stimulation of eye endothelial cells in our laboratory, produced only a +0.5 stimulation of the keratinocyte cultures; the data from Fisher et al.[15] suggesting that maximal stimulations are not seen until 1 μ*M* PAF require further investigation. The values shown are relative to control (CONT), which is defined as "0" in the table. The control values of PGE_2 were approximately 0.5 ng/ml in the 0.5-h period, with an increase to approximately 1 ng/ml in the 2- to 5-h time period and 1 to 5 ng/ml in the 16 to 24-h time period.

a rapid and perhaps better-defined pathway for certain mediators to explore specific mechanisms for these mediators.

As shown in Figure 3, 2- to 2.5-fold stimulations in IP metabolism were observed with histamine, bradykinin, and thrombin and 4-fold stimulations with calcium ionophore. TPA and EPP had little effect in the 30-min incubation time. These effects appeared similar to those observed using PGE_2 measurements and are in general agreement with recent reports by Talwar et al.[13,14] Although these researchers did not present CI data, they reported a stimulation of IP release in response to increased Ca^{2+} in the keratinocyte medium. The observed stimulation by CI also suggests a role for Ca^{2+} in regulating IP/DAG metabolism, but also may represent nonspecific membrane damage. Additional agonists known to stimulate IP metabolism in other target cells (carbachol, norepinephrine, vasopressin, angiotensin II, α-melanocyte stimulating hormone [α-MSH], prostaglandin $F_{2\alpha}$, or substance P, all at 1 μ*M*) were also screened, with little or no observed response (<50%).

Certain effectors which showed clear stimulation of the IP/DAG pathway were examined

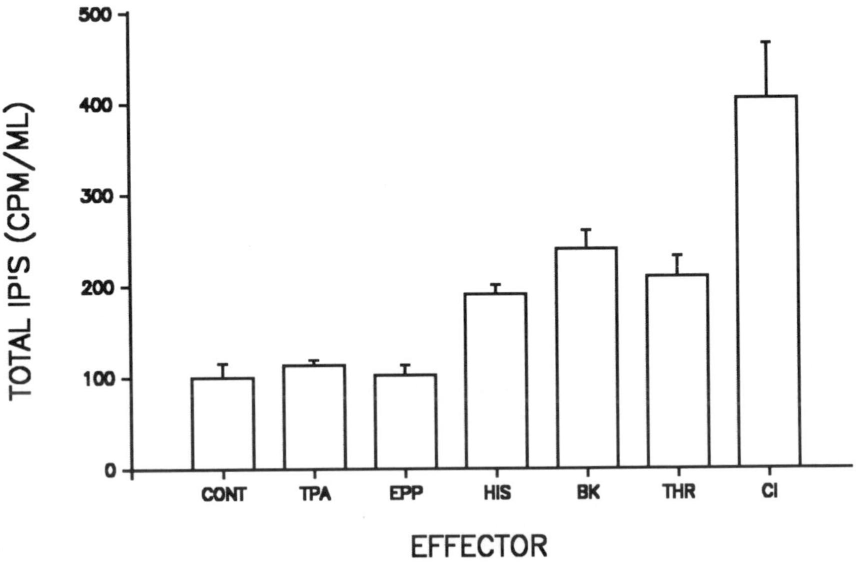

FIGURE 3. Effect of various effectors on inositol phosphate release in cultured keratinocytes. Cells were preincubated with 15 μCi/ml [^{3}H]-myoinositol for 24 h (see Section II). The concentrations of the effectors were the same as in Figure 1.

in greater detail to better understand the specificity of the effects as well as to define inhibitory agents. Dose-response studies for histamine, bradykinin, and thrombin were performed (see Figures 4A to C). Histamine was maximally effective at 10μ*M* ($EC_{50} \approx$ 2 to 3 μ*M*), and bradykinin was maximally effective at 100 n*M* ($EC_{50} \approx$ 20 n*M*). The bradykinin results agree with those of Talwar et al.,[13] who characterized the keratinocyte bradykinin receptor as the B_2 subtype. Thrombin maximally stimulated at 3 U/ml (20 n*M*), with an EC_{50} of 0.2 to 0.3 U/ml (1.3 to 2 n*M*).

Since inositol 1,4,5-triphosphate (Ins 1,4,5-P_3) and diacylglycerol are considered to be active second messengers generated by receptor-mediated hydrolysis of phosphatidylinositol 4,5-bisphosphate by phospholipase C,[21] it was asked whether $InsP_3$ was produced by histamine treatment of keratinocytes. Figure 5 shows that 1 min of treatment with 10 μ*M* histamine resulted in an elevation of all three classes of IPs: a 40% increase in radioactivity in $InsP_3$, a 40% increase in inositol bisphosphate ($InsP_2$), and a 30% increase in $InsP_1$.

Inhibitor studies were next conducted to characterize the histamine, bradykinin, and thrombin effects (see Figure 6). To characterize the keratinocyte histamine receptor, cells were treated with histamine plus various concentrations of diphenhydramine, an H_1-specific antagonist, or metiamide, an H_2-specific antagonist.[22] As shown, IP stimulation induced by 30 μ*M* histamine was inhibited >95% by 100 n*M* diphenhydramine ($EC_{50} \approx$ 1 n*M*). Metiamide showed little effect at concentrations <10 μM, suggesting H_1 histamine receptor specificity. Bradykinin effects on the keratinocytes were examined using the bradykinin antagonist [D-phe^7]-bradykinin,[36] which at 20 μ*M* inhibited 100 n*M* bradykinin-stimulated IP metabolism by >50%. Diisofluorophosphate-treated thrombin (DIP-thrombin), whose serine active site is irreversibly inactivated without affecting its ability to bind to cell membranes,[23] showed little ability to stimulate inositol phosphates at concentrations where thrombin is maximally active (100 n*M*). This finding suggests that the active site of thrombin is required for its stimulation of the IP/DAG pathway in keratinocytes.

The thrombin effect was of considerable interest due to its presence in wound healing and skin inflammation. Also, as a serine protease, it shares characteristics with several other proteases. Stimulation of inositol phosphates by other proteases has been reported in other

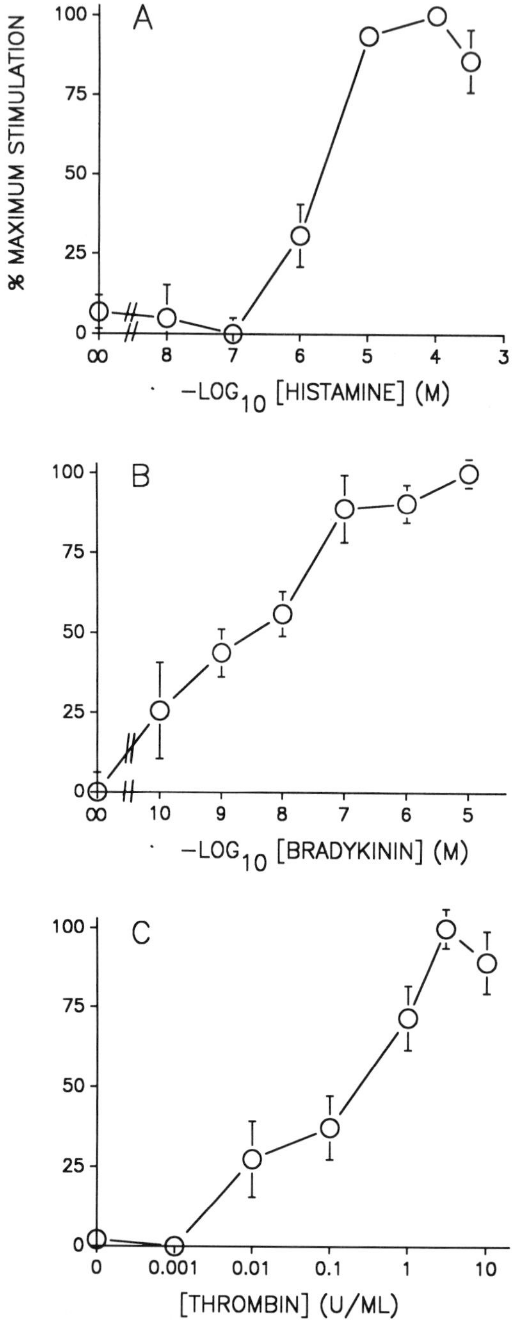

FIGURE 4. Inositol phosphate release as a function of agonist concentration. (A) Bradykinin, (B) histamine, (C) thrombin. Cells preincubated with 15 μCi/ml [^{3}H]-myoinositol for 24 hours were given 10 n*M* $LiCl_2$ plus various concentrations of either bradykinin, histamine, or thrombin and incubated for 30 min at 37°C. Incubations were terminated and analyzed as described in Section II. Maximal stimulations for the agents tested were: histamine, 2-fold; bradykinin, 2.5-fold; thrombin, 2-fold.

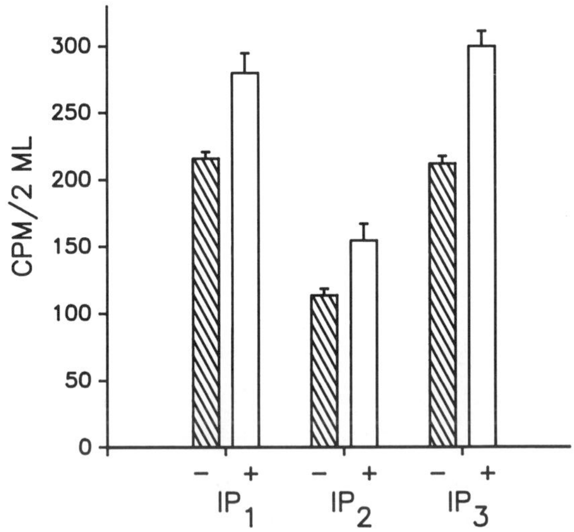

FIGURE 5. Effect of histamine on formation of $InsP_1$, $InsP_2$, and $InsP_3$. Cells preincubated with 15 μCi/ml [^3H]-myoinositol for 24 h were exposed for 1 min to 10 n*M* $LiCl_2$ and 30 μ*M* histamine. Incubations were stopped, inositol phosphates extracted, and extracts from three wells pooled and chromatographed as described in Section II. Following elution of inositol with 10 ml of H_2O and glycerophosphate with 6 ml of 5 m*M* NaB_4O_7/60 n*M* NaCOOH, inositol phosphates were step-eluted as follows: $InsP_1$ — 8 ml of 0.2 *M* $HCOONH_4$; $INSP_2$ — 8 ml of 0.4 *M* $HCOONH_4$/0.1 *M* HCOOH; $InsP_3$ — 8 ml of 1.0 *M* $HCOONH_4$/0.1 *M* HCOOH. A 2-ml volume of each fraction as counted in 14 ml of scintillant. Hatched bars = no histamine; open bars = 10 μ*M* histamine.

cell types;[24,25] hence, a variety of proteases potentially may be involved in skin pathologies. Accordingly, the effects of additional proteases were assessed to help define the specificity of the keratinocyte response to thrombin. It was found that trypsin, α-chymotrypsin, and plasmin also stimulated inositol phosphate formation (see Figure 7), but papain, carboxypeptidase A, and thermolysin had little or no effect at the concentrations tested. To determine whether thrombin, trypsin, and α-chymotrypsin are acting through the same sites or mechanisms, IP synthesis in response to various combinations of thrombin (at maximally stimulating levels) and other proteases or agonists was measured. Although thrombin plus another agonist (either histamine or bradykinin) gave stimulation levels greater than either agonist alone, combinations of thrombin plus any of the effective proteases stimulated only about the same as the proteases alone (data not shown). This is consistent with the concept that thrombin, trypsin, plasmin, and α-chymotrypsin act at the same sites on the cell.

Next, a variety of protease inhibitors (hirudin, leupeptin, soybean trypsin inhibitor) were employed to examine potential blocking effects on the thrombin response. Hirudin is a thrombin-specific inhibitor,[26] while leupeptin and soybean trypsin inhibitor are selective inhibitors of serine and cysteine proteases such as trypsin, plasmin, and papain.[27] Various concentrations of either hirudin, leupeptin, or soybean trypsin inhibitor, together with 1 U/ml thrombin, were added to keratinocytes and incubated for 30 min. Figure 8 shows that hirudin at 30 to 50 n*M* ($EC_{50} \approx 4$ to 6 n*M*) completely inhibited the thrombin effect, whereas leupeptin ($EC_{50} \approx 5$ μ*M*) and soybean trypsin inhibitor (only minimal inhibition observed) were at least 1000 times less effective. This supports the thrombin specificity of the IP stimulation observed with this agent.

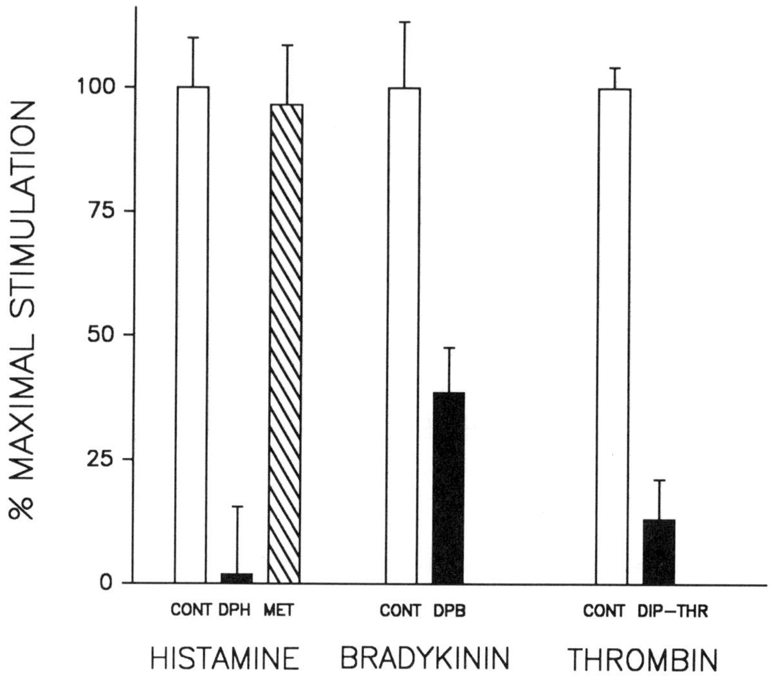

FIGURE 6. Effect of blocking agents on histamine-, bradykinin-, and thrombin-stimulated inositol phosphate release in cultured keratinocytes. The cells were briefly preincubated with either (1) 1 μ*M* diphenhydramine (DPH) or 1 μ*M* metiamide (MET) followed by 30 μ*M* histamine, or (2) 20 μ*M* [D-Phe^7]-bradykinin (DBP) followed by 100 n*M* bradykinin. The histamine effect was inhibited 46% by 10 μ*M* metiamide (not shown). For thrombin, cells were incubated with either 100 n*M* DIP-inactivated thrombin (DIP-THR) or 100 n*M* active thrombin (see text). Maximal stimulations were ~2.5-fold for all three agonists. Because units measure thrombin activity rather than absolute concentration, thrombin and DIP-thrombin concentrations are expressed here as molarities. For active thrombin, 1 U/ml = 7 n*M*; for DIP-thrombin, 1 U/ml = 140 n*M*.

Evaluations of second messenger responses to various agents involved in irritant or inflammatory effects provide information regarding the characteristics of the human keratinocyte which may be relevant to damage and repair processes in the skin.

IV. DISCUSSION

Stimulation of PGE_2 production and changes in inositol phosphate (IP) formation in cultured keratinocytes appear to provide useful markers for the exploration of possible irritant and toxic effects in exogenous dermatoses. The examination of keratinocyte responses over time, using living, isolated cells, suggests interesting approaches to both initial mechanisms of the inflammatory response as well as later effects of agents which induce pathological changes in the skin. By studying keratinocytes isolated from other cells in the skin, it should be possible to define cellular responses which can then be compared to responses of other cell types in the skin and in human skin organ culture models. The PGE_2 data presented above showed delayed cellular responses which could be quite important in understanding irritant or inflammatory effects on this cell type. These findings regarding stimulation of the IP pathway are in general agreement with recent reports by Talwar et al.[13,14] from Vorhees' laboratory. Together, the data strongly suggest that the keratinocyte may play an important role in mediating initial responses to agents such as thrombin, bradykinin, and histamine and in delayed responses to agents such as TPA and EPP. Keratinocyte responses

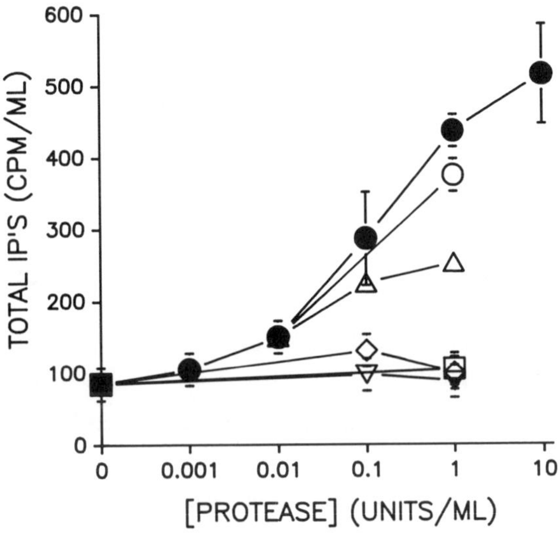

FIGURE 7. Effect of various proteases on inositol phosphate formation. Cells preincubated with 15 μCi/ml [^{3}H]-myoinositol overnight were given 10 m*M* $LiCl_2$ and the indicated activities of various proteases for 30 min at 37°C. Incubations were terminated and cell extracts analyzed (as in the Figure 1 legend). Units of each protease are those given by the manufacturer. ● = trypsin; ○ = α-chymotrypsin; △ = plasmin; ◇ = papain; ▽ = carboxypeptidase A; □ = thermolysin. Trypsin was the only protease tested which did not produce loss or destruction of the cells at concentrations >1 U/ml.

to potentially toxic substances (such as hydrogen peroxide and chemical toxins and irritants such as phenol, toluene, and chloroform) are currently under investigation in our laboratory using PGE_2 levels and PI metabolism, along with alterations which may be produced in other biochemical and cellular functions.

We have previously observed in studies of irritants in the mouse inflammatory ear and human skin organ culture models that PGE_2 provides a useful marker by which to follow effects over time. While activation of the IP/DAG signaling pathway is expected to produce some stimulation of the PG pathway, other mechanisms (such as activation of phospholipases) may be involved in the stimulation of PGE_2 and other eicosanoids by the cells. Also, the time course of irritant effects is more readily followed using PGE_2.

The stimulation of keratinocyte inositol phosphate formation by bradykinin, histamine, and thrombin supports the view that the IP/DAG pathway may be an important means by which inflammatory mediators exert direct effects. Studies of IP stimulation may be useful to characterize the mechanisms of response and inhibition of specific receptors.

The bradykinin dose response ($EC_{50} \approx 20$ n*M*) is similar to that found in other human epithelial cells[38] and is consistent with the observations of other investigators in keratinocytes[14] and other cell types.[28] These findings suggest that bradykinin effects on keratinocytes may be physiologically relevant. Inhibition of bradykinin effects by the antagonist [D-phe^7]-bradykinin is consistent with the observations of Vavrek and Stewart,[36] who found this compound to be an antagonist in the guinea pig ileum myotropic activity test. In contrast, Brass et al.[37] found [D-phe^7]-bradykinin to be a full agonist for IP turnover in murine neuroblastoma cells, suggesting possible species or cell type specificity. The data also emphasize the importance of testing bradykinin antagonists in human keratinocytes.

Histamine was maximally active at a concentration (10 to 100 μ*M*) characteristic of this

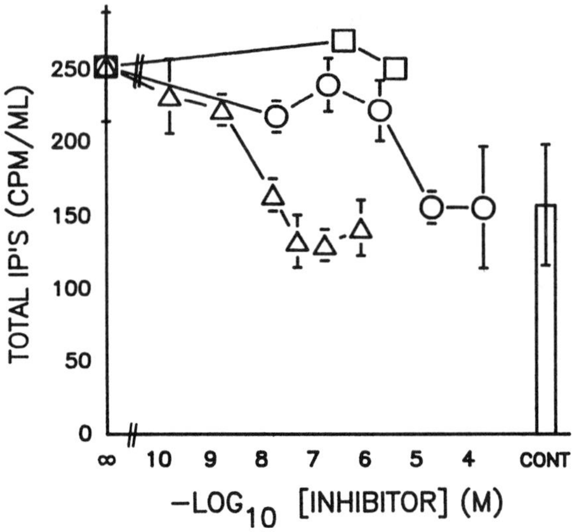

FIGURE 8. Effect of protease inhibitors on the stimulation of inositol phosphates by thrombin. Cells preincubated with 15 μCi/ml [^{3}H]-myoinositol overnight were treated with 10 m*M* $LiCl_2$, 1 U/ml thrombin, and various concentrations of either hirudin (△), leupeptin (○), or soybean trypsin inhibitor (□) for 30 min. The incubations were terminated and the cells treated and analyzed as described in Section II. Maximal stimulation by thrombin was 1.7-fold.

bioactive amine in other cells types.[20,29] The complete inhibition of histamine-mediated IP stimulation by 100 n*M* diphenhydramine suggests that human keratinocytes possess H_1 histamine receptors. This is substantiated by the absence of an inhibitory effect by metiamide, a specific H_2 antagonist. The involvement of H_1 receptors is in accord with the stimulation of IP formation by histamine in other cell types.[20,29]

Thrombin stimulation of IP formation in keratinocytes as described in our study has also been reported recently.[13,14] The response to thrombin is intriguing, since this agent is a mitogen for human splenocytes,[30] chick embryo fibroblasts,[31] and mouse embryo cells.[32] Thrombin has also been postulated to stimulate wound healing via cell division at the site of injury. Thrombin may play a role in the inflammatory response, since Bar-Shavit et al.[33] have reported that thrombin stimulates macrophage proliferation. In mouse keratinocytes, increased mitotic activity is linked to the IP/DAG pathway. An interesting possibility is that thrombin may stimulate keratinocyte mitogenesis via this pathway.

Our data show the thrombin effect also may be evoked by three additional proteases, two of which (trypsin and α-chymotrypsin) are serine proteases and, thus, have active sites similar to thrombin.[34] Plasmin and thrombin both have important roles in blood coagulation and, hence, may be exposed to keratinocytes *in vivo* under conditions of inflammation, clot formation, or wound healing. In addition to thrombin, trypsin has also been shown to stimulate cell proliferation.[35] We cannot rule out the possibility, however, that these proteases may mimic the effect of another, unknown protease operating *in vivo*.

Overall, the findings that inflammatory mediators show specific effects on PG and IP/DAG pathways in human keratinocytes raise the possibility that these pathways may be involved in eliciting physiological responses in the skin and, thus, may be involved in a number of manifestations of exogenous dermatoses. Evaluations of drug effects on these pathways in keratinocytes and other skin cells may lead to improved clinical treatments, including therapies for inflammation and wound healing.

ACKNOWLEDGMENT

Ernest Bloom is supported by the Colgate-Palmolive Fellowship in *In Vitro* Toxicology.

REFERENCES

1. **Goldyne, M.,** Human fibroblast and keratinocyte synthesis of eicosanoids in response to interleukin 1, *Ann. N.Y. Acad. Sci.*, 518, 108, 1988.
2. **Kartasova, D., Ponec, M., and Van de Putte, P.,** UV irradiation of human epidermal keratinocytes, *Exp. Cell Res.*, 174, 421, 1988.
3. **Kvedar, J. and Levine, L.,** Modulation of arachidonic acid production in a cultured newborn rat keratinocyte cell line, *J. Invest. Dermatol.*, 88, 124, 1987.
4. **DeLeo, V. A., Scheide, S., Horlich, H., Hanson, D., et al.,** The effect of 12-*O*-tetradecanoylphorbal-13-acetate (TCA) on phospholipid metabolism of human epidermal keratinocytes in culture, *Carcinogenesis*, 7, 937, 1986.
5. **Furstenberger, G., Richter, H., Fujenis, N. E., and Marks, F.,** Arachidonic acid and prostaglandin E_2 release and enhanced cell proliferation induced by the phorbal ester TPA in a murine epidermal cell line, *Cancer Lett.*, 11, 191, 1981.
6. **Bloom, E., Maibach, H. I., Tammi, R., and Polansky, J. R.,** *In vitro* models for cutaneous effects of glucocorticoids using human skin organ and cell culture, in *Models in Dermatology*, Maibach, H. I. and Lowe, N. J., Eds., S. Karger, Basel, 1989, 2.
7. **Bloom, E., Goldyne, M., Maibach, H. I., Tammi, R., and Polansky, J. R.,** In vitro effects of irritants using human skin cell and organ culture models, *Clin. Res.*, 35, 670A, 1987.
8. **Bloom, E., Maibach, H. I., Tammi, R., and Polansky, J. R.,** In vitro models of irritant effects on human skin, *Toxicologist*, 9(1), 6, 1989.
9. **Bloom, E., Maibach, H. I., Jones, H., Goldyne, M., and Polansky, J. R.,** *In vitro* studies of human keratinocytes and skin fibroblasts: TPA and dexamethasone responses, *Clin. Res.*, 36, 634A, 1988.
10. **Evans, M. G. and Marty, A.,** Potentiation of muscarine and α-adrenergic responses by an analogue of guanosine 5′-triphosphate, *Proc. Natl. Acad. Sci. U.S.A.*, 83, 4099, 1986.
11. **Paris, S., Chambard, J.-C., and Pouyssegur, J.,** Coupling between phosphoinositide breakdown and early mitogenic events in fibroblasts, *J. Biol. Chem.*, 262, 1977, 1987.
12. **Rijzewijk, J. J., Bauer, F. W., Boezeman, J. B., Happle, R., and Mier, P. O.,** Recruitment of quiescent (G_0) cells following epidermal injury is initiated by activation of the phosphoinositol cycle, *J. Invest. Dermatol.*, 90, 44, 1988.
13. **Talwar, H. S., Elder, J. T., Voorhees, J. J., and Fisher, G. J.,** Bradykinin causes phosphoinositide turnover, arachidonic acid release and c-myc expression in adult human keratinocytes, *Clin. Res.*, 37, 703A, 1989.
14. **Talwar, H. S., Fisher, G. J., Harris, V. A., and Voorhees, J. J.,** Agonist-induced hydrolysis of phosphoinositol and formation of 1,2-diacylglycerol in adult human keratinocytes, *J. Invest. Dermatol.*, 93, 242, 1989.
15. **Fisher, G. J., Talwar, H. S., Ryder, N. S., and Voorhees, J. J.,** Differential activation of human skin cells by platelet activating factor: stimulation of phosphoinositide function and arachidonic acid mobilization in keratinocytes but not in fibroblasts, *Biochem. Biophys. Res. Commun.*, 103, 1344, 1989.
16. **Oliw, E., Granström, E., and Änggård, E.,** The prostaglandins and essential fatty acids, in *Prostaglandins and Related Substances*, Pace-Asciak, C. and Granström, E., Eds., Elsevier, Amsterdam, 1983, 1.
17. **Berridge, M. J., Downes, C. P., and Hauley, M. R.,** Lithium amplifies agonist-dependent phosphatidylinositol responses in brain and salivary gland, *Biochem. J.*, 206, 587, 1982.
18. **Schacht, J.,** Extraction and purification of polyphosphoinositides, *Methods Enzymol.*, 72, 626, 1981.
19. **Berridge, M. J.,** Rapid accumulation of inositol triphosphate reveals that agonists hydrolyze polyphosphoinositides instead of phosphatidylinositol, *Biochem. J.*, 212, 849, 1983.
20. **Crook, R. B., Bazan, N. G., Alvarado, J. A., and Polansky, J. R.,** Histamine stimulation of inositol phosphate metabolism in cultured human non-pigmented ciliary epithelial cells, *Curr. Eye Res.*, 8, 415, 1989.
21. **Berridge, M. J.,** Inositol triphosphate and diacylglycerol: two interacting second messengers, *Annu. Rev. Biochem.*, 56, 159, 1987.
22. **Black, J. W., Duncan, W. A. M., Durant, C. J., Ganellin, C. R., and Parsons, E. M.,** Definition and antagonism of histamine H_2-receptors, *Nature (London)*, 236, 385, 1972.

23. **Aubrey, B. J., Hoak, J. C., and Owen, W. G.,** Binding of human thrombin to cultured human endothelial cells, *J. Biol. Chem.,* 254, 4092, 1979.
24. **Cunningham, D. D., Carney, D. H., and Glenn, K. C.,** Initiation of cell division by thrombin and trypsin, in *Chemistry and Biology of Thrombin,* Lundblad, R. L., Fenton, J. W., and Mann, K. G., Eds., Ann Arbor Science, Ann Arbor, MI, 1977, 545.
25. **Martin, B. M., Feinman, R. D., and Detwiler, T. C.,** Platelet stimulation by thrombin and other proteases, *Biochemistry,* 14, 1308, 1975.
26. **Low, D. A. and Cunningham, D. D.,** A novel method for measuring cell surface-bound thrombin, *J. Biol. Chem.,* 257, 850, 1982.
27. **Powers, J. C. and Harper, W. C.,** Inhibitors of serine proteinases, in *Proteinase Inhibitors,* Barrett, A. J. and Salveson, G., Eds., Elsevier, New York, 1986, 55.
28. **Yano, K., Higashida, H., Inoue, R., and Nozawa, Y.,** Bradykinin-induced rapid breakdown of phosphatidylinositol 4,5 bis-phosphate in neuroblastoma × glioma hybrid NG108-18 cells, *J. Biol. Chem.,* 259, 10201, 1984.
29. **Lo, W. W. Y. and Fan, T.-P. D.,** Histamine stimulates inositol phosphate accumulation via the H_1-receptor in cultured human endothelial cells, *Biochem. Biophys. Res. Commun.,* 148, 47, 1987.
30. **Chen, L. B., Teng, N. N. H., and Buchanan, J. R.,** Mitogenicity of thrombin and surface alterations on mouse splenocytes, *Exp. Cell Res.,* 101, 41, 1976.
31. **Chen, L. B. and Buchanan, J. M.,** Mitogenic activity of blood components. I. Thrombin and prothrombin, *Proc. Natl. Acad. Sci. U.S.A.,* 72, 131, 1975.
32. **Zetter, B. R., Sun, T. T., Chen, L. B., and Buchanan, J. M.,** Thrombin potentiates the mitogenic response of cultured fibroblasts to serum and other growth promoting agents, *J. Cell. Physiol.,* 92, 233, 1977.
33. **Bar-Shavit, R., Kahn, A. J., Mann, K. G., and Wilner, G. D.,** Identification of a thrombin sequence with growth factor activity on macrophages, *Proc. Natl. Acad. Sci. U.S.A.,* 83, 976, 1986.
34. **Barrett, A. J.,** An introduction to the proteinases, in *Proteinase Inhibitors,* Barrett, A. J. and Salveson, G., Eds., Elsevier, New York, 1986, 7.
35. **Teng, N. H. and Chen, B.,** The role of surface proteins in cell proliferation as studied with thrombin and other proteases, *Proc. Natl. Acad. Sci. U.S.A.,* 72, 413, 1975.
36. **Vavrek, R. J. and Stewart, J. M.,** Competitive antagonists of bradykinin, *Peptides,* 6, 161, 1985.
37. **Brass, K. M., Manning, D. C., Perry, D. C., and Snyder, S. H.,** Bradykinin analogues: differential agonist and antagonist activities suggesting multiple receptors, *Br. J. Pharmacol.,* 94(1), 3, 1988.
38. **Crook, Richard B. and Polansky, J. R.,** unpublished data.

Section IV. Clinical Aspects

Chapter 21

TESTIFYING IN MEDICOLEGAL CASES INVOLVING DERMATOSES OF POSSIBLE EXOGENOUS ORIGIN

Alexander A. Fisher

TABLE OF CONTENTS

I. INTRODUCTION

Dermatologists are frequently called upon to give expert testimony on the causal relationship of certain dermatoses with exposure to various environmental chemicals, including surgical procedures. The testifying "expert" must given an unbiased opinion based on personal experience, a thorough review of the pertinent literature, and, when feasible, consultation with not only dermatologic colleagues, but with "experts" on the effect of various environmental factors on the skin.

What follows are the opinions that were provided by this writer in specific instances of medicolegal cases involving the possibility of environmental factors producing various dermatoses.

II. THE "SAFETY" OF STAINLESS STEEL SUTURES IN NICKEL-SENSITIVE INDIVIDUALS

This author previously wrote that "several nickel-sensitive individuals have had stainless steel sutures inserted for surgical procedures without a reaction."[1] As a result of this statement, a nickel-sensitive patient was referred to prove whether or not the use of stainless steel sutures had produced certain symptoms and signs in a nickel-sensitive person who was suing a surgeon for malpractice. The patient insisted that although the surgeon had been informed that the patient was allergic to nickel and that the stainless steel sutures contained nickel, stainless steel sutures had been used nevertheless and had produced certain symptoms and signs. The patient and her attorney agreed to allow me to perform certain tests and procedures.

A. TEST PROCEDURES

1. The Patch Test with Nickel Sulfate

No patient can be classified as being nickel sensitive unless a positive patch test reaction to nickel sulfate can be obtained.[2] Experience has shown that some patients who claim that they are allergic to nickel, but have negative patch test reactions, upon further investigation show other causes for their dermatitis or symptoms. However, this patient did have a strong positive patch test reaction to nickel sulfate.

2. The Dimethylglyoxime Test

The stainless steel sutures that the surgeon used in this case contained 10% nickel. However, the results of the dimethylglyoxime test were negative. In this author's experience, stainless steel objects that give a negative reaction to the dimethylglyoxime test are safe for use in nickel-sensitive persons.

3. The Suture Insertion Skin Test

The litigant consented to have a stainless steel suture inserted into the skin and left in place for 1 week and to have a silk suture inserted as a control. After 1 week the stainless steel suture produced very little (if any) reaction, less than that produced by the silk suture.

The dimethylglyoxime test performed on the stainless steel suture after it was removed from the skin showed negative results. This "postoperative" test was performed since some investigators state that "body fluids can leach out nickel from stainless steel."[3,4]

As a result of these tests it was concluded that the stainless steel sutures did not produce the alleged symptoms and signs in this nickel-sensitive patient, and the suit was dropped.[5] When stainless steel sutures are to be used in a patient with alleged "metal sensitivity", or "nickel allergy" in particular, a patch test for nickel sensitivity should be performed. The dimethylglyoxime test should be done on the sutures and, in this litigious era, a preliminary "suture insertion skin test" should be done.

III. THE "MISUSE" OF THE PATCH TEST TO DETERMINE MERCURY "POISONING"

A 45-year-old dentist claimed that he acquired "mercury poisoning" from exposure to mercury used in his dental practice. The only "objective" sign of the alleged mercury poisoning was said to be a positive patch reaction containing mercuric chloride. The so-called "patch test kit" contains 0.02% mercuric chloride. A positive reaction is said to consist not only of erythema, edema, and papules at the patch test site, but of changes in heart rate, blood pressure, temperature, and various "toxic" symptoms such as gastrointestinal upset, headache, fatigue, and depression, all during the 48-h patch-test period. On the basis of such "positive" results, many dentists urge their patient to have their mercury amalgam fillings replaced with gold or composite resin fillings.

It was emphasized that patch testing for nonallergic "hypersensitivity" with mercury chloride is completely irrational for the following reasons:

1. Patch tests may reveal *allergic* hypersensitivity to mercury, but do not reveal mercury poisoning or so-called nonallergic hypersensitivity.
2. Patch tests with mercury chloride are notoriously unrealiable. Even such dilute solutions as 0.01% aqueous mercury chloride may show "irritant" patch test reactions in control patients.
3. A proper test for allergic mercury hypersensitivity is 5% ammoniated mercury. Even a positive reaction to this patch test is meaningless as far as mercury "poisoning" is concerned, as is a positive reaction to the patch test with 0.5% metallic mercury in petrolatum, which is used by some European investigators. Hence, the claim of mercury poisoning or "toxicity" based on such patch testing is completely irrational.

It was emphasized that allergic hypersensitivity to mercury, as proved by a properly applied and correctly interpreted patch test reaction, may result in dermatitis, stomatitis, or urticaria. Toxic reactions or "poisoning" by mercury manifested by nausea, diarrhea, tremors, depression, insomnia, nephritis, pneumonitis, dark pigmentation of marginal gingiva, and loosening of the teeth cannot be determined by a patch test.[6] An opinion obtained from the American Dental Association confirmed these conclusions.[7]

IV. SEVERE REACTIONS TO DIMETHYL SULFOXIDE USED AS A VEHICLE FOR FOOD ALLERGIC PATCH TESTS

A 46-year-old asthmatic woman acquired a generalized dermatitis following the application of a 3-d patch test in which food had been dissolved in dimethyl sulfoxide (DMSO). The exact concentration of the solvent DMSO was not known. The patient claimed that she had been "poisoned". Her lawyer pointed out that the U.S. Food and Drug Administration (FDA) and the American Medical Association (AMA)[9,10] stated that DMSO could cause many undesirable effects.

It was pointed out that the therapeutic uses of topical DMSO are the subject of controversy. The FDA has warned that, because of its power as a solvent, DMSO may act as a "carrier" for many internally poisonous substances that ordinarily do not permeate the skin.[8] Many arthritis preparations, for example, contain camphor and menthol, which are safe when applied topically, but may cause unpleasant and dangerous effects when transported through the skin and into the bloodstream.

The AMA's Council on Scientific Affairs reports[9,10] that in recent years much controversy has been generated about the safety and efficacy of DMSO. The agent has been proclaimed by the media as a miracle drug and a medical panacea. However, it has also been suggested to be a hoax.

When DMSO is applied to the skin in concentrations as low as 20% it causes transient erythema at the site of application in most subjects. Contact urticarial lesions are frequently caused at higher concentrations. Since the response may occur on first contact, the mechanism is classified as nonimmunologic.

A few cases of generalized dermatitis have occurred following application of DMSO. A wheal and histamine type of erythematous reaction may occur at sites distant from the area of application. This generalized dermatitis occasionally may be severe.

A. THE USE OF DMSO IN PATCH TESTING FOR FOOD ALLERGY

The test uses a DMSO vehicle to introduce fat and oil as well as water-soluble foods to subcutaneous immune structures. It is reported to detect many of the non-IgE-mediated food allergies that aqueous food extracts often miss.

Breneman,[11] the author of the test, claims that patch testing with DMSO has a sensitivity "almost 14 times higher than that of conventional RAST® - (radioallergosorbent test) based screens". He added that the specificity of the test was also high, since false negative reactions were "quite rare".

This DMSO patch test is recommended for Type I reactions due to the ingestion of foods, but not for the Type IV delayed eczematous allergic reactions caused by the handling of foods. Although DMSO has been used in the past to enhance Type IV delayed contact allergy and to enhance the penetration of various substances through the skin *in vivo*, this is a "new" use for a "food" allergy test.[12-14]

It was pointed out that there is a great difficulty in interpreting coverd DMSO patch tests for the following reasons:

1. Kligman[15] reported that when patch tests were occluded with plastic tape, the primary irritant effects of DMSO were greatly enhanced. Epidermal vesiculation, histologic evidence of epidermal cell death, and a perivascular dermal infiltrate were observed.
2. This author's experiences confirm that when skin is treated with DMSO at concentrations of 20 to 90% and occluded with a bandage or a patch test for 2 to 3 d, marked erythema and vesiculation often occur under the bandaged area at the patch test site.[16]
3. It therefore seems difficult or impossible to differentiate the nonspecific irritant vesicular reaction that usually occurs with a 3-d closed DMSO patch test from the alleged specific allergic reactions to food or other substances patch tested using DMSO as the vehicle.
4. The explanation by the investigator who suggested the use of DMSO as a vehicle for food patch testing was that "the so-called 'non-specific irritant' reaction appears to be immune, also — mostly IgE. We use a DMSO patch so we know the amount of irritation from the DMSO alone. We then compare this with the results of the food test patch."

This author gives the opinion that the DMSO patch test should not be used routinely, since this type of testing cannot be controlled easily, even with the use of a DMSO control patch because of lack of standardization.

V. DOES CHRONIC ALLERGIC CONTACT DERMATITIS PLAY A ROLE IN MYCOSIS FUNGOIDES?

A 46-year-old bricklayer had a chronic hand dermatitis presumably due to cement, since he had a strongly positive patch test to potassium dichromate. Subsequently, the patient developed a generalized mycosis fungoides. The claim was made that the chronic allergic hand dermatitis was a precipitating factor in the causation of the mycosis fungoides. It is obvious that the implication of such a relationship would have far-reaching medicolegal consequences.

The patient's lawyer and physician cited the following reports in the literature to bolster their claim.

In 1983, Schlenzka et al.[23] stated that 20 patients who had clinical and histologic diagnoses of mycosis fungoides were subjected to patch tests with chromate, nickel, and cobalt between 1971 and 1982. Positive patch test results appeared in 12 patients (60%). The prevalence of such contact allergies in these patients is consequently much higher than that found in other patients tested in recent years. These results are claimed to support the hypothesis that mycosis fungoides is an "immunoma".

Winkelmann,[17] in comparing mycosis fungoides with the Sezary syndrome, states that in many instances when studying "immune" reactions, delayed patch test reactions are frequently positive in patients with mycosis fungoides, but that in patients with the Sezary syndrome, delayed patch test reactions are negative. However, this statement appears to be at variance with an earlier report by Ecker and Winkelmann,[18] in which it is stated that five elderly patients with recurrent and progressive erythematous patchy dermatitis had temporary flares of generalized erythema and associated positive patch test reactions. The histologic findings were of a distinct pattern of lymphocytic dermal reticulosis and chronic epithelial response. They believed this represents a potential pre-Sezary state that can be recognized by the course and histologic pattern.

In rebuttal, it was pointed out that Ackerman et al.[19] warn that eruptions of spongiotic foci with atypical mononuclear cells can be misdiagnosed as being Pautrier microabscesses, with resultant mismanagement of the patient's condition. Ackerman et al.[19] reported on several patients with spongiotic dermatitis, most commonly allergic contact dermatitis, that could not be distinguished histologically from mycosis fungoides. This observation is consonant with the concept that the microscopic structure does not always parallel biologic behavior. They further state that there is a tendency for the histopathologist to overdiagnose mycosis fungoides. The patient whose condition has been misdiagnosed as mycosis fungoides may be subjected to unnecessary and potentially injurious treatment, in addition to the burden of living with a potentially fatal disease.

In addition, the report of Orbaneja[20] was cited in which there is a description of a 54-year-old man who had noticed a dry, pruritic, erythematous plaque (about the size of the palm of his hand) on his right thigh. The plaque progressively grew larger and deeper. This histologic picture was similar to that of "infiltrated mycosis fungoides". The dermatitis, however, was discovered to be an allergic reaction to the "striker" area of a box of matches. Three other cases were studied that were clinically suggestive of mycosis fungoides because of their infiltrated plaque-like lesions. Topical hypersensitivity was suspected, however, when a positive patch test reaction was obtained using the striker area of a box of matches.

Finally, the recent report of Turp et al.,[21] which was a case-controlled study of the possible causative factors in mycosis fungoides, was cited. These authors state that a detailed case control study was carried out on 53 patients (33 men and 20 women) with histologically proven mycosis fungoides and on an age- and sex-matched control population. The possible causative factors that were investigated included occupation, recreation, and exposure to petrochemicals, pesticides, insecticides, and potential carcinogens. Exposure to plants of the Compositae family, tanning, and chronic sun exposure were also investigated, as were smoking, drug ingestion, and other skin diseases. Personal and family histories of other malignancies were also investigated. The investigators concluded that these factors played "*no part in the cause of mycosis fungoides*".

The opinion was given that the cause of mycosis fungoides is as yet unknown and that prolonged allergic contact dermatitis may be confused clinically and histologically with mycosis fungoides. Contact allergens apparently do not play a role in causing mycosis fungoides.[22]

REFERENCES

1. **Fisher, A. A.,** *Contact Dermatitis,* 3rd ed., Lea & Febiger, Philadelphia, 1986.
2. **Fisher, A. A.,** Problems attending patch testing for nickel sensitivity, *Cutis,* 29, 148, 1982.
3. **Katz, S. A. and Samitz, M. H.,** Leaching of nickel from stainless steel consumer commodities, *Acta Derm. Venereol.,* 55, 113, 1975.
4. **Fisher, A. A.,** Somewhat controversial reports of metal reactions, *Cutis,* 24, 28, 1979.
5. **Fisher, A. A.,** Medico-legal aspects of the use of stainless steel sutures in nickel-sensitive persons, *Cutis,* 41, 25, 1988.
6. **Fisher, A. A.,** The misuse of patch testing to determine hypersensitivity to mercury amalgams, *Cutis,* 35, 110, 1985.
7. Council on Dental Materials, Instruments, and Equipment and Council on Dental Therapeutics, Patch tests for sensitivity to mercury on nickel, *J. Am. Dent. Assoc.,* 108, 381, 1984.
8. Use of DMSO for unapproved indication, *F.D.A. Drug Bull.,* 10, 20, 1980.
9. DMSO (dimethyl sulfoxide), AMA Drug Evaluations, 5th ed., Division of Drugs; American Medical Association, Chicago, 1983, 785.
10. Council on Scientific Affairs of the American Medical Association, Dimethyl sulfoxide — controversy and current status, 1981, *J. Am. Med. Assoc.,* 1369, 1982.
11. **Breneman, J. C.,** DMSO Food Allergy Test, *Allergy Observer,* 2, 1, 1985.
12. **Maibach, H. J. and Feldman, R. J.,** The effect of DMSO on percutaneous penetration of hydrocortisone and testosterone in man, *Ann. N.Y. Acad. Sci.,* 141, 423, 1967.
13. **Thormann, J. and Wildenhoff, K. E.,** Contact allergy to idoxuridine, *Contact Dermatitis,* 6, 170, 1980.
14. **Roeleveld, C. G. and van Ketel, W. G.,** Allergic reaction to a hair dye elicited by an ointment containing DMSO, *Contact Dermatitis,* 1, 311, 1975.
15. **Kligman, A. M.,** Topical pharmacology and toxicology of dimethyl sulfoxide. I, *J. Am. Med. Assoc.,* 193, 796, 1965.
16. **Fisher, A. A.,** Dimethyl sulfoxide as a vehicle for food allergy patch tests, *Cutis,* 35, 109, 1985.
17. **Winkelmann, R. K.,** The nature of Sezary syndrome (mycosis fungoides) controversies in T cell skin disease: mycosis fungoides and Sezary syndrome, in *Controversies in Dermatology,* Epstein, E., Ed., W. B. Saunders, Philadelphia, 1984, 399.
18. **Ecker, R. I. and Winkelmann, R. K.,** Lymphatoid contact dermatitis, *Contact Dermatitis,* 7, 84, 1981.
19. **Ackerman, A. B., Prez, T. S. and Capland, L.,** Spongiotic simulants of mycosis fungoides, *Arch. Dermatol.,* 109, 218, 1975.
20. **Orbaneja, J. G., Diez, L. I., Lozano, J. L. S., et al.,** Lymphomatoid contact dermatitis — a syndrome produced by epicutaneous hypersensitivity with clinical features and a histopathologic picture similar to that of mycosis fungoides, *Contact Dermatitis,* 2, 139, 1976.
21. **Tuyp, E., Burgoyne, A., Aitchison, T., et al.,** A case-control study of possible causative factors in mycosis fungoides, *Arch. Dermatol.,* 123, 196, 1987.
22. **Fisher, A. A.,** Allergic contact dermatitis mimicking mycosis fungoides, *Cutis,* 40, 19, 1987.
23. **Schlenzka, Quednow, and Magdeburg,** paper presented at Int. Contact Dermatitis Symp., Erfurt, East Germany, 1983.

Chapter 22

SYSTEMICALLY INDUCED ALLERGIC CONTACT DERMATITIS

Niels K. Veien, Torkil Menné, and Howard I. Maibach

TABLE OF CONTENTS

I. INTRODUCTION

Clinically speaking, the primary event in allergic contact dermatitis is an eczematous reaction at the site of contact with the hapten which caused the sensitization. Eczema may occur at any time later in life at any site on the body which again comes into contact with the same hapten.

II. CAN ALLERGIC CONTACT DERMATITIS BE ELICITED AFTER ORAL INGESTION OF THE HAPTEN?

In answering this question, it is helpful to review cases in which patients have developed cellular immune reactions and dermatitis due to substances not found ubiquitously. A number of case reports have described patients who had adverse skin reactions to systemically administered medicaments to which they had patch-test-proven cellular immune reactions. Bernard et al.[1] described widespread maculopapular eruptions, fever and purpuric lesions, and leukocytosis in two patients who were given 1 g pristamycine orally.

An oral dose of 1.4 g pivampicillin gave rise to a rash in the skin folds, the anogenital region, and the palms and soles.[2] A similar rash appeared after an ampicillin-impregnated sponge was placed in a patient's middle ear. Biopsy showed leukocytoclastic vasculitis, and an intracutaneous test with ampicillin showed evidence of an immune-complex reaction.[3]

Hours after an intravenous dose of gentamicin was given, a patient developed eczema of the arms and neck.[4] Generalized dermatitis, including the palms and soles, was seen in a patient after a daily dose of 100 mg phenobarbitone, given orally for 1 month.[5] Even a poorly absorbed drug like nystatin has given rise to eruptions in the skin folds and the anogenital area after oral ingestion.[6] Urticaria in a patient given oral codeine was associated with a positive patch test.[7] Dimethyl sulfoxide placed intravesically produced eczematous eruptions on the fingers and arms of a patient.[8] Tagami[9] described a patient with toxic epidermal necrolysis after systemic administration of ampicillin and reviewed ten other patch-test-proven cases of dermatitis of similar morphology caused by various medicaments.

Generalized eczema followed the subcutaneous injection of merthiolate in 1 of 45 merthiolate-sensitive patients,[10] and widespread eczematous eruptions were seen following the ingestion of alprenolol in an occupationally sensitized person.[11] Oral aspirin reproduced an eczematous eruption in a patient sensitized by topical application of methyl salicylate.[12]

Cashew nuts may be contaminated by bits of the shells which contain an oleoresin closely related to that of poison ivy. A total of 54 persons developed widespread maculopapular eruptions, largely on the extremities and trunk and in intertriginous areas. A few had stomatitis, perianal itching, and desquamation of the head and feet. Nine patients had positive patch tests to cashew shell extract and to poison ivy urushiol.[13] A similar oleoresin is found in the fruit of the Ginkgo tree, and one patient developed perioral and perianal dermatitis after eating this fruit.[14] The cinnamon in vermouth caused a flare of dermatitis in a patient sensitive to balsam of Peru, and vanilla sugar caused a similar reaction.[15]

A. DELIBERATE ORAL CHALLENGE WITH MEDICAMENTS

If substances such as those mentioned above are deliberately given orally at a time when a patient's dermatitis is quiescent, various characteristic skin reactions may occur. Ekelund and Möller[16] challenged 12 neomycin-sensitive patients with this poorly absorbed substance and noted a relapse of the original dermatitis in 5 patients; in 6 patients the neomycin patch test site flared. Vesicular dermatitis of the palms was seen in three of these patients, and some of them experienced generalized dermatitis. Ekelund and Möller[16] also reported flares, most commonly of the original dermatitis, in six of nine hydroxyquinoline-sensitive patients. Leifer and Steiner[17] noticed a flare of dermatitis in a hydroxyquinoline-sensitive patient after oral challenge with this substance.

During World War II, topical sulfonamides were used extensively in treating wounds. Systemic administration of sulfonamides to persons who had become sensitized after this topical use caused dose-dependent flares of dermatitis. The patients who had the most pronounced reactions were those most sensitive to the sulfonamides. Practically all persons in this category had positive patch tests to sulfonamide, while most patients with less severe reactions often had negative patch tests.[18]

Symptoms similar to those presented above have also been seen after inhalation of a substance to which a person has developed cellular immunity.[19]

The above-mentioned examples of reactions to substances not regularly encountered by the patients suggest that an eczematous reaction may indeed be elicited as a specific event after oral administration or inhalation of a substance to which one has a cellular immune reaction.

III. WILL COMMON CONTACT ALLERGENS PRODUCE FLARES OF DERMATITIS IF THEY ARE ADMINISTERED SYSTEMICALLY?

If it is established that systemically induced contact dermatitis is a clinical entity, the next question is: would a patient who is sensitized to a common contact allergen react with a flare of dermatitis after systemic administration of the hapten? Such reactivity would mean that systemically induced contact dermatitis has more than theoretical interest.

The clinical evidence suggesting that reactions of this type do take place is mostly anecdotal, but some of the case stories are rather convincing. A patient with a history of nickel dermatitis developed vesicular hand eczema and eczema of the arms 24 h after she had swallowed a coin. The dermatitis cleared after the coin was removed by gastroscopy.[20] A similar case was described by Lacroix et al.,[21] whose patient had widespread, oozing, vesicular eruptions which did not clear until a coin was removed from the gastric mucosa. This patient also had eosinophilia which returned to normal after the coin was removed.

Vesicular hand eczema was seen in three of four patients whose nickel dermatitis flared after the insertion of skin clips.[22] Nickel-plated retractors used for abdominal surgery caused flares of dermatitis on the ear lobes and thighs of seven patients with a history of nickel dermatitis. Five of these patients had positive nickel patch tests.[23]

A vitallium (cobalt, chromium, and molybdenum alloy) screw used for osteosynthesis caused widespread dermatitis which began at sites of garter clasps and flared when the screw was removed, only to subside completely 3 weeks after removal. The patient in question had nickel, cobalt, and chromium allergies.[24] A similar case was caused by a stainless steel screw.[25] Urticaria was associated with the insertion of a vitallium nail in a nickel-sensitive patient.[26] Severe bilateral hand eczema was seen in a nickel- and cobalt-sensitive patient after the use of an orthopedic plate made of vitallium.[27] A nickel-sensitive patient had widespread dermatitis after a metal implant containing nickel was inserted in the left elbow and again after placement of a dental plate containing nickel. All cutaneous symptoms disappeared after the two metal items were removed.[28] A patient developed oozing dermatitis on the abdomen at the site of contact with a minichain containing nickel used in connection with treatment for a fractured mandible.[29] Waterman and Schrik[30] suggested that sensitization to nickel and cobalt, chromium, and methacrylate might be a consequence of metal-to-polyethylene hip prostheses.

Hand eczema was seen in a chromate-sensitive patient after placement of a metal denture containing chromium[31] and in a chromate-sensitive patient after insertion of dentures made of vitallium.[32] Circumstantial evidence suggested that vesicular hand eczema which flared after the insertion of a stainless steel pacemaker was caused by nickel in the stainless steel.[33]

Four patients with intrauterine devices (IUDs) containing copper had widespread cutaneous eruptions and positive patch tests to copper;[34] a similar patient was described by Barranco.[35] A patient with a pacemaker developed generalized dermatitis, presumably caused by incompletely hardened epoxy resin in the pacemaker.[36] A mercury-sensitive patient developed facial eczema after dental restoration with amalgam.[37]

The evidence presented here suggests that substances implanted in the body may release amounts of a hapten sufficient to produce various cutaneous eruptions in patients who have positive patch tests to these haptens. This view is supported by studies of nickel release under experimental conditions. The amount of nickel released from coins has been studied by Menné and Solgaard,[38] who demonstrated that, over a 24-h period at 35°C, used coins released an average of 600 μg nickel in synthetic sweat, or four times the average daily oral intake of nickel.[39] Hemingway and Molokhia[40] studied the release of nickel from metallic nickel in relation to time, acidity, and temperature and found that it was correlated to all three parameters. The nickel and cobalt release from dental alloys in physiological saline was studied by Magnussen et al.,[41] who found that up to 365 μg nickel and 780 μg cobalt was released over a period of 5 d. *In vivo* studies in rabbits showed that the highest concentration of nickel from stainless steel prostheses was released in the vicinity of the transplant.[42]

A. DELIBERATE ORAL CHALLENGE

Deliberate oral challenge with 200 mg parabens of a paraben-sensitive patient who had widespread dermatitis after ingesting foods containing parabens resulted in a similar eruption within a few hours.[43] Two patients with positive patch tests to butylated hydroxyanisole (BHA) and butylated hydroxytoluene (BHT) were challenged orally with 20 and 40 mg BHA, respectively, and both developed acute vesicular hand dermatitis.[44]

Oral challenge with 20 mg tetrametylthiuram given daily for 2 d to a thiuram-sensitive patient with dermatitis of the face caused a recurrence of this dermatitis.[45] Flares of the original dermatitis, either alone or in combination with reactivation of a previously positive patch test site, were seen in 11 patients who were sensitive to *para*-amino compounds after oral challenge with sulfonylurea compounds in amounts ranging from 250 to 500 mg.[46]

Of 38 patients with positive patch tests to propylene glycol, 15 experienced widespread exanthema and also, in some cases, vertigo after oral challenge with 2 to 15 ml propylene glycol.[47] Of 20 patients with positive patch tests to azo dyes who were challenged orally with the dyes, 12 experienced pruritus and/or flares of their dermatitis; 2 of 20 controls had similar flares.[48]

A detailed description of a course of oral hyposensitization therapy in a patient who was very sensitive to poison ivy noted pronounced anogenital pruritus after the ingestion of the oleoresin. The same patient had generalized scarlatiniform eruptions and vesicular hand and foot eczema, as well as flares at sites of previous poison ivy patch tests.[49] Kligman[50] injected poison ivy oleoresin into two highly sensitive patients who developed generalized dermatitis, leukocytosis, fever, and eosinophilia. Of five other persons who took part in the test, three developed urticaria, one experienced a dyshidrotic eruption, four had pruritus and fever, and two had exacerbation of their dermatitis.

IV. THE CLINICAL FEATURES OF SYSTEMIC CONTACT DERMATITIS

The studies listed above indicate that a wide range of cutaneous eruptions is seen as a result of the systemic administration of a hapten in patients with cellular immunity to this hapten. The eruptions seen in these studies were usually symmetrical. Widespread maculopapular eruptions were seen primarily in drug-related cases or after oral challenge with propylene glycol. Erythematous rashes in the skin folds and in the anogenital area were also seen when the eruptions were drug related. Vesicular eruptions of the palm and soles occurred after oral challenge with anti-oxidants, after the ingestion of oleoresins from poison ivy, and after the administration of drugs.

Most of the substances mentioned caused flares of previously positive patch test sites as well as flares in areas of previous dermatitis. Anal pruritus was seen especially after hyposensitization with poison ivy extracts. Vasculitis and erythema multiforme were most often drug related. Metal implants causing dermatitis of the overlying skin were especially associated with osteosynthesis and the implantation of pacemakers.[51]

V. IS SYSTEMICALLY INDUCED NICKEL DERMATITIS OF MORE THAN THEORETICAL INTEREST?

Most cases of systemically caused contact dermatitis have been associated with oral ingestion or percutaneous absorption of the hapten. Intravenous administration has also led to severe flares of dermatitis, as had the release of the hapten from items implanted in the mouth and around fractures or in IUDs. Reactions following intravesical instillation of dimethyl sulfoxide have also been seen.[8] The respiratory tract is yet another route of administration, as the hapten can also be inhaled.

In previously published detailed descriptions of groups of nickel-allergic patients, it has been noted that their nickel dermatitis often spread from areas of direct contact with nickel-plated items to other areas of the body. Calnan[52] described such secondary eruptions in 232 of 312 nickel-allergic patients. The most common sites of secondary eruptions were the elbow flexures, the eyelids, the sides of the neck, and the inner thighs, as well as the entire body. He also noted that 20% of his patients had hand eczema. Noneczematous patterns, such as erythema multiforme, urticaria, and generalized pruritus, were also listed as secondary eruptions.

Marcussen[53] examined 622 nickel-sensitive adult patients, 517 (80%) of whom had eczema which had spread to a site other than the primary site of nickel contact. Of the 517, 38% had hand eczema, 47% had dermatitis of the elbow flexures (this was more frequent among strongly sensitive persons than among the weakly sensitive), 28% had facial eczema, and 32% had eczema of the shoulders and upper arm.

Christensen[54] described secondary eruptions in 34 of 66 nickel-allergic patients with hand eczema. The secondary eruptions were usually in the elbow flexures and on the neck, the eyelids, and the soles of the feet. The chronicity of the hand eczema among these patients and the flares of hand eczema which occurred without obvious external contact with nickel led Christensen to consider the hands a secondary rather than a primary site of dermatitis. Of the 66 patients, 51 had hand eczema classified as pompholyx eczema.

Most of the types of secondary eruptions described among nickel-sensitive persons after topical exposure to nickel are comparable to the eruptions seen in persons sensitized to rarely encountered substances after these substances have been given orally or parenterally.

The use of topical drugs or balsam of Peru may also be followed by widespread erythema, multiforme-like eruptions, or other types of vasculitis when the patients have become sensitized to these drugs.[55-57] Such eruptions correspond to some of the secondary eruptions among nickel-sensitive patients noted by Calnan.[52]

Pompholyx hand eczema was common among Christensen's[54] patients. This morphology of dermatitis has also been described as part of the reaction pattern seen after systemic challenge with a hapten in patients sensitized to the same hapten. It was, therefore, a natural next stop to formulate the hypothesis that this type of dermatitis exemplifies a systemically induced allergic contact dermatitis. Christensen and Möller[58] explored this hypothesis by means of a placebo-controlled oral challenge with 5.6 mg nickel given as nickel sulfate; in 9 of 12 patients the pompholyx eczema flared after challenge with nickel, but not after the

TABLE 1
Review of the Results of Oral Challenge with Nickel in Patients with Positive Patch Tests to Nickel

Method	Amount of nickel (given as $NiSO_4$)	Total no. tested	Flare of usual eczema	Ref.
Double-blind, placebo controlled	5.6 mg	12	9 patients	58
Single-blind, placebo controlled	0.6 mg 1.2 mg 2.5 mg	14	1 patient 1 patient 9 patients	59
Double-blind, placebo controlled	2.4 mg	20	13 patients	60
Single-blind, placebo controlled, + blind evaluation of photographs	2.5 mg	28	17 patients	61
Double-blind, placebo controlled, cross-over	0.5 mg daily for 2 d	10	1 patient	62
Double-blind, placebo controlled, cross-over	2 mg daily for 2 d 4 mg daily for 2 d	22	5 to Ni; 5 to placebo; 4 to both 6 to Ni; 2 to placebo; 2 to both	63
Placebo controlled, repeated if in doubt	2.24 mg	49	31 patients; of the 31 patients, 15 had recurrent vesicular hand eczema, 10 periorbital eczema, and 6 had other reactions	64
Open	0.6 mg 1.2 mg 2.5 mg	5 5 5	1 patient 3 patients 4 patients	65
Double-blind, placebo controlled	0.4 mg daily for 2 d 2.5 mg daily for 2 d 5.6 mg	10 10 6	3 to Ni, 2 to placebo, 2 to both 2 to Ni, 1 to placebo, 3 to both 4 to Ni, 0 to placebo, 2 to both	66
Double-blind, placebo controlled	2.5 mg; also 1 mg cobalt	144	55 of 131 with nickel allergy 23 of 131 to placebo or metals + placebo (see Section V of text for details)	67
Double-blind, placebo controlled	2.5 mg	19	5 of 12 with recurrent vesicular hand eczema; 3 of 7 with other types of nickel dermatitis	68

placebo. Seven of the nine patients also had flares at previous sites of dermatitis or previous nickel patch test sites. Widespread maculopapular eruptions were also seen, and one patient had tinnitus and dizziness.

This study encouraged others to use similar challenge procedures with varying doses of nickel. In general, the use of single doses of <1 mg nickel produced few flares of dermatitis, while 2.5 mg nickel gave rise to flares in approximately one half of the patients tested, and doses of 5 mg induced flares in 75% of the patients tested (Table 1).

Roduner et al.[68] found no difference in the rate of reactivity to oral challenge with nickel between those with vesicular hand eczema and those with other types of nickel dermatitis. All the patients in Roduner's study who had vesicular hand eczema prior to the challenge and who experienced flares of dermatitis after the challenge had flares of vesicular hand eczema. None of the patients who did not have vesicular hand eczema prior to the challenge experienced this particular type of flare-up reaction after the oral challenge. Veien et al.[67] challenged nickel-allergic patients with eczema other than vesicular hand eczema and saw flares of the original dermatitis in patients with nonvesicular hand eczema, eyelid dermatitis, perianal dermatitis, and vasculitis-like lesions.

Most challenge studies have been performed in groups of patients with vesicular hand eczema using doses of nickel which are higher than the expected daily intake in food. Nickel has usually been given as a single dose of an inorganic nickel salt. In considering this procedure the following questions arise:

- Is recurrent vesicular hand eczema associated with nickel allergy?
- Is it likely that the concentrations of nickel in the serum and in the skin can be altered "spontaneously" to the same degree as these concentrations are altered by oral challenge with milligram doses of nickel?
- Are there other reasons which make it clinically relevant to carry out an oral challenge with milligram doses of an inorganic nickel salt?
- What is a reasonable dose of nickel for the single-dose oral challenge?

Both recurrent vesicular hand eczema and nickel allergy are common entities. In one hospital dermatology department in Sweden, 1% of all first consultations were due to recurrent vesicular hand eczema.[69] One runs the risk of making a false association between vesicular hand eczema and nickel allergy because of their common joint occurrence. Menné[70] carried out epidemiological studies to determine whether nickel allergy is associated with hand eczema and concluded that such an association does exist and also that hand eczema predisposes to nickel allergy. The hand eczema in nickel-allergic patients is due to environmental rather than genetic factors.

Similarly, Peltonen[71] reported that hand eczema was common among nickel-allergic females and that palmar eczema was more common than eczema on the dorsum of the hands, and Christensen[54] noted that the majority of his patients with nickel allergy and hand eczema had eczema of the pompholyx variety. Edman[72] examined 153 patients with vesicular palmar eczema and found a statistically significant association between this type of eczema and nickel allergy.

The concentration of nickel in palmar skin and the variations in the concentrations of nickel during quiescent phases of dermatitis and periods of active vesicular eruptions are largely unknown. Likewise, it is uncertain whether the local stimulation of the immune system directed against nickel is higher in palmar skin among patients with vesicular hand eczema and nickel allergy than among those without nickel allergy.

There are those who argue against an association between recurrent vesicular hand eczema and the oral ingestion of nickel. Their argument is based on the results of oral challenge with nickel in doses of the magnitude expected to be present in food[62] and the lack of reactivity among some groups of nickel-allergic patients challenged with 2.5 to 4 mg nickel.[63,66] Two studies also showed that the levels of nickel in the serum and urine of one very nickel-sensitive female who was followed closely could not be well correlated to the activity of her dermatitis.[73,74]

It is also difficult to explain how it is that reactivation of nickel patch test sites and flares of dermatitis at previous sites of nickel contact, while frequently seen after oral challenge with nickel in doses of 2.5 to 5 mg, are rarely, if ever, seen during "spontaneous" flares of nickel dermatitis. In order to resolve the question of the clinical relevance of the doses of nickel most commonly used for oral challenge, it is necessary that large numbers of patients with different degrees of sensitivity be tested using varying doses of nickel. This would help to determine whether a reactivation of the relevant dermatitis can occur without creating other symptoms.

In this context, it is interesting to consider the concentration of nickel used for patch testing. The accepted procedure for determining whether a patient has nickel allergy is the application of 5% nickel sulfate in petrolatum for 48 h under occlusion. This procedure results in relatively few false positive or false negative reactions. For some patients, however, it is necessary to carry out intradermal testing in addition to patch testing in order to arrive at a diagnosis of nickel allergy.[75]

Approximately 0.2 mg nickel is applied when 5% nickel sulfate is used for patch testing.[76] Strongly positive patch tests result, however, from the release of nickel in amounts of 0.3 μg per 48 h from alloys containing nickel.[77]

Emmett et al.[78] showed that some patients with nickel allergy reacted to 0.01% nickel in petrolatum and that the individual variations in the threshold for reaction to nickel placed on the skin of nickel-allergic persons ranged from 0.47 to 118 μg. Wahlberg[79] found similar variability in the reactivity of patients with positive patch tests to nickel. While the most sensitive patients reacted to 0.019% in distilled water, there was a mean reactivity to 0.5% $NiSo_4$ in petrolatum. Dooms-Goossens[80] described two females with histories of nickel allergy who had positive patch tests to 10% nickel sulfate, but not to lower concentrations.

There is a 1000-fold difference between the reactivity of the most sensitive of Emmett's patients and the two women described by Dooms-Goossens. The choice of a standard patch test concentration of nickel sulfate is 5%, or 500 times the lowest concentration to which Emmett's patients reacted and 10 times the mean concentration to which Wahlberg's patients reacted.

During the patch test procedure the nickel is placed directly in contact with the target organ. The oral challenge procedure is, however, a much more complicated means of presenting nickel to the skin's immunocompetent cells, and a number of individual factors could influence the amount of nickel which actually reaches the skin. It is reasonable to carry out the oral challenge procedure with a dose of nickel which is substantially higher than the expected natural daily intake for the following reasons:

1. The bioavailability of nickel, particularly if it is given in the form of an inorganic salt of the metal, may vary considerably.
2. Depending on when it is ingested, the absorption of nickel may also vary significantly. Nickel ingested after fasting is absorbed much more completely than nickel ingested during a meal,[81] and it has also been shown that nickel sulfate added to a breakfast is incompletely absorbed.[82] One study showed that patients with terminal ileitis appeared to absorb more nickel than control persons.[83]
3. The individual degree of nickel sensitivity could influence the reaction to ingested nickel.
4. The concentration of nickel in the target organ may vary due to factors other than ingested nickel, such as the mobilization of nickel from the tissues or sweating. Significant amounts of nickel may be excreted in sweat, and drugs like disulfiram and indomethacin may cause nickel to be released from the tissues.[84,85]
5. Even though an oral challenge is carried out when the dermatitis is quiescent, endogenous factors other than nickel may cause aggravation. Edman[72] found an association between vesicular hand eczema, the intake of acetylsalicylic acid and oral contraceptives, and smoking.
6. External contact with irritants and/or nickel also may influence the dermatitis of a patient during the oral challenge procedure.
7. The eruptive nature of recurrent, vesicular hand eczema, which is the most common type of dermatitis, is an indication that intermittent high intake of nickel may be more likely to precipitate flares of dermatitis than the regular, average intake.

Any number of combinations of the above factors are possible. It is, therefore, illogical to base the dose of nickel to be used for an oral challenge procedure on the mean daily intake of nickel. The dose of nickel chosen should provide a reactivation of the clinical problem in a reasonable number of patients without causing too many nonphysiological reactions. This was the method used in arriving at the concentration of nickel sulfate to be used for the patch test.

One oral challenge study showed that 5.6 mg nickel caused a number of reactions other than flares of the dermatitis. These included flares of old patch test sites and widespread erythema.[58] Doses of 0.5 to 1 mg nickel caused flares in only a few nickel-sensitive patients (Table 1).

In a recent study by Nielsen,[86] 12 nickel-allergic females were challenged orally with a diet naturally rich in nickel for five consecutive days (490 μg nickel per day). On day 11 after initiation of the study, the vesicular hand eczema of 10 of the 12 patients was aggravated.

A dose of 2.5 mg nickel used for routine investigation of nickel-sensitive patients suspected to have endogenous reactions to nickel caused flares of dermatitis in more than one half of those tested, with few nonphysiological reactions.[67] Therefore, it is suggested that this dose can be used for the oral challenge until more sophisticated test procedures are developed.

Nickel dermatitis can serve as a model for other types of systemically induced contact dermatitis caused by ubiquitous haptens. In cases of drug-induced systemic contact dermatitis, the diagnosis is established both by challenge with and depletion of the hapten. This is a relatively simple procedure when the substance in question is not found in the natural environment.

In the case of nickel, the mouth is the main port of entry into the body, and depletion experiments have included low-nickel diets. It has been shown that nickel intake can be reduced to one half of the regular daily intake when following what can be thought of as a normal diet.[87]

Depletion studies carried out before detailed knowledge of the occurrence of nickel in food was available showed that the dermatitis of approximately one half of the patients who followed a "low nickel" diet improved after 1 to 2 months. A concomitant reduction of nickel excretion was demonstrated.[88,89] The dermatitis of a very nickel-sensitive patient improved after she had followed a low-nickel diet for 2 weeks and remained clear for the 6 months that she followed the diet. During this time, the excretion of nickel in the urine decreased to one half the prediet level and the amount of nickel in whole blood decreased to one third of the original level.[90] Depletion studies are difficult to carry out in a properly controlled manner, and the above results are based on open studies.

The results of treatment of nickel-allergic patients with disulfiram, which has a profound effect on nickel homeostasis, are also of interest, as diethyldithiocarbamate (a split product of disulfiram) is the drug of choice for nickel carbonyl poisoning.[91] In a controlled study, disulfiram had a significantly better effect on several parameters of nickel dermatitis than a placebo.[92] During the initial days of disulfiram therapy, some patients experienced a severe flare of the dermatitis complicated by a leukocytoclastic vasculitis. At the same time, the nickel levels in serum increased to 10 to 40 times the usual level.[93] Such flares indicate that alterations of the nickel homeostasis do have some influence on the course of nickel dermatitis.

Similarly, intravenous release of microgram doses of nickel from an infusion cannula may give rise to severe flares of nickel dermatitis, including bullous eruptions on the hands, generalized eruptions, and fever.[94] A dramatic increase of nickel levels in plasma was associated with such clinical symptoms in a dialysis patient.[95]

VI. SUBSTANCES WHICH CAUSE SYSTEMICALLY INDUCED CONTACT DERMATITIS

A. MEDICAMENTS

Largely on the basis of case reports, a number of drugs have been listed as having caused systemic contact dermatitis. Maibach[96] studied a group of patients who had discontinued the use of the clonidine transdermal therapeutic system due to dermatitis. Of 52 patients with positive patch tests, 29 were challenged orally with a therapeutic dose of clonidine. Only one patient reacted with a flare-up at the site of the original dermatitis. It was concluded that there is a dose relationship when dermatitis results from the oral administration of a substance suspected of causing dermatitis in patients with cellular immunity to this substance. Other examples of reactions to medicaments are given in Section II of this chapter.

B. OLEORESINS

Oleoresins in cashew nut shells have also given rise to epidemics of systemically induced contact dermatitis in patients sensitive to poison ivy.[13]

C. NICKEL

The evidence presented in Section V of this chapter indicates that nickel dermatitis may be aggravated by situations other than topical cutaneous contact with nickel. Likewise, there is some evidence that other ubiquitous haptens which commonly cause cellular immune reactions and eczema may also cause such reactions if the hapten is administered systemically.

D. COBALT

Cobalt allergy is common and is often associated with nickel allergy, particularly in females. In contrast to nickel, cobalt is well absorbed in the gastrointestinal tract. An oral challenge procedure carried out with nickel, cobalt, and a placebo among 146 patients with positive patch tests to nickel and/or cobalt showed a statistically significant relationship between positive cobalt patch tests and the results of the oral challenge with cobalt. Of 13 patients with positive patch tests only to cobalt, 7 had flares of dermatitis after oral challenge with 1 mg cobalt.[67]

E. CHROMIUM

Mali[19] demonstrated a positive Prausnitz-Küstner reaction in a chromate-sensitive patient whose vesicular hand eczema flared after inhalation of paint fumes containing chromate. Oral challenge with potassium dichromate in doses of from 50 μg to 10 mg produced flares of dermatitis in a number of chromate-sensitive patients.[97-99] In a placebo-controlled study Kaaber and Veien[100] showed that 11 of 31 chromate-sensitive patients experienced flares of their dermatitis after oral challenge with 2.5 mg chromium given as potassium dichromate. In another study one of three patients challenged orally with 50 μg potassium dichromate had a positive reaction to the challenge.[101]

F. BALSAM OF PERU

Oral challenge with 20 to 50 mg of balsam of Peru and challenge with individual spices have occasionally given rise to flares of dermatitis.[102-106] A double-blind, placebo-controlled study showed that 10 of 17 patients with positive patch tests to balsam of Peru had flares of dermatitis after oral challenge with 1 g balsam of Peru, compared with 6 of 26 with positive patch tests to wood tar, perfume mixture, and/or colophony. One in each group reacted to the placebo and 2 of the 26 to balsam of Peru as well as the placebo.[107]

The clinical relevance of the above oral challenge results remains to be confirmed by, for example, well-controlled depletion trials. An open trial to reduce the intake of balsams in balsam-sensitive patients led to an improvement in the dermatitis of approximately one half of the patients who followed a low-balsam diet for at least 1 month.[108]

VIII. MECHANISMS INVOLVED IN SYSTEMICALLY INDUCED CONTACT DERMATITIS

Most investigations of the mechanisms of reaction of systemically induced dermatitis have involved nickel-allergic patients. Histological examination of patch test sites which flared after oral challenge with nickel showed eczematous lesions,[109] and immune histochemical examinations showed that cellular immune reactions, rather than antibody-mediated reactions, had taken place.[110] Möller[111] considered flares of dermatitis seen in nickel-sensitized mice after epicutaneous application of nickel at another site to be due to cellular immunity. An increase in the levels of lymphocyte transformation after oral challenge with nickel supports the above findings.[112]

Sjövall et al.[113] attempted to hyposensitize nickel-allergic patients by repeated oral administration of nickel. Nickel sensitivity declined with doses of 5 mg nickel per week, while tuberculin sensitivity was unaffected, thus indicating a nickel-specific mechanism.

Oral ingestion of the hapten in 21 patients sensitive to poison ivy or poison oak produced hyposensitization significantly more often than did a placebo. The hyposensitization appeared to be antigen specific.[114]

Other mechanisms also may be involved in the reactions seen in nickel-allergic patients after oral challenge. Christensen[109] noted widespread erythema which appeared a few hours after a nickel-allergic patient was challenged with nickel, and Veien et al.[115] detected antibodies against a nickel-albumin complex (which had produced antibodies in rabbits) in the serum of similar patients.

Eruptions seen when nickel-allergic patients treated with disulfiram also had a high level of nickel in their plasma were histologically similar to leukocytoclastic vasulitis.[93] This would suggest an immune-complex-mediated reaction. Such reactivity was also suggested by Polak and Turk.[116] Excoriated lesions in nickel-allergic patients were described as nickel vasulitis by Hjorth;[117] similar lesions were studied by Veien and Krogdahl,[118] who saw leukocytoclastic vasculitis in two patients and perivascular lymphohistiocytic infiltrates in the upper dermis of nine others. None of the patients had spongiosis or other evidence of classical allergic contact dermatitis.

A study by Meneghini et al.[55] showed that the clinical and histological features of systemic reactions to various drugs in patients with positive patch tests to the same substances were similar to erythema multiforme and other types of vasculitis. A search for circulating immune complexes was also carried out among these patients, but none were found.

There are various types of immune reactions, as evidenced by patients with adverse reactions to nickel who show no evidence of cellular immune reactions and in whom nickel nonetheless causes asthma, rhinitis, and immediate-type reactions.[119-121] One of the common clinical features of patients suspected to suffer from systemically induced contact dermatitis is recurrent vesicular hand eczema, often called pompholyx. This morphology of dermatitis has been associated with a number of causes and should probably be regarded as a nonspecific reaction pattern on the palms, fingers, soles, and toes.

Although external contact with a hapten in individuals sensitized to it can cause vesicular hand eczema,[122,123] it is commonly considered to be an endogenous dermatosis. In some situations, dermatophytoses, commonly seen on the feet, have caused "id" reactions on the hands. Psychological stress and premenstrual tension have also been associated with vesicular hand eczema, and a significant relationship to atopy has also been noted. Menné and Hjorth[124] reviewed the currently known causes of vesicular hand eczema and concluded that there is still insufficient knowledge of its causes.

Since then, Schwanitz[125] has studied a group of patients with vesicular eruptions on the hands, and he found many smokers among them. He discussed the possible presence of irritants in sweat. Edman[72] also found an association between vesicular hand eczema and smoking, as well as an association between this type of dermatitis and the intake of acetylsalicylic acid and oral contraceptives.

Patch-test-negative patients with vesicular hand eczema have been seen to react with flares of dermatitis after oral challenge with nickel, cobalt, chromium, and balsam of Peru.[107,126] Some of the patients who reacted to nickel also had flares at previous nickel contact sites, and a careful history of that stage revealed that they were, in fact, probably nickel allergic and had had false-negative patch tests. For this group of patients, the vesicular hand eczema could be both a specific and a nonspecific reaction. The reaction could be specific to circulating proteins associated with the immunological reactions, and a possible specificity could be hapten coupled to these proteins. Only sophisticated investigations of the immunological processes in the area of vesicular hand dermatitis itself can disclose whether this clinical response is a specific event. Until such investigations have been carried out, we are left with classical methods such as the history of the patient, the results of cutaneous testing (possibly supplemented by *in vitro* tests), the results of oral challenge procedures, and the results of depletion studies in order to arrive at a correct diagnosis of systemically induced contact dermatitis.

REFERENCES

1. **Bernard, P., Fayol, J., Bonnafoux, A., Bedane, D., Delrous, J. L., Catanzano, G., and Bonnetblanc, J. M.,** Toxidermies après prise orale de pristinamycine, *Ann. Dermatol. Venereol.,* 115, 63, 1988.
2. **Rasmussen, L. P. and Menné, T.,** Systemisk kontakteksem — "baviansyndromet" — ved ampicillin-allergi, *Ugeskr. Laeg.,* 147, 1341, 1985.
3. **Andersen, K. E.,** Systemic contact dermatitis, *Acta Derm. Venereol.,* 68 (Suppl. 135), 62, 1988.
4. **Ghadially, R. and Ramsay, C. A.,** Gentamicin: systemic exposure to a contact allergen, *J. Am. Acad. Dermatol.,* 19, 428, 1988.
5. **Pigatto, P. D., Morelli, M., Polenghi, M. M., Mozzanica, N., and Altomare, G. F.,** Phenobarbital-induced allergic dermatitis, *Contact Dermatitis,* 16, 279, 1987.
6. **Lechner, T., Grytzmann, B., and Bäurle, G.,** Hämatogenes allergisches Kontaktekszem nach oraler Gabe von Nystatin, *Mykosen,* 30, 143, 1987.
7. **de Groot, A. C. and Conemans, J.,** Allergic urticarial rash from oral codeine, *Contact Dermatitis,* 14, 209, 1986.
8. **Nishimura, M., Takano, Y., and Toshitani, S.,** Systemic contact dermatitis, medicamentosa occurring after intravesical dimethyl sulfoxide treatment for interstitial cystitis, *Arch. Dermatol.,* 124, 182, 1988.
9. **Tagami, H., Tetauta, K., Iwataki, K., and Yamada, M.,** Delayed hypersensitivity in ampicillin-induced toxic epidermal necrolysis, *Arch. Dermatol.,* 119, 910, 1983.
10. **Förström, L., Hannuksela, M., Kousa, M., and Lehmuskallio, E.,** Merthiolate hypersensitivity and vaccination, *Contact Dermatitis,* 6, 241, 1980.
11. **Ekenvall, L. and Forsbeck, M.,** Contact eczema produced by a β-adrenergic blocking agent (alprenolol), *Contact Dermatitis,* 4, 190, 1978.
12. **Hindson, C.,** Contact eczema from methyl salicylate reproduced by oral aspirin (acetylsalicylic acid), *Contact Dermatitis,* 3, 348, 1977.
13. **Marks, J. G., Jr., DeMelfi, T., McCarthy, M. A., Witte, E. J., Castagnoli, N., Epstein, W. L., and Aber, R. C.,** Dermatitis from cashew nuts, *J. Am. Acad. Dermatol.,* 10, 627, 1984.
14. **Becker, L. E. and Skipworth, G. B.,** Ginkgo-tree dermatitis, stomatitis, and proctitis, *J. Am. Med. Assoc.,* 231, 1162, 1975.
15. **Fisher, A. A.,** The clinical significance of positive patch test reactions to balsam of Peru, *Cutis,* 13, 909, 1974.
16. **Ekelund, A.-G. and Möller, H.,** Oral provocation in eczematous contact allergy to neomycin and hydroxyquinolines, *Acta Derm. Venereol.,* 49, 422, 1969.
17. **Leifer, W. and Steiner, K.,** Studies in sensitization to halogenated hydroxyquinolines and related compounds, *J. Invest. Dermatol.,* 17, 233, 1951.

18. **Park, R. G.,** Cutaneous hypersensitivity to sulphonamides, *Br. Med. J.*, 2, 69, 1943.
19. **Mali, J. W. H.,** Über einen Fall von dyshidrotischem Ekzem durch Chromat, *Hautarzt,* 11, 27, 1960.
20. **Boonk, W. J., and Van Ketel, W. G.,** Nickel dermatitis after swallowing a nickel coin (Dutch florin), *Contact Dermatitis,* 5, 341, 1979.
21. **Lacroix, J., Morin, C. L., and Collin, P.-P.,** Nickel dermatitis from a foreign body in the stomach, *J. Pediatr.*, 95, 428, 1979.
22. **Oakley, A. M. M., Ive, F. A., and Carr, M. M.,** Skin clips are contraindicated when there is nickel allergy, *J. R. Soc. Med.*, 80, 1987.
23. **Kvorning, S. A.,** Post-operative dermatitis following non-epidermal nickel contact, *Contact Dermatitis,* 1, 327, 1975.
24. **Oleffe, J. and Wilmet, J.,** Generalized dermatitis from an osteosynthesis screw, *Contact Dermatitis,* 6, 365, 1980.
25. **Barranco, V. P. and Soloman, H.,** Eczematous dermatitis from nickel, *J. Am. Med. Assoc.*, 220, 1244, 1972.
26. **McKenzie, A. W., Aitken, C. V. E., and Ridsdill-Smith, R.,** Urticaria after insertion of Smith-Petersen Vitallium nail, *Br. Med. J.*, 4, 36, 1967.
27. **Meyrick Thomas, R. H., Rademaker, M., Goddard, N. J., and Munro, D. D.,** Severe eczema of the hands due to an orthopaedic plate made of Vitallium, *Br. Med. J.*, 294, 106, 1987.
28. **Grimalt, F. and Romaguera, C.,** Acute nickel dermatitis from a metal implant, *Contact Dermatitis,* 6, 441, 1980.
29. **Busch, H. P. and Wendt, V.,** Hämatogene Kontaktdermatitis gegen Nickelionen durch Septopal-Minikette, *Dtsch. Z. Mund. Kiefer GesichtsChir.*, 11, 247, 1987.
30. **Waterman, A. H. and Schrik, J. J.,** Allergy in hip arthroplasty, *Contact Dermatitis,* 13, 294, 1985.
31. **Hubler, W. R., Jr. and Hubler, W. R., Sr.,** Dermatitis from a chromium dental plate, *Contact Dermatitis,* 9, 377, 1983.
32. **Glendenning, W. E.,** Allergy to cobalt in metal denture as cause of hand dermatitis, *Contact Dermatitis Newsl.*, 10, 225, 1971.
33. **Landwehr, A. J. and van Ketel, W. G.,** Pompholyx after implantation of a nickel-containing pacemaker in a nickel-allergic patient, *Contact Dermatitis,* 9, 147, 1983.
34. **Romaguera, C. and Grimalt, F.,** Contact dermatitis from a copper-containing intrauterine contraceptive device, *Contact Dermatitis,* 7, 163, 1981.
35. **Barranco, V. P.,** Eczematous dermatitis caused by internal exposure to copper, *Arch. Dermatol.*, 106, 386, 1972.
36. **Romaguera, C. and Grimalt, F.,** Pacemaker dermatitis, *Contact Dermatitis,* 7, 333, 1981.
37. **White, I. R. and Smith, B. G. N.,** Dental amalgam dermatitis, *Br. Dent. J.*, 156, 259, 1984.
38. **Menné, T. and Solgaard, P.,** Temperature-dependent nickel release from nickel alloys, *Contact Dermatitis,* 5, 82, 1979.
39. **Flyvholm, M.-S., Dalsgaard Nielsen, G., and Andersen, A.,** Nickel content of food and estimation of dietary intake, *Z. Lebensm. Unters. Forsch.*, 179, 427, 1984.
40. **Hemingway, J. D. and Molokhia, M. M.,** The dissolution of metallic nickel in artificial sweat, *Contact Dermatitis,* 16, 99, 1987.
41. **Magnusson, B., Bergman, M., Bergman, B., and Söremark, R.,** Nickel allergy and nickel-containing dental alloys, *Scand. J. Dent. Res.*, 90, 163, 1982.
42. **Samitz, M. H. and Katz, S. A.,** Nickel dermatitis hazards from prostheses, *Br. J. Dermatol.*, 92, 287, 1975.
43. **Kleinhans, D. and Knoth, W.,** Praben-Kontaktallergie mit enteraler Provokation, *Z. Hautkr.*, 48, 699, 1973.
44. **Roed-Petersen, J. and Hjorth, N.,** Contact dermatitis from antioxidants, *Br. J. Dermatol.*, 94, 233, 1976.
45. **Goitre, M., Bedello, P. G., and Cane, D.,** Allergic dermatitis and oral challenge to tetramethylthiuram disulphide, *Contact Dermatitis,* 7, 272, 1981.
46. **Angelini, G. and Meneghini, C. L.,** Oral tests in contact allergy to *para*-amino compounds, *Contact Dermatitis,* 7, 311, 1981.
47. **Hannuksela, M. and Förström, L.,** Reactions to peroral propylene glycol, *Contact Dermatitis,* 4, 41, 1978.
48. **Baer, R. L. and Leider, M.,** The effects of feeding certified food azodyes in paraphenylenediamine-hypersensitive subjects, *J. Invest. Dermatol.*, 13, 223, 1949.
49. **Shelmire, B.,** Cutaneous and systemic reactions observed during oral poison ivy therapy, *J. Allergy,* 12, 252, 1941.
50. **Kligman, A. M.,** Poison ivy *(Rhus)* dermatitis, *Arch. Dermatol.*, 77, 149, 1958.
51. **Tilsley, D. A. and Rotstein, H.,** Sensitivity caused by internal exposure to nickel, chrome and cobalt, *Contact Dermatitis,* 6, 175, 1980.
52. **Calnan, C. D.,** Nickel dermatitis, *Br. J. Dermatol.*, 68, 229, 1956.

53. **Marcussen, P. V.,** Spread of nickel dermatitis, *Dermatologica,* 115, 596, 1957.
54. **Christensen, O. B.,** Nickel Allergy and Hand Eczema in Females, Doctoral thesis, Lund University, Malmö, Sweden, 1981, 28.
55. **Meneghini, C. L. and Angelini, G.,** Secondary polymorphic eruptions in allergic contact dermatitis, *Dermatologica,* 163, 63, 1981.
56. **Degreef, H., Bonamie, A., Van Derheyden, D., and Dooms-Goossens, A.,** Mephenesin contact dermatitis with erythema multiforme features, *Contact Dermatitis,* 10, 220, 1984.
57. **Bruynzeel, D. P., Van Den Hoogenband, H. M., and Koeduk, F.,** Purpuric vasculitis-like eruption in a patient sensitive to balsam of Peru, *Contact Dermatitis,* 11, 207, 1984.
58. **Christensen, O. B. and Möller, H.,** External and internal exposure to the antigen in the hand eczema of nickel allergy, *Contact Dermatitis,* 1, 136, 1975.
59. **Kaaber, K., Menné, T., Tjell, J. C., and Veien, N. K.,** Antabuse® treatment of nickel dermatitis. Chelation — a new principle in the treatment of nickel dermatitis, *Contact Dermatitis,* 5, 221, 1979.
60. **Sertoli, A., Lombardi, P., Francalanci, S., Gola, M., Girogini, S., and Panconesi, E.,** Effetto della somministrazione orale de apteni in soggetti sensibilizatti affetti da eczema allergico da contatto, *G. Ital. Dermatol. Venereol.,* 120, 213, 1985.
61. **Kaaber, K., Veien, N. K., and Tjell, J. C.,** Low nickel diet in the treatment of patients with chronic nickel dermatitis, *Br. J. Dermatol.,* 98, 197, 1978.
62. **Jordan, W. P., Jr. and King, S. E.,** Nickel feeding in nickel-sensitive patients with hand eczema, *J. Am. Acad. Dermatol.,* 1, 506, 1979.
63. **Burrows, D., Creswell, S., and Merret, J. D.,** Nickel, hands and hip prostheses, *Br. J. Dermatol.,* 105, 437, 1981.
64. **Bedello, P. G., Goitre, M., Cane, D., and Roncarolo, G.,** Nickel: aptene ubiquitario, *G. Ital. Dermatol. Venereol.,* 120, 293, 1985.
65. **Cronin, E., de Michiel, A., and Brown, S. S.,** Oral nickel challenge in nickel-sensitive women with hand eczema, in Abstr. Swansea Conf. Nickel Toxicology, Swansea, Wales, September 3 to 6, 1980, 91.
66. **Gawkrodger, D. J., Cook, S. W., Fell, G. S., and Hunter, J. A. A.,** Nickel dermatitis: the reaction to oral nickel challenge, *Br. J. Dermatol.,* 115, 33, 1986.
67. **Veien, N. K., Hattel, T., Justesen, O., and Nørholm, A.,** Oral challenge with nickel and cobalt in patients with positive patch tests to nickel and/or cobalt, *Acta Derm. Venereol.,* 67, 321, 1987.
68. **Roduner, J., Haudenschild-Falb, E., Kunz, E., Hunziker, T., and Krebs, A.,** Perorale Nickelprovokation bei nichtdyshidrosiformem und dyshidrosiformem Nickeleksem, *Hautarzt,* 38, 262, 1987.
69. **Thelin, I. and Agrup, G.,** Pompholyx — a one year series, *Acta Derm. Venereol.,* 65, 214, 1985.
70. **Menné, T.,** Nickel Allergy, Thesis, University of Copenhagen, Copenhagen, Denmark, 1983, 69.
71. **Peltonen, L.,** Nickel sensitivity in the general population, *Contact Dermatitis,* 5, 27, 1979.
72. **Edman, B.,** Palmar eczema: a pathogenetic role for acetylsalicylic acid, contraceptives and smoking?, *Acta Derm. Venereol.,* 68, 402, 1988.
73. **de Jongh, G. J., Spruit, D., Bongaarts, P. J. M., and Duller, P.,** Determination of factors influencing a case of nickel contact dermatitis. I. Presentation of data, *Contact Dermatitis,* 4, 142, 1978.
74. **de Jongh, G. J. and Spruit, D.,** Factors influencing nickel dermatitis. II., *Contact Dermatitis,* 4, 149, 1978.
75. **Marcussen, P. V.,** Comparison of intradermal test and patch test using nickel sulfate and formaldehyde, *J. Invest. Dermatol.,* 40, 263, 1963.
76. **Fischer, T. and Maibach, H.,** Amount of nickel applied with a standard patch test, *Contact Dermatitis,* 11, 285, 1984.
77. **Menné, T., Brandrup, F., Thestrup-Pedersen, K., Veien, N. K., Andersen, J. R., Yding, F., and Valeur, G.,** Patch test reactivity to nickel alloys, *Contact Dermatitis,* 16, 255, 1987.
78. **Emmett, E. A., Risby, T. H., Jiang, L., Ng, S. K., and Feinman, S.,** Allergic contact dermatitis to nickel: bioavailability from consumer products and provocation threshold, *J. Am. Acad. Dermatol.,* 19, 314, 1988.
79. **Wahlberg, J. E. and Skog, E.,** Nickel allergy and atopy, *Br. J. Dermatol.,* 85, 97, 1971.
80. **Dooms-Goossens, A., Naert, C., Chrispeels, M. T., and Degreef, H.,** Is a 5% nickel sulfate patch test concentration adequate?, *Contact Dermatitis,* 6, 232, 1980.
81. **Sunderman, F. W., Jr., Hopfer, S. M., Swift, T., Ziebka, L., Marcus, A. H., Most, B. M., and Creason, J.,** Nickel absorption and elimination in human volunteers, in *Trace Elements in Human Health and Disease,* 2nd Nordic Symp., Odense University, Odense, Denmark, August, 1987.
82. **Solomons, N. W., Viteri, F., Shuler, T. R., and Nielsen, F. H.,** Bioavailability of nickel in man: effects of foods and chemically-defined dietary constituents on the absorption of inorganic nickel, *J. Nutr.,* 112, 39, 1982.
83. **Schöfer, H. Von, Rosenberger, G., Hottenrott, C., Strohm, W. D., Burmester, H. J., Gonnermann, B., and Hensel, O.,** Sensibilisierung gegen Nickelsulfat bei Patienten mit Ileitis terminalis (Morbus Crohn), *Dermatosen,* 36, 157, 1988.

84. **Christensen, O. B., Möller, H., Andrasko, L., and Lagesson, V.,** Nickel concentration of blood, urine and sweat after oral administration, *Contact Dermatitis,* 5, 312, 1979.
85. **Andersen, K. E., Benezra, C., Burrows, D., Camarasa, J., Dooms-Goossens, A., Ducombs, G., Frosch, P., Lachapelle, J.-M., Lahti, A., Menné, T., Rycroft, R., Scheper, R., White, I., and Wilkinson, J.,** Contact dermatitis. A review, *Contact Dermatitis,* 16, 55, 1987.
86. **Nielsen, G. D.,** Oral challenge of nickel-allergic patients with hand eczema, in *Nickel and Human Health: Current Perspectives,* Nieboer, E. and Aitio, A., Eds., John Wiley & Sons, in press.
87. **Veien, N. K. and Andersen, M. R.,** Nickel in Danish food, *Acta Derm. Venereol.,* 66, 502, 1986.
88. **Kaaber, K., Veien, N. K., and Tjell, J. C.,** Low nickel diet in the treatment of patients with chronic nickel dermatitis, *Br. J. Dermatol.,* 98, 197, 1978.
89. **Veien, N. K., Hattel, T., Justesen, O., and Nørholm, A.,** Dietary treatment of nickel dermatitis, *Acta Derm. Venereol.,* 65, 138, 1985.
90. **Gawkrodger, D. J., Shuttler, I. L., and Delves, H. T.,** Nickel dermatitis and diet: clinical improvement and a reduction in blood and urine nickel levels with a low-nickel diet, *Acta Derm. Venereol.,* 68, 453, 1988.
91. **Sunderman, F. W., Jr., Aitio, A., Morgan, L. G., and Norseth, T.,** Biological monitoring of nickel, *Toxicol. Ind. Health,* 2, 17, 1986.
92. **Kaaber, K., Menné, T., Veien, N., and Hougaard, P.,** Treatment of nickel dermatitis with Antabuse®, a double-blind study, *Contact Dermatitis,* 9, 297, 1983.
93. **Kaaber, K., Menné, T., Tjell, J. C., and Veien, N.,** Antabuse® treatment of nickel dermatitis. Chelation — a new principle in the treatment of nickel dermatitis, *Contact Dermatitis,* 5, 221, 1979.
94. **Smeenk, G. and Teunissen, P. C.,** Allergische reacties op nikkel uit infusie-toedieningssystemen, *Ned. Tijdschr. Geneeskd.,* 121, 4, 1977.
95. **Olerud, J. E., Lee, M. Y., Uvelli, D. A., Goble, G. J., and Babb, A. L.,** Presumptive nickel dermatitis from hemodialysis, *Arch. Dermatol.,* 120, 1066, 1984.
96. **Maibach, H. I.,** Oral substitution in patients sensitized by transdermal clonidine treatment, *Contact Dermatitis,* 16, 1, 1987.
97. **Fregert, S.,** Sensitization to hexa- and trivalent chromium, in *Pemphigus: Occupational Dermatosis due to Chemical Sensitization,* Liebner, E. and Flórián, E., Eds., Budapest, 1965, 50.
98. **Schleiff, P.,** Provokation des Chromatekzems zu Testzwecken durch interne Chromzufuhr, *Hautarzt,* 19, 209, 1968.
99. **Goitre, M., Bedello, P. G., and Cane, D.,** Chromium dermatitis and oral administration of the metal, *Contact Dermatitis,* 8, 203, 1982.
100. **Kaaber, K. and Veien, N. K.,** The significance of chromate ingestion in patients allergic to chromate, *Acta Derm. Venereol.,* 57, 321, 1977.
101. **Sertoli, A., Lombardi, P., Francalanci, S., Gola, M., Giorgini, S., and Panconesi, E.,** Effetto della somministrazione oral de apteni in soggetti sensibilizzati affetti da eczema allergico da contatto, *G. Ital. Dermatol. Venereol.,* 120, 213, 1985.
102. **Fisher, A. A.,** Allergic eczematous contact dermatitis due to foods, *Cutis,* 16, 603, 1975.
103. **Forsbeek, M. and Skog, E.,** Immediate reactions to patch tests with balsam of Peru, *Contact Dermatitis,* 3, 201, 1977.
104. **Niinimäki, A. and Hannuksela, M.,** Immediate skin test reactions to spices, *Allergy,* 36, 487, 1981.
105. **Bedello, P. G., Goitre, M., and Cane, D.,** Contact dermatitis and flare from food flavouring agents, *Contact Dermatitis,* 8, 143, 1982.
106. **Warin, R. P. and Smith, F. J.,** Chronic urticaria investigations with patch and challenge tests, *Contact Dermatitis,* 8, 117, 1982.
107. **Veien, N. K., Hattel, T., Justesen, O., and Nørholm, A.,** Oral challenge with balsam of Peru, *Contact Dermatitis,* 12, 104, 1985.
108. **Veien, N. K., Hattel, T., Justesen, O., and Nørholm, A.,** Reduction of intake of balsams in patients sensitive to balsam of Peru, *Contact Dermatitis,* 3, 270, 1985.
109. **Christensen, O. B., Lindström, C., Löfberg, H., and Möller, H.,** Micromorphology and specificity of orally induced flare-up reactions in nickel-sensitive patients, *Acta Derm. Venereol.,* 61, 505, 1981.
110. **Christensen, O. B., Beckstead, J. H., Daniels, T. E., and Maibach, H. I.,** Pathogenesis of orally induced flare-up reactions at old patch sites in nickel allergy, *Acta Derm. Venereol.,* 65, 298, 1985.
111. **Möller, H.,** Flare-up of allergic contact dermatitis in the mouse after topical distant provocation, *Acta Derm. Venereol.,* 64, 125, 1984.
112. **Veien, N. K., Svejgaard, E., and Menné, T.,** *In vitro* lymphocyte transformation to nickel: a study of nickel-sensitive patients before and after epicutaneous and oral challenge with nickel, *Acta Derm. Venereol.,* 59, 447, 1979.
113. **Sjövall, P., Christensen, O. B., and Möller, H.,** Oral hyposensitization in nickel allergy, *J. Am. Acad. Dermatol.,* 17, 774, 1987.
114. **Epstein, W. L., Byers, V. S., and Frankart, W.,** Induction of antigen specific hyposensitization to poison oak in sensitized adults, *Arch. Dermatol.,* 118, 630, 1982.

115. **Veien, N. K., Christiansen, A. H., Svejgaard, E., and Kaaber, K.,** Antibodies against nickel-albumin in rabbits and man, *Contact Dermatitis,* 5, 378, 1979.
116. **Polak, L., Turk, J. L., and Frey, J. R.,** Studies on contact hypersensitivity to chromium compounds, *Prog. Allergy,* 17, 145, 1973.
117. **Hjorth, N.,** Nickel vasculitis, *Contact Dermatitis,* 2, 356, 1976.
118. **Veien, N. K. and Krogdahl, A.,** Is nickel vasculitis a clinical entity?, in *Current Topics in Contact Dermatitis,* Forsch, P. J., Dooms-Goøsens, A., Lachapelle, J.-M., Rycroft, F. J. G., and Scheper, R. J., Eds., Springer-Verlag, Berlin, 1989, 172.
119. **Mitchell, J. C.,** Type I without type III and type IV hypersensitivity reaction to nickel, *Contact Dermatitis,* 7, 270, 1981.
120. **Niordson, A.-M.,** Nickel sensitivity as a cause of rhinitis, *Contact Dermatitis,* 7, 273, 1981.
121. **Block, G. T. and Yeung, M.,** Asthma induced by nickel, *J. Am. Med. Assoc.,* 247, 1600, 1982.
122. **Hjorth, N. and Roed-Petersen, J.,** Occupational protein contact dermatitis in food handlers, *Contact Dermatitis,* 2, 28, 1976.
123. **Schorr, W. F.,** The skin and chemical additives to food, *Arch. Dermatol.,* 105, 131, 1972.
124. **Menné, T. and Hjorth, N.,** Pompholyx — dyshidrotic eczema, *Semin. Dermatol.,* 2, 75, 1983.
125. **Schwanitz, H. J.,** *Das Atopische Palmoplantarekzem,* Springer-Verlag, Berlin, 1986.
126. **Veien, N. K., Hattel, T., Justesen, O., and Nørholm, A.,** Oral challenge with metal salts. I. Vesicular patch-test negative hand eczema, *Contact Dermatitis,* 9, 402, 1983.

Chapter 23

KATHON® CG — AN UNUSUAL CONTACT SENSITIZER

Magnus Bruze, Birgitta Gruvberger, and Bert Björkner

TABLE OF CONTENTS

I. INTRODUCTION

Preservatives are required in water-based products/systems to prevent growth of bacteria, fungi, yeasts, and algae. These microorganisms may be pathogenic for humans, but even when harmless for humans they may be undesirable because they degrade constituents of water-based products/systems. Degradation of constituents may imply loss or change of activity, discoloration, and malodor. Preferably, the preservatives themselves should not constitute any hazard to human health or the environment. In recent years some preservatives, and in particular formaldehyde, have been questioned due to suspicions that they possess carcinogenic properties. This may, in part, explain why the new preservative Kathon® CG has become widely distributed so rapidly. Also contributing to this development are other factors such as effectiveness, degradation in the environment,[1,2] and the marketing claim that Kathon® CG is safe to human health when used as recommended by the manufacturer.

However, preservatives are biologically active, and most preservatives seem to possess the capacity to sensitize. Thus, it is not surprising that Kathon® CG is a sensitizer, but some of the features in its sensitizing capacity are more remarkable. Within the last few years, more articles seem to be published on Kathon® CG as a sensitizer than on any other allergen. There has been and still is some controversy about Kathon® CG, particularly concerning its allergic and irritant properties, patch test technique for diagnostic purpose, and clinical relevance of demonstrated hypersensitivity. Based on our own experience and data published in the literature, this chapter is an attempt to clarify what is known about Kathon® CG as a sensitizer, emphasizing the controversial issues. It is these authors' conviction that our understanding will increase if the information is given in those steps in which it has been achieved.

II. KATHON® PRESERVATIVES

Kathon® preservatives are manufactured by Rohm and Haas Company, Philadelphia, and contain either 2-octyl-4-isothiazolin-3-one or methylisothiazolinones. Of these Kathon® CG (cosmetic grade) is the most frequently discussed in dermatological literature. It consists of two active ingredients (a.i.), stabilizing salts, and water. 5-Chloro-2-methyl-4-isothiazolin-3-one (5243-K-CG) and 2-methyl-4-isothiazolin-3-one (243-K-CG) (Figure 1) constitute the a.i., and due to conditions present during synthesis the chlorinated a.i. is formed at a three times higher concentration than the nonchlorinated one. The CTFA-adopted names for 5243-K-CG and 243-K-CG are methylchloroisothiazolinone and methylisothiazolinone, respectively. Rohm and Haas also manufactures other methylisothiazolinone-based preservatives with the same principal composition as Kathon® CG (Table 1). These preservatives are corrosive at high concentrations, but maintain effectiveness against bacteria, fungi, yeasts, and algae at low concentrations; they have a broad spectrum of applications, some of which are shown in Table 2. The recommended use concentrations vary between 1 and 5 ppm a.i. in industrial water treatment and up to 35 ppm in an electrodeposition paint system.[3,4] For cosmetics and toiletries a maximum concentration of 15 ppm a.i. is recommended.[5]

III. ESTABLISHING AS A SENSITIZER

A. THE FIRST PATCH TEST REACTIONS

In 1981, two workers with hand dermatitis suspected to be occupation related were referred to the Department of Occupational Dermatology, General Hospital, Malmö. Both individuals were exposed to cutting fluids which, according to data sheets, contained the preservative Kathon® 886 MW (metal working). At this time Kathon® 886 MW was com-

A

B

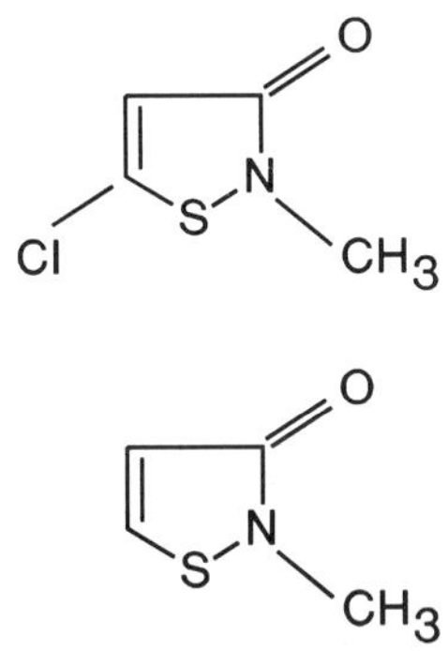

FIGURE 1. Structures of (A) 5-chloro-2-methyl-4-isothiazolin-3-one and (B) 2-methyl-4-isothiazolin-3-one.

TABLE 1
Various Methylisothiazolinone-Based Preservatives Manufactured by Rohm and Haas Company

Kathon® CG
Kathon® CG/ICP
Kathon® LX
Kathon® WT
Kathon® 886 MW

TABLE 2
Products and Processes Where Methylisothiazolinones Can Be Used

Cosmetics	Paints
Cutting fluids	Swimming pool water
Cooling tower water	Toiletries
Detergents	In commercial photoprocessing
Fountain solutions (offset)	In paper mill production
Glues	

pletely unknown to Swedish dermatologists. When patch tested with Kathon® 886 MW in water at 1000 ppm a.i., both workers reacted positively. At this time nothing was published on the sensitizing capacity of Kathon® 886 MW, in humans or in animals; nor did the manufacturer know of any sensitizing capacity of the preservative. Furthermore, the morphology of the positive patch test reactions had features of both an allergic and an irritant reaction. The elevated infiltration with a papular-vesicular surface indicated an allergic reaction, while a sharp demarcation to macroscopically normal skin was evidence of an irritant nature. To further elucidate the nature of the reaction, it was decided to perform patch testing with serial dilutions and patch testing of controls.[6]

B. PATCH TESTING OF CONTROLS AND WITH SERIAL DILUTIONS

According to the manufacturer, Kathon® CG contains the fewest contaminants of the preservatives listed in Table 1. Because of this, Kathon® CG was inserted into our routine patch test series at the Departments of Occupational Dermatology in Malmö and Lund.[6] Kathon® CG was diluted in water, and the intended patch test concentration was 300 ppm a.i. Initially, positive patch test reactions were obtained only in Malmö. The Finn chamber technique is used in Malmö for patch testing, while the Al-test technique is used in Lund.

TABLE 3
Result of Patch Testing with Serial Dilutions in Kathon® CG-Reacting Individuals

Test concentration in ppm a.i.[a]	Number of reactors
300	34
100	17
30	8
10	2

[a] Active ingredients.

The discrepancy in results was therefore believed to be due to differences in technique. The aluminum foil in the Al-test is covered by a polyethylene film, while the aluminum cup in the Finn Chamber® is uncovered. Irritant degradation products of Kathon® CG were believed to be formed in Finn Chambers® with aluminum as a catalyst. However, the concentration of the a.i. of Kathon® CG was found to be too low in Lund; when corrected, positive patch test reactions also were obtained in Lund.[6] Surprisingly, a great number of positive patch test reactions were obtained, with 4.4% of consecutively tested patients reacting. The morphology of the reactions was similar for all patients and did not differ from the positive reactions to Kathon® 886 MW that were noted in the two metal workers. This high prevalence of positive patch test reactions in the controls was indicative of an irritant nature of the reactions, but did not exclude an allergic nature.[6]

Neither did the result of patch testing with serial dilutions of Kathon® CG facilitate the interpretation of the patch test reactions.[6] Starting with the concentration of an allergen giving moderate patch test reaction, it is usually possible to decrease this concentration 100 times without losing the capacity to elicit a positive patch test reaction.[7,8] On the other hand, the ability to elicit an irritant reaction is usually lost with minute dilution steps from the concentration giving a moderate reaction. Thus, the result shown in Table 3 with loss of reactivity with a tenfold dilution in the majority of patients was supportive of an irritant nature of the reactions.

In summary, extensive investigations to elucidate the nature of the patch test reactions had not been conclusive. Most facts suggested an irritant nature, but an allergic nature could not be ruled out entirely. Still, nothing was published on any sensitizing capacity of Kathon® CG, and the manufacturer did not disclose any information on such a capacity. At this point it was decided to patch test with Kathon® CG at a higher concentration. If the test reactions in the controls represented irritant reactions, it was believed that a concentration increase of more than threefold would at least double the frequency of positive patch test reactions, while it was considered to be unlikely to find a prevalence of hypersensitivity to a virtually unknown preservative at around 10%.

C. IRRITANT REACTIONS AND PATCH TEST SENSITIZATION

A total of 40 patients were simultaneously patch tested with Kathon® CG at 1000 ppm and 300 ppm a.i.[6] No reaction of the previously mentioned type was noted when the tests were read after 3 d. However, at this time a positive reaction of a different type was noted in ten patients. Macroscopically, these reactions were sharply demarcated and brownish red, with a slightly shiny and dry surface. Thus, this morphology was consistent with what usually is considered to be a typical irritant reaction. However, the other type of reaction appeared on the back of some patients 10 to 30 d after the application of the first patch test.

Furthermore, these reactions were located on that part of the back where Kathon® CG had been applied. Kathon® CG and the other test preparations which had been patch tested on that part of the back were retested; positive reactions were noted in all retested patients at the regular reading time (after 3 d), but only to Kathon® CG. By definition, a negative patch test reaction followed by a flare-up reaction after 10 to 30 d and then a positive reaction after 3 d at retesting means that sensitization was induced by the patch test procedure;[9] i.e., Kathon® CG was shown to be a sensitizer for the first time.

The fact that Kathon® CG was shown to be a contact sensitizer did not mean that the other positive patch test reactions to Kathon® CG at 300 ppm a.i. necessarily represented allergic reactions. However, two facts strongly argued in favor of this interpretation. First, the morphology of the flare-up reactions to Kathon® CG and of the positive patch test reactions to Kathon® CG after 3 d in those who were retested because of a flare-up reaction, as well as of the positive reactions after 3 d in those reacting when patch tested the first time, was the same. Second, the strength of reactivity was the same for the patch-test-sensitized individuals and the patients reacting positively when routinely patch tested the first time; i.e., it was impossible to elicit a positive reaction also in the patch-test-sensitized individuals when patch testing with the patch-test-sensitizing test solution diluted tenfold.

The frequency of contact allergy to Kathon® CG in our region was obviously as high as 4 to 5%. However, how had these patients become sensitized, and did this Kathon® CG hypersensitivity have any clinical relevance?

IV. EXPLANATION OF SENSITIZATION

Except for the two workers whose reactions to Kathon® 886 MW initiated the studies on Kathon®, none of the patients reacting positively to Kathon® CG when routinely patch tested had been exposed to cutting fluids. However, according to Table 2, Kathon® CG may be present in cosmetics and toiletries, and almost everyone is exposed daily to either one or both of these types of products. Labeling of ingredients of cosmetics and toiletries is not required by law in Sweden and most other European countries, as it is in the U.S.[10] Therefore, patients with hypersensitivity to Kathon® CG were asked to bring their own cosmetics and toiletries for chemical analyses.[11] It was in this way possible to demonstrate Kathon® CG (more precisely, the a.i.) in approximately one third of the products. Only a minor portion of the Kathon® CG-containing products were labeled. Perhaps a more important finding was that every Kathon® CG-hypersensitive person had been or was being exposed to at least one product containing Kathon® CG. The concentrations of the a.i. were within the recommended range. Occupational exposure to preservatives based on 5243-K-CG and 243-K-CG was also examined for these persons but not found for anyone; therefore, it is obvious that exposure to cosmetics and toiletries with Kathon® CG at recommended concentrations can induce sensitization.

V. PRESENT EXPOSURE

Since methylisothiazolinone-based preservatives were introduced in European countries in the mid-1970s and then in the U.S. in 1980, they have gradually gained in importance. During a period of 4 years, the number of uses for Kathon® CG increased from 38 to 222 and then, during the following 3-year period, to 512.[12] In 1980, according to the manufacturer, Kathon® CG was used in 55,000 tons of cosmetic products in Europe and in 20,000 tons in the U.S.[13] In the Netherlands, approximately 25% of all cosmetics and toiletries are preserved by methylisothiazolinone-based products.[14] The current figure for Sweden is similar, according to our experience, including information from manufacturers and suppliers and also from chemical analyses of about 110 products annually. In one region in the U.S., 57% of over-the-counter shampoos and conditioners contained Kathon® CG.[15]

VI. PATCH TESTING WITH KATHON® CG

At present, the only way to determine if a person is hypersensitive to Kathon® CG is by performing a patch test. There has been some controversy regarding the manner of performing patch testing with Kathon® CG. This controversy is, in our opinion, partly due to two factors: (1) the major purpose of patch testing is forgotten, and (2) the factors of importance for elicitation of a positive patch test reaction in a hypersensitive individual are not fully considered. The major purpose of patch testing is to establish (or rule out) contact allergy, rather than to diagnose allergic contact dermatitis. However, one of the advantages of the patch test technique over *in vitro* techniques (when available) for establishing contact allergy is that it is also a provocation test for allergic contact dermatitis. In a particular patient other conditions are also required to be present in order to conclude that the patient's dermatitis is an allergic contact dermatitis from the sensitizer just disclosed. The obvious goal of patch testing with a particular sensitizer is to trace as many individuals as possible with hypersensitivity and with as few adverse reactions as possible. With the same patch test technique a higher concentration of the sensitizer usually means that hypersensitivity in more persons will be traced, but also at a higher risk of induction of irritant reactions and patch test sensitization.

The main exogenous factors that may help provoke a patch test reaction in a certain region in a hypersensitive individual may be considered to belong to one of the following three groups:[16] (1) surface concentration and total number of molecules applied, (2) penetration, and (3) allergen exposure time. Provided that the same patch test technique, vehicle, and occlusion time are used, the number of molecules of the sensitizer will be decisive for the result. Thus, comparison of the results of patch testing with Kathon® CG at two departments where, for example, concentrations at 200 and 100 ppm a.i., respectively, have been used may be a comparison of the same thing, provided that one half the volume/amount of the test preparation was applied with the 200 ppm preparation. Variation in the number of molecules applied also may occur within the same patch test laboratory, particularly when petrolatum is the vehicle.[17] Thus, there is always an inter- and intradepartmental variation in the number of molecules applied for any sensitizer patch tested in petrolatum or when no fixed volumes are used for solutions. However, for most routinely patch-tested sensitizers this is of minor importance, since the routine patch test concentrations permit a variation in the number of molecules applied without a great risk of giving false negative reactions, irritant reactions, or induction of patch test sensitization. On the other hand, the corresponding concentration range for Kathon® CG is very narrow, while the number of molecules applied is of crucial significance for the patch test result. For example, patch testing with serial dilutions of Kathon® CG showed that Kathon® CG hypersensitivity was not traced in one half the number of persons when the patch test concentration was changed from 300 ppm a.i. to 100 ppm (Table 3). In our departments, the patch testing with aqueous Kathon® CG has always been based on fixed volumes of 15 μl for the Finn Chamber® technique and 30 μl for the Al-test technique.

In most articles on contact allergy to Kathon® CG, information is missing on the volumes/amounts applied on the patch unit. Therefore, it is difficult to compare and discuss the results of such studies. Despite this, certain general aspects can be discussed about the requirements for a proper patch test preparation.

Patch testing with Kathon® CG in water at 300 ppm a.i. and with a fixed volume meant detection of many contact allergies without any irritant reactions.[6] On the other hand, patch testing with Kathon® CG under equivalent conditions, but at 1000 ppm a.i., gave many irritant reactions.[6] Thus, to avoid irritant patch test reactions, the concentration can be more than 300 ppm but should be lower than 1000 ppm a.i. Internationally, an aqueous test preparation at 100 ppm a.i. has been recommended to avoid irritant reactions. The main

TABLE 4
Patch Test Concentrations of Chromate, Neomycine and Some Preservatives and Their Concentrations in Cement, Antibacterial Creams and Cosmetics/Toiletries, Respectively

Substance	Patch test concentration	Use concentration	Patch test concentration/use concentration
Chromate	0.5%	0.0056%	90
Formaldehyde	2%	0.05—0.2%	10—40
Kathon® CG	200 ppm	5—15 ppm	13—40
Neomycin	20%	0.5—1%	20—40
Parabenes	15%	0.2—0.4%	37—75

reason for this recommendation seems to be the result of an investigation where aqueous Kathon® CG preparations at various concentrations were given repeatedly to the upper arms.[18] Reactions indicating an irritant nature were scored daily for the respective concentration, and the sums for the whole period were calculated and compared. A small positive sum was obtained for the 100 ppm a.i. solution, and increasing irritant scores were recorded with increasing concentrations. However, no irritant reactions were reported after the first exposure, and irritant reactions first obtained of multiple applications do not give any guidance for the choice of a nonirritant patch test preparation. There have been remarks on irritant patch test reactions to 100 ppm a.i. in some reports.[15,19-20] Of course, irritant reactions to 100 ppm may occur, but the morphology of these putative irritant reactions was not described; in fact, sometimes erythematous reactions seem to have been interpreted as irritant reactions. Almost all of our patients with just an erythematous reaction to Kathon® CG have been retested with the routine preparation and with a solution containing Kathon® CG at a higher concentration. All of these patients have then reacted to the more highly concentrated preparation with a morphological allergic reaction. Thus, a more likely explanation of these erythematous reactions is that they represent threshold reactions for allergic reactions.

It has also been argued that a patch test concentration of 100 ppm a.i. is preferable to higher concentrations because it is closer to the recommended use concentration of Kathon® CG in cosmetics and toiletries. However, this argument indicates that the establishment of contact allergy to Kathon® CG is confused with the diagnosis of allergic contact dermatitis from Kathon® CG. It is obvious that hypersensitivity in persons traced by patch testing with Kathon® CG at 15 ppm a.i. will more likely be clinically relevant than the hypersensitivity requiring patch testing at 300 ppm. a.i. Still, contact allergy to Kathon® CG just traced by patch testing with Kathon® CG at higher concentrations than 100 ppm a.i. may be clinically relevant.[6,22] Furthermore, the difference in Kathon® CG concentration for the patch test preparation and cosmetics and toiletries is similar or lower compared to the difference of the corresponding item for other sensitizers in the standard test series, as shown in Table 4.

Patch test sensitization rather than irritancy is the limiting factor for the routine patch test concentration for Kathon® CG. Although the true patch test sensitization rates for various concentrations are not known and are difficult to ascertain,[23-26] the minimum figure of approximately 1% noted for patch test sensitization to 300 and 250 ppm a.i. is unacceptably high.[6] In Sweden, therefore, 200 ppm a.i. is recommended for routine patch testing, and after having patch tested some 1000 patients only one person has become patch test sensitized (as shown by a flare-up reaction). Patch test sensitization to 150 and 100 ppm a.i. has also been reported.[21,27] With regard to present knowledge and the wish to trace hypersensitivity in as many persons as possible yet avoid getting unacceptable numbers of irritant reactions and patch test sensitization, patch testing with aqueous Kathon® CG at 200 ppm a.i. seems appropriate.

TABLE 5
Frequency Figures for Kathon® CG Hypersensitivity in Various Countries

Country	Selection of patients	Test concentration (ppm a.i.)	Vehicle[a]	No. of patients	Positive reactions No.	Positive reactions %	Ref.
Belgium	Contact dermatitis	100	aq	786	7	0.9	52
Denmark	Contact dermatitis	100	aq	1511	13	0.9	31
Denmark	Contact dermatitis	100	aq	1396	18	1.3	21
England	Contact dermatitis	100	aq	1836	5	0.3	50
England	Contact dermatitis	100	aq	1309	12	0.9	23
Finland	Preservative allergy	100	aq	443	3	0.7	22
Finland	Contact dermatitis	100	aq	591	27	4.6	22
Hungary	Cosmetic allergy	15	aq	300	0	0	56
Italy	Cosmetic allergy	100	aq	1600	8	0.5	37
Italy	Face contact dermatitis	100	aq	98	6	6.1	34
Italy	Contact dermatitis	100	aq	3744	50	1.3	41
Italy	Contact dermatitis	100	aq	620	52	8.3	49
The Netherlands	Cosmetic allergy	150	pet	179	6	3.4	29
The Netherlands	Contact dermatitis	100	pet	501	7	1.4	35
The Netherlands	Contact dermatitis	100	aq	243	8	3.3	46
Switzerland	Contact dermatitis	15	aq	625	11	1.8	43
Switzerland	Contact dermatitis	100	aq	420	23	5.5	53
Sweden	Contact dermatitis	300	aq	976	43	4.4	6
Sweden	Contact dermatitis	250	aq	170	10	5.9	6
Sweden	Contact dermatitis	100	aq	210	4	1.9	6
Sweden	Work-related skin disease	100	aq	124	0	0	20
Sweden	Work-related skin disease	150	aq	534	4	0.7	20
USA	Contact dermatitis	100	aq	327	2	0.6	52
USA	Contact dermatitis	100	aq	365	13	3.6	15
West Germany	Contact dermatitis	100	aq	671	23	3.4	45
West Germany	Cosmetic allergy	50	aq	49	5	10.2	54

[a] aq = aqueous; pet = petrolatum.

VII. EPIDEMIOLOGY

A. CONTACT ALLERGY

Since the first reports on hypersensitivity to Kathon® CG[27,28] there have been many reports on contact allergy, mainly from European countries but also from the U.S.A. (Table 5).[6,14,15,20-22,29-55] The hypersensitivity frequency varies between 0 and 10.2%. The reasons for variation are partly due to differences in patient groups, indications for patch testing, patch test concentrations, and national differences in distribution, i.e., the total amount of methylisothiazolinone-based preservatives used and the types of products in which the preservatives are ingredients. A particularly high frequency was reported in patients with suspected cosmetic dermatitis,[54] and a low frequency (0%) was reported when a low patch test concentration was used.[56]

B. ALLERGIC CONTACT DERMATITIS

Kathon® CG in cosmetics and toiletries such as moisturizing creams, shampoos, and sunscreens have been the most common causes of allergic contact dermatitis from Kathon® CG.[22,27-29,38,39,45,46,52] Most of these patients have been exposed to products with Kathon® CG at recommended concentrations, but occasionally exposure has been at higher concentrations.[22,49] Occupational allergic contact dermatitis has been described in a truck driver[52] and in chemists[32,57] handling Kathon® CG as well as in other workers after exposure to

Kathon® CG in cutting fluids,[6,32] milk sample pots,[33] products for skin and hair care,[22,49,52] and radiography liquids.[52]

VIII. CLINICAL FEATURES

Although the allergic patch test reaction to Kathon® CG has a somewhat different macroscopic morphology than allergic patch test reactions to other sensitizers,[6,41] allergic contact dermatitis from Kathon® CG is indistinguishable from allergic contact dermatitis from other sensitizers. We have seen allergic contact dermatitis from Kathon® CG located on all parts of the body, but predominately on the hands and face. This is in agreement with the results of other studies.[6,32,33,39,41,46,49,50] Women outnumber men in most studies.[22,31,34,41,45,46,49-52] It has been pointed out that Kathon® CG hypersensitivity frequently is an aggravating factor of a preceding dermatitis.[41] When figures are given for preexisting dermatitis, these vary between 54%[50] and 89%.[22]

IX. CLINICAL RELEVANCE

Sometimes it is very difficult to assess whether the hypersensitivity to Kathon® CG is clinically relevant in a particular patient. Besides establishing contact allergy, the process of making the diagnosis of allergic contact dermatitis includes two major steps: (1) demonstration of present exposure to preservatives based on methylisothiazolinones and with known present exposure and (2) determination of whether this type of exposure explains the clinical picture and course of the dermatitis. The crucial step is the demonstration of present exposure. This is particularly difficult in countries where labeling of daily-life products such as cosmetics and toiletries is not required by law. Sometimes, these products may not be suspected to contain Kathon® CG and therefore are not investigated further. Even when these unlabeled products are investigated further and patch tested, the concentration of Kathon® CG, when present, is usually too low to permit a positive patch test reaction to the product as such. The possibility of obtaining a list of the ingredients of unlabeled cosmetics and toiletries from manufacturers and suppliers varies in countries where labeling is not a prerequisite. Inquiry on ingredients is a laborious and time-consuming procedure when the patients are exposed to many cosmetics and toiletries.

However, exposure to Kathon® CG also may be overlooked when these products are labeled with the ingredients. Sometimes, the manufacturers of cosmetics and toiletries think that there is a temporary need for a stronger preservative in their products. In this way Kathon® CG may appear in the products, but not on the label, which says that another preservative is present in the product. A similar source of error sometimes arises in the metal-working industry when a nonmethylisothiazolinone-based preservative is present in the cutting fluid. Occasionally, the metal-working system needs to be cleaned, and for that purpose a strong biocide (e.g., methylisothiazolinone-based preservatives) may be used at a high concentration. This preservative may then contaminate the common cutting fluid for a period of time. We have also become aware of another type of exposure to methylisothiazolinone-based preservatives in finished products which usually is unknown to the manufacturer and supplier of these products. As Kathon® CG can be used as a slimicide in paper mills, it can contaminate the paper used for napkins.[72] In the same way the raw materials used for cosmetics and toiletries may be preserved with Kathon® CG. According to our experience, it is very uncommon that a manufacturer knows that the products contain Kathon® CG at low concentrations because of preservation of the raw materials used. There are also other occasions when it is difficult or impossible to obtain appropriate information on the presence of methylisothiazolinones in various products.

Obviously, there is a great need for chemical investigations. Unfortunately, at present

TABLE 6
Results of Use Testing

Product	Concentration of Kathon® CG (ppm a.i.)[a]	No. of applications daily	No. of applications total	Concentration of Kathon® CG × total no. of applications	No. tested	Positive No.	Positive %	Ref.
Lotion	7.7	1	5	39	2	0	0	31
Lotion	8.6	1	5	43	9	0	0	31
Cream	7	2	14	98	10	5	50	22
Cream	15	2	14	210	2	1	50	22
Cream	9	2	14	126	7	3	43	45
Lotion	15	2	14	210	7	4	57	45
Cream	15	2	14	210	20	8	40	41
Lotion	7	2	28	196	7	7	100	49
Cream	12	2	14	168	5	3	60	46
Cream	15	2	14	210	13	7	54	6

[a] Active ingredients.

no simple spot test for demonstration of methylisothiazolinones is available that can be carried out at any dermatological laboratory. Some investigations using high-performance liquid chromatography (HPLC) for demonstration of methylisothiazolinones in products have been reported.[11,38,49,50,58] However, most dermatologists do not have access to a chemical laboratory; hence, conclusions stating that a hypersensitivity to Kathon® CG is clinically nonrelevant have to be considered carefully.

Although it might be known that a Kathon® CG-hypersensitive individual is exposed to cosmetics and toiletries containing Kathon,® it may be difficult to ascertain the clinical relevance of the exposure to these products, particularly when patch testing has yielded a negative result. To explore this issue, use tests have been carried out and generally by applying a cream or lotion containing around 15 ppm a.i. in the antecubital fossa twice daily for up to 2 weeks.[6,15,19,22,31,41,44,46,49] The results of such tests are presented in Table 6. To summarize the results, a positive use test is noted in approximately one half the number of patients, which is understandable considering the conditions for the use tests and the strength of reactivity in Kathon® CG-hypersensitive individuals (Table 3). The lowest total dose of Kathon® CG was given in the use test which was negative for all investigated persons.[31] The only reported use test carried out with rinse-off products such as shampoos and hair conditioners gave a negative result.[59] It shall also be pointed out that positive use tests have been reported in persons not reacting to patch test concentrations of <200 to 300 ppm a.i..[6,22,60]

Initially, many patients with contact allergy to Kathon® CG seem to have a dermatitis which is not an allergic contact dermatitis from Kathon® CG but another exogenous or endogenous dermatitis. Because of this there might be a greater need for certain cosmetics and moisturizing creams and, hence, a higher risk of exposure to Kathon® CG. Under these circumstances it is very difficult to assess the clinical relevance of the exposure to Kathon® CG, but it is likely that it will aggravate a preexisting dermatitis. Apparently, this is an area that should be explored further.

X. PREVENTION

A. INDIVIDUAL PREVENTION

For a Kathon® CG-hypersensitive individual it is important to know what to avoid to stay free of dermatitis. From a general point of view, it is obvious that products containing

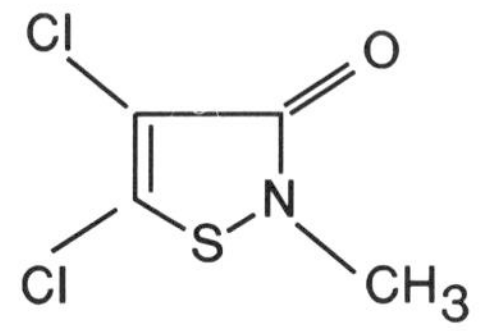

FIGURE 2. Structure of 4,5-dichloro-2-methyl-4-isothiazolin-3-one.

TABLE 7
Preservatives Containing 5-Chloro-2-Methyl-4-Isothiazolin-3-One and 2-Methyl-4-Isothiazolin-3-One

Acticide AZ	Mitco CC 31 L
Acticide SPX	Mitco CC 32 L
Algucid CH 50	Mitrol 933
Amerstat 250	Mx 323
Bodoxin	Nalcon® 7647
Celkil 90	Parmetol A 23
Euxyl K 100	Parmetol DF 12
Fennosan IT 21	Parmetol DF 18
GR 856 Izolin	Parmetol DF 35
Grotan® TK 2	Parmetol K 40
Kemtox® IT	Parmetol K 50
Magnacide® 26 S	P 3 Multan D
Mergal K 7	Piror® P 109
Metatin GT	

Note: Rohm and Haas preservatives are not included and are listed in Table 1.

Kathon® CG should be avoided. The results of the use tests shown in Table 6 indicate that there is a great risk of developing a dermatitis if leave-on products containing Kathon® CG are applied regularly and frequently at recommended concentrations. It is likely that even more than 50% of the patients will develop a dermatitis if such leave-on products containing Kathon® CG are applied on damaged skin or during a period exceeding the common use test time, particularly since results indicate that Kathon® CG can accumulate within the skin.[61,62] The risk of developing a dermatitis when exposed to rinse-off products containing Kathon® CG is, of course, less, but elicitation of dermatitis has been reported in Kathon® CG-hypersensitive individuals.[49]

Kathon® CG is a preservative consisting of two a.i. but also other substances. When Kathon® CG unambiguously was shown to be a contact sensitizer in humans,[6] any of its ingredients could have been proven to be the sensitizer(s), but the a.i. were considered to be the most likely ones. More recent investigations have also demonstrated the a.i. to be the main sensitizers and the chlorinated one to be the more potent of the two.[32,52] A contaminant in Kathon® CG, a dichlorinated methylisothiazolinone (Figure 2), has also been shown to be a sensitizer in Kathon® CG of the same potency as 5243-K-CG, but it is clinically less relevant since it is present in the preservative at a much lower concentration.[64] These findings mean that exposure to these methylisothiazolinones should be avoided to enable healing of dermatitis and relapses. Currently, methylisothiazolinones are not known to be present in the environment except in preservatives. However, there are many other preservatives besides those from Rohm and Haas (Table 7) which contain or are based on

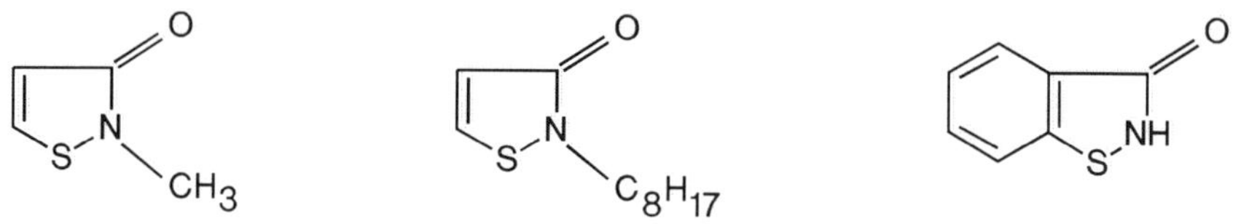

A B C

FIGURE 3. Structures of (A) 2-methyl-4-isothiazolin-3-one, (B) 2-octyl-4-isothiazolin-3-one, and (C) benzisothiazolin-3-one.

the methylisothiazolinones constituting the a.i. in the Kathon® CG preservatives listed in Table 1. Thus, Kathon® CG-hypersensitive persons should also avoid exposure to products containing the preservatives listed in Table 7. Some of these may appear in cosmetics and toiletries, but most of them are in occupational settings, which makes tracing difficult.

To avoid relapses of allergic contact dermatitis, exposure to the primary sensitizer should be avoided as well as exposure to compounds cross-reacting with it. The methylisothiazolinones were first reported to be sensitizers when they were demonstrated to be the allergens in Kathon® CG. Therefore, the cross-reacting patterns for these methylisothiazolinones had to be established. The a.i. are chemically related to two other preservatives: benzisothiazolin-3-one and 2-octyl-4-isothiazolin-3-one (Figure 3). Simultaneous reactions to Kathon® CG and these latter preservatives have been reported,[28] but we have not been able to demonstrate any cross-reactivity between methylisothiazolinones and octylisothiazolinone or benzisothiazolinone in either man or animals.[65,66]

In summary, a Kathon® CG-hypersensitive person should be advised to avoid exposure to methylisothiazolinones, which means preservatives based on (or containing) these compounds (Tables 1 and 7).

B. GENERAL PREVENTION

Preferably, sensitization to and allergic contact dermatitis from any sensitizer should be prevented. Therefore, knowledge about the sensitizing capacity of a compound or product before its introduction into the market is one prerequisite. In this way, potent sensitizers may never be introduced or their use may be restricted to avoid skin exposure and sensitization. There are many predictive sensitization tests using guinea pigs for evaluation of the sensitizing capacity of a compound or a product.

In connection with the introduction of Kathon® CG into the market, the manufacturer conducted a guinea pig maximization test (GPMT) with Kathon® CG which gave a negative result.[67] Guidelines for the GPMT require induction and challenge with separate ingredients of a formulated product rather than with the formulated product itself.[68] The negative result in the mentioned GPMT may be due to the use of a formulated product or the use of exceedingly low concentrations for induction and challenge. However, a later study using Kathon® CG as such demonstrated the preservative to be a contact sensitizer in the guinea pig.[69] In this sensitization study, various concentrations of Kathon® CG were used for induction and challenge. Based on the results for the various concentrations and statistical calculations, the authors claim that there is a "no response concentration" zone, where Kathon® CG can be used without concern for clinically significant delayed contact dermatitis. The GPMT with a.i. have demonstrated that 5243-K-CG is a strong sensitizer and that 243-K-CG is a weak one.[65] In another GPMT, the dichlorinated methylisothiazolinone was shown to possess a sensitizing capacity comparable to 5243-K-CG.[66]

There have also been studies in humans investigating sensitization to Kathon® CG present in leave-on and rinse-off products. No individual was sensitized when using a lotion containing 15 ppm a.i. once daily on nondamaged skin of the legs, arms, and trunk for 13

TABLE 8
Risk Assessment of the Use of Kathon® CG-Containing Cosmetics and Toiletries

	Sensitization	Elicitation
Rinse-off products		
On normal skin	(*)	(*)
On damaged skin	*	**
Leave-on products		
On normal skin	*	**
On damaged skin	***	***

Note: *** = high risk; ** = moderate risk; * = low risk; (*) = minimal risk.

weeks.[69] On the other hand, sensitization has been noted to rinse-off products containing Kathon® CG at 12.5 ppm a.i. and 20 ppm, respectively.[61] However, these rinse-off products were applied under occlusion.

The sensitizing capacity is not the only factor of importance for risk assessment concerning allergic contact dermatitis from Kathon® CG. However, when strong sensitizers are used in cosmetics and toiletries of the leave-on type which are used daily by almost everybody (sometimes also on damaged skin), a high frequency of allergic contact dermatitis is expected. Thus, the development of hypersensitivity to Kathon® CG and allergic contact dermatitis from exposure to it was predictable.[70]

Apparently, the introduction of Kathon® CG into the market unfortunately was made without the results of unequivocally performed sensitization studies in animals. However, a frequency of contact allergy to a preservative exceeding 1% in routinely patch-tested dermatitis patients is considered to be unacceptably high and merits measures to decrease it.[52] We think that a great number of the problems with Kathon® CG and other methylisothiazolinone-based preservatives would disappear if these preservatives were not used in leave-on products[55] and in industrial applications, meaning frequent exposure of skin to methylisothiazolinones at high concentrations. Based on the results of various reported investigations and our own experience, we have summarized our opinion on individual and general prevention concerning allergic contact dermatitis from Kathon® CG in Table 8.

XI. SUMMARY

Kathon® CG and other products containing methylisothiazolinones are effective preservatives at low concentrations. Unfortunately, methylisothiazolinones are potent sensitizers when used at recommended concentrations in cosmetics and toiletries. The allergic patch test reaction to Kathon® CG has features of both irritant and allergic morphology. Generally, it is impossible in a Kathon® CG-hypersensitive individual to elicit a positive patch test reaction with a dilution step involving just a few dilutions of the concentration that gives a moderate test reaction. For the choice of patch test concentration, patch test sensitization is the limiting factor. Use tests have shown that clinically relevant hypersensitivities require patch testing with Kathon® CG at concentrations which are unacceptable because of a high risk of patch test sensitization. The tracing of methylisothiazolinones is difficult since they are used in many preservatives which are widely distributed in the environment.

REFERENCES

1. **Krzeminski, S. F., Brackett, C. K., and Fisher, J. D.,** Fate of microbicidal 3-isothiazolone compounds in the environment: modes and rates of dissipation, *J. Agric. Food Chem.*, 23, 1060, 1975.
2. **Krzeminski, S. F., Brackett, C. K., Fisher, J. D., and Spinnler, J. F.,** Fate of microbicidal 3-isothiazolone compounds in the environment: products of degradation, *J. Agric. Food Chem.*, 23, 1068, 1975.
3. **Rohm and Haas Co.,** Kathon WT Safety Guidelines, SC-WTSG-July 1988.
4. **Rohm and Haas Co.,** Kathon LX Safety Guidelines, SC-LX-SG-September 1988.
5. **Rohm and Haas Co.,** Kathon CG Safety Guidelines, SC-CGSG-February 1989.
6. **Björkner, B., Bruze, M., Dahlquist, I., Fregert, S., Gruvberger, B., and Persson, K.,** Contact allergy to the preservative Kathon® CG, *Contact Dermatitis,* 14, 85, 1986.
7. **Fregert, S.,** Publication of allergens, *Contact Dermatitis,* 12, 123, 1985.
8. **Bruze, M.,** Contact sensitizers in resins based on phenol and formaldehyde, *Acta Derm. Venereol. Suppl.*, 119, 1, 1983.
9. **Cronin, E.,** *Contact Dermatitis,* Churchill Livingstone, Edinburgh, 1980, 15.
10. **Larsen, W. G.,** Why is the U.S.A. the only country with compulsory cosmetic labeling?, *Contact Dermatitis,* 20, 1, 1989.
11. **Gruvberger, B., Persson, K., Björkner, B., Bruze, M., Dahlquist, I., and Fregert, S.,** Demonstration of Kathon® CG in some commercial products, *Contact Dermatitis,* 15, 24, 1986.
12. **Decker, R. L., Wenninger, J. A.,** Frequency of preservative use in cosmetic formulas as disclosed to FDA — 1987, *Cosmet. Toiletries,* 102, 21, 1987.
13. **Wright, C., Gingold, E., Venitt, S., and Crofton-Sleigh, C.,** Mutagenic activity of Kathon, an industrial biocide and cosmetics preservative containing 5-chloro-2-methyl-4-isothiazolin-3-one and 2-methyl-4-isothiazolin-3-one, *Mutat. Res.*, 119, 35, 1983.
14. **de Groot, A. C., Weyland, J. W.,** Kathon CG: a review, *J. Am. Acad. Dermatol.*, 18, 350, 1988.
15. **Fransway, A. F.,** Sensitivity to Kathon CG: findings in 365 consecutive patients, *Contact Dermatitis,* 19, 342, 1988.
16. **Wester, R. C. and Maibach, H. I.,** Percutaneous absorption relative to occupational dermatology, in *Occupational and Industrial Dermatology,* 2nd ed., Maibach, H. I., Ed., Year Book Medical Publishers, Chicago, 1987, 241.
17. **Fischer, T. and Maibach, H.,** Amount of nickel applied with a standard patch test, *Contact Dermatitis,* 11, 285, 1984.
18. **Maibach, H. I.,** Diagnostic patch test concentration for Kathon CG, *Contact Dermatitis,* 13, 242, 1985.
19. **Hannuksela, M. and Salo, H.,** The repeated open application test (ROAT), *Contact Dermatitis,* 14, 221, 1986.
20. **Färm, G. and Wahlberg, J. E.,** Positive test reactions to Kathon® in patients with work-related skin disease, *Contact Dermatitis,* 16, 228, 1987.
21. **Menné, T. and Hjorth, N.,** Kathon® CG reactivity in 1396 consecutively patch tested patients in the Copenhagen area, *Contact Dermatitis,* 19, 260, 1988.
22. **Hannuksela, M.,** Rapid increase in contact allergy to Kathon® CG in Finland, *Contact Dermatitis,* 15, 211, 1986.
23. **Shuster, S. and Spiro, J.,** Measurement of risk of sensitisation and its application to Kathon, *Contact Dermatitis,* 17, 299, 1987.
24. **Cox, N. H. and Shuster, S.,** Risk of sensitisation to Kathon®, *Contact Dermatitis,* 18, 54, 1988.
25. **Bruze, M. and Assennato, G.,** Risk of sensitization to Kathon® CG, Letter to the Editor, *Contact Dermatitis,* 20, 76, 1989.
26. **Shuster, S.,** Risk of sensitization to Kathon® CG, Letter to the Editor, *Contact Dermatitis,* 20, 78, 1989.
27. **de Groot, A. C., Liem, D. H., and Weyland, J. W.,** Kathon® CG: cosmetic allergy and patch test sensitization, *Contact Dermatitis,* 12, 76, 1985.
28. **Foussereau, J., Brändle, I., and Boujnah-Khouadja, A.,** Allergisches Kontaktekzem durch Isothiazolin-3-on-Derivate, *Dermatosen,* 32, 208, 1984.
29. **de Groot, A. C., Liem, D. H., Nater, J. P., and van Ketel, W. G.,** Patch tests with fragrance materials and preservatives, *Contact Dermatitis,* 12, 87, 1985.
30. **Emmons, W. W. and Marks, J. G., Jr.,** Immediate and delayed reactions to cosmetic ingredients, *Contact Dermatitis,* 13, 258, 1985.
31. **Hjorth, N. and Roed-Petersen, J.,** Patch test sensitivity to Kathon® CG, *Contact Dermatitis,* 14, 155, 1986.
32. **Pilger, C., Nethercott, J. R., and Weksberg, F.,** Allergic contact dermatitis due to a biocide containing 5-chloro-2-methyl-4-isothiazolin-3-one, *Contact Dermatitis,* 14, 201, 1986.
33. **Grattan, C. E. H., Harman, R. R. M., and Tan, R. S. H.,** Milk recorder dermatitis, *Contact Dermatitis,* 14, 217, 1986.

34. **Tosti, A., Manuzzi, P., and de Padova, M. P.,** Contact dermatitis to Kathon CG, *Contact Dermatitis,* 14, 326, 1986.
35. **de Groot, A. C., Bos, J. D., Jagtman, B. A., Bruynzeel, D. P., van Joost, T., and Weyland, J. W.,** Contact allergy to preservatives. II, *Contact Dermatitis,* 15, 218, 1986.
36. **Fuchs, Th.,** Aktuelle kontaktallergene und ihre perspektiven. Formaldehyde; isothiazoline (Kathon® CG), *Die Medizinische Welt,* 37, 1305, 1986.
37. **Angelini, G., Vena, G. A., and Meneghini, C. L.,** Dermatite da contatto con Kathon® CG, *Boll. Dermatol. Allergol. Prof.,* 1, 20, 1986.
38. **van Joost, Th., Habets, J. M. W., Stoltz, E., and Geursen-Reitsma, A. M.,** Sodium hypochlorite sensitization, *Contact Dermatitis,* 16, 114, 1987.
39. **de Groot, A. C.,** Contact allergy to cosmetics: causative ingredients, *Contact Dermatitis,* 17, 26, 1987.
40. **de Groot, A. C. and Bruynzeel, D. P., Kathon CG,** Letter to the Editor, *Contact Dermatitis,* 17, 189, 1987.
41. **Meneghini, C. L., Angelini, G., and Vena, G. A.,** Contact allergy to Kathon® CG, *Contact Dermatitis,* 17, 247, 1987.
42. **Favet, J., Griffiths, W., Hunziker, N., Amacker, P.-A., and Schorer, E.,** Sensibilité aux tests épicutanés pour le Kathon® CG et le Germall® II, *Pharm. Acta Helv.,* 62, 282, 1987.
43. **Hunziker, N., Favet, J., and Griffiths, W.,** Contact sensitization to Kathon® CG in Geneva, *Boll. Dermatol. Allergol. Prof.,* 2, 145, 1987.
44. **Tosti, A., Melino, M., and Veronesi, S.,** The clinical relevance of patch test positivity to some preservatives, *Boll. Dermatol. Allergol. Prof.,* 2, 185, 1987.
45. **Frosch, P. J. and Schulze-Dirks, A.,** Kontaktallergie auf Kathon CG, *Hautarzt,* 38, 422, 1987.
46. **de Groot, A. C.,** Istothiazolinone preservative as important contact allergen in cosmetics, *Dermatosen,* 35, 169, 1987.
47. **de Groot, A. C. and Bos, J. D.,** Preservatives in the European standard series for epicutaneous testing, *Br. J. Dermatol.,* 116, 289, 1987.
48. **Marston, S.,** Kathon CG in cosmetics, *Contact Dermatitis,* 18, 58, 1988.
49. **Tosti, A.,** Prevalence and sources of Kathon CG sensitization in Italy, *Contact Dermatitis,* 18, 179, 1988.
50. **Cronin, E.,** Allergy to cosmetics, *Acta Derm. Venereol.,* Suppl., 134, 77, 1987.
51. **de Groot, A. C.,** Adverse Reactions to Cosmetics, Thesis, State University of Groningen, The Netherlands, 1988, chap. 4.
52. **Cronin, E., Hannuksela, M., Lachapelle, J.-M., Maibach, H. I., Malten, K., and Meneghini, C. L.,** Frequency of sensitisation to the preservative Kathon® CG, *Contact Dermatitis,* 18, 274, 1988.
53. **Pasche, F. and Hunziker, N.,** Sensitization to Kathon CG in Geneva and Switzerland, *Contact Dermatitis,* 20, 115, 1989.
54. **Kleinhans, D.,** Kontaktallergie gegen Isothiazolinone (Kathon® CG, Euxyl® K 100) in kosmetischen Produkten, *Derm. Kosmet.,* 28, 27, 1987.
55. **de Groot, A. C. and Herxheimer, A.,** Isothiazolinone preservative: cause of a continuing epidemic of cosmetic dermatitis, *Lancet,* 2, 314, 1989.
59. **Husz, S. and Simon, N.,** Epicutaneous patch test with the preservative Kathon® CG, *Contact Dermatitis,* 15, 245, 1986.
57. **O'Driscoll, J. B. and Beck, M. H.,** Occupational allergic contact dermatitis from Kathon® WT, *Contact Dermatitis,* 19, 63, 1988.
58. **Rastogi, S. C.,** Kathon CG content in cosmetic products, *Contact Dermatitis,* 19, 263, 1988.
59. **Weaver, J. E., Cardin, C. W., and Maibach, H. I.,** Dose-response assessments of Kathon® biocide. I. Diagnostic use and diagnostic threshold patch testing with sensitized humans, *Contact Dermatitis,* 12, 141, 1985.
60. **Björkner, B., Bruze, M., Dahlquist, I., Fregert, S., Gruvberger, B., and Persson, K.,** Contact allergy to the preservative Kathon® CG, Letter to the Editor, *Contact Dermatitis,* 14, 198, 1986.
61. **Cardin, C. W., Weaver, J. E., Bailey, P. T.,** Dose-response assessments of Kathon® biocide. II. Threshold prophetic patch testing, *Contact Dermatitis,* 15, 10, 1986.
62. **deBethizy, J. D., Longacre, S. L., Steigerwalt, R. B., Deckert, F. W., Moss, J. N., Hayes, A. W., Smith, J. M., and Scribner, H. E.,** Absorption and disposition of ^{14}C-labelled Kathon® biocide, a mixture of 5-chloro-2-methyl-4-isothiazolin-3-one and 2-methyl-4-isothiazolin-3-one, following intravenous or dermal administration to male Sprague-Dawley rats, *Fd. Chem. Toxic.,* 24, 43, 1986.
63. **Bruze, M., Dahlquist, I., Fregert, S., Gruvberger, B., and Persson, K.,** Contact allergy to the active ingredients of Kathon® CG, *Contact Dermatitis,* 16, 183, 1987.
64. **Bruze, M., Dahlquist, I., and Gruvberger, B.,** Contact allergy to dichlorinated methylisothiazolinone, *Contact Dermatitis,* 20, 219, 1989.
65. **Bruze, M., Fregert, S., Gruvberger, B., and Persson, K.,** Contact allergy to the active ingredients of Kathon® CG in the guinea pig, *Acta Derm. Venereol.,* 67, 315, 1987.
66. **Bruze, M., Gruvberger, B., and Persson, K.,** Contact allergy to a contaminant in Kathon® CG in the guinea pig, *Dermatosen,* 35, 165, 1987.

67. **Rohm and Haas Co.,** Kathon CG Bulletin SC-465-ER/0983, 1983.
68. **Wahlberg, J. E. and Boman, A.,** Guinea pig maximization test, in *Contact Allergy Predictive Tests in Guinea Pigs,* Anderson, K. E. and Maibach, H. I., Eds., S. Karger, Basel, 1985.
69. **Chan, P. K., Baldwin, R. C., Parsons, R. D., Moss, J. N., Stiratelli, R., Smith, J. M., and Hayes, A. W.,** Kathon biocide: manifestation of delayed contact dermatitis in guinea pigs is dependent on the concentration for induction and challenge, *J. Invest. Dermatol.,* 81, 409, 1983.
70. **Schwartz, S. R., Weiss, S., Stern, E., Morici, I. J., Moss, J. N., Goodman, J. J., and Scarborough, N. L.,** Human safety study of body lotion containing Kathon® CG, *Contact Dermatitis,* 16, 203, 1987.
71. **Bruze, M., Gruvberger, B., and Agrup, G.,** Sensitization potential to Euxyl K 400, Letter to the Editor, *Contact Dermatitis,* 19, 392, 1988.
72. **Bruze, M.,** unpublished data.

Chapter 24

CONTACT DERMATITIS TO TOPICAL CORTICOSTEROIDS: DIAGNOSTIC PROBLEMS

A. Dooms-Goossens

TABLE OF CONTENTS

I. INTRODUCTION

The use of topical corticosteroids has increased enormously in recent years, as has the number of publications concerning their contact allergic potential. In the literature, we found 13 such papers dated between 1959 and 1968,[1-13] 15 dated between 1969 and 1978,[14-29] and 53 dated between 1979 and the present,[30-82] 42 of which were published in the last 5 years. Dermatologists are becoming aware that such reactions are not at all as exceptional as had been previously thought, and recent statistics[76,83] even indicate that the frequency of positive reactions to corticosteroids is in the range of the frequency of some "standard" allergens. Moreover, even cross-reactions seem to be quite common.

The pitfalls associated with the detection of corticosteroid allergies may well be why they went undetected so long. In 1986[63] we described some of these problems. We will discuss them again here in the light of more recent findings.

II. PROBLEMS RELATED TO THE CORTICOSTEROID ITSELF

A. STRUCTURE AND PHARMACOLOGICAL EFFECT

Corticosteroids have long remained above suspicion in medical circles regarding their sensitizing capacity, and perhaps this is often still the case. The reasons for this are the very characteristics of the corticosteroids themselves:

- Corticosteroids are structurally related to cortisol, a substance already present in the body. However, cortisol (hydrocortisone) gives rise to contact allergic reactions, perhaps because hydrocortisone is converted to a metabolite when applied cutaneously. This metabolite is no longer experienced as a part of the body and so can function as an allergen like any other chemical substance.
- Corticosteroids have an anti-inflammatory effect, which is why they are used to treat allergic skin conditions. Indeed, topical glucocorticosteroids modulate the delayed-type hypersensitivity reaction.[84,85] In fact, contact sensitivity reactions (as well as many other side effects) are maked by this anti-inflammatory effect.

B. THE PRESENCE OF IMPURITIES AND METABOLITES

Corticosteroids, like all other pharmaceutical substances,[86] may contain impurities[4,63] or metabolites formed during storage[38] that could have a role in contact allergic reactions. In 1960, Church[4] demonstrated that 21-diol acetate, a precursor present as an impurity in hydrocortisone acetate, was the actual allergenic culprit for one of his patients. Therefore, if a patient reacts to a corticosteroid product, but not to a reference standard or any of the vehicle ingredients, it may be advisable to check for impurities or metabolites in the corticosteroid itself.

III. PROBLEMS RELATED TO THE CORTICOSTEROID-ALLERGIC PATIENT

A. "MASKED" CONTACT DERMATITIS REACTIONS

The anti-inflammatory activity of corticosteroid preparations may mask a contact allergy to the corticosteroid itself or to another ingredient of the preparation. Indeed, the clinical picture of such patients is rarely spectacular, as the lesions are generally chronic and do not respond to local corticosteroid treatment. The inability of the skin lesions to heal often leads the attending physician to prescribe other, still stronger corticosteroid preparations. This often brings no relief of the symptoms and causes other adverse corticosteroid effects (e.g., rosacea, perioral dermatitis) that may even dominate the clinical picture. In exceptional cases, the contact allergy may express itself as an id-like spread elsewhere on the body.

Masking effects also occur in patch test reactions: corticosteroids, especially in higher concentrations, may actually suppress a sensitivity reaction or delay the response.[38,80] Hence, the test conditions are critical, and late patch test readings — often up to 7 d[66,87] — must be made.

B. MULTIPLE SENSITIVITIES

Patients with contact eczema reactions to topical preparations often acquire new sensitivities deriving from the particular immunological microenvironment in lesional skin, which predisposes it to further immunological reactions at the same site.[88] In fact, it has been shown that contact allergic patients develop multiple contact allergies more often than is predicted on the basis of the frequency of each allergy occurring in isolation[89] and that this phenomenon is especially prominent when rare sensitizers are involved.[90] We need not be surprised, then, that corticosteroid sensitivity is found to be particularly common among patients with chronic eczema, who use vast numbers of topical agents and so tend to develop concurrent hypersensitivity to several ingredients of topical medications.[23,27,38,51,70,80,87,91] However, this too is a pitfall for the dermatologist searching for allergens present in such preparations. One is apt to be satisfied when one can identify one allergenic ingredient when, in fact, the patient could be sensitive to several,[92] including the corticosteroid itself.

IV. PROBLEMS OF DEMONSTRATING A CORTICOSTEROID CONTACT ALLERGY

A. THE CUTANEOUS PARADOX

A weak concentration of a corticosteroid in a pharmaceutical product can cause a contact allergy on the treated eczema site while giving a negative patch test reaction[63] or even a negative usage test[93] on an intact test site. This "cutaneous paradox" is not restricted to corticosteroids, as Fisher's "paraben paradox"[94] and other examples[86] demonstrate, but may be particularly significant with corticosteroids because of their anti-inflammatory properties. In any event, the dermatologist must keep it in mind.

B. REDUCED BIOAVAILABILITY IN PATCH TESTING

The bioavailability of a corticosteroid diluted in a test vehicle like petrolatum is generally lower than in a commercial preparation in which the vehicle has been formulated specially to enhance skin penetration. Substances like urea, salicylic acid, and propylene glycol (and also certain emulsified forms, being solvent systems of an aqueous solution of a surface-active agent) are more efficient in this regard than is petrolatum.[95] Consequently, a corticosteroid may well cause an allergic eczema in the commercial preparation, but give a negative patch test when present in a neutral test vehicle. Unfortunately, penetration-enhancing chemicals often elicit irritant reactions, which precludes their general use in patch testing. Nevertheless, we found that the addition of dimethyl sulfoxide (DMSO) was helpful in some cases.[87]

C. THE LACK OF OPTIMAL TEST CONDITIONS

The epicutaneous testing of corticosteroids has not been — and perhaps never will be — standardized because of the particularly wide divergence in individual patient reactions. Inadequate test conditions may well be the most important cause of underestimating corticosteroid sensitivity. It has been suggested that increasing the test concentration could overcome the problem of false negative reactions.[17,25] However, the critical bioavailability factors are relative solubility in the vehicle and affinity for the skin,[96] so merely increasing the concentration of the corticosteroid would hardly enhance the penetration. As several authors have reported this experience, a change of vehicle is called for.[20,38] While there is as yet no agreement on an optimum test vehicle or concentration, alcoholic solutions seem generally to be more adequate than petrolatum for testing corticosteroids,[38,63,76,93] although some patients may already react strongly to some of the petrolatum suspensions.[87] Moreover, since dissolving a drug at its maximum concentration is necessary for its greatest release from a vehicle,[96] a saturated solution in alcohol might well be optimal. We have been able to demonstrate this for hydrocortisone: 1% ethanol solutions sometimes failed to cause a reaction, while a saturated, 2.5% solution did.[63] This is also the case for hydrocortisone-17-butyrate,[93] for which a concentration of 0.1 to 1% in ethanol seems to be optimal. This finding is supported by the results of *in vitro* penetration studies, which have shown that the percutaneous absorption of this molecule is improved in ethanol up to a saturation solubility of 0.07%.[97]

However, tests with saturated alcoholic solutions do not always work. For example, most of our patients[63,87] who reacted to a cream containing 1% hydrocortisone acetate reacted neither to a 25% concentration in petrolatum nor to a 0.5% saturated solution in ethanol (at its maximum solubility). The relatively greater ability of hydrocortisone than hydrocortisone acetate to penetrate the skin might account for this. Thus, other vehicles need to be examined also.

D. PRESERVATION OF TEST SUBSTANCES

It has been shown that many corticosteroids degrade when stored in ethanolic solutions. Förström and co-workers[38] found that most of the 17-butyric ester of hydrocortisone had been converted into the 21-butyric ester within 2 months of refrigerated storage in ethanol. Similarly, after 1 month of storage under the same conditions, the chromatographic purity of fluocinolone acetonide, triamcinolone acetonide, and betamethasone-17-valerate varied from 95 to 75%. However, Reitamo and co-workers[70] found that stored and freshly prepared solutions of hydrocortisone-17-butyrate gave the same patch-test results.

Although it has been shown that alcohol has a stabilizing effect on hydrocortisone,[98] refrigeration of ethanolic solutions of this corticosteroid, in our experience, often gives rise to the formation of precipitates that do not always redissolve at room temperature, thus distorting the test concentration. We found that this precipitating effect differs considerably as a function of the brand and the batch number. This is probably due to certain impurities or metabolites, since the U.S.P. hydrocortisone reference standard, which approaches 100% purity, remains "clear" much longer.

Therefore, it would be ideal to test patients with freshly prepared solutions, but this is difficult in practice.

V. PROBLEMS RELATED TO THE CHEMICAL RELATIONSHIPS BETWEEN CORTICOSTEROID MOLECULES

A. DETERMINATION OF CROSS-SENSITIVITY PATTERNS

There has been a great deal of speculation about associated positive patch tests and possible cross-reactions between various corticosteroids. Corticosteroid-sensitive patients have generally been exposed to several corticosteroid preparations, so concomitant sensitization often cannot be ruled out. However, associated positive reactions to "pure" corticosteroids to which the patients could not have been previously exposed[63] constitute irrefutable proof of cross-sensitivity.

Several hypotheses have been formulated to account for the cross allergic patterns observed. The formation by enzymatic oxidation or hydrolysis of a common allergenic metabolite in the skin, such as a 21-dihydrocortisone, a so-called glyoxal, has been suggested.[99] However, testing with 21-dihydrocortisone in corticosteroid-sensitive subjects has remained negative.[104] Some authors have postulated that the total steroid molecule is the

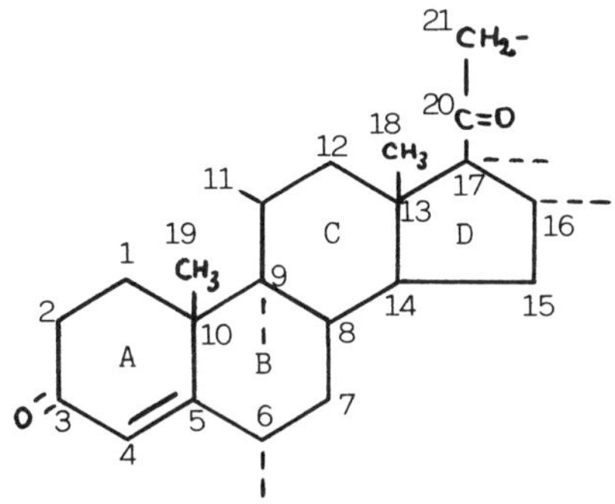

FIGURE 1. The structure and nomenclature of adrenocorticosteroids. The prototype is cortisol (hydrocortisone).

immunological requisite,[17,77] while others have considered the immunological requisite to be, in some cases, very specific structures, as could be demonstrated for the C_{16},C_{17}-*cis*-ketal and -diol structures.[75]

On the basis of literature data, to which we applied rigorous technical criteria for data selection,[100,101] and our own patch test data,[87] we have concluded that at least 85% of the positive results of simultaneously tested substances can be placed in well-determined structural classes, even when account is taken of possible selective tests of certain corticosteroids by some case report authors.

Indeed, we were able to demonstrate that an analogous substitution pattern on the steroid D-ring or the carbon side chain (C_{20}, C_{21}) (cf. Figure 1) seems to be important, while this is not the case for other structural variables, such as the presence of a double bond in the steroid A-ring or fluor substitutions on the B-ring. While a strict delineation between the various structures is not always possible, our results suggest four classes of sensitizing corticosteroids (this classification is independent of potency):

Class A:	Hydrocortisone and tixocortol pivalate type
Class B:	Triamcinolone acetonide type
Class C:	Betamethasone type
Class D:	Hydrocortisone-17-butyrate and clobetasone-17-butyrate type

Figures 2 to 5 present these four classes and the corticosteroids assigned to them on the basis of our study.

B. THE SEARCH FOR POTENTIAL MARKERS OR SCREENING AGENTS

Cross-allergy studies are interesting from the point of view of pathophysiological theory for obtaining information on the molecular structure of antigenic determinants. However, since there are dozens of corticosteroids on the market and since it is impracticable to screen patients with all the corticosteroid molecules to which he or she could have been exposed, one hopes that optimal screening agents will be found to detect contact allergy to the various classes of cross-reacting corticosteroids for diagnostic purposes.

The following markers have thus far been identified: hydrocortisone-17-butyrate (0.1 to 1% ethanol)[70,93] and tixocortol pivalate (1% petrolatum).[76,87] Our findings indicate that the former may well be a good marker for Class D and the latter for Class A. However, other corticosteroid contact-allergy markers still need to be identified, and their test conditions still need to be optimized, since the most reliable strategy for detecting corticosteroid allergy may well consist of patch testing certain representatives of the four classes.

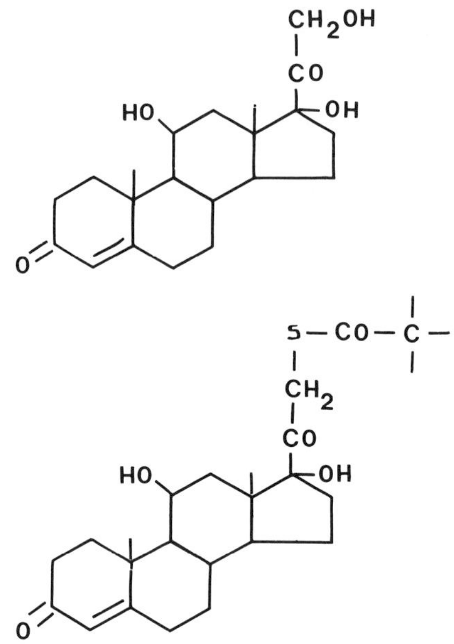

FIGURE 2. Class A — hydrocortisone and tixocortol pivalate type: corticosteroids without substitution on the corticosteroid D-ring or C_{20}, C_{21}-carbon chain, but including C_{17}- and/or C_{21}-short chain (acetate) esters *and* the C_{21}-thioester, tixocortol pivalate. Corticosteroids in Class A: cloprednol, cortisone (and 21-acetate), hydrocortisone (and 21-acetate), fludrocortisone, methylprednisolone (and 21-acetate), prednisolone (and 21-acetate), prednisone, and tixocortol-21-pivalate.

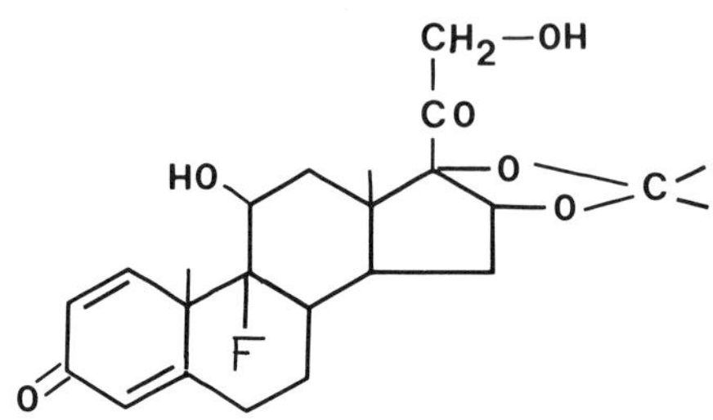

FIGURE 3. Class B — triamcinolone acetonide type: corticosteroids with a C_{16}, C_{17}-*cis*-ketal or -diol structure. Corticosteroids in Class B: amcinonide, budenoside, desonide, fluocinolone acetonide, flucinonide, halcinonide, triamcinolone acetonide, and triamcinolone alcohol.

Moreover, one must bear in mind that not all corticosteroids necessarily fall into one of the particular classes of cross-allergens.[101] This could be due to individual differences in skin metabolism, such as corticosteroid-esterase activity in the skin,[102] that could be responsible for the conversion of hydrocortisone-17-butyrate to hydrocortisone and fluocortolone hexanoate or of pivalate to fluocortolone. Furthermore, an intrinsically limited specificity of the cross-allergy phenomenon could also account for this. For example, a certain continuity in the transition between the various defined classes cannot be denied. This is illustrated by

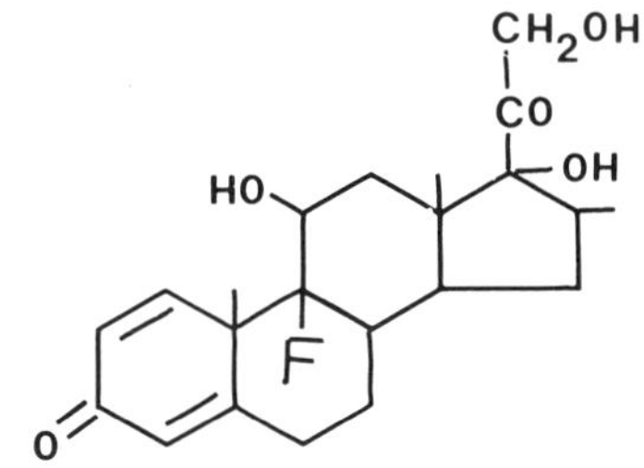

FIGURE 4. Class C — betamethasone type: corticosteroids with a C_{16}-methyl substitution. Corticosteroids in Class C: betamethasone (and 21-disodium phosphate), dexamethasone (and 21-disodium phosphate), and fluocortolone.

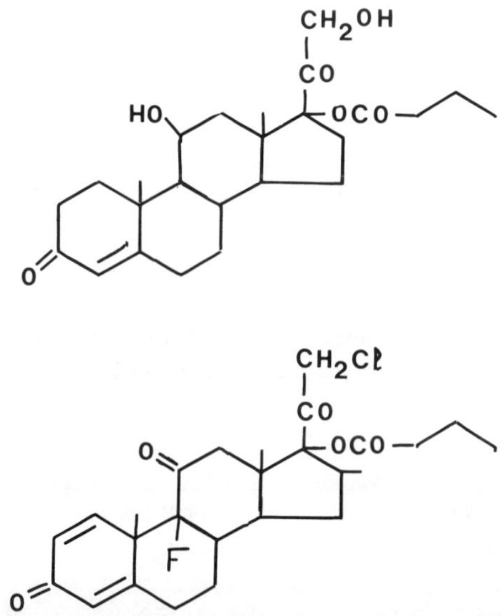

FIGURE 5. Class D — hydrocortisone-17-butyrate and clobetasone-17-butyrate type: corticosteroids with a long-chain ester at C_{17} and/or C_{21}, and with or without a C_{16}-methyl substitution. Corticosteroids in Class D: (A) Without a C_{16}-methyl substitution — hydrocortisone-17-butyrate and hydrocortisone-17-valerate. (B) With a C_{16}-methyl substitution — alclomethasone-17,21-dipropionate, betamethasone-17-valerate, betamethasone-17,21-dipropionate, clobetasol-17-propionate, clobetasone-17-butyrate, fluocortolone-21-hexanoate, fluocortolone-21-pivalate, and fluprednidene acetate.

the difficult position of the short-chain esters and by some corticosteroids that combine different substitutes, such as amcinonide, which is a 16,17-*cis*-ketal-21-acetate. It could well be that the classes as we have defined them will have to be modified or expanded as research advances.

C. TREATMENT OF CORTICOSTEROID-ALLERGIC PATIENTS

Another reason why cross-allergy studies are interesting and necessary is to facilitate the choice of corticosteroids if the condition of the patient indicates their use.

In a patient in which a corticosteroid contact allergy is a possible diagnosis, it is best to avoid systemic corticosteroid therapy because of the risk of generalized skin and other adverse reactions as well as local corticosteroid therapy. Indeed, it has been shown that allergic reactions may be caused by topical, oral, intralesional, and intra-articular administration of corticosteroids.[14,28,32,53,54,58,73,82] We ourselves have observed two such cases.

Once the corticosteroid allergen is identified, corticosteroid therapy is still possible if an alternative therapeutic agent of suitable potency can be selected from classes that do not contain the offending corticosteroid. With the aid of our computer product data file,[86] we have been able to classify all commercially available corticosteroids within the four classes given above. After having determined the corticosteroids to which our patients were allergic, we have noted that most of the lesions cleared upon application of a corticosteroid of a class to which no reaction had been obtained by patch testing.

VI. PRACTICAL GUIDELINES

Patients who suffer from a chronic eczema and who do not respond to corticosteroid therapy or have a known contact allergy to local therapeutics should be suspected of having contact allergy to one or more corticosteroids. All local corticosteroid therapy must then be stopped immediately and systemic corticosteroid therapy reduced and also stopped. Hospital admission may be necessary if the patient has to be removed from his possibly allergenic environment. The most ''inert'' possible therapy must be commenced (such as wet dressings and systemic antihistamines).

Once the lesions have cleared, the patient must be tested with all the topical pharmaceutical products used along with their individual ingredients, including the corticosteroid itself. It is important here that no prejudgments be made regarding the sensitization power of particular ingredients, since such patients may well manifest multiple sensitivities, often to unexpected substances.

There seem to be no general ''ideal'' test conditions for all corticosteroids, except perhaps for hydrocortisone (2% ethanol), tixocortol pivalate (1% petrolatum), and hydrocortisone-17-butyrate (1% ethanol). If no suitable markers or test conditions are available, we suggest testing first with petrolatum for practical reasons and then with ethanol if a false-negative result is suspected. If a corticosteroid contact allergy is still suspected after all tests remain negative, usage tests or ROAT (repeated open application test)[103] and testing with various concentrations and penetration-enhancing agents such as DMSO may be useful. There are simply too many unknowns at present about the test concentrations, storage conditions, and, indeed, the vehicles themselves.

Late readings, up to at least 5 d after application, should be taken in order not to miss delayed positive reactions. Moreover, erythematous reactions and slightly edematous reactions should not be ignored, but should be further investigated because the test conditions are critical.

If a corticosteroid tests positively, it may be beneficial to test corticosteroids — ideally all of them — of the various corticosteroid classes. Of course, the patient must be carefully informed about the corticosteroids that could possibly be safe to use should corticosteroid therapy ever be necessary. In our experience, many patients who have suffered for years from chronic eczema with intermittent acute exacerbations have been able to be made lesion free in this way.

At first sight, the development of a ''corticosteroid mix'' for use in the standard series to identify all corticosteroid-sensitive patients would seem to be a practicable goal. However, even if we were to develop an optimal screening mix, we would still have to cope with all the problems peculiar to mixes (particularly if the vehicle is ethanol), such as stability, chemical interactions, and false positive and false negative results. No breakthrough that could simplify the sometimes overwhelmingly complex problem of corticosteroid sensitivity seems to be in the offing.

REFERENCES

1. **Burckhardt, W.,** Kontactekzem durch Hydrocortison, *Hautarzt,* 10, 42, 1959.
2. **Dorn, H.,** Kontaktallergie gegenüber Salben-Konservierungsmitteln und Hydrocortison, *Z. Hautkr.,* 27, 305, 1959.
3. **Kooij, R.,** Hypersensitivity to hydrocortisone, *Br. J. Dermatol.,* 71, 392, 1959.
4. **Church, R.,** Sensitivity to hydrocortisone acetate ointment, *Br. J. Dermatol.,* 72, 341, 1960.
5. **Sönnichsen, N.,** Beitrag zur Hydrocortison-Überempfindlichkeit, *Hautarzt,* 13, 226, 1962.
6. **Vermeulen, C. W. and Malten, K. E.,** Contact-eczeem door 6-alpha-chloorprednison en neomycin (bestanddelen van Adremycine-zalf), *Ned. Tijdschr. Geneeskd.,* 107, 548, 1963.
7. **Coskey, R. J.,** Contact dermatitis due to topical hydrocortisone and prednisolone, *Mich. Med.,* 64, 669, 1965.
8. **Bandmann, H.-J., Huber-Riffeser, G., and Woyton, A.,** Kontaktallergie gegen Triamcinolonacetonid, *Hautarzt,* 17, 183, 1966.
9. **Zina, G. and Bonu, G.,** Contact sensitivity to corticosteroids, *Contact Dermatitis Newsl.,* 2, 26, 1967.
10. **Coskey, R. J. and Bryan, H. G.,** Contact dermatitis due to methylprednisolone, *J. Am. Med. Assoc.,* 199, 172, 1967.
11. **Wilkinson, R. D., McGarry, E. M., and Solomon, S.,** Allergic contact dermatitis to hydrocortisone, *J. Invest. Dermatol.,* 43, 295, 1967.
12. **Wulf, K.,** Beitrag zur Triamcinolon-Kontaktallergie, *Z. Hautkr.,* 42, 765, 1967.
13. **Wiegel, O.,** Kontaktallergie durch Kortikosteroid-haltige Externa (Triamcinolon-Acetonid und Dexametason), *Med. Welt,* 13, 828, 1968.
14. **Comaish, S.,** A case of hypersensitivity to corticosteroids, *Br. J. Dermatol.,* 81, 919, 1969.
15. **Edwards, M. and Rudner, E. J.,** Dermatitis venenata due to hydrocortisone alcohol, *Cutis,* 6, 757, 1970.
16. **Foussereau, J., Liman-Mestiri, S., Khochnevis, A., and Basset, A.,** L'allergie à l'association thérapeutique locale "nystatine, néomycine et acétonide de triamcinolone", *Bull. Soc. Fr. Dermatol.,* 78, 457, 1971.
17. **Alani, M. D. and Alani, S. D.,** Allergic contact dermatitis to corticosteroids, *Ann. Allergy,* 30, 181, 1972.
18. **Bunney, M. H.,** Contact dermatitis due to betamethasone-17-valerate (Betnovate), *Contact Dermatitis Newsl.,* No. 12, 318, 1972.
19. **Malten, K. E.,** Betnelan V lotion contact sensitivity, *Contact Dermatitis Newsl.,* No. 13, 360, 1973.
20. **Gutzwiller, P.,** Zum Problem der Kortikosteroid-Allergie, *Dermatologica,* 148, 253, 1974.
21. **Van Ketel, W. G.,** Allergy to Utralan preparations, *Contact Dermatitis Newsl.,* No. 15, 427, 1974.
22. **Krook, G.,** Contact dermatitis due to Ficortril (hydrocortisone 1% ointment, Pfizer), *Contact Dermatitis Newsl.,* No. 15, 460, 1974.
23. **Alani, S. D. and Alani, M. D.,** Allergic contact dermatitis and conjunctivitis to corticosteroids, *Contact Dermatitis,* 2, 301, 1976.
24. **Calnan, C. D.,** Use and abuse of topical steroids, *Dermatologica,* 152 (Suppl. 1), 247, 1976.
25. **Tegner, E.,** Contact allergy to corticosteroids, *Int. J. Dermatol.,* 15, 520, 1976.
26. **Menné, T. and Andersen, K. E.,** Allergic contact dermatitis from fluocortolone, fluocortolone pivalate and fluocortolone caproate, *Contact Dermatitis,* 3, 337, 1977.
27. **Coskey, R. J.,** Contact dermatitis due to multiple corticosteroid creams, *Arch. Dermatol.,* 114, 115, 1978.
28. **Shigemi, F., Tanaka, M., Ohtsuka, T.,** A case of sensitization to oral corticosteroids, *J. Dermatol.,* 5, 231, 1978.
29. **Rietschel, R. L.,** Photocontact dermatitis to hydrocortisone, *Contact Dermatitis,* 4, 334, 1978.
30. **Soesman-van Waadenoijen Kernekamp, A. and Van Ketel, W. G.,** Contact allergy to hydrocortisone-17-butyrate, *Contact Dermatitis,* 5, 268, 1979.
31. **Brandao, F. M. and Camarasa, J. G.,** Contact allergy to hydrocortisone-17-butyrate, *Contact Dermatitis,* 5, 354, 1979.
32. **Fisher, A. A.,** Allergic reactions to intralesional and multiple topical corticosteroids, *Cutis,* 23, 564, 1979.
33. **Kark, E. C.,** Sensitivity to fluorinated steroids presenting as a delayed hypersensivity, *Contact Dermatitis,* 6, 214, 1980.
34. **Romaguera, C. and Grimalt, F.,** Five cases of contact dermatitis from pyrrolnitrine, *Contact Dermatitis,* 6, 352, 1980.

35. **Brown, R.**, Allergy to hydrocortisone-17-butyrate, *Contact Dermatitis,* 6, 504, 1980.
36. **Van Hecke, E. and Temmerman, L.**, Contact allergy to the corticosteroid budesonide, *Contact Dermatitis,* 6, 509, 1980.
37. **Van Ketel, W. G. and Swain, A. F.**, Allergy to clobetasol-17-propionate (Dermovate), *Contact Dermatitis,* 7, 278, 1981.
38. **Förström, L., Lassus, A., Salde, L., and Niemi, K.-M.**, Allergic contact eczema from topical corticosteroids, *Contact Dermatitis,* 8, 128, 1982.
39. **Brown, R.**, Simultaneous hypersensitivity to 3 topical corticosteroids, *Contact Dermatitis,* 8, 339, 1982.
40. **Hausen, B. M.**, Dinitrochlorbenzol (DNCB)/Dermoxin-Kontaktallergie, *Z. Hautkr.*, 57, 1840, 1982.
41. **Bachmann-Buffle, B.**, Allergy to clobetasol-17-propionate (Dermovate), *Dermatologica,* 167, 104, 1983.
42. **Chalmers, R. J. G., Beck, M. H., and Muston, H. L.**, Simultaneous hypersensitivity to clobetasone butyrate and clobetasol propionate, *Contact Dermatitis,* 9, 317, 1983.
43. **Pasricha, J. S. and Gupta, R.**, Contact sensitivity to betamethasone-17-valerate and fluocinolone acetonide, *Contact Dermatitis,* 9, 330, 1983.
44. **Dooms-Goossens, A., Vanhee, J., Vanderheyden, D., Gevers, D., Willems, L., and Degreef, H.**, Allergic contact dermatitis to topical corticosteroids: clobetasol propionate and clobetasone butyrate, *Contact Dermatitis,* 9, 470, 1983.
45. **Sturtz, R. P. and Rau, R. C.**, Contact dermatitis to desonide, *Arch. Dermatol.*, 119, 1023, 1983.
46. **Esser, B.**, Beitrag zur Kortison-Allergie, *Z. Hautkr.*, 58, 29, 1983.
47. **Kuhlwein, A., Hausen, B. M., and Hoting, E.**, Kontaktallergie durch halogenierte Kortokosteroide, *Z. Hautkr.*, 58, 794, 1983.
48. **Seitz, P.**, Kontakdermatitis durch ein Glukokortikoid, *Z. Hautkr.*, 58, 1338, 1983.
49. **Boyle, J. and Peachey, R. D. G.**, Allergic contact dermatitis to Dermovate and Eumovate, *Contact Dermatitis,* 11, 50, 1984.
50. **Boujnah-Khouadja, A., Brändle, I., Reuter, G., and Foussereau, J.**, Allergy to 2 new corticoid molecules, *Contact Dermatitis,* 11, 83, 1984.
51. **Guin, J. D.**, Contact sensitivity to topical corticosteroids, *J. Am. Acad. Dermatol.*, 10, 773, 1984.
52. **Ohi, T., Yoshii, T., and Tanioku, K.**, A case of contact dermatitis to amcinonide, *Skin Res. (Hifu),* 26, 648, 1984.
53. **De Boer, E. M., Van Den Hoogenband, H. M., and Van Ketel, W. G.**, Allergie voor inspuitbare corticosteroïden, *Ned. Tijdschr. Geneeskd.*, 128, 1486, 1984.
54. **Pambor, M.**, Kortikosteroid-allergie, *Dermatol. Monatsschr,* 170, 608, 1984.
55. **Yoshikawa, K., Watanabe, K., and Mizuno, N.**, Contact allergy to hydrocortisone-17-butyrate and pyridoxine hydrochloride, *Contact Dermatitis,* 12, 55, 1985.
56. **Koch, E. M. W.**, Contact allergic reaction to valerate esters of betamethasone and hydrocortisone, *Contact Dermatitis,* 12, 58, 1985.
57. **Hayakawa, R., Matsunaga, K., Ukei, C., and Hosokawa, K.**, Allergic contact dermatitis from amcinonide, *Contact Dermatitis,* 12, 213, 1985.
58. **Lübbe, D.**, Epidermale Prednisolonsensibilisierung — ein Erfahrungsbericht über 7 klinische Beobachtungen, *Wiss. Z. Univ. Halle,* 34, 63, 1985.
59. **Hosokawa, K., Ukei, C., Matsunaga, K., and Hayakawa, R.**, Amcinonide contact dermatitis, *Skin Res. (Hifu),* 27, 564, 1985.
60. **Hausen, B. M. and Kulenkamp, D.**, Kontaktallergie auf Fludroxycortid und Cetylalcohol, *Derm. Beruf Umwelt,* 33, 27, 1985.
61. **Junghans, C.**, Allergische Kontaktdermatitis gegen drei verschiedene topische Corticosteroide, *Derm. Kosmet.*, 26, 92, 1985; 26, 126, 1985.
62. **Jelen, G., Grosshans, E., and Foussereau, J.**, Une autre forme d'allergie au pivalate de tixocortol, *Lett. Gerda,* 5, 16, 1985.
63. **Dooms-Goossens, A., Verschaeve, H., Degreef, H., and Van Berendoncks, J.**, Contact allergy to hydrocortisone and thixocortol pivalate: problems in the detection of corticosteroid sensitivity, *Contact Dermatitis,* 14, 94, 1986.
64. **Spiro, J. G. and Lawrence, C. M.**, Contact sensitivity to clobetasol propionate, *Contact Dermatitis,* 14, 116, 1986.
65. **Meding, B. and Dahlberg, E.**, Contact allergy to budesonide in a nasal spray, *Contact Dermatitis,* 14, 253, 1986.
66. **Nakajima, H., Mohri, S., and Nagai, R.**, A case of contact dermatitis due to betamethasone valerate, *J. Dermatol.*, 13, 467, 1986.
67. **Foussereau, J. and Jelen, G.**, Tixocortol pivalate, an allergen closely related to hydrocortisone, *Contact Dermatitis,* 15, 37, 1986.
68. **Kubo, Y., Nonaka, S., and Yoshida, H.**, Contact allergy to amcinonide, *Contact Dermatitis,* 15, 109, 1986.

69. **Junghans, C. Kuhlwein, A., and Hausen, B. M.,** Merfachsensibilisierung gegen topisch anngewandte Kortikosteroide, *Derm. Beruf Umwelt,* 34, 68, 1986.
70. **Reitamo, S., Lauerma, A. I., Stubb, S., Käyhkö, K., Visa, K., and Förström, L.,** Delayed hypersensitivity to topical corticosteroids, *J. Am. Acad. Dermatol.,* 15, 582, 1986.
71. **Jelen, G.,** Un nouveau cas d'allergie au pivalate de tixocortol. Allergie croisée avec l'hydrocortisone, *Lett. Gerda,* 3, 27, 1986.
72. **Hayakawa, R., Matsunaga, K., Suzuki, M., and Hosokawa, K.,** Contact dermatitis from amcinonide, *Contact Dermatitis,* 16, 48, 1987.
73. **Maucher, O. M., Faber, M., Knipper, H., Kirchner, S., and Schopf, E.,** Kortikoidallergie, *Hautarzt,* 38, 577, 1987.
74. **Tosti, A., Guerra, L., Manuzzi, P., and Lama, L.,** Contact dermatitis from clobetasol propionate, *Contact Dermatitis,* 17, 256, 1987.
75. **Sasaki, E., Shinya, M., Fujioka, M., Sato, M., Hata, M., Yajima, J., Hattori, S., and Honda, M.,** Contact dermatitis due to amcinonide, *Contact Dermatitis,* 18, 61, 1988.
76. **Dooms-Goossens, A.,** Identification of undetected corticosteroid allergy, *Contact Dermatitis,* 18, 124, 1988.
77. **Lozinsky, A. Z. and Mitchel, J. C.,** Allergic contact dermatitis from multiple topical corticosteroids, *Contact Dermatitis,* 18, 244, 1988.
78. **Lindelöf, B.,** Contact allergy to hydrocortisone, *Contact Dermatitis,* 18, 309, 1988.
79. **Camarasa, J. G., Malet, A., Serra-Baldrich, E., and Lluch, M.,** Contact allergy to tixocortol pivalate, *Contact Dermatitis,* 19, 147, 1988.
80. **Feldman, S. B., Sexton, F. M., Buzas, J., and Marks, J. G.,** Allergic contact dermatitis from topical steroids, *Contact Dermatitis,* 19, 226, 1988.
81. **Cox, N. H.,** Contact allergy to clobetasol propionate, *Arch. Dermatol.,* 124, 911, 1988.
82. **Voss, M.,** Eine Prednisolon-Allergie mit pustulösem Exanthem, *Dermatol Monatsschr.,* 174, 221, 1988.
83. **Dooms-Goossens, A., Andersen, K., Burrows, D., Camarasa, J., Ducombs, G., Frosch, P., Lachapelle, J.-M., Lahti, A., Menné, T., Rycroft, R., White, I., and Wilkinson, J.,** A survey of results of patch tests with tixocortol pivalate, a marker for contact allergy to certain corticosteroids, *Contact Dermatitis,* 20, 158, 1989.
84. **Ashworth, J., Booker, J., and Breathnach, S. M.,** Effects of topical corticosteroid therapy on Langerhans cell antigen presenting function in human skin, *Br. J. Dermatol.,* 118, 457, 1988.
85. **Lauerma, A. I., Visa, K., Pekonen, M., Förström, L., and Reitamo, S.,** Cellular kinetics of delayed hypersensitivity test reactions to topical glucocorticosteroids, *Arch. Dermatol. Res.,* 279, 379, 1987.
86. **Dooms-Goossens, A.,** *Allergic Contact Dermatitis to Ingredients Used in Topically Applied Pharmaceutical Products and Cosmetics,* Leuven University Press, Leuven, 1983.
87. **Dooms-Goossens, A., Degreef, H., Mariën, K., and Coopman, S.,** Contact allergy to corticosteroids: a frequently missed diagnosis?, *J. Am. Acad. Dermatol.,* 21, 538, 1989.
88. **Smolle, J. and Kresbach, H.,** Anmerkungen zu den Kontaktekzemen — Ätiopathogenetische Überlegungen anhand von immunhistochemischen Untersuchungen: Contact dermatitis — immunohistological findings and etiopathogenic considerations, *Z. Hautkr.,* 62, 1681, 1987.
89. **Paramsothy, Y., Collins, M., and Smith, A. G.,** Contact dermatitis in patients with leg ulcers: the prevalence of late positive reactions and evidence against systemic ampliative allergy, *Contact Dermatitis,* 18, 30, 1988.
90. **Moss, C., Friedman, P. S., Shuster, S., and Simpson, J. M.,** Susceptibility and amplification of sensitivity in contact dermatitis, *Clin. Exp. Immunol.,* 61, 232, 1985.
91. **Dooms-Goossens, A.,** Corticosteroid contact allergy, in Abstr. Proc. Eur. Symp. Contact Dermatitis, Heidelberg, May 27 to 29, 1988, 45.
92. **Dooms-Goossens, A.,** Sensibilisatie ten opzichte van corticoidhoudende topica, in *Topische Corticotherapie, Corticothérapie Locale,* Lapière, C. M. and Piérard, G. E., Eds., Symposium Publications Division, Brocades-Belge S.A., Brussels, Belgium, 1976, 49.
93. **Reitamo, S., Lauerma, A. I., and Förström, L.,** Optimal testing for contact allergy to topical corticosteroids, poster (no. 26) presented at the Eur. Symp. Contact Dermatitis, Heidelberg, May 27 to 29, 1988.
94. **Fisher, A. A.,** The paraben paradox, *Cutis,* 12, 830, 1973.
95. **Poelman, M. C., Leveque, J. L., and Le Gall, F.,** Objective determination of the bioavailability of dermocorticoids — influence of the formulation, *Br. J. Dermatol.,* 111 (Suppl. 27), 158, 1984.
96. **Schaefer, H., Zesch, A., and Stüttgen, G.,** *Skin Permeability,* Springer-Verlag, Berlin, 1982, 657.
97. **Ponec, M. and Polano, M. K.,** Hydrocortisone butyrate penetration through epidermis *in vitro*: influence of vehiculum, concentration in the vehiculum, and the amount applied, *Arch. Dermatol. Forsch.,* 245, 381, 1972.
98. **Das Gupta, V.,** Effect of vehicles and other active ingredients on the stability of hydrocortisone, *J. Pharm. Sci.,* 67, 299, 1978.

99. **Bundgaard, H.,** The possible implication of steroid-glyoxal degradation products in allergic reactions to corticosteroids, *Arch. Pharm. Chem. Sci.,* 8, 83, 1980.
100. **Coopman, S. and Dooms-Goossens, A.,** Cross-reactions in topical corticosteroid contact dermatitis, *Contact Dermatitis,* 19, 145, 1988.
101. **Coopman, S., Degreef, H., and Dooms-Goossens, A.,** Identification of cross-reaction patterns in allergic contact dermatitis to topical corticosteroids, *Br. J. Dermatol.,* 121, 27, 1989.
102. **Tauder, V. and Toda, T.,** Transformation of diflucortolone valerate in the skin of rat, guinea pig and man, *Arzneim. Forsch.,* 26, 1484, 1976.
103. **Hannuksela, M. and Salo, H.,** The repeated open application test (ROAT), *Contact Dermatitis,* 14, 221, 1986.
104. **Bundgaard, H.,** personal communication, 1988.

Chapter 25

SKIN PROBLEMS IN THE FISH PROCESSING INDUSTRY

Kristian Thestrup-Pedersen and Lars Halkier-Sørensen

TABLE OF CONTENTS

I. INTRODUCTION

The fishing industry contains working conditions which are known to be able to induce skin irritation. They include low temperatures, wet work, trauma-like cuts or stings, and contact with biological material which can induce skin irritation or allergies.

This chapter describes skin problems in the Danish fishing industry, which can be considered representative of similar industries in Northern Europe, including the Scandinavian countries (Denmark, Sweden, Norway, Finland, Iceland, Greenland), England, Scotland, Ireland, Belgium, The Netherlands, Germany, Poland, and the Soviet Union. Fishermen from these countries often work within the same geographical areas, catching the same sort of fish (which, of course, can vary a lot worldwide).

II. EPIDEMIOLOGICAL DATA

A. FISH AND FISH PRODUCTS

Most of the fish dealt with in this investigation are "roundfish" such as codfish (71.7%), haddock (22.7%), coalfish (5.3%), and whiting (0.3%) as opposed to "flatfish" such as plaice or flounder. The fish were caught in the Baltic Sea, Danish seas (Kattegat and Skagerak), the North Sea, or the North Atlantic. Immediately after the catch, their intestines are removed by the fishermen and the fish are then placed among crushed ice in boxes. By law, the addition of chemicals to inhibit putrefaction is not allowed. The fishing boats may be at sea for up to 2 weeks before landing their catch.

The fishermen come into contact with intestines, intestinal fluids, and slime on the surface of the fish, all of which may contain bacteria or algae. The skin contact in the fish-processing factories is mostly with slime, skin, and "fish juice", which is the juice surrounding the fillets on their arrival for processing. The workers control the fillets for bones, cut them into suitable sizes, and then send them further on for packing. During packing, salts or polyphosphates may be added as required by the customer. The temperature of the fish in the factories is around 6°C or colder.

The protein content of cod varies between 15 and 19%, the oil content between 0.1 and 0.9%, and the water content between 78 and 83%. Just following spawning there is a reduction in the protein content due to an increased activity of muscle enzymes (cathepsin A, C, and D), making the fillets soft.

Fish meat and juices from fish meat contain volatile nitrogen-rich substances such as ammonium, trimethylamine, dimethylamine, trimethylaminoxide, hypoxanthine, inosine monophosphate, adenosine monophosphate, adenosine diphosphate, adenosine triphosphate, and formaldehyde. In six samples of fish juice known to have caused skin irritation, we found very low concentrations of histamine in one sample (<0.01 μmol/ml) and cadaverine in four samples (0.02—0.05 μmol/l).

We also studied bacteria and algae from the fish. Algae were not found, and we did not observe any unusual bacterial growth, any difference between samples which gave rise to skin irritation, or samples without skin-irritative properties.

B. SKIN PROBLEMS AMONG FISHERMEN

Fishing at sea is one of the most dangerous occupations and, in Denmark, carries a mortality rate about 50 times higher than the normal population. The average fisherman does not regard dermatological problems as being of major significance for his health, although they are rather frequent. Thus, a recent questionnaire among 280 Danish fishermen showed that 24% had experienced skin irritancy or eczema; two thirds of the workers believed that their condition was due to contact with fish.[1] The men had worked for an average of 24 years as fisherman, with a range between 2 and 54 years. Skin disease was present in

TABLE 1
Distribution of Skin Symptoms among 156 Employees with Subjective Symptoms

Symptoms	Number	Percentage
Itching	136	87%
Redness	100	64%
Stinging	61	39%
Papules	24	15%
Burning	16	10%
Facial rash	13	8%

TABLE 2
Localization of Skin Symptoms among 156 Employees

Localization	Number	Percentage
Lower arms	109	70%
Inner side	105	67%
Outer side	4	3%
Face and neck	94	60%
Back of hands	41	26%
Between fingers	8	5%
On fingers	7	4%
Palms	5	3%

40% of 115 fishermen in Norway;[2] two thirds exhited contact eczema on the hands and one fifth had infections.

C. SKIN PROBLEMS AMONG EMPLOYEES IN THE FISHING INDUSTRY

We have performed a clinical study among 196 randomly selected employees in four fish-processing factories. A total of 156 (80%) had experienced skin symptoms during their work with the fish products over a 5-year period. All employees went through a clinical investigation and several measurements of their skin physiology (transepidermal water loss [TEWL]; skin conductivity; skin temperature).[3] Table 1 shows the subjective symptoms and Table 2 the localization of symptoms.

1. Clinical Symptoms

The symptoms were present for short time periods (e.g., hours). They could arise suddenly with employees experiencing that the fish were ''biting''. Some days certain persons had complaints; others had them on ensuing days. The symptoms disappeared within a few hours after stopping work.

The clinical investigation by a dermatologist revealed that 11% had urticarial-type redness at the time of investigation. Only 2% had mild eczema, and nail changes were uncommon.

2. Cause of Symptoms

When asked what caused the skin symptoms, 89% said that they were due to skin contact with ''fish juice''. ''Fish juice'' is the ooze from the fish or fish fillets that blends with melted ice and collects in the fish boxes. The employees knew that fish kept through the weekend often give rise to a yellow discoloration of ''fish juice''. The protein content was around 10 g/l, which is one seventh of the protein content in human plasma.

TABLE 3
The Table Shows Redness Induced by Scratch Testing Volunteers with Various Fish Products and Their Dependence upon Postmortal Age

Item	Postmortal age (days)	No. tested	Positive reactions
Fish juice	1—2	23	0
Fish meat	1—2	23	0
Fish juice	5—7	23	15 (65%)
Fish meat	5—7	23	2 (9%)
Fish juice	10—12	12	7 (58%)
Fish meat	10—12	12	0
Fish skin	1—2	8	0
Fish skin	10—12	8	6 (75%)
Slime	4—5	7	0
Slime	10—12	7	6 (86%)
Juice from box	10—12	7	6 (86%)
Control, 0.9% NaCl		39	0
Control, 0.3% histamine		24	24 (100%)

3. Skin Physiology

Temperature measurements revealed that the employees had a temperature less than 20°C on the fingers and palms, while the temperature on the back of the hands was 25°C and 30°C on the forearms. Their finger temperature is, thus, on the average 12.2°C lower than in control persons.[3] We have shown that cooling of the skin to below 20°C abolishes the histamine-induced itch and reduces redness by one half, but does not interfere with the wheal reaction.[4] This could explain why urticarial itching is not found on the fingers and palms.

The TEWL was low in the employees compared to control persons. This low TEWL changes after work, with a significant increase 1 to 2 h following work, when it normalized and left a "dried-out" sensation in many employees. TEWL is significantly correlated to skin temperature, explaining the low TEWL at work.

II. EXPERIMENTAL STUDIES OF SKIN IRRITANCY FROM FISH PRODUCTS

A total of 145 volunteers in our hospital without previous skin problems following contact with fish were scratch tested using different products from roundfish. Following a scratch on the volar forearm, the fish material was applied for 20 min, after which we recorded itching, stinging, and redness. The redness was recorded using area measurements and a laser Doppler flowmeter in order to ensure objective results.

The most common symptoms following scratch testing were redness and itching. Table 3 shows that items from fish with a high postmortal age were the main cause of symptoms, especially substances accumulating in "fish juice".

It should be stressed that a subclinical trauma to the epidermis — such as the induction of an invisible scratch — could lead to the recognition of itching, redness, and stinging in the test persons following exposure to fish products.

IV. BIOCHEMICAL INVESTIGATIONS INCLUDING FRACTIONATION

When performing lipid or protein extraction, we showed that the skin-irritant capacity was confined to the protein fraction larger than 10,000 kDa in molecular weight. There was no difference in preliminary studies of sodium dodecyl sulfate (SDS)-polyacrylamide gel electrophoresis of protein fractions known to be skin irritants or those without this capacity. We have not investigated the cathepsin activity in our test material, but have noticed that overnight incubation of human epidermal sheets in "fish juice" leads to various degrees of degeneration of the tissue. The volatile amines were not found to be skin irritative when tested in a pure chemical form.

V. CONCLUSIONS

The fish-processing industry in Denmark is ninth in a list of industries where the frequency of dermatological occupational diseases (i.e., mostly hand eczema) was rated.[5] Approximately 80% of the employees do, however, experience skin irritancy from time to time with urticarial reactions. These symptoms are annoying, but do not prevent continued work. Very few were found with eczema (2%), and among 4000 employees only 0.7% had left the industry due to skin problems over a 10-year period.

Previous studies of workers in the fishing industry have described erysipeloid, which is still a rather common, acute, time-limited skin infection in fishermen,[6,7] and other forms of dermatosis.[8] Depending upon the fish products, some investigators have found up to 11% with eczema in the fishing industry[9] as compared to our 2%. Dogger bank itch was not found to be relevant in our situation and is still rather rare even among fishermen, but is severely disabling.[10,11] The antigen does not come from fish, but from a marine animal (*Alcyonidium gelatinosum* [Bryozoa]).[10,11] The allergy is a type IV allergy, but may also give rise to symptoms of asthma. Asthma has recently been described to occur among fisherman exposed to cuttlefish, but proper allergic studies were not carried out.[14]

The most well-described skin symptoms due to fish are protein contact urticaria.[8,14-18] Type I-like reactions also may be induced using raw fish material in both healthy and atopic subjects, although to a larger extent in the atopics.[19,20]

In our investigation, the localization of the urticarial symptoms was mainly on the volar side of the lower arm, in the face and neck from splashes of fish juice, and on the back of the hands. Surprisingly, very few symptoms occur on the fingers and palms, and this seems to be related to the low skin temperature in these locations; itching is particularly prevented.[3,4] Also, minor trauma predisposes to the development of symptoms.

It was clearly demonstrated that the fractions with skin irritant properties were the protein fractions of high molecular weight. This may support the impression that protein-rich fish, i.e., those caught in the early months of the year before spawning, can rather commonly give rise to symptoms. However, the postmortal age of the fish at landing seems to be more important.[21]

A possible cause for the occurrence of skin symptoms may be the damage that enzymes from fish intestine may induce on keratin. It has recently been shown that trypsin and pepsin in the fish gut can dissolve human keratin.[22,23] Other enzymes may be involved, too, such as cathepsins. This sort of cumulative damage could be one of several damaging events upon the epidermis, thus preparing the background for irritative symptoms to occur.

VI. PREVENTIVE MEASUREMENTS

As preventive measures, the industry should encourage that the postmortal age of landed fish be as low as possible, that "fish juice" in the boxes be rinsed from the fillets before processing, that contamination of fish meat with intestinal juices be avoided, and that damage to the epidermal barrier be prevented in order to inhibit access of high-molecular-weight substances to dermal tissue. Pollutants, bacteria, algae, or volatile amines arising during putrefaction do not seem to play a role in the elicitation of skin irritancy from fish.

It is our experience that if employees are given this information and they use protective clothing when exposed to fish products of a postmortal age of more than 1 week, annoying skin-irritancy problems in the industry will be virtually eliminated.

REFERENCES

1. **Halkier-Sørensen, L. and Thestrup-Pedersen, K.,** Subjective skin symptoms among Danish fishermen, unpublished.
2. **Tellus, G.,** Skin diseases among fisherman, *Tidsskr. Nor. Laegeforen.*, 107, 566, 1987 (in Norwegian).
3. **Halkier-Sørensen, L. and Thestrup-Pedersen, K.,** Skin temperature and skin symptoms among workers in the fish processing industry, *Contact Dermatitis,* 19, 206, 1988.
4. **Halkier-Sørensen, L. and Thestrup-Pedersen, K.,** The relevance of low skin temperature inhibiting histamine-induced itch to the location of contact urticarial symptoms in the fish processing industry, *Contact Dermatitis,* 21, 179, 1989.
5. **Veien, N. K., Heydenreich, G., Kaaber, K., and Villumsen, P.,** Occupational Skin Diseases Diagnosed by Dermatologists in Jutland, Institute of Occupational Medicine, Copenhagen, 1986 (in Danish).
6. **Klauder, J. V., Righter, L. L., and Harkins, M. J.,** A distinctive and severe form of erysipeloid among fish handlers, *Arch. Dermatol. Syphilol.*, 14, 662, 1926.
7. **Proctor, D. M. and Richardson, I. M.,** A report on 235 cases of erysipeloid in Aberdeen, *Br. J. Ind. Med.*, 11, 175, 1954.
8. **Beer, W. E., Jones, M., and Jones, W. E.,** Dermatoses in lobster fishermen, *Br. Med. J.*, 1, 807, 1968.
9. **Kavli, G. and Moseng, D.,** Contact urticaria from mustard in fish-stick production, *Contact Dermatitis,* 17, 153, 1987.
10. **Fraser, J. H. and Lyell, A.,** Dogger bank itch, *Lancet,* 1, 61, 1963.
11. **Newhouse, M. L.,** Dogger bank itch: survey of trawlermen, *Br. Med. J.*, 1, 1142, 1966.
12. **Carlé, J. S., Thybo, H., and Christophersen, C.,** Dogger bank itch. 3. Isolation, structure, determination and synthesis of a hapten, *Contact Dermatitis,* 8, 43, 1982.
13. **Carlé, J. S. and Christophersen, C.,** Dogger bank itch. An eczema-causing sulfoxonium ion from the marine animal, *Alcyonidium gelatinosum* (Bryozoa), *Toxicon,* 1, 307, 1982.
14. **Tomaszunas, S., Weclawik, Z., and Lewinski, M.,** Allergic reactions to cuttlefish in deep-sea fishermen, Letter to the Editor, *Lancet,* 1, 1116, 1988.
15. **Hjorth, N. and Roed-Petersen, J.,** Occupational protein contact dermatitis in food handlers, *Contact Dermatitis,* 2, 28, 1976.
16. **Görannson, K.,** Contact urticaria to fish, *Contact Dermatitis,* 7, 282, 1981.
17. **Lombardi, P., Campolmi, P., Giorgini, S., Spallanzani, P., and Sertoli, A.,** Contact urticaria from fish, honey and peach skin, *Contact Dermatitis,* 9, 422, 1983.
18. **Melino, M., Toni, F., and Riguzzi, G.,** Immunologic contact urticaria to fish, *Contact Dermatitis,* 17, 182, 1987.
19. **Beck, H. I. and Nissen, B. K.,** Type-I reactions to commercial fish in non-exposed individuals, *Contact Dermatitis,* 9, 219, 1983.
20. **Beck, H. I. and Nissen, B. K.,** Contact urticaria to commercial fish in atopic persons, *Acta Derm. Venereol.*, 63, 257, 1982.
21. **Halkier-Sørensen, L. and Thestrup-Pederson, K.,** Skin irritancy from fish is related to its postmortem age, *Contact Dermatitis,* 21, 172, 1989.
22. **Bjelland, S., Gildberg, A., and Volden, G.,** Degradation of human keratin by fish pepsin, *Arch. Dermatol. Res.*, 280, 119, 1989.
23. **Bjelland, S., Hjelmeland, K., and Volden, G.,** Degradation of human epidermal keratin by cod trypsin and extracts of fish intestines, *Arch. Dermatol. Res.*, 280, 469, 1989.

Chapter 26

CONTACT URTICARIA TO SURGICAL AND HOUSEHOLD RUBBER GLOVES

Kristiina Turjanmaa and Timo Reunala

TABLE OF CONTENTS

I. INTRODUCTION

The fact that various rubber products such as rubber (latex) gloves can cause allergic contact dermatitis has been known for decades. This delayed type of contact allergy is not caused by the natural rubber latex, i.e., by the milky liquid of the rubber tree, but by the rubber-compounding chemicals added to natural rubber during the manufacture of rubber products. The diagnosis of rubber glove contact dermatitis is easily verified with patch tests, and most patients are allergic to thiuram chemicals.[1] An immediate type of contact allergy is immunologic contact urticaria, in which the allergens are proteins and the reaction is mediated by immunoglobulin E (IgE) antibodies. Immunologic contact urticaria in the intact or eczematous skin of sensitized people has been ascribed to a wide range of plants, vegetables, and animal products.[2] In 1979 Nutter[3] reported the first case of contact urticaria from household latex gloves, and the following year Förström[4] described the case of a nurse who contracted contact urticaria from surgical gloves. In the 1980s several reports, mostly from European countries, have focused attention on latex glove contact urticaria (LGCU), which can readily go undiagnosed.[5-16] It is also now well established that, in addition to local cutaneous symptoms, this allergy can cause systemic symptoms such as generalized urticaria, asthma, and even anaphylaxis.[14,17,18] Several skin tests have been carried out in an effort to diagnose LGCU, but standardized skin tests are still unavailable because the allergen has not been accurately defined. Even so, there is abundant evidence that the allergen is a latex protein(s) which elutes from the gloves into the skin and may sensitize atopic people in particular.[11,19]

II. FREQUENCY AND PREDISPOSING FACTORS

In Table 1, 40 cases of latex glove contact urticaria reported from various countries are compiled; 5 (13%) of the patients are doctors and 14 (35%) are nurses, showing that about half of the patients with LGCU are in the medical profession and have obviously been sensitized by surgical latex gloves. The other half of the patients contracted LGCU from household gloves used either at home or at work. In our series of 27 patients with LGCU admitted to our department during a 2-year period, the ratio was about the same; 13 were doctors and nurses sensitized by surgical gloves, and 14 were patients sensitized by household gloves.[20]

The only epidemiologic study on LGCU to date was carried out on 512 Finnish glove-using medical employees.[21] All were skin tested with latex glove pieces, and 15 (3%) were found to have LGCU. The frequency was highest (6%) in operating units, where 7% of the doctors and 6% of the nurses were sensitized (Table 2). In nonoperating units, the frequency was still high in doctors (6%), but lower in nurses (1%). Nilsson[22] examined 142 hospital wet-workers in Sweden and found that as many as 41 exhibited work-related contact urticaria symptoms. These occurred mostly in kitchen work involving food handling; in only four (3%) were the symptoms caused by household gloves. We used skin tests to examine 36 food handlers who wore latex household gloves in a food factory; one was found to have LGCU.[20] The results of these screening studies show that the frequency of LGCU is as high as 3% in occupations where either surgical or household latex gloves are worn in daily work. In addition to routine prick test series, we carried out a latex prick test on 851 patients admitted to the Allergy Laboratory of our Department with suspected atopy and found LGCU in 9 (1%) patients (Table 2). All of these had a history of LGCU, and eight were sensitized by using household gloves. The ninth patient was a young boy who had had several attacks of facial edema caused by contact with balloons. Later, when a plastic catheter was inserted into his urethra, he contracted generalized urticaria from the surgical gloves worn by the doctor.[20]

TABLE 1
Reported Cases of Latex Glove Contact Urticaria

Year	Country	Number of patients	Glove type		Number with generalized symptoms	Tests used for diagnosis								Ref.
			Surgical	Cleaning		OPT	RT	S-CT	PT	UT	P-KT	RAST	BHRT	
1979	England	1	0	1	0	+1			+1,2					3
1980	Finland	1	1	0	1			+1						4
1982	FRG	2	0	2	1		+1	+2						5
1983	Finland	4	0	4	0	+1,2			+2	+1	+2			6
1984	Finland	2	2	0	2				+1,2		+1,2	+1		7
1984	FRG	3	1	2	1	+1	+1	+2						8
1984	Sweden	3	3	0	1	+1,2			+2	+1				9
1986	Sweden	6	NG	NG	0			+1		+1		+2		10
1986	Spain	1	1	0	1				+1,2			+1	+1,2	11
1986	FRG	3	3	0	3	+1,2				+1		+1,2		12
1987	FRG	3	0	3	2		+1					+1,2		13
1987	Sweden	5	0	5	5				+1	+2		+2		14
1987	Finland	4	1	3	1			+1	+2					15
1988	Canada	2	2	0	2				+2			+2		16

Note: 1 = glove material, 2 = natural rubber, OPT = open patch test, RT = rub test, S-CT = scratch-chamber test, PT = prick test, UT = use test, P-KT = Prausnitz-Küstner test, RAST = radioallergosorbent test, BHRT = basophil histamine release test.

TABLE 2
Frequency of Latex Contact Urticaria among 512 Glove-Using Hospital Employees, 44 Medical Students, 36 Cleaning-Glove Using Food Handlers, and Latex Prick-Tested Control Patients

Subjects	Total frequency	Frequency in different occupations		
		Doctors	Nurses	Others
Hospital employees				
Total	15/512 (3%)	7/108 (7%)	7/325 (2%)	1/79 (1%)
In operating units	9/145 (6%)	4/54 (7%)	4/71 (6%)	1/29 (3%)
In other units	6/367 (2%)	3/54 (6%)	3/254 (1%)	0/59 (0%)
Medical students	1/44 (2%)	—	—	—
Food handlers in a food factory	1/36 (3%)	—	—	—
Prick-tested control patients	9/851 (1%)	—	—	—

TABLE 3
Hand Eczema, Atopy, and Routine Prick Test Findings for 42 Subjects with Latex Glove Contact Urticaria

Subjects	Hand eczema		Atopy, positive	
	Past	Present	Personal history	Prick test[a]
Surgical glove users (n = 28)	9 (32%)	16 (57%)	23 (82%)	17 (61%)
Cleaning glove users (n = 14)	2 (14%)	12 (86%)	13 (93%)	12 (86%)
Total (n = 42)	11 (26%)	28 (67%)	36 (86%)	29 (69%)

[a] At least one positive reaction in a series of 20 common inhalant allergens.

Most of the patients with LGCU are females. In our series of 42 subjects only 6 (14%) were males, all of whom were medical practitioners and had been sensitized by contact with the surgical gloves.[20] The same was true for many of our female patients and also for those reported previously (Table 1). Interestingly, we could not find any correlation between the total length of time that surgical gloves were worn and the development of LGCU among the 512 hospital employees examined.[21] In allergic employees the length of time gloves had been worn varied from 3 to 20 years, suggesting that LGCU does not necessarily appear during the early years of work in the medical profession and surgical glove usage. However, we have diagnosed one medical student with LGCU, indicating that this allergy may also appear within a short time of wearing surgical gloves.[23]

A prominent feature of LGCU is that most of the patients present with symptoms of atopic disorders or at least have a positive personal history of atopy such as atopic dermatitis, allergic rhinitis, and asthma. Of our 42 patients with LGCU, 36 (86%) had a personal history of atopy (Table 3). Among the patients reported in the literature (Table 1), a possible history of personal atopy was found in 25 (81%) of 31 patients with LGCU. Whether atopy is a real predisposing factor or is indirectly associated with the development of LGCU is not yet known. Atopics tend to suffer from irritant hand eczema and need to wear household gloves for personal protection.[24] Moreover, in such atopics the latex allergen eluting from the gloves can penetrate the eczematous skin more easily than the intact skin, and this may be an additional factor contributing to the development of LGCU. In agreement with this, as many as 67% of our 42 patients with LGCU had hand eczema, and another 26% had a past history of the condition.[20]

TABLE 4
Symptoms in 27 Subjects with Latex Glove Contact Urticaria

Subjects	Redness and itching or exacerbation of HE[a]	Contact urticaria syndrome stage[b]			
		I	II	III	IV
Using surgical gloves (n = 13)	1 (8%)	6 (46%)	2 (15%)	2 (15%)	2 (15%)
Using cleaning gloves (n = 14)	2 (14%)	5 (36%)	3 (22%)	2 (14%)	2 (14%)
Total (n = 27)	3 (11%)	11 (40%)	5 (19%)	4 (15%)	4 (15%)

[a] 9/27 had present hand eczema.
[b] According to von Krogh and Maibach.[2]

III. SYMPTOMS

The symptoms of LGCU may vary from mild, uncharacterized itching to severe systemic reactions (Table 4). The symptoms are similar for surgical and household gloves, and most patients soon become aware of the skin irritation caused by the gloves. However, only a few patients present with a typical history of immediate localized urticarial response. Itching and burning or a history of recurrent hand eczema are common. Immediate symptoms appear 1 to 15 min after the gloves have been put on and disappear in 1 to 2 h. The patients, even if they are surgeons, frequently suspect that glove powder or hand washing, not the latex gloves, causes these symptoms. About half of our patients with LGCU had experienced extracutaneous symptoms (Table 4) including generalized urticaria and facial edema (stage II), rhinoconjunctivitis, orolaryngeal and gastrointestinal symptoms, and asthma (stage III) or anaphylaxis (stage IV).[2] Five of our female patients with LGCU had experienced anaphylactic or severe stage III reactions during either delivery or gynecological surgery as a result of the surgical gloves worn by doctors.[18,20] Of the cases reported previously in the literature (Table 1), 55% had had local symptoms, a figure that is very similar to the 52% reported for our patients with LGCU.

In a few cases there are no immediate symptoms or they are only very mild, but during continuous use of gloves the patient begins to notice the emergence or reappearance of hand eczema. As many as 28 (67%) of the patients in our series with LGCU presented with simultaneous symptoms of LGCU and hand eczema; in 9 of these the hand eczema disappeared or imporved markedly once they avoided wearing latex gloves. Whether the eczema in these patients is a consequence of repeated urticarial lesions being rubbed and scratched or is a separate irritant contact dermatitis is unclear. However, the healing of hand eczema is similar to that observed for protein contact dermatitis among veterinary surgeons and food handlers.[25,26] The hypothesis that protein allergens may cause eczematous reactions in the skin of patients with atopic eczema is interesting in terms of LGCU and associated hand eczema. Some other factors, e.g., irritant chemicals and an atopic background, may be the reason why only one third of the hand eczema cases among our LGCU subjects healed as a result of avoiding latex gloves.

IV. DIAGNOSIS

Several types of skin and *in vitro* tests are available for diagnosing LGCU (Table 1). These include open and closed application of latex glove pieces to either intact or slightly eczematous skin, with subsequent development of wheals in positive cases. Moistening of the glove pieces or the skin permits the allergen to dissolve more quickly from the glove into the skin. The scratch chamber technique improves the penetration of the allergen into

TABLE 5
Latex Prick Test, Use Test, and RAST Results and Serum IgE Levels in 42 Subjects with Latex Glove Contact Urticaria

Subjects	Prick test: Positive n	Prick test: Grade 2+	Grade 3+	Grade 4+	Use test positive n (%)	RAST: Positive n (%)	RAST Score 1	Score 2	Score 3	Score 4
Surgical glove users (n = 28)	28	7	9	12	21[a] (95%)	17 (61%)	2	7	6	2
Cleaning glove users (n = 14)	14	2	6	6	14 (100%)	10 (71%)	1	4	3	2
Total (n = 42)	42	9	15	18	35 (97%)	27 (64%)	3	11	9	4

[a] Six subjects not tested.

the skin, but false positive reactions are not uncommon with this test.[21] Prick testing seems to be the best way to confirm the diagnosis of LGCU, but as long as commercial latex allergen preparations do not exist, the test solutions should be made correctly. We use the following procedure: about 20 pieces (1 cm^2) are cut from an allergenic latex glove (Exona®, Semperit, Austria), incubated for 30 min in 5 ml of sterile physiological saline (taken from a plastic bag to avoid contamination from rubber cork) at room temperature, and used as a fresh eluate or after storage at +4°C in prick tests.[27] The allergenicity of the eluates of various surgical and household gloves can vary considerably, and in order to avoid false negative prick test results it is important to use a well-characterized glove brand as allergen source.[19] When available, rubber tree sap can be used at the same time as glove eluate in prick tests.[27] We have never observed any adverse reactions, even when highly allergic patients have been simultaneously prick tested with several glove eluates made with the above technique. In contrast to prick testing, scratch testing with concentrated natural liquid latex and use testing with whole latex gloves have been implicated as causes of systemic, and even life-endangering, reactions; these test techniques should therefore be abandoned.[12,23] At least weak positive prick tests should be confirmed by the one finger latex glove use test.[27] Our use tests were judged positive only when some wheals were seen. Unspecific erythema and dermographism can occur in the use test, but these are easily excluded if a simultaneous test with a polyvinylchloride (PVC) glove is performed. We have never seen positive use test or prick test reactions from PVC gloves, although it has been claimed that such reactions occur.[28,29]

The diagnostic sensitivity of the prick test with latex glove eluate seems to be better than that of the commercially available latex radioallergosorbent test (latex RAST, k82, Pharmacia Diagnostics AB, Uppsala, Sweden), in which rubber sap is used as allergen. We have examined 42 patients with LGCU (Table 5). All had positive prick tests, but latex RAST was positive in only 27 (64%) of the subjects. However, the RAST detected all the highly allergic subjects; thus, it is a valuable *in vitro* test for LGCU when confirming and grading LGCU. Previously, one patient with LGCU has been examined with the basophil histamine release test (BHRT).[11] This patient showed positive histamine release with both latex glove eluate and natural rubber (liquid latex). In our study, BHRT gave positive results in 93% of the 14 subjects with LGCU who were evaluated. Moreoever, the amount of histamine released at the optimal latex glove allergen concentration correlated very well with the size of the prick test reactions. These results indicate that BHRT is a reliable *in vitro* test for LGCU and also seems to give more positive results than the RAST.[30]

V. ALLERGENICITY OF DIFFERENT LATEX GLOVES

Household latex gloves are worn for personal protection in cleaning and industrial work. Examination (either sterile or unsterile) gloves are worn during medical examinations and in diagnostic and therapeutic procedures, and surgical gloves are worn during surgical work. These different types of latex gloves are all made by dipping a mold into liquid, natural rubber latex. However, the allergenic capacity of different surgical and household latex gloves varies considerably. After prick testing the eluates of 17 surgical and 2 household latex gloves, we found that 4 glove brands were highly allergenic.[19] These gave positive prick tests in 91 to 100% of the subjects, whereas four other brands were positive in only 8 to 21% of the patients. The only gloves with no positive reactions were the nonlatex gloves used as controls; one of these was made from PVC and the other from synthetic rubber. Our results show that all latex surgical and household gloves contain the allergen, but that differences exist either in the composition or amount of latex allergen eluting from the gloves. We performed protein measurements on the eluates of three different latex gloves and found the lowest total protein concentration (0.025 mg/g glove) in the glove causing only a few positive prick tests (Ansell Gammex®, Ansell Rubber Co. Ltd., Thailand). This is in agreement with our previous clinical observation that almost half of the hospital employees allergic to latex surgical gloves can continue with their work if they wear this special glove. It was, however, surprising to detect a six times higher protein concentration (0.28 mg/g glove) in the household glove (Natura®, Vascongada de Gaucho y Latex, Spain) than in another surgical glove (0.046 mg/g Triflex®, Travenol Laboratories, Belgium) because both gloves caused positive prick tests at the same high frequency. Therefore, the allergenicity of a particular glove probably cannot be predicted from the total amount of protein eluting from that glove.[19]

Both direct and indirect evidence now exists that the allergen in the latex gloves is certainly a protein or proteins. Carrillo et al.[11] used BHRT to show that both latex glove eluate and natural rubber (liquid latex) contained the allergen. Serial filtrations indicated that the allergenity was strongest in the fraction with >30,000 mol wt. We used high-pressure liquid chromatography (HPLC) to analyze the proteins eluting from one surgical glove and one household glove causing a high number of positive prick tests.[19] The HPLC curves were similar except for one prominent protein peak at 30,000 mol wt in the household glove. The allergenic fractions from the gloves had distinct peaks at 2,000, 5,000 and 30,000 mot wt. These results suggest that there are no major qualitative differences between the proteins of these gloves. Interestingly, the eluate made from the rubber tree leaf had a similar HPLC protein profile, confirming that some proteins derived from natural rubber persist in latex gloves.

Gloves are also powdered, generally with corn starch. This starch has been reported to cause contact urticaria in some rare cases.[31,32] We tested commercial Biosorb® (Johnson & Johnson Arlington, TX) glove powder in 42 subjects and Elastyren® glove powder (Remy Industries, Wijgmaal-Leuven Belgium) in 32 subjects with LGCU, but our results were negative.[20] This indicates that glove powder is not an important allergen among subjects with LGCU.

The coexistence of LGCU and allergic contact dermatitis due to rubber chemicals in the gloves seems to be a more frequent occurrence than expected. Among 35 patients with previously diagnosed LGCU, we found 5 (14%) who had positive patch tests to thiurams and some other rubber chemicals. In contrast, using latex allergen we prick tested 18 patients with allergic contact dermatitis due to rubber gloves and found 1 patient who also had LGCU.[33] The frequent coexistence of type I and type IV contact allergy as a result of wearing latex gloves supports the view that simultaneous prick and patch tests should be performed when latex glove contact urticaria or dermatitis is suspected.

VI. MANAGEMENT OF LATEX GLOVE CONTACT URTICARIA

Treatment of LGCU is simple when the diagnosis is verified with skin tests and the RAST. The major goal is to avoid latex products, gloves in particular. Other latex products besides gloves are cuffs, masks, tubes, catheters in medicine, cofferdams in dentistry, dummies, teats, balloons, condoms, and elastic thread. Latex gloves can be avoided by wearing gloves made from PVC or synthetic rubber or, in some instances, by wearing cotton or plastic undergloves. Special attention should be paid to informing patients with LGCU and their doctors about risks that arise during surgery, deliveries, vaginal examinations, and dental treatment. To avoid serious systemic reactions, operating teams must use PVC or synthetic rubber gloves, and any latex tubes and catheters used during anaesthesia should be exchanged for nonlatex products when the patient has a diagnosed or suspected latex contact urticaria.

VII. CONCLUSION

For 10 years latex gloves have been known to cause contact urticaria, and more than 100 patients with LGCU sensitized by contact with either surgical or household gloves have been reported.[20,34,35] About 3% of the hospital employees who wear latex surgical or examination gloves at work seem to have LGCU, and as many as 1% of the patients prick tested for suspected atopy may show LGCU. The incidence of LGCU may increase among medical workers in the near future because of the more frequent use of protective latex gloves in the modern era of AIDS. The local symptoms of LGCU are often unspecific or may be masked by coexisting hand eczema. LGCU can also cause systemic, even anaphylactic, symptoms not readily connected with glove use. The diagnosis of LGCU is, however, easy to make with latex prick tests by eluting the allergen from the gloves. The one-finger use test and latex RAST are useful additional diagnositic tests. *In vivo* and *in vitro* results strongly indicate that the latex allergen in the gloves is a protein, or proteins, and derives from natural rubber latex. However, the exact nature and molecular weight of latex allergen have not been defined. Better characterization of the latex allergen and the determination of its chemical structure are needed. Standardized latex allergen preparation is needed, especially for prick testing, and better manufacturing methods should be developed to eliminate the latex allergen from the gloves.

REFERENCES

1. **Cronin, E.,** *Contact Dermatitis,* Churchill Livingstone, Edinburgh, 1980, 714.
2. **von Krogh, G. and Maibach, H. I.,** The contact urticaria syndrome — an updated review, *J. Am. Acad. Dermatol.,* 5, 328, 1981.
3. **Nutter, A. F.,** Contact urticaria to rubber, *Br. J. Dermatol.,* 101, 597, 1979.
4. **Förström, L.,** Contact urticaria from latex surgical gloves, *Contact Dermatitis,* 6, 33, 1980.
5. **Galinski, T. and Kleinhans, D.,** Kontakt-Urtikaria durch Gummihandschuhe, *Dermatosen,* 30, 118, 1982.
6. **Köpman, A. and Hannuksela, M.,** Contact urticaria to rubber, *Duodecim,* 3, 39, 1983.
7. **Turjanmaa, K., Reunala, T., Tuimala, R., and Kärkkäinen, T.,** Severe IgE mediated allergy to surgical gloves, *Allergy (Suppl.),* 2 (Abstr.) 35, 1984.
8. **Kleinhans, D.,** Contact urticaria to rubber gloves, *Contact Dermatitis,* 10, 124, 1984.
9. **Meding, B. and Fregert, S.,** Contact urticaria from natural latex gloves, *Contact Dermatitis,* 10, 52, 1984.
10. **Wrangsjö, K., Mellström, G., and Axelsson, G.,** Discomfort from rubber gloves indicating contact urticaria, *Contact Dermatitis,* 15, 79, 1986.
11. **Carrillo, T., Cuevas, M., Munoz, T., Hinojosa, M., and Moneo, I.,** Contact urticaria and rhinitis from latex surgical gloves, *Contact Dermatitis,* 15, 69, 1986.

12. **Frosch, P. J., Wahl, R., Bahmer, F. A., and Maasch, H. J.**, Contact urticaria to rubber gloves is IgE-mediated, *Contact Dermatitis,* 14, 241, 1986.
13. **Seifert, H. U., Seifert, B., Wahl, R., Vocks, E., Borelli, S., and Maasch, H. J.**, Immunoglobulin E-vermittelte Kontakturtikaria bzw. Asthma bronchiale durch Latex-enthaltende Haushaltgummihandschuhe, *Dermatosen,* 35, 137, 1987.
14. **Axelsson, J. G. K., Johansson, S. G. O., and Wrangsjö, K.**, IgE-mediated anaphylactoid reactions to rubber, *Allergy,* 42, 46, 1987.
15. **Estlander, T., Jolanki, R., and Kanerva, L.**, Contact urticaria from rubber gloves: a detailed description of four cases, *Acta Derm. Venereol. Suppl.,* 134, 98, 1987.
16. **Spanner, D., Sussman, G., Tarlo, S., Buttoo, K., Evans, S., and Dolovich, J.**, Latex allergy: two case reports, *J. Allergy Clin. Immunol.,* 81, 179, 1988.
17. **Axelsson, I. G. K., Eriksson, M., and Wrangsjö, K.**, Anaphylaxis and angioedema due to rubber allergy in children, *Acta Paediatr. Scand.,* 77, 314, 1988.
18. **Turjanmaa, K., Reunala, T., Tuimala, R., and Kärkkäinen, T.**, Allergy to latex gloves: unusual complication during delivery, *Br. Med. J.,* 297, 1029.
19. **Turjanmaa, K., Laurila, K., Mäkinen-Kiljunen, S., and Reunala, T.**, Rubber contact urticaria, allergenic properties of 19 brands of latex gloves, *Contact Dermatitis,* 19, 362, 1988.
20. **Turjanmaa, K.**, Latex Glove Contact Urticaria, Thesis, University of Tampere, Tampere, Finland, 1988.
21. **Turjanmaa, K.**, Incidence of immediate allergy to latex gloves in hospital personnel, *Contact Dermatitis,* 17, 270, 1987.
22. **Nilsson, E.**, Contact sensitivity and urticaria in "wet" work, *Contact Dermatitis,* 13, 321, 1985.
23. **Turjanmaa, K. and Reunala, T.**, Contact urticaria from rubber gloves, in *Occupational Dermatoses, Dermatologic Clinics,* Vol. 6, Taylor, J. S., Ed., W. B. Saunders, Philadelphia, 1988, 47.
24. **Rystedt, I.**, Atopic background in patients with occupational hand eczema, *Contact Dermatitis,* 12, 247, 1985.
25. **Hjorth, N. and Roed-Petersen, J.**, Occupational protein contact dermatitis in food handlers, *Contact Dermatitis,* 2, 28, 1976.
26. **Kalveram, K.-J., Kästner, H., and Forck, G.**, Nachweis von spezifischen IgE-Antikörpern bei Tierärzten mit Kontakturticaria, *Z. Hautkr.,* 61, 75, 1986.
27. **Turjanmaa, K., Reunala, T., and Räsänen, L.**, Comparison of diagnostic methods in latex surgical glove contact urticaria, *Contact Dermatitis,* 19, 241, 1988.
28. **Estlander, T., Jolanki, R., and Kanerva, L.**, Dermatitis and urticaria from rubber and plastic gloves, *Contact Dermatitis,* 14, 20, 1986.
29. **Frosch, P. J., Born, C. M., and Schutz, R.**, Kontaktallergien auf Gummi-, Operations- und Vinylhandschuhe, *Hautarzt,* 38, 210, 1987.
30. **Turjanmaa, K., Räsänen, L., Lehto, M., Mäkinen-Kiljunen, S., and Reunala, T.**, Basophil histamine release and lymphocyte proliferation tests in latex contact urticaria, *Allergy,* 44, 181, 1989.
31. **Fisher, A. A.**, Contact urticaria and anaphylactoid reaction due to corn starch surgical glove powder, *Contact Dermatitis,* 16, 224, 1987.
32. **van der Meeren, H. L. M. and van Erp, P. E. J.**, Life-threatening contact urticaria from glove powder, *Contact Dermatitis,* 14, 190, 1986.
33. **Turjanmaa, K. and Reunala, T.**, Latex contact urticaria associated with delayed allergy to rubber chemicals, in *Current Topics in Contact Dermatitis,* Frosch, P. J., Dooms-Goossens, A., Lachapelle, J.-M., Rycroft, R. J. G., and Scheper, R. J., Eds., Springer-Verlag, Heidelberg, 1989, 460.
34. **Dooms-Goossens, A.**, Contact urticaria caused by rubber gloves, *J. Am. Acad. Dermatol.,* 18, 1360, 1988.
35. **Wrangsjö, K., Wahlberg, J. E., and Axelsson, I. G. K.**, IgE-mediated allergy to natural rubber in 30 patients with contact urticaria, *Contact Dermatitis,* 19, 264, 1988.

Chapter 27

SKIN PROBLEMS IN THE OFFICE ENVIRONMENT

J. E. Wahlberg and B. Stenberg

TABLE OF CONTENTS

I. INTRODUCTION

The proportion of people employed in clerical work has gradually increased. In Sweden the change was from less than 15% in 1870 to 60% in 1980, while those involved in farming decreased from 72 to 5% during the same period (Figure 1). A similar redistribution is probable for most industrialized countries. According to the 1980 U.S. census, approximately 40 million Americans work in office buildings.[1]

The risk of accidents among clerical workers is extremely small; however, for example, skin, eye, mucous membrane, neck, and shoulder complaints are steadily increasing, especially after the introduction of new equipment such as visual display terminals (VDTs) and new materials such as carbonless copy paper (CCP). In our experience skin symptoms among office personnel are often accompanied by symptoms from the mucous membranes, which is in contrast to many other employment situations where the skin alone is affected. This limits the dermatologist's chance of making a correct diagnosis and suggesting adequate prophylactic means; frequently, other specialists have to be consulted.

Table 1 indicates — somewhat schematically and according to our experience — the frequency of various skin diseases to stress that irritation and nonspecific symptoms from the skin dominate and that the symptoms are frequently more prominent than the signs. The ailments are rarely well defined, and overlapping among the various conditions seems to be common. Non-office-related skin diseases such as impetigo, scabies, urticaria, and psoriasis (Table 1) in persons in key positions sometimes cause much public attention, and uncritical journalists contribute to the spread of the "epidemic". Ailments and skin diseases among office workers have been reviewed by Foussereau et al.,[2] Rycroft,[3] and Marks.[4]

II. THE SICK BUILDING SYNDROME

A. BACKGROUND

During the last 25 years, health problems related to indoor air quality have gained increasing interest. "Sick" buildings were recognized as early as the 1950s, but have become a main occupational health problem in many Western countries during the last 10 years. By 1987 the National Institute for Occupational Safety and Health (NIOSH) in the U.S. had conducted approximately 450 field investigations of indoor air quality problems.[5]

In 1982 a World Health Organization (WHO)[6] expert group defined the "sick building syndrome" (SBS). Their definition, based mainly on reports from the Scandinavian countries and the U.S., is still in widespread use and includes these symptoms:

- Eye, nose, and throat irritation
- Sensation of dry mucous membranes and skin
- Erythema
- Mental fatigue
- Headache, high frequency of airway infections, and cough
- Hoarseness, wheezing, itching, and nonspecific hypersensitivity
- Nausea, dizziness

The syndrome has several names; "stuffy office syndrome" and "tight building syndrome" are often used in British and American literature. In 1986 the WHO[7] estimated that 10 to 30% of newly built office buildings in Western countries had indoor air problems of this kind.

However, clinical findings are sparsely described. In a Swedish study[8] of children in day-care centers with indoor air problems, dry, itching, erythematous skin on the face and the extremities and eye and airway irritation were significantly more prevalent than in control

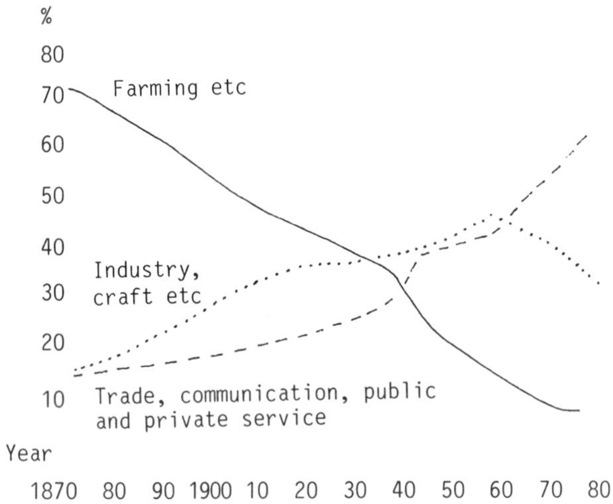

FIGURE 1. Changes in occupation of the Swedish population from 1870 (4 million inhabitants) to 1980 (8 million inhabitants).

TABLE 1
Ailments and Skin Diseases among Office Personnel

Allergic contact dermatitis	Rare
Irritant contact dermatitis	Uncommon
Photo contact dermatitis	Rare
Irritation of skin and mucous membranes sometimes accompanied by general symptoms such as headache and lethargy. Skin symptoms are nonspecific, i.e., tingling, stinging of facial skin, erythema, edema, dryness and scaling, papules/pustules, etc.	Common
Aggravation of hand eczema, especially atopic dermatitis	Not uncommon
Impetigo, scabies, urticaria, psoriasis, pustulosis, etc. incorrectly blamed on the office environment	Not uncommon

buildings. Dermatological findings among workers in ''sick'' buildings were reported in 1987.[9] Scaling dermatitis of the scalp, ears, and face, facial erythema, rosacea, urticaria/pruritis, and ''itching folliculitis'' were common work-related findings.

B. PREVALENCE

The prevalence of the full syndrome (i.e., general symptoms and mucosal and skin symptoms coexisting) is not known. In an interview study by Finnegan et al.[10] comparing naturally and mechanically ventilated buildings (with or without humidification and recirculation of air), symptoms were found in prevalences shown in Table 2.

In a Danish study,[11] work-related skin reactions were reported by 0 to 31% of the women in 28 buildings investigated. The corresponding figures for men were 0 to 20%. Higher figures were claimed for ''general symptoms'' (up to 69%) and for irritation of the mucous membranes (up to 62%) in women.

In a questionnaire study[12] of an office building with known indoor air problems (inadequate ventilation rates, water damage to the fabric) and a control building, the prevalences shown in Figure 2 were found. There is a prominent sex difference in the prevalences. Different reasons for this common finding have been discussed. Different job category among males and females has been shown to be an important factor.[11] It can be speculated that females are more prone to reporting this kind of symptom than men. A difference in susceptibility is also possible. When comparing prevalences from different studies, it is of

TABLE 2
Prevalence of Skin Symptoms (%) in Relation to Method of Air Supply

	Number of subjects			
	Natural ventilation	**Mechanical ventilation only**	**Humidified, no recirculation**	**Humidified with recirculation**
Symptoms	259	73	354	477
Dry skin	5.7	5.5	16.2***	14.9***
Rash	1.9	2.7	3.1	2.9
Itchy skin	2.9	2.7	7.4*	7.2*

Note: Significance of difference when compared with natural ventilation — * signifies $p < 0.05$, ** signifies < 0.01, and *** signifies < 0.001.

From Finnegan, M. J., Pickering, C. A. C., and Burge, P. S., *Br. Med. J.*, 289, 1573, 1984. With permission.

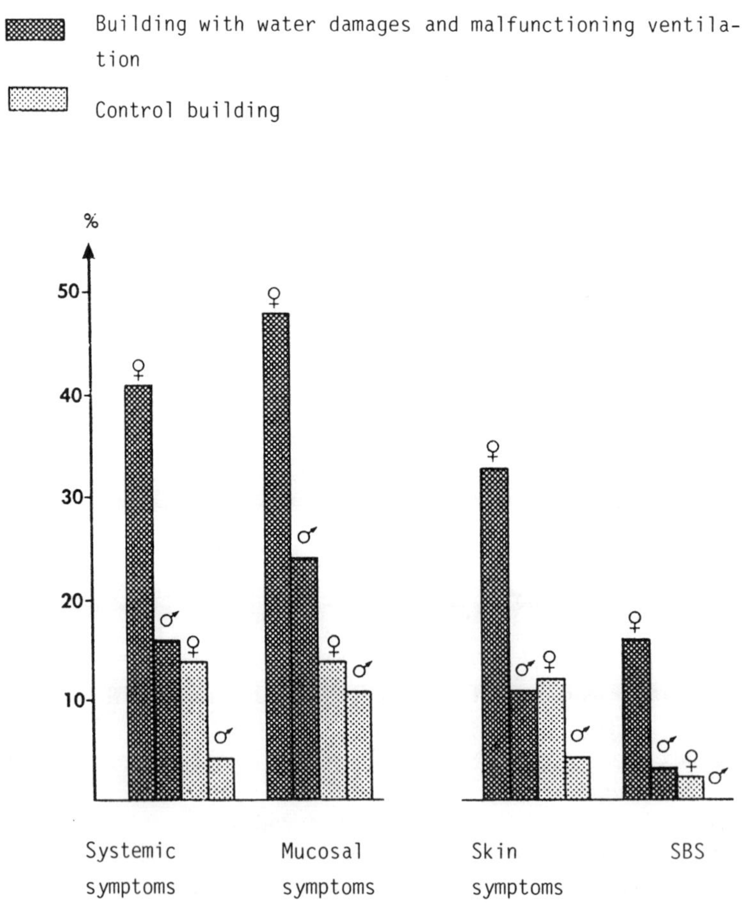

FIGURE 2. Prevalence of systemic symptoms (headache, lethargy); mucosal symptoms (eye, nose, throat irritation); skin symptoms (dryness, erythema, irritation of facial skin); and SBS (symptoms from all three groups) appearing every week reported in a questionnaire study (Stenberg[12]).

utmost importance to know the precise construction of the questions as well as to have a definition of the term "work related".

C. SUSPECTED OFFENDING FACTORS

The exact cause of the SBS is unknown. Formaldehyde[13,14] is one of few substances reported to cause similar symptoms, yet today formaldehyde is rarely found in sufficient amounts to explain the symptoms. It is widely accepted that ventilation rate (i.e., the amount of outdoor air flow) is critical.[15] Indoor air pollutants — volatile and particle-bound — can accumulate as a result of insufficient flow rate and/or recirculation of air. The relative importance of each factor varies between different buildings.

It can hardly be expected that these symptoms appear as a result of inadequate ventilation as such. The fact that "sick" buildings have become more numerous since the 1970s is often explained as a result of lowered ventilation rates due to energy saving and of the introduction of new building materials with a complex emission to indoor air.[6] Besides building materials, the possible sources of pollution are furniture, heating systems, smoking, chemicals produced by processes inside the building, emissions from humans, etc. In Scandinavia, degradation products from water-damaged flooring materials represent a problem of current interest. Under alkaline conditions, water can hydrolyze phtalates in polyvinylchloride (PVC) floors into alcohols, leading to odor problems. As with many other proposed causes of "building sickness", the cause-effect relationship has not been established in provocation studies. Glass fibers[16] from ceiling insulation materials have caused both skin and mucosal symptoms. SBS-like symptoms have been attributed to fumes from copying machines[17] in badly ventilated work areas.

As the cause of the symptoms is mainly unknown, it can be speculated that perhaps some of the symptoms are primary reactions to irritation and some the results of mental stress.

In a study by Finnegan et al.,[10] persons working in mechanically ventilated buildings had significantly more symptoms than those in naturally ventilated buildings. This is not a constant finding. In the Danish study,[11] female sex, job category (clerks and social workers), photocopying, working with VDTs, and handling of CCP correlated significantly with the prevalence of work-related mucosal and general symptoms.

III. CARBONLESS COPY PAPER

CCP has been used since the 1940s, extensively during the last 20 years. Calnan and Connor[18] reported on an episode of skin, mucosal, and general symptoms appearing in 1965 where intensive handling of CCP in a small, badly ventilated room was stated to be the cause. Increased ventilation solved the problem. Calnan[19] reviewed problems related to working with CCP. Typical are skin irritation, rashes on the hands and face, and irritation in the eyes, nose, and upper respiratory tract. Associated systemic symptoms are headaches, drowsiness, and lethargy. The color former of CCP is suggested to be the agent responsible for the irritative reactions.[19]

Reports on allergic contact dermatitis from CCP are extremely rare. Marks[20] reported a positive patch test to *p*-toluene sulfinate of Michler's hydrol (PTSMH) and contact urticaria[21] and airway obstruction from CCP with a concomitant increase in prostaglandin F_2 (PGF_2) and thromboxane B_2.

Menné et al.[22] reported positive prick tests with Idem paper in ten affected subjects. Jeansson et al.,[23] however, failed to reproduce these results. In a study of 151 subjects referred to a Department of Occupational Dermatology because of symptoms associated to working with CCP, they could not confirm that the paper was the cause.

The contradiction may arise from the fact that the symptoms are caused by different exposures. Thus, irritation on the hands is logical as a result of handling the paper, and symptoms from lips and face may, to some extent, be caused by contamination from the hands. However, most symptoms are almost identical to those reported in SBS. One possible explanation for SBS symptoms appearing in work with CCP is that the paper may act as an indoor air pollutant. Particles and volatile substances such as solvents are released.[24] Reports on the relationship between symptoms and small, hot, poorly ventilated rooms and a beneficial effect of increased ventilation support this view.

IV. VISUAL DISPLAY TERMINALS

A. BACKGROUND

Skin symptoms among VDT operators have been investigated and described mainly from Norway, Sweden, and England. Findings and a review of the literature have been published recently.[25]

Symptoms were first observed in Bergen on the west coast of Norway, where Nilsen[26] reported on six cases. He noted erythematous macules, papules, and teleangiectasia on the zygomatic areas in four cases, while two had cleared when the persons affected came for examination. All investigations, including provocation tests, were negative, but Nilsen summarized that the VDTs were the only probable eliciting factor of the skin rashes.

Other case reports appeared from Norway in 1981 and 1984 by nondermatologists.[27,28] Recent consultations with Norwegian dermatologists have confirmed that no new cases have been referred and that there is no public discussion of the subject. In 1984, Rycroft and Calnan[29] reported on four cases of facial rash among VDT operators. The rashes had been observed by the medical officer, but when the cases were examined by the authors the rashes had cleared. From the U.S., single cases have been reported by Feldman et al.[30] and by Fisher.[31]

B. "RADIATION" FROM VDTs

Experts have repeatedly confirmed that there is no hazardous radiation such as ultraviolet (UV) or X-rays from VDTs, but the Swedish public seems not to have been convinced by their statements.[32] Static electricity was initially blamed for the skin rashes stated to be associated with VDT work.[28] Electrostatic charges are caused by friction between clothing materials, but also emanate from VDTs. However, elimination of static electricity from VDTs with sophisticated grounding devices has not solved the problem.[33] The role of electromagnetic radiation in the elicitation of skin rash is still unclear. In a Swedish study not yet published, skin symptoms were provoked/reappeared even when the VDT was disconnected; this implies a psychological mechanism of elicitation.

C. REVIEW OF SWEDISH VDT STUDIES

In several studies, Swedish dermatologists have tried to elucidate the question of skin disease allegedly due to VDTs. The first was an epidemiological study[34] where 536 subjects (395 VDT operators and 141 controls) answered a questionnaire. Clinical examinations were performed on 74 of 96 individuals; skin lesions were recorded as objective (visible during the examination) or anamnestic (reported as having been present previously). Altogether, 15 types of skin lesion were found; the most frequent were eczema, dry skin, redness, itching and/or dryness, papules, seborrheic dermatitis, rosacea, acne, and psoriasis. This varied pattern of diagnoses speaks against one single factor originating from the VDT. There was a tendency toward a higher frequency of seborrheic dermatitis, acne, and rosacea (one diagnosis group) among the VDT-exposed subjects not found among the controls ($p < 0.05$). There was also a more positive electrostatic charge in this diagnosis group than in the rest of the population. The cohort is at present — 5 years after the first examination — being reexamined to observe the outcome of the ailments and diseases previously found and to record whether any new diseases have appeared.

In a second study[35] a questionnaire was sent to all 179 patients, 16 to 66 years old, with rosacea or perioral dermatitis diagnosed at a Department of General Dermatology in 1982. Answers were obtained from 166 subjects, 42 of whom had worked with a VDT. Of the 42, 8 suspected aggravation due to VDT work, but aggravation was also reported among office workers not exposed to VDTs.

In a third study[36] 100 consecutive VDT operators, referred to a Department of Occupational Dermatology, received the same diagnoses — namely, acne (16 cases), rosacea (15 cases), and seborrheic dermatitis (22 cases). Telangiectases (11 cases) with no signs of accompanying rosacea was common. In nine cases dry skin, erythema, and a few papules were noticed, but no specific dermatological diagnosis could be made. In six cases the skin was judged normal even at repeated visits. The patients were treated as appropriate for the diagnosis. At reexamination it was noticed that the therapies given had had a favorable effect in spite of continuous VDT exposure. Since then, approximately 150 new cases have been referred and examined with the same diagnoses as in the previous study. No specific VDT skin disease has been observed so far.

In a recent study by Berg[33] of 201 referred VDT operators (90% women), 46% were diagnosed as having rosacea, 14% seborrheic eczema, 10% atopic dermatitis, and 18% nonspecific skin problems (e.g., itching and redness). An association to migraine was observed. Light-testing of 123 patients showed no difference compared to a control group. Of the 52 Electrostatic shields tested, 32 were defective.

A fourth study[37] reported 9 cases of skin complaints (5 rosacea, 3 erythema, 1 subjective irritation) appearing in 12 office employees soon after introducing VDTs; 12 workmates not engaged in VDT work developed no similar skin symptoms during the period. This clustering of cases and the observation that symptoms could be aggravated later in non-VDT work support the hypothesis of a synergistic effect of VDTs and other workplace factors.

D. ONGOING STUDIES

Several studies are in progress in Sweden, and some were presented at the Second International Conference on Work with Display Units in Montreal in 1989. Results from provocation studies on patients exposed to magnetic and electrical fields will be reported. A prevalence study and a case-referent study on SBS- and VDT-related skin disease in Northern Sweden will be reported in 1990.

E. COMMENTS

Our studies[34-37] indicate that there may be a relation between VDT work and aggravation of skin disorders such as rosacea, seborrheic and atopic dermatitis, and acne. These are common skin diseases with multiple causes and variable courses. We do not know if the aggravation reported depends on physical, chemical or psychological factors.

It is our impression that patients suffer from subjective irritation on the face, often mimicking "sunburn", before any objective signs appear. Erythema on the flush area and periorbital edema are early findings. Later on, diverging signs of skin disease are noted, and rosacea seems to be the most common reaction pattern.

It is unlikely that the different skin diseases found among VDT operators would be caused or aggravated by a common exposure factor originating from the VDT. We believe that other factors in the office affecting the indoor environment (i.e., air pollution, paper dust, ventilation, humidity, temperature, floorings, flowers, tobacco smoke, cleansing, or stress) contribute in some of the cases where there seems to be a connection between the dermatoses and office work. However, it has not been proved that these skin diseases are overrepresented among office workers or VDT operators.[33] The relative role of VDTs in the elicitation of symptoms remains to be determined.

At the 1986 International Conference on VDT Work[28,37] we invited reports on skin complaints, but only a few cases were presented and no specific diagnoses were given. We do not know why skin complaints in association with VDT work at present seem to be concentrated in Sweden. Climatic factors have been suggested.

V. LOW HUMIDITY OCCUPATIONAL DERMATOSES

Rycroft and co-workers,[38,39] in two reports, have described a phenomenon which they name ''low-humidity occupational dermatoses''. Although not primarily described in an office environment, these dermatoses are listed by Rycroft[3] as a problem affecting office personnel. In one case, workers installing a telephone exchange were affected; another case occurred in a contact lens factory and a third in a company producing silicon chips. The skin symptoms were pruritus and urticaria on covered areas and a scaling dermatitis on the face, scalp, and ears. The common exposure factor found was a relative humidity at the lower end of the comfort range, around 35%. The problems were solved when the relative humidity was raised to 45 to 50%.

Gaul[40] monitored skin symptoms in eight subjects during 1 year parallel to meteorological observations. A questionnaire study of 577 subjects was carried out simultaneously. A quick change from high to low humidity led to chapping of the skin, first on the back of the hands, then on lips and other parts of the face. Ichthyosis appeared first on the lower legs, then on the arms, and circular erythematous plaques were seen on the thighs and hips. Similar findings were reported by Chernosky[41] in patients living and working in air-conditioned houses when air humidity was lowered drastically.

It is not our experience that symptoms of the kind reported by Rycroft et al.[38,39] appear in Northern Scandinavia, as relative humidity quickly falls in autumn and winter. Interestingly, however, the symptoms of ''low humidity occupational dermatoses'' strongly resemble our findings in ''sick'' buildings where low air humidity is not a feature separating these from buildings with low prevalences of skin complaints.

As low air humidity and temperature vary with many other factors affecting the quality of the indoor environment, it is an open question what role humidity itself plays in evoking skin symptoms.

Andersen,[42] in experiments with exposures of up to 78 h in dry air, did not succeed in evoking any objective skin symptoms in the subjects. Interestingly, they could not accurately judge whether they were exposed to low (20%) or to high air humidity (70%) when the temperature was held constant. The experiments support the view that the subjective experience of dry air is best correlated to the degree of pollution in the air.[42]

VI. MASS PSYCHOGENIC ILLNESS

Because of the nonspecific nature of many symptoms appearing in the office environment, psychological factors are often suggested as causes for the complaints. They may even be used as an excuse for incomplete investigations.[43] Solid investigations revealing psychological factors as a cause are rare. Colligan[44] has reviewed the present knowledge of the psychological effects of indoor air pollution. Ozone, for example, is reported[45] to induce drowsiness and concentration difficulties, as well as mucosal irritation. Odors[46] may induce drastic alterations in mood, state, and demeanor, independent of any acute toxic effects. Low levels of offensive pollutants have been shown to increase aggressiveness. Colligan[44] summarizes that indoor air pollutants may have direct and specific neurobehavioral effects as well as a diffuse impact on psychological functions via the autonomic system.

Boxer[47] has summarized experience from 19 episodes of probable mass psychogenic illness investigated by NIOSH. Typical symptoms are listed in Table 3. No dermatological

TABLE 3
Symptoms Most Frequently Reported in Probable "Mass Psychogenic Illness"

Symptom
Headache
Dizziness/light-headedness
Nausea
Dry mouth/throat
Eye/nose/throat irritation
Sleepiness/drowsiness
Weakness
Numbness/tingling
Chest tightness/discomfort

From Boxer, P. A., *J. Occup. Med.*, 27, 867, 1985. With permission.

TABLE 4
Signs Suggestive of "Mass Psychogenic Illness"[a]

1. Relatively severe symptoms of sudden onset among a number of people, particularly if symptoms begin after leaving the alleged source of exposure
2. Moderate to severe symptoms without the usual symptoms caused by the alleged contaminant
3. Diversity of symptoms with scarcity of physical signs or laboratory findings
4. Occasional conversion symptoms, i.e., loss or alteration of physical functioning, not under voluntary control, that cannot be explained by any physical disorder or known pathophysiological mechanism
5. A higher attack rate among females than males when the population at risk includes both sexes
6. Becoming affected not necessarily because of proximity to the source of the putative exposure, but rather only after learning of either the suspected exposure or the fact that someone else is exhibiting signs of illness
7. Recurrences in affected individuals when they congregate
8. Benign morbidity with lack of sequelae

[a] According to Boxer.[47]

problems were reported, but skin complaints might be missing due to the construction of the questionnaires; most of these studies use questionnaires to record symptoms. Some symptoms are obviously common to SBS and complaints from handling CCP. These should be looked upon as an indication of the nonspecific and invariable nature of human response to different stimuli. Boxer[47] also lists signs that are suggestive of mass psychogenic illness (see Table 4). Most of these signs make it easy to separate this phenomenon from real indoor air problems.

VII. CONTACT DERMATITIS

A. ALLERGIC CONTACT DERMATITIS

In our experience allergic contact dermatitis is rare, and most sensitivities seem to have been acquired outside the office (e.g., to nickel, perfumes, or ornamental flowers); however, relapses can occur from occupational exposure. Allergens specific for the office environment are rare. Textbooks and previous reviews[2-4,48-51] list possible allergen sources. Chemicals present in rubber objects such as finger stalls, bands, erasers, and sponges should especially be mentioned. Contact allergy from inks, dyes (e.g., in carbon paper[53]), glues, colophony, and formaldehyde resins in paper have also been reported.

TABLE 5
Suggested Investigations and Consultations when Exploring Symptoms Appearing among Office Personnel

By the dermatologist
- Patch tests
- Prick/scratch tests with suspected materials — control necessary
- Provocation tests with suspected materials and placebos
- IgE, RAST
- Biopsies and cultivations usually of limited value unless for research purposes
- Photographic pictures to document the clinical appearance, repeatedly, before and after provocation at work, sick leave, etc.

By others
- Information on temperature, humidity, ventilation, flooring materials, building and furniture materials, damage caused by water leakage, smoking habits, cleansing, flowers, use of herbicides, etc. are great value.
- Examination by specialists in allergy, ear-nose-throat (ENT), occupational medicine
- Experts in epidemiology, questionnaire investigation.

B. IRRITANT CONTACT DERMATITIS

Since irritant contact dermatitis is mainly a diagnosis of exclusion, the frequency is difficult to state. Overlapping with the conditions described above (Sections II to IV) is probable. It is usually the skill of the dermatologist, the outcome of supervised provocation tests, and the possibility of work-site observations that will lead to the diagnosis.

Among offending agents, different types of paper, paper dust, and solvents in typewriter cleansers should be mentioned. Soap and detergents not used at home can, of course, induce an irritant contact dermatitis. Several patients with this condition have a history of atopic dermatitis, and it is mainly a semantic question how they should be classified — irritant contact dermatitis in an atopic individual or atopic hand dermatitis?

C. PHOTO CONTACT DERMATITIS

The fact that a dermatitis may be caused by exposure to artificial light may be overlooked, and we do not know the importance of, or the contribution from, artificial light sources in offices. A few cases caused by thioureal — a copy machine antioxidant — have been described.[53]

VIII. SUGGESTED INVESTIGATIONS AND CONSULTATIONS

Besides taking a detailed history and performing a clinical examination of affected skin sites — usually hands and face and allegedly affected distant sites — some tests and other investigations are recommended to exclude other diagnoses such as contact allergy or atopy (Table 5).

Patch, prick, scratch, or provocation tests with materials from the office environment are rarely positive. In cases of positive reaction, control tests in unexposed subjects are indispensable. When testing paper, it is recommended to test with the paper ''as is'' (both sides) and with aqueous and solvent extracts.

Photographs to document the clinical appearance of affected and allegedly affected skin sites at each visit, before and after provocation tests, etc., have been found to be valuable to support observations made by the dermatologist. Where there are several cases from the same office, we usually ask the industrial physician to carry out a questionnaire investigation, do an examination, and then refer the most representative cases to us. For the assessment of reported symptoms, knowledge of some physical factors, materials, cleaning, etc. (see Table 5) in the office is of great value. Work-site visits are recommended. However, consultations with other specialists are sometimes necessary, since it is rare that only the skin is affected.

IX. CONCLUSIONS

Traditional investigations are of limited value when exploring and examining complaints, symptoms, ailments, and diseases of the skin in office workers. However, contact allergy, photosensitivity, and atopy should be excluded and common facial dermatoses such as acne, rosacea, and seborrheic dermatitis diagnosed and treated. The complaints are often accompanied by symptoms from the mucous membranes, headache and lethargy/fatigue. To be able to make a correct diagnosis and suggest prophylactic means, the dermatologist sometimes must consult other specialists.

REFERENCES

1. **Hodgson, M. J., Block, G. D., and Parkinson, D. K.,** Organophosphate poisoning in office workers, *J. Occup. Med.,* 28, 434, 1986.
2. **Foussereau, J., Benezra, C., and Maibach, H.,** *Occupational Contact Dermatitis,* Munksgaard, Copenhagen, 1982, 336.
3. **Rycroft, R. J. G.,** Occupational dermatoses among office personnel, *Occup. Med.,* 1, 323, 1986.
4. **Marks, J. G., Jr.,** Dermatological problems of office workers, *Dermatol. Clin.,* 6, 75, 1988.
5. Guidance for Indoor Air Quality Investigations, Hazard Evaluations and Technical Assistance Branch Division of Surveillance, Hazard Evaluations and Field Studies, National Institute for Occupational Safety and Health, Cincinnati, Ohio, January 1987.
6. **WHO,** Indoor Air Pollutants: Exposure and Health Effects, EURO Reports and Studies 78, World Health Organization, Copenhagen, 1983.
7. **Alimenko, V. V., Anderson, L., Lebowitz, M. D., and Lindvall, T.,** The "sick" building syndrome, in *Indoor Air,* Vol. 6, Evaluations and Conclusions for Health Sciences and Technology, Berglund, B., Berglund, U., Lindvall, T., and Sundell, J., Eds., Swedish Council for Building Research, Stockholm, 1986, D13.87.
8. Stockholms-undersökning: barn på problemdaghem har oftare hud- och ögonsymptom, *Läkartidningen,* 79, 4544, 1982 (in Swedish).
9. **Stenberg, B.,** Skin complaints in buildings with indoor climate problems, *Environ. Int.,* 15, 81, 1989.
10. **Finnegan, M. J., Pickering, C. A. C., and Burge, P. S.,** The sick building syndrome: prevalence studies, *Br. Med. J.,* 289, 1573, 1984.
11. **Skov, P. and Valbjorn, O.,** The "sick" building syndrome in the office environment: the Danish town study, *Environ. Int.,* 13, 339, 1987.
12. **Stenberg, B.,** unpublished data.
13. **Godish, T.,** Formaldehyde and building-related illness, *J. Environ. Health,* 44, 116, 1981.
14. **Olsen, J. H. and Dossing, M.,** Formaldehyde induced symptoms in day care centers, *Am. Ind. Hyg. Assoc. J.,* 43, 366, 1982.
15. **Fanger, P. O.,** A comfort equation for indoor air quality and ventilation, in *Healthy Buildings 88,* Vol. 1, Berglund, B. and Lindvall, T., Eds., State of the Art Reviews, Swedish Council for Building Research, Stockholm, 1988, D19.39
16. **Verbeck, S. J. A., Buise-van Unnik, E. M. M., and Malten, K. E.,** Itching in office workers from glass fibres, *Contact Dermatitis,* 7, 354, 1981.
17. **Jensen, M. and Roed-Petersen, J.,** Itching erythema among post office workers caused by a photocopying machine with wet toner, *Contact Dermatitis,* 5, 389, 1979.
18. **Calnan, C. D. and Connor, B. L.,** Carbon paper dermatitis due to nigrosine, *Berufs-Dermatosen,* 20, 248, 1972.
19. **Calnan, C. D.,** Carbon and carbonless copy paper, *Acta Derm. Venereol.,* 59 (Suppl. 85), 27, 1979.
20. **Marks, J. G.,** Allergic contact dermatitis from carbonless copy paper, *J. Am. Med. Assoc.,* 245, 2331, 1981.
21. **Marks, J. G., Trautlein, J. J., Zwillich, C. W., and Demers, L. M.,** Contact urticaria and airway obstruction from carbonless copy paper, *J. Am. Med. Assoc.,* 252, 1038, 1984.
22. **Menné, T., Asneas, G., and Hjorth, N.,** Skin and mucous membrane problems from "no carbon required" paper, *Contact Dermatitis,* 7, 72, 1981.
23. **Jeansson, I., Löfström, A., and Lindblom, A.,** Complaints relating to the handling of carbonless copy paper in Sweden, *Am. Ind. Hyg. Assoc. J.,* 45, B-24, 1984.

24. **Molhave, L. and Grunnet, K.,** Addendum: headspace analysis of gases and vapors emitted by carbonless paper, *Contact Dermatitis,* 7, 76, 1981.
25. **Wahlberg, J. E. and Lidén, C.,** Is the skin affected by work at visual display terminals?, *Dermatol. Clin.,* 6, 81, 1988.
26. **Nilsen, A.,** Facial rash in visual display unit operators, *Contact Dermatitis,* 8, 25, 1982.
27. **Tjonn, H. H.,** Report of facial rashes among VDT operators in Norway, in *Health Hazards of VDT's?,* Pearce, B. G., Ed., John Wiley & Sons, New York, 1984, 17.
28. **Wedberg, W. C.,** Facial particle exposure in the VDU environment, the role of static electricity, in *Work with Display Units 86,* Knave, B. and Widebäck, P.-G., Eds., Elsevier/North-Holland, Amsterdam, 1987, 151.
29. **Rycroft, R. J. G. and Calnan, C. D.,** Facial rashes among visual display unit operators, in *Health Hazards of VDT's?,* Pearce, B. G., Ed., John Wiley & Sons, New York, 1984, 13.
30. **Feldman, L., Eaglstein, W. H., and Johnson, R. B.,** Terminal illness, Letter to the Editor, *J. Am. Acad. Dermatol.,* 12, 366, 1985.
31. **Fisher, A. A.,** "Terminal" dermatitis due to computers (visual display units), *Cutis,* 37, 153, 1986.
32. **Lagerholm, B.,** Bildskärmar och hudförändringar, Ingående undersökningar motiverade, *Lakartidningen,* 83, 60, 1986 (in Swedish).
33. **Berg, M.,** Skin problems in workers using visual display terminals, a study of 201 patients, *Contact Dermatitis,* 19, 335, 1988.
34. **Lidén, C. and Wahlberg, J. E.,** Work with video display terminals among office employees. V. Dermatologic factors, *Scand. J. Work. Environ. Health,* 11, 489, 1985.
35. **Lidén, C. and Wahlberg, J. E.,** Does visual display terminal work provoke rosacea?, *Contact Dermatitis,* 13, 235, 1985.
36. **Wahlberg, J. E.,** Någon specifik bildskärmsrelaterad hudsjukdom har inte observerats bland 100 undersökta operatörer, *Lakartidningen,* 83, 3328, 1986 (in Swedish).
37. **Stenberg, B.,** A rosacea-like skin rash in VDU-operators, in *Work with Display Units 86,* Knave, B. and Widebäck, P.-G., Eds., Elsevier/North-Holland, Amsterdam, 1987, 160.
38. **Rycroft, R. J. G. and Smith, W. D. L.,** Low humidity occupational dermatoses, *Contact Dermatitis,* 6, 488, 1980.
39. **White, I. R. and Rycroft, R. J. G.,** Low humidity occupational dermatosis — an epidemic, *Contact Dermatitis,* 8, 287, 1982.
40. **Gaul, L. E. and Underwood, G. B.,** Relation of dew point and barometric pressure to chapping of normal skin, *J. Invest. Dermatol.,* 19, 9, 1952.
41. **Chernosky, M. E.,** Prutitic skin disease and summer air conditioning, *J. Am. Med. Assoc.,* 179, 1005, 1962.
42. **Andersen, I. B., Lundqvist, G. R., Jensen, P. L., and Proctor, D.,** Human response to 78 hour exposure to dry air, *Arch. Environ. Health,* 29, 319, 1974.
43. **Halley, S. F. and Brilliant, L. B.,** Is the diagnosis of "mass hysteria" an excuse for incomplete investigation of low-level environmental contamination?, *J. Occup. Med.,* 23, 22, 1981.
44. **Colligan, M. J.,** The psychological effects of indoor air pollution, *Bull. N.Y. Acad. Med.,* 57, 1014, 1981.
45. **Key, M. M., Henschel, A. E., and Butler, J., Eds.,** Occupational Diseases: A Guide to Their Recognition, DHEW (NIOSH) Publ. No. 77-181, U.S. Government Printing Office, Washington D.C., 1977.
46. **Rotton, J., Frey, J., and Barry, T.,** The air pollution experience and physical aggression, *J. Appl. Soc. Psychol.,* 9, 397, 1979.
47. **Boxer, P. A.,** Occupational mass psychogenic illness, *J. Occup. Med.,* 27, 867, 1985.
48. **Cronin, C.,** *Contact Dermatitis,* Churchill Livingstone, Edinburgh, 1980, 883.
49. **Fregert, S.,** *Manual of Contact Dermatitis,* 2nd ed., Year Book Medical Publishers, Chicago, 1981.
50. **Adams, R. M.,** *Occupational Skin Disease,* Grune & Stratton, New York, 1983, 422.
51. **Fisher, A. A.,** *Contact Dermatitis,* 3rd ed., Lea & Febiger, Philadelphia, 1986, 503.
52. **Shehade, S. A., Beck, M. H., and Chalmers, R. J. G.,** Allergic contact dermatitis to crystal violet lactone in carbonless copy paper, *Contact Dermatitis,* 17, 310, 1987.
53. **Van der Leun, J. C., de Kreek, E. J., Deenstre-van Leeuwen, M., and van Weelden, H.,** Photosensitivity owing to thiourea, *Arch. Dermatol.,* 113, 1611, 1977.

Chapter 28

MISCHIEVOUS METALS (CHROMATE, COBALT, NICKEL, AND MERCURY)*

Desmond Burrows

TABLE OF CONTENTS

* Modified from Burrows, D., *Clin. Exp. Dermatol.*, 14, 265, 1989. With permission.

I. INTRODUCTION

It seems especially appropriate that someone with an interest in contact dermatitis should give this oration because of Prosser-White's preeminent reputation in that subject. It is therefore a special honor to give this oration. There are many metals which can be mischievious. These include aluminum, antimony, arsenic, beryllium, boron, cadmium, copper, gold, palladium, platinum, selenium, tin, zinc, and zirconium. This chapter will concentrate on four of them — chromate, cobalt, nickel, and mercury — and discuss the propaganda, the problems, and the prospects.[1]

II. PROPAGANDA

What is the propaganda about metals?

1. Millions of people are disabled by nickel allergy.
2. Patch testing is not worthwhile.
3. Get rid of nickel in earrings and the problem is solved.
4. The dimethylglyoxime test is not accurate enough.
5. Nickel allergy predisposes to hand dermatitis.
6. Nickel allergy predisposes to other allergens.
7. Clear nickel from the diet and the problem will disappear.
8. Chromate patch testing is inaccurate.
9. We should worry about mercury.

III. PROBLEMS

A. MILLIONS OF PEOPLE ARE DISABLED BY NICKEL ALLERGY

Nickel allergy is doubling every 10 years;[2-4] perhaps this is related to the doubling of nickel production. In a test population, Peltonen[5] in Finland showed that 0.9% of the female population had nickel allergy and hand eczema, which would mean that about a half million people are affected in Britain. Looking at it another way, around 12% of women have nickel allergy, and probably slightly less than 10% of that group will have hand eczema. This would produce the same sort of figure. Chromium and cobalt have become rare causes of hand dermatitis.

B. PATCH TESTING IS NOT WORTHWHILE

Those who express this view argue that it is perfectly obvious when a person has nickel allergy because the person will give a history of intolerance to inexpensive jewelry and that a strong nickel-positive patch test will increase the chance of other false positive results. There are five reasons for continuing to patch test with nickel:

1. There may be another metal involved, such as cobalt.
2. There may be other allergens.
3. It is important from a legal point of view to have confirmation.
4. A person may be allergic to nickel without any history of nickel allergy.[6] We found about 4% positive patch tests in those with no nickel history on routine testing.
5. The patch test may be negative even with a strong history of nickel allergy.

A particularly interesting group includes those who have a strong history of nickel allergy — that is, intolerance to inexpensive jewelry with a negative patch test — i.e., Kieffer 47%,[7], Menne 29%,[8] Lantigna et al. 42%,[9] Burrows 50%,[62] Schmiel 48%,[10] Fischer and

TABLE 1
Effects of Ear Piercing on the Development of Nickel Hypersensitivity in Relation to Sex in Young Patients (Age <30 Years)

	Female patients			Male patients		
		Nickel positive			Nickel positive	
Piercing	Group size	No.	%	Group size	No.	%
−	48	3	6.3%	149	4	2.7%
+	568	225	39.6%	41	8	19.5%

From van Hoogstraten, I. M. W., Anderson, K. E., von Blomberg, B. M. E., Boden, D., Bruynzeel, D. P., Burrows, D., Camarasa, J. M. G., Dooms-Goossens, A., Kraal, G., Lahti, A., Menné, T., Rycroft, R. J. G., Todd, D., Vreeburg, K. J. J., Wilkinson, J. D., and Scheper, in *Current Topics in Contact Dermatitis,* Frosch, P. J., Dooms-Groossens, A., Lachapelle, J.-M., Rycroft, R. J. G., and Scheper, R. J., Eds., Springer-Verlag, Heidelberg, 1990, 178. With permission.

Rystedt 17%,[11] and Gilboa et al. 58%.[12] There are two reasons for this. One is that patch testing may not be accurate enough. Lachapelle[13] tested with identical materials on either side of the back, producing discordant results in 3 of 70 positive tests. Testing by *in vitro* methods still failed to demonstrate nickel allergy in these patients.[12,14] Secondly, it has been suggested that this is a finding which particularly occurs in atopics.[6,7,11,13] We have looked at three statistically matched samples: (1) those with normal skin, (2) those with a nickel history who are patch test positive, and (3) those with a nickel history who are patch test negative. No statistically significant difference was found in atopy between these three groups, measuring personal history, family history, patch tests, and skin tests.

C. GET RID OF NICKEL IN EARRINGS AND THE PROBLEM IS SOLVED

Larsson-Stymne and Widstrom[3] looked at 960 consecutive patients. Of those with a history of ear piercing 13% had positive patch tests for nickel, compared with only 1% of those who had not had their ears pierced. This has been confirmed by a recent joint European contact dermatitis study[14] (Table 1). Larsson-Stymne and Widstrom looked at whether a history of irritation at the time of ear piercing was important. Of 43 girls who had their ears pierced, 32 had a history of oozing ears (all gave a positive nickel patch test) and 11 had no history of oozing ears (all were patch test negative). This certainly corresponds with clinical experience.

It is odd that in spite of increasing use of earrings by men there does not seem to be a corresponding increase in nickel allergy. Widstrom and Erikssohn[15] studied 216 18-year-old men who were entering the Swedish Army. Ear piercing was found to be common (22%). In spite of that, nickel allergy was found to be much lower than in females, and this also applied to those with pierced ears. The Joint European Study also confirmed this (Table 1). Why is this so? It may be related to the later onset of ear piercing in men and, perhaps more importantly, to the fact that men tend to wear more expensive jewelry containing less nickel.

D. THE DIMETHYLGLYOXIME TEST IS NOT ACCURATE ENOUGH

Two questions involving the dimethylglyoxime (DMG) test are:

1. What is the lowest concentration of nickel which the DMG test will detect?
2. What is the lowest concentration of nickel that will produce a flare-up?

Solutions of various dilutions of nickel sulfate will produce a reaction as low as 10 ppm.[62] If the DMG test will detect 10 ppm of nickel sulfate, is this sufficient, or will patients react to lower concentrations? With patch tests, Allenby and Goodwin[16] found that only one person reacted below 11.2 ppm. Patients allergic to nickel appear to react to lower concentrations of metallic nickel than nickel salts. For instance, Fischer et al.[17] showed that four of ten people reacted to 0.05 ppm, and Menné et al.[18-19] showed that most people react to amounts greater than 1 ppm of nickel sulfate.

White gold, which contains small amounts of nickel (<0.5 ppm), produced a reaction in all 18 nickel-sensitive patients tested, even when the DMG test was negative.[20] Hence, there is no doubt that patients will react to lower amounts than the DMG test can detect. Therefore, just because the DMG test is negative it cannot be said that the metal is not responsible for exacerbating the dermatitis. However, it is unlikely that this occurs frequently. Why do patients react to metallic nickel in much lower concentrations than to metallic salts? It may be that nickel sulfate penetrates poorly. It certainly penetrates more slowly than nickel chloride, but this probably is not the whole answer and is the opposite of chromate, where the metal will not produce any reaction. Some have suggested that metallic nickel is an irritant.[12]

E. NICKEL ALLERGY PREDISPOSES TO CONTACT DERMATITIS

This is important because of assessment of suitability for employment. If it can be shown that women who have nickel allergy, but whose skin is completely clear, are more prone to develop an irritant contact dermatitis to detergents and wetness, then it may be a reasonable policy not to employ such people if they would be exposed to irritants. Although there is some conflict of evidence, there does seem to be a higher incidence of nickel allergy in those with hand dermatitis caused by irritants (wetness and detergents), but one cannot draw the conclusion from this that nickel dermatitis predisposed them to it. It may be that skin which is damaged allows penetration of nickel more easily and so facilitates sensitization. Menne et al.[21] and Peltonen[5] showed that there was considerable increase of hand eczema in those with nickel allergy. For instance, in the general population, 19% of women have hand eczema, but if there was a history of nickel allergy, this would increase to 43%. In a metal-working factory, this went up from 26% (in those without allergy) to 46% (in those with nickel allergy).[22] Nilsson and Back[23] showed that a history of nickel allergy increased the chance of getting hand eczema by about 1.8%. To put this into perspective, if there was previous history of hand eczema, then the chances of being nickel allergic were 13 times greater. Menné et al.[21] in a prospective study showed that nickel allergy did predispose to a hand eczema.[21]

F. NICKEL ALLERGY PREDISPOSES TO OTHER ALLERGENS

It has been suggested that nickel ought to be eliminated from the battery because it leads to an increase in the angry back syndrome and an increase in the number of other positives. Lammintausta and Kalimo,[24] however, have shown that allergy to nickel does not increase the number of positive patch tests. Paramsothy et al.[25] confirmed this in a study of varicose ulcers. Cobalt may be an exception. A patient with a positive nickel test is 20 times more likely to have a cobalt allergy than one with a negative nickel test. A patient with a strong nickel patch test is 50 times more likely to have a strong cobalt allergy.[26]

Copper and palladium are two other metals which seem to be associated with a higher incidence of nickel allergy. It has often been suggested that this means that these metals are very often commonly associated in metal; therefore, if a person is frequently in contact with nickel and gets sensitized to it, the same can occur with copper and palladium.

However, there are two findings which would suggest that it is more than that. First, Lammintausta et al.[27] found that guinea pigs which were sensitized to nickel were more

TABLE 2
Results of Six Double Blind Clinical Trials with Oral Nickel

Nickel (mg)	No. patients	Ref.
5.6	12	37
0.5	10	38
2.0	22	39
4.0	22	39
0.4	26	40
2.5	26	40
2.5	26	40
2.5	131	41
2.5	19	42

Note: Total number of patients = 280; total number of tests = 420.

easily sensitized to cobalt than those not sensitized to nickel. Second, it has been shown that lymphocytes with a monoclonal sensitivity to nickel will also react to copper and palladium, but not cobalt (R. Scheper, personal communication).

G. CLEAR NICKEL FROM THE DIET AND THE PROBLEM WILL DISAPPEAR

This is not a new concept. In 1926, Walthard[28] showed that a flare-up in a nickel patch test could be produced with oral nickel in experimental animals. There are a number of reasons why this idea has gained favor:

1. Anecdotal: cases have been described where dermatitis flared up after swallowing a coin[29] or eating margarine with a high nickel content[30] or due to metal retractors used during an operation.[31]
2. Increased excretion of nickel during flare-up has been reported, suggesting that for some reason increased absorption of nickel occurs; however, numbers are small, and there is neither agreement nor a statistically significant trial.[32]
3. Low-nickel diets have been used. The two difficulties are conducting a double-blind trial and producing a low-nickel diet. Veien[33] showed that it was very difficult to decrease a normal nickel diet which varied from 0.06 to 0.25 mg daily for a high-nickel diet to 0.01 to 0.14 mg for a low-nickel diet. Sunderman et al.[34] showed that absorption varies greatly depending on how the nickel was taken and on the food with which it was taken.
4. Animal experiments: Ziegler and Hohndorf[35] implanted guinea pigs with metal containing chromium and nickel and sensitized them to both. Some developed dermatitis over the metal implant. This has been seen clinically in patients. Thomas et al.[36] reported a patient who was allergic to cobalt and had a plate inserted containing cobalt; dermatitis developed over the plate, but it cleared when the plate was removed.
5. The best method of testing this hypothesis is by double-blind trials of oral nickel. Up to 1986 there had been studies in four centers which were double blind both for the patient and the observing doctor (Table 2).[37-42] In these trials, when 5.6 mg of nickel was given, a difference was suggested between those receiving the placebo and those receiving the nickel. A smaller dose produced no effect. In 1987 there were two further trials; one[42] was not statistically significant, and in the other Veien et al.[41] gave 2.5 mg nickel vs. placebo. They showed a difference in flares between nickel and placebo.

TABLE 3
Relevance of Positive Patch Tests to Chromate (European Contact Dermatitis Group Study)

Patients tested	2939
Positive reaction to both percentages	143
Number relevant	96
Positive to 0.5%	95
Positive to 0.25%	84
Chromate allergy missed by 0.25%	11
Chromate allergy missed by 0.5%	1

From Andersen, K. E., Burrows, D., Cronin, E., Dooms-Groossens, A., Rycroft, R., and White, I., *Contact Dermatitis,* 19, 389, 1988. With permission.

There now have been a total of 210 patients on trial with oral nickel in 280 cases, which suggests that doses of about 2.5 mg or above may produce flares. When one considers that the normal diet is about 0.5 mg, an enormous intake of food would be required to produce flares with nickel.

6. Chelating agents: Antabuse and Trientene are excellent chelators of nickel. There have been a number of trials treating patients with nickel dermatitis with these agents, none of which have been conclusive.[43,44]

The history of metal allergy in hip prostheses tends to confirm these conclusions. Initially, hips were made with a metal head on a metal cup which produced a significant amount of metal in the tissues. There were many reports of systemic aggravation of skin disease and sensitization from these artificial hip prostheses, with rejection of the prostheses due to metal allergy. Since the change to metal-on-plastic hips, sensitization problems and hip rejection to metal allergy have disappeared. This has been confirmed recently by a very long-term study by Moller and Carlsson.[45] All this evidence suggests that systemic nickel effects are dose related, and the amounts encountered in the everyday diet are unlikely to cause flares.

H. CHROMATE PATCH TESTING IS NOT ACCURATE

It is remarkable the great diversity of the incidence of chromate allergy reported in different parts of the world. This may be more apparent than real, and the answer lies not in different levels of allergy in different centers, but in interpretation of patch tests. The problem is that 0.5% potassium dichromate is irritant and will produce false positive patch tests, but if 0.25% is used, many significant patch tests will be missed. The European Environmental Contact Dermatitis Group carried out a multicenter study on this, and these figures show that if 0.5% is used, a substantial number of false positives will occur, but that about 8% of all true positive patch tests will be missed (Table 3).

I. WE SHOULD WORRY ABOUT MERCURY

There are two problems with mercury — sensitization and toxicity. Sensitization is not a big problem and is now much lower than previously reported for three reasons:

1. The percentages used to patch test with mercury in the past were often highly irritant concentrations with many false positives.
2. It is used much less commonly in local applications than previously.
3. Many patients who are allergic to thimersal are not allergic to mercury, but to the nonmercury part of the molecule.

TABLE 4
Effects of Orthodontic Treatment on the Development of Nickel Hypersensitivity in Young Patients (Age <30 Years)

Orthodontic treatment	Piercing	Group size	Nickel positive No.	Nickel positive %
Brace	—	57	2	2.5%
—	—	143	5	3.5%
—	+	431	168	39.0% (p <0.05)
Brace before piercing	+	86	24	27.9%

Allergy to mercury in amalgam is probably not a large problem. White and Smith,[46] in a review of the literature, found only 28 cases where mercury amalgam produced allergic reactions. The exception to this may be lichen planus of the mouth, where a higher incidence of mercury allergy is found than in the general population.

Toxicity is a much greater problem. Large quantities of mercury are used by dentists, and there has been very little research into the harmful effects of amalgam. There is no doubt that mercury can be released from amalgam, particularly when chewing, and if mercury fillings are removed the concentration in the mouth is 20 times higher than that which would be allowed in a factory. It is a concern that metallic mercury can be transformed to the more dangerous methyl mercury. One case report describes changes in lymphocytes, but there were only three cases reported.[47]

We compared 28 medical students and 25 final-year dental students, paying particular attention to allergy to mercury and lymphocyte subsets. None showed any positive patch tests to mercury, but there was a statistically significant higher total lymphocyte count, total T-cell numbers, T-helper/inducer cells, and T-suppressor/cytotoxic cells in dental students compared to medical students.[48]

V. PROSPECTS

These, then, are my views on the propaganda and problems. It has been said that the notion that as a man grows older his illusions leave him is not quite true. What is true is that his early illusions are supplanted by new and, to him, equally convincing illusions. This is probably true, for I have changed my mind over the years on a number of these, but as Dean Swift said, "To acknowledge that you were wrong yesterday is simply to let the world know that you are wiser today than you were."

What about the prospects? There are many interesting prospects for the future.

A. IMMUNOTOLERANCE

Vreeburg et al.[49] showed that guinea pigs that were fed nickel prior to being sensitized were much more difficult to sensitize to nickel. Van der Burg et al.[50] demonstrated that there was a suggestion of a lower incidence of nickel allergy in those who had orthodontic treatment prior to ear piercing. The European Contact Dermatitis Group has conducted a large multicenter study on this (Table 4).[14] The figures of our own department (Table 5)[16] within this study are in agreement with the main study and show a reduction in nickel allergy in those who had a brace inserted before, rather than after, having their ears pierced.

B. NEW DRUGS

In the future new drugs which will have a specific action on contact dermatitis may be developed. For instance, cyclosporin A has been shown topically to inhibit contact sensitivity in rabbits,[51] and in 1986 these same workers showed that it inhibited patch test reactions to

TABLE 5
Relevance of Positive Patch Tests to Chromate (Belfast Figures)

		Nickel	
		Positive	**Negative**
Brace:	Before	9 (25%)	27 (75%)
	After	13 (50%)	13 (50%)

From Todd, D. J. and Burrows, D., *Ulster Med. J.*, 58(2), 168, 1989. With permission.

nickel.[52] It is difficult, however, to get a suitable vehicle — either very little is absorbed or so much is absored that there is danger of a systemic effect.

C. BAN NICKEL IN JEWELRY

No country has yet taken this drastic action, but the Ministry of Environmental Affairs in Denmark is recommending a law that would make merchants selling jewelry containing more than 0.05 $\mu g/cm^2$/week of nickel guilty of a criminal offense. From what was shown earlier, this is probably a little unrealistic; perhaps 0.5% $\mu g/cm^2$/week would be more acceptable. The work of Larsson-Styme and Widstrom[3] would confirm that, at a stoke of the law, nickel dermatitis could be very drastically reduced.

D. FERROUS SULFATE IN CEMENT

Ferrous sulfate is a strong reducing agent and will reduce hexavalent chromate to trivalent chromate. Trivalent chromate is relatively harmless and does not have a high sensitizing or dermatitis-elicitating potential, mainly because it does not penetrate the skin. Ferrous sulfate has been added to cement in three European countries — Denmark, Sweden, and Finland. Its cost in Denmark is 10 million kroner per year in a population of 5 million. It adds about 1% to the cost of cement. In Britain, about 13 million tons of cement are used per year; at £50 a ton, this is £650 million and 1% of the cost would be about £6 million. With the Danish figures, on a population basis, the cost in Britain would be about £10 million, so the figure is somewhere between about £6 and 10 million. There are three arguments against adding ferrous sulfate to cement:

1. There is no proof that it is effective. Cement dermatitis is decreasing very rapidly in countries which do not use ferrous sulfate;[53] even in those which do use ferrous sulfate, the decrease in cement dermatitis was occurring long before its addition and made no difference to the downward curve.[54]
2. Even if it were proven to be effective, surely a better and more effective use of £5 million a year could be found in dermatology.
3. Builders will be resistant to its use in cement because of problems in the past with additives.

E. BETTER DIAGNOSIS

1. *IN VITRO* TESTS

Patch testing is rather an inexact science. There can be quite a discordance, not only among repeating patch tests, but also patch tests done at the same time. Lachapelle[13] showed that 4% of 73 identical positive patch tests were discordant; that is, they were positive on one side and negative on the other. There has therefore been a need for a more accurate method of diagnosing contact allergy. Since 1962 there have been 86 reports of *in vitro*

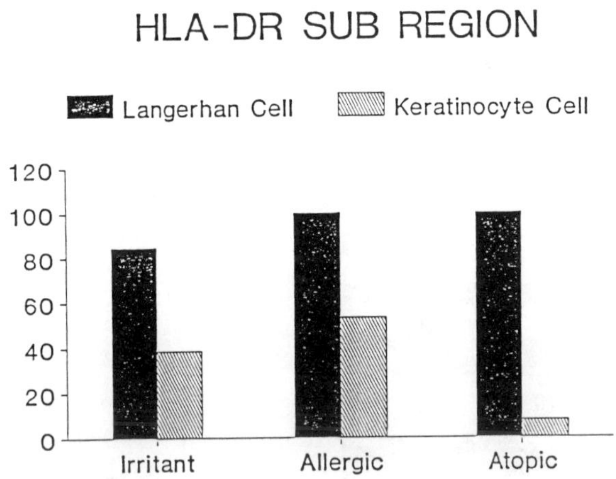

FIGURE 1. Expression of HLA-DR antigen on Langerhans' cells and keratinocytes in allergic contact dermatitis, irritant contact dermatitis, and atopic eczema.

tests in the literauture. These have mainly been lymphocyte transformation tests (70) and either lymphocyte migration or macrophage migration inhibition tests. The disadvantages of *in vitro* tests are that they are expensive, still subjective, and time-consuming; there are also toxic problems with dyes and irritant chemicals, and accuracy is still unsatisfactory. As Von Blomberg[55] has said, "*In vitro* tests are a delicacy, not a routine food."[55]

2. DR CLASS 2 ANTIGENS

Histocompatibility antigens of class 2 (DR antigens) are present on Langerhans' cells and spread to keratinocytes in many skin diseases, particularly those which are lymphocyte rich, such as graft vs. host (GVH), lichen planus, mycoses fungoides, and contact dermatitis. Barker et al.[56] have shown that expression on keratinocytes is very low or absent in atopic eczema.[56] We have studied 13 patients with atopic eczema, 13 patients with irritant contact dermatitis, and 15 patients with allergic contact dermatitis.[57] In the atopic eczema, the involved skin was biopsied; in the irritant and allergic cases, biopsies were taken of patch tests, either of the allergen to which they were allergic or an irritant patch test with dithanol (0.2%). The keratinocytes and Langerhans' cells were stained for HLA DR, DQ, and DP. The highest expression was shown for HLA DR. Figure 1 shows the results of the expression of class 2 antigen DR on keratinocytes; 40% stained in allergic contact dermatitis, 84% in irritant dermatitis, and only 6% in atopic dermatitis. This would not appear to be sufficiently diagnostic to be a useful test, but it does perhaps point the way for the future. Katz has shown that these DR-staining keratinocytes had a down-regulation function on lymphocyte proliferation,[58] which is the reverse of the effect of normal keratinocytes, which have a stimulating effect on lymphocytes.[59] In other words, they seem to occur in situations where there is a strong lymphocyte production and appear to be an attempt to suppress it. One wonders whether this might be of significance in the etiology of atopic eczema and its unusual responses. It also might be an explanation for persistent positive patch tests. Perhaps a more useful test in the long term might be the presence of immunoglobulin E (IgE) on Langerhans' cells.[60]

F. OXYGEN RADICALS

Oxygen radicals are increased, and may cause damage, in various conditions such as hyperoxemia, ischemia, inflammation, drugs and toxins, neoplasms, and radiation. Free radical scavengers mop them up (i.e., vitamin E or Dapsone).

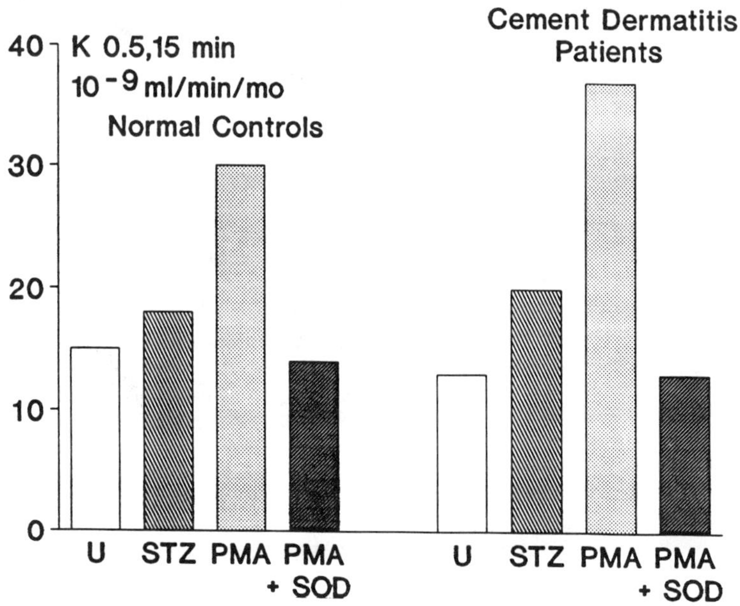

FIGURE 2. Levels of oxygen radicals in monocytes in normal controls and cement patients. U = unstimulated, STZ = S T Zymosan, PMA = phormal myristate acetate, PMA + SOD = phorbol myristate acetate and superoxide dismutase.

Normally, enzymes (e.g., superoxide dismutase [SOD]) catalyze their breakdown. Miyachi et al.[61] showed that oxygen radicals in polymorphonuclear leukocytes were markedly raised in patients with cement dermatitis. SOD, which in this situation should be increased to mop these up, was normal, whereas in cement workers who were not allergic to chromate the SOD was raised. Dapsone recently was shown to decrease oxygen radical levels and was found to be clinically effective in the treatment of cement dermatitis. We have examined oxygen radicals in monocytes in chromate dermatitis without stimulation and stimulated with SR zymosan (SRZ) or phorbol myristate acetate (PMA). Figure 2 shows that oxygen radicals were much more prevalent in the stimulated monocytes in chromate dermatitis than in stimulated monocytes of controls. It was aslo demonstrated that these oxygen radicals decreased with the addition of Dapsone. However, a double-blind trial of Dapsone in five patients did not show any significant improvement in chromate dermatitis.

Contact dermatitis is far from a dying specialty or a ''blister which has burst''. There are many questions to be answered, and the door is opening just sufficiently on some of these to suggest that what lies beyond is very interesting. In the preface to his book Prosser White said, ''To a reader a book is a tool to use, a tilt to abuse, or a toy to amuse: but mirrors the soul of the writer.'' I would imagine he would have felt the same about his oration. I hope you have found it so.

REFERENCES

1. **Oliver, M. I.,** Prevention of coronary heart disease: propaganda, promises, problems and prospects, *Circulation,* 73, 1, 1986.
2. **Cronin, E.,** *Contact Dermatitis,* Churchill Livingstone, Edinburgh, 1980.

3. **Larsson-Stymne, B. and Widstrom, L.,** Ear piercing — a cause of nickel allergy in schoolgirls, *Contact Dermatitis,* 13, 289, 1985.
4. **Romaguera, C., Grimalt, F., and Vilaplana, J.,** Contact dermatitis from nickel: investigation of its source, *Contact Dermatitis,* 19, 52, 1988.
5. **Peltonen, L.,** Nickel sensitivity in the general population, *Contact Dermatitis,* 5, 27, 1979.
6. **Nilsson, E.,** Individual and experimental risk factors for hand eczema in hospital workers, *Acta Derm. Venereol. Suppl.,* 128, 1986.
7. **Kieffer, M.,** Nickel sensitivity: relationship between history and patch test reaction, *Contact Dermatitis,* 5, 398, 1979.
8. **Menne, T.,** Nickel allergy — reliability of patch test evaluated in female twins, *Derm. Beruf Umwelt,* 29, 156, 1981.
9. **Lantinga, H., Nater, J. P., and Coenraads, P. J.,** Prevalence, incidence and course of eczema in the hands and forearm in a sample of the general population, *Contact Dermatitis,* 10, 135, 1984.
10. **Schmiel, G.,** Haufigkeit von Nickel-Kontaktallergien am unaustewahlten Patientengut im Raum Munchen, *Derm. Beruf Umwelt,* 33, 92, 1985.
11. **Fischer, T. and Rystedt, I.,** False positive follicular and irritant patch test reactions to metal salts, *Contact Dermatitis,* 12, 93, 1985.
12. **Gilboa, R., Al-Tawil, N. G., and Marcusson, J. A.,** Metal allergy in cashiers: an *in vitro* and *in vivo* study for the presence of metal allergy, *Acta Derm. Venereol.,* 68, 317, 1988.
13. **Lachapelle, J. M.,** 1st Eur. Symp. Contact Dermatitis, Heidelberg, May 27 to 29, 1988.
14. **Moller, H. and Svensson, A.,** Metal sensitivity: positive history but negative test indicates atopy, *Contact Dermatitis,* 14, 57, 1986.
15. **Widstrom, L. and Erikssohn, I.,** Nickel allergy and ear piercing in young men, 1st Eur. Symp. Contact Dermatitis, Heidelberg, May 27 to 29, 1988.
16. **Allenby, C. F. and Goodwin, B. F. S.,** Influence of detergent washing powders on minimal elicitating patch test concetration of nickel and chromium, *Contact Dermatitis,* 9, 491, 1983.
17. **Fischer, T., Fregert, S., Gruvberger, B., and Rystedt, I.,** Nickel release from ear piercing kits and earrings, *Contact Dermatitis,* 10, 39, 1984.
18. **Menne, T., Bandrup, F., Thestrup-Pedersen, K., Veien, N. K., Andersen, J. R., Yding, F., and Valeur, G.,** Patch test reactivity to nickel alloys, *Contact Dermatitis,* 16, 255, 1987.
19. **Menne, T., Andersen, K. E., Kaaber, K., Osmundsen, P. E., Andersen, J. R., Yding, F., and Valeur, G.,** Evaluation of dimethylglyoxime stick tests for detection of nickel, *Derm. Beruf Umwelt,* 35, 128, 1987.
20. **Fischer, T., Fregert, S., Gruvberger, B., and Rystedt, I.,** Contact sensitivity to nickel in white gold, *Contact Dermatitis,* 10, 23, 1984.
21. **Menne, T., Borgan, O., and Green, A.,** Nickel allergy and hand dermatitis in a stratified sample of the Danish female population: an eipdemiological study including statistic appendix, *Acta Derm. Venereol.,* 62, 35, 1982.
22. **Rydtedt, I. and Fischer, T.,** Relationship between nickel and cobalt sensitization in hard metal workers, *Contact Dermatitis,* 9, 195, 1983.
23. **Nilsson, E. and Back, O.,** The importance of anamnestic information of atopy, metal dermatitis in earlier hand eczema in the development of hand dermatitis in women in hospital work, *Acta Derm. Venereol.,* 66, 45, 1986.
24. **Lammintausta, K. and Kalimo, K.,** Do positive nickel reactions increase nonspecific patch test reactivity?, *Contact Dermatitis,* 16, 160, 1987.
25. **Paramsothy, Y., Collins, M., and Smith, A. G.,** Contact dermatitis in patients with leg ulcers: the presence of late positive reactions and evidence against systemic ampliative allergy, *Contact Dermatitis,* 18, 30, 1988.
26. **van Joost, T. and van Everdingen, J. J.,** Sensitization to cobalt associated with nickel allergy: clinical and statistical studies, *Acta Derm. Venereol.,* 62, 525, 1982.
27. **Lammintausta, K., Pitkanen, O. P., Kalimo, K., and Jansen, C. T.,** Interrelationship of nickel and cobalt contact sensitization, *Contact Dermatitis,* 13, 148, 1985.
28. **Walthard, B.,** Die Erzeugung experimentaller Nickelidiosynkrasie bei Laboratoriumstieren, *Schweiz. Med. Wochenschr.,* 56, 603, 1926.
29. **Lacroix, J., Morin, C. L., and Collin, P. P.,** Nickel dermatitis from a foreign body in the stomach, *J. Pediatr.,* 95, 428, 1979.
30. **Rudzki, E. and Crzywa, Z.,** Exacerbation of dermatitis by margarine, *Contact Dermatitis,* 3, 344, 1977.
31. **Kvorning, S. A.,** Postoperative dermatitis following non-epidermal nickel contact, *Contact Dermatitis,* 1, 327, 1975.
32. **Menne, T. and Thorboe, A.,** Nickel dermatitis: nickel excretion, *Contact Dermatitis,* 2, 353, 1976.
33. **Veien, N. K.,** Nickel diet, in 8th Int. Contact Dermatitis Symp., Cambridge, England, 1986.

34. **Sunderman, F. W., Hopfer, S. M., Swift, T., et al.,** Nickel absorption and elimination in human volunteers, in *Trace Elements in Human Health and Disease,* 2nd Nordik Symp., Odense, Denmark, August 1987.
35. **Ziegler, V. and Hohndorf, H.,** Implantation of metal in guinea pigs, in Int. Symp. Contact Dermatitis, Erfurt, East Germany, June 1983.
36. **Thomas, R. H. M., Rademaker, M., Goddard, N. J., and Munro, D. D.,** Severe eczema of the hands due to an orthopaedic plate made of Vitallium, *Br. Med. J.,* 294, 106, 1987.
37. **Christensen, O. B. and Moller, H.,** Nickel allergy and hand eczema, *Contact Dermatitis,* 1, 129, 1975.
38. **Jordan, W. P. and King, S. E.,** Nickel feeding in nickel sensitive patients with hand eczema, *J. Am. Acad. Dermatol.,* 1, 506, 1979.
39. **Burrows, D., Creswell, S., and Merrett, J. D.,** Nickel, hands and hip prostheses, *Br. J. Dermatol.,* 105, 437, 1981.
40. **Gawkrodger, D. J., Vestey, J. P., Wong, W.-K., and Buxton, P. K.,** Contact clinic survey of nickel sensitive subjects, *Contact Dermatitis,* 14, 165, 1986.
41. **Veien, N. K., Hattel, T., Justesen, O., and Nørholm, A.,** Oral challenge with nickel and cobalt in patients with positive patch test to nickel and/or cobalt, *Acta Derm. Venereol.,* 67, 321, 1987.
42. **Roduner, J., Haudenschild-Falb, E., Kuntz, E., Hunziker, T., and Krebs, A.,** Oral nickel-challenge in non-pompholyx and pompholyx-type nickel eczema, *Hautarzt,* 38, 262, 1987.
43. **Burrows, D., Rogers, S., Beck, M., McMaster, D., Merrett, D., and Eedy, D. J.,** Treatment of nickel dermatitis with trientine, *Contact Dermatitis,* 15, 55, 1986.
44. **Veien, N. K., Hattel, T., Justensen, O., and Nørholm, A.,** Dietary treatment of nickel dermatitis, *Acta Derm. Venereol.,* 65, 138, 1985.
45. **Moller, H. and Carlsson, A.,** Bioimplantation of metals in patients with metal allergy: the late outcome, 1st Eur. Symp. Contact Dermatitis, Heidelberg, May 27 to 29, 1988.
46. **White, I. R. and Smith, B. G. N.,** Dental amalgam dermatitis, *Br. Dent. J.,* 156, 259, 1984.
47. **Eggleston, D. W.,** Effect of dental amalgam and nickel alloys on T lymphocytes: preliminary report, *J. Prosthet. Dent.,* 15, 617, 1984.
48. **Eedy, D. J., Burrows, D., Clifford, T., and Fay, A.,** Alterations in the Immunological Status of Dental Students, Society of Dental Research, Belfast, April 1988.
49. **Vreeburg, K. J., de Groot, K., von Blomberg, M., and Scheper, R. J.,** Induction of immunological tolerance by oral administration of nickel and chromium, *J. Dent. Res.,* 63, 124, 1984.
50. **Van den Burg, C. K. H., Bruynyzeel, D. P., Vreeburg, K. J. J., von Blomberg, M. E., and Scheper, R. J.,** Hand eczema in hairdressers and nurses: a prospective study, *Contact Dermatitis,* 14, 275, 1986.
51. **Aldridge, R. D., Thompson, A. W., Rankin, R., Whiting, P. H., Cunningham, C., and Simpson, J. G.,** Inhibition of contact sensitivity reactions to DNFB by topical cyclosporin application in the guinea-pig, *Clin. Exp. Immunol.,* 59, 23, 1985.
52. **Aldridge, R. D., Sewell, H. F., King, G., and Thomson, A. W.,** Topical cyclosporin A in nickel contact hypersensitivity: results of a preliminary clinical immunohistochemical investigation, *Clin. Exp. Immunol.,* 66, 582, 1986.
53. **von Gailhofer, G. and Ludvan, M.,** Zur anderung des allergenspekturums bei kontaktekzenmen in den jahren 1975 — 1984, *Derm. Beruf Umwelt,* 35, 12, 1987.
54. **Farm, G.,** Changing patterns in chromate allergy, *Contact Dermatitis,* 15, 298, 1986.
55. **von Blomberg, B. M. E.,** Contact allergy *in vitro,* 1st Eur. Symp. Contact Dermatitis, Heidelberg, May 27 to 29, 1988.
56. **Barker, J. N. W. N. and MacDonald, D. M.,** Epidermal class 2 human lymphocyte antigen expression in atopic dermatitis: a comparison with experimental allergic contact dermatitis, *J. Am. Acad. Dermatol.,* 16, 1175, 1987.
57. **McKenna, K., Burrows, D., and Walsh, M.,** A comparison of expression of human lymphocyte class 2 antigens by cutaneous Langerhan cells and keratinocytes between patients with allergic, irritant and atopic dermatitis, 1st Eur. Symp. Contact Dermatitis, Heidelberg, May 27 to 29, 1988.
58. **Katz, S.,** Derm-Path meeting, British Association of Dermatologists, London, July 1988.
59. **Sauder, D., Carter, C., Katz, S. I., and Oppenheim, J. J.,** Epidermal cell derived lymphocyte activating factor, *J. Invest. Dermatol.,* 79, 34, 1982.
60. **Barker, J. N. W. N., Alegre, V. A., and MacDonald, D. M.,** Surface-bound immunoglobulin E on antigen presenting cells and cutaneous tissue of atopic dermatitis, *J. Invest. Dermatol.,* 90, 117, 1988.
61. **Miyachi, Y., Uchida, K., Komura, J., Asada, Y., and Niwa, Y.,** Auto-oxidative damage in cement dermatitis, *Arch. Dermatol. Res.,* 277, 288, 1985.
62. **Burrows, D.,** unpublished data.

Chapter 29

EXOGENOUS DERMATOSES IN THE TROPICS

C. L. Goh

TABLE OF CONTENTS

I. INTRODUCTION

The tropical climate varies minimally throughout the year. The ambient temperature (usually >25°C) and relative humidity (usually >70%) remain high throughout the year. Most countries in the tropics are either developing or are newly developed nations. In general, the socioeconomic status and the level of health care in these countries are low compared to developed nations elsewhere. Their industries tend to produce low-value-added products and are usually labor intensive. In this chapter the features of exogenous dermatoses seen in the tropics are discussed.

II. CONTACT DERMATITIS

Contact dermatitis is a common exogenous dermatosis in the tropics. Contact dermatitis accounted for about 10% of all dermatological and about 25% of all dermatitis cases seen in a skin hospital in tropical Singapore; it also accounted for 55% of hand eczema cases.[1]

A. IRRITANT CONTACT DERMATITIS

We believe that irritant dermatitis is more prevalent than allergic dermatitis in most tropical countries. For example, it was reported that irritant dermatitis contributed to more than 55% of patients with contact dermatitis of the hands and more than 60% of patients with occupational dermatitis in tropical Singapore.[2]

Topical medicaments are the most common causes of nonoccupational irritant dermatitis in the tropical countries. Self-medication with over-the-counter (OTC) medicaments is a common practice in many tropical countries.[3] Most of these medicaments contain counter-irritants (e.g., menthol, camphor, eucalyptus oil, salicylates, phenols, turpentine, peppermint, etc.). It is often difficult to differentiate irritant from allergic contact dermatitis caused by such medicaments because their exact constituents are usually unknown and are not available for patch testing. Lee and Lam[4] suggested that patch testing with the constituents in such medicaments (when available) can help elucidate causative irritants in these products. However, patch testing may not detect weak irritants.

Hand eczema is common in tropical countries, and most cases are due to irritant contact dermatitis. For example, Goh[1] reported that in more than one half of his patients with exogenous hand eczema the condition was due to irritant contact dermatitis in Singapore; a high proportion was caused by occupational agents. Occupational dermatitis accounted for 30% of exogenous hand eczema, but for only 10% of exogenous eczema elsewhere.

Occupational hand eczema is also less prevalent in females in tropical countries. The female:male ratio of patients with nonoccupational exogenous hand eczema in Singapore was 1:1, but for occupational hand eczema it was 1:2. Similarly, occupational irritant hand eczema was reported to be less prevalent in females in Nigeria.[5] One explanation for the lower prevalence in female workers in tropical countries may be because most females do domestic chores from a young age (for socioeconomic reasons), and they tend to develop hardening (tolerance). They are therefore less susceptible to occupational irritants compared to their male counterparts when they work in industries.[5]

Susceptibility to irritants seems to relate to differences in ethnic goups in some tropical countries. Goh and Chia[6] reported that the light-colored skin of Chinese appeared to be more susceptible to irritants than the dark-colored skin of Inidans. In their study on skin irritability to sodium lauryl sulfate using skin water vapor loss as an objective measurement, they reported that the skin of Chinese individuals appeared to be slightly more irritable than Malays and the latter more so than the Indians.[6]

B. ALLERGIC CONTACT DERMATITIS

The prevalent contact allergens in tropical countries differ slightly from those elsewhere. Table 1 compares the prevalence of contact allergy to some common allergens in some tropical and temperate countries.[7] There are some differences in the prevalence of contact allergy to topical medicaments and occupational allergens. The prevalence of allergy to different types of medicaments is influenced by availability, the prescribing habits of physicians of a country, and the occupational allergens to its prevalent industries.[8-10] It can be seen that the prevalence of allergy to ubiquitous allergens such as nickel and chromate does not differ very much in different countries in different regions of the world.

1. Allergic Contact Dermatitis to Medicaments

Topical medicaments are the most common contact allergens in many tropical countries.[8,10,11] OTC medicaments are the most common group. This is because self-medication with OTC medicaments is a common practice in tropical countries. OTC medicaments are inexpensive and readily available.

Among topical medicaments, antimicrobials are the most common contact allergens in tropical countries. Many of these are OTC antimicrobials such as proflavine and nitrofurazone. They are widely used in these countries and are well-known sensitizers. Proflavine and nitrofurazone are reported to be common skin sensitizers in Singapore[13] and India,[10] respectively.

We recently compared the sensitizing potential of some of these OTC antimicrobials and confirmed that many of the less expensive OTC antimicrobials are potent sensitizers. Seven OTC topical antimircobials (namely, proflavine, iodine, *p*-chlorometaxylenol, benzalkonium chloride, propamidine, chlorhexidine, and cetrimide were compared using a modified Beuhlers guinea pig study. Four out of ten guinea pigs were sensitized to proflavine compared to two of ten for propamidine, and for *p*-chlorometaxylenol, one of ten for iodine, and zero of ten for chlorhexidine and cetrimide.[44]

In the case of controlled topical antimicrobials (those requiring a doctor's prescription), it again appears that less expensive antibiotics are more sensitizing. Using the same animal study method, we also compared the sensitizing potential of ten controlled topical antibiotics commonly prescribed in tropical countries. The results showed that neomycin sensitized two of ten guinea pigs, gentamicin sensitized one of ten guinea pigs, and kanamycin, bacitracin, polymyxin, tetracycline, fusidic acid, and sodium fusidate sensitized zero of ten guinea pigs. In addition, the overall results indicated that controlled topical antimicrobials are less sensitizing than OTC topical antimicrobials.[44]

In addition to OTC antimicrobials, traditional herbal preparations, which are concoctions of plants extracts, are also commonly used as medicaments by people in tropical countries. Some of these preparations are reported to cause allergic contact dermatitis.[11] Contact allergy to such preparations is seldom reported, as the causative allergen is usually unknown.

2. Plant Dermatitis

A group of plants called "rengas" are the most common cause of phytodermatitis in the tropics.[13,14] "Rengas" is a name coined by the indigenous people of tropical Southeast Asia, where the plants flourished. The rengas consists of five genera of plants — namely, *Gluta, Melanochyla, Melanorrhoea, Semecarpus,* and *Swintonia*. All belonged to the Anacardaceae family. These plants contain a toxic resinous sap which has the characteristic property of blackening when exposed to air and of becoming lacquer-like. The resin is notoriously known to cause severe skin problems. The chemical nature of the resin is unknown. Past reports appear to suggest that the resin is a potent skin irritant and sensitizer.[15,16] One observation noted that soldiers developed eruptions on their feet after wading in rivers into which the fruit had fallen.[15] It has also been observed that it was undesirable

TABLE 1
Comparison of Prevalence of Contact Allergy to Common Allergens in Some Tropical Countries and Temperate Countries[7]

Allergen	Tropical countries			Temperate countries		
	Singapore %	Thailand %	Indonesia %	Europe %	North America %	Japan %
Nickel sulfate 5 %	13.9	na	7.0	6.7	11.0	4
Cobalt chloride 1%	4.7	4.1	3.0	6.9	na	3.7
Potassium dichromate 0.5%	6.3	12.2	9.0	6.6	8	6.8
Proflavine 0.5%	6.3	na	na	na	na	na
Neomycin 20%	5.7	6.4	6.0	3.7	6.0	5.8
Clioquinol 5%	2.6	na	na	1.7	na	1.3
Turpentine 1%	1.1	4.4	na	5.9	6.0	1.5
Colophony 20%	4.9	7.0	5.0	3.3	—	2.9
p-Phenylenediamine 1%	3.4	6.7	9.0	4.9	8.0	7.1
Balsam of Peru 25%	3.1	6.4	3.0	6.3	na	5.7
Lanolin/wool alcohol 30%	3.1	na	1.0	2.6	3.0	1.2
Parabens mix 15%	0.8	4.1	na	1.9	3.0	0.8
Ethylenediamine 1%	1.4	na	na	na	7.0	1.9
Formalin 2%	0.7	na	na	3.5	na	3.4
Ref.	9	25	26	41	42	43

Note: na = not applicable.

to shelter under a rengas tree during a tropical rainstorm because drops of water could carry down the resin and cause dermatitis.[15] Carpenters working with rengas woods are known to risk sensitization to the resin.[13,14] Unlike rengas, other tropical woods used in furniture making, e.g., teak (*Tectona grandis*), meranti (*Shorea* sp.), nyatoh (*Palaquiu* sp., *Ganua* sp., *Payena* sp.), kapor (*Dryobalanops* sp.), balau (*Shorea* sp.), rubber (*Hevea brasiliensis*), and chengal (*Balanocarpus neimii*) appeared to be weak skin sensitizers. In a survey of 479 woodworkers in 33 furniture-making factories for contact allergy to these woods in Singapore, none were found to be allergic to any of these woods. Only 1.6% of the workers were found to have irritant contact dermatitis from various wood dusts.[17]

Another plant reported to cause phytodermatitis in the tropics is the mango plant (*Mangifera* sp.). It belongs to the Anacardaceae family. Outbreaks of dermatitis from the plant saps often occur during fruiting seasons. The allergen comes from the sap of the stem. The exact allergen remains unknown. The sap is also an irritant. Fruit pickers develop acute dermatitis on the hands and arms, whereas patients sensitive to the saps may develop perioral dermatitis when they eat the fruit.[13]

3. Cosmetic Dermatitis

Contact allergy to cosmetics appears to be less prevalent in tropical countries than elsewhere. This is probably because usage of cosmetics is less prevalent in tropical countries. In Singapore, *p*-phenylenediamine was reported to be the most common cosmetic sensitizer; the prevalence was highest in males over 40 years of age. The higher prevalence in older males was attributed to the fact that males tend to dye their own hair, whereas females tend to go to hairdressers. In this way males risk sensitization from the dye due to poor dyeing technique.[18]

Contact allergy to cosmetic preservatives, such as Kathon® CG and Dowicil® 200, which are common in Europe and the U.S., is uncommon in the tropical countries.[18]

A characteristic type of pigmented contact dermatitis is seen in tropical countries. It is caused by an Indian cosmetic called ''kumkum''.[19] ''Kumkums'' are used traditionally by Hindu women. They come in different colors and are applied on the mid-forehead. Pigmented contact dermatitis from ''kumkum'' presents with mild erythema and irritation on the mid-forhead (where the ''kumkum'' was applied). This is followed over a variable period by dark gray pigmentation. The causative allergen is Sudan I, an allergen which was previously reported to cause outbreaks of pigmented contact dermatitis from Japanese cosmetics.[19,20] Using gas chromatography-mass spectrophotometry analysis, Kozuka et al.[21] have confirmed the presence of Sudan I in ''kumkums''.

III. CONTACT ALLERGY AND ETHNICITY

The sensitizing capacity of different ethnic groups in tropical countries appears to be the same. Goh[22] compared the prevalence of contact allergy among different ethnic groups (Chinese, Malays, and Indians) attending a contact dermatitis clinic in Singapore and reported that the rates of positive patch test reactions to common contact allergens were not significantly different for each ethnic group.

IV. PATCH TESTING IN THE TROPICS

There is little climatic variation throughout the year in the tropics; thus, there is no specific period of the year when it is most ideal to perform the patch test. The patch test is an uncomfortable procedure for the patient because of the high heat and humidity in the tropics. Patients should be briefed on the procedure carefully before patch testing. The Finn Chambers® on Scanpor tapes appear to be a convenient patch test method for the tropics. We suggest some modification to the standard patch test procedure in the tropics:

1. We allow the patient to shower after application of the patch test chambers, but he is instructed to keep his back dry during the first 48 h. He is allowed to use moist towels to clean his back daily; he is instructed to dab but not to wipe.
2. All patients engaged in heavy manual labor (e.g., construction work) are given medical leave while the patch test chambers are in place. After removal of the chambers (i.e., at 48 h) the patients are allowed to return to normal duties, as most workers prefer to return to work for socioeconomic reasons.
3. Skin marking is done with marking inks. We have found the preparation containing gential violet 1%, methyl alcohol 50%, silver nitrate 20%, and distilled water 29% w/w to be most suitable. The preparation may cause irritation in some individuals. Freshly prepared ink is needed, as the marker becomes too concentrated due to rapid evaporation of the solvent in the hot climate. Concentrated marker inks tend to cause skin irritation. Commercial skin markers and fluorescent markers do not stay on the skin longer than 48 h in the tropics due to perspiration. Marking with tapes is impractical, as the tapes tend to dislodge easily due to the perspiration.
4. It is necessary to examine patch test reactions in a well-lit room, as positive patch test reactions are easily missed in dark-colored individuals.

A. STORAGE OF PATCH TEST MATERIALS IN THE TROPICS

The homogeneity of patch test materials can be affected by the heat in the tropics. For example, Goh and Kowk[23] showed that nickel sulfate and potassium dichromate patch test materials become less homogeneous (after storing for 2 months) when they are stored at room temperature (mean temperature of 28°C), but remained relatively homogeneous when stored (refrigerated) at 4°C. The coefficients of variation of concentration of nickel sulfate and potassium dichromate in different aliquots of the two allergens in their respective tubes were higher when stored at room temperature (25.3 and 9.6%, respectively) than when kept refrigerated (3.1 and 4.6%, respectively). This change in homogeneity was attributed to the reduction in the viscosity resulting in the redistribution of these allergens in the tubes because of the high ambient temperature in the tropics. They recommend that patch test materials dispersed in petrolatum be stored in refrigerators.

V. OCCUPATIONAL DERMATOSES

In general, the construction, metal/engineering, and electronics industries are the predominant industries in the newly industrialized countries, whereas agriculture remains the predominant industry in the developing countries in the tropics. The high heat and humidity in the tropics tend to promote heat-related skin disorders among workers.

A. OCCUPATIONAL CONTACT DERMATITIS

As in other regions of the world, contact dermatitis is the most common presentation of occupational dermatosis. In tropical Singapore, more than 90% of all cases of occupational dermatosis were contact dermatitis. Irritant contact dermatitis is more common than allergic contact dermatitis, accounting for more than 60% of all workers with occupational dermatitis.[2]

Cement in the construction industry, cutting fluids, oils and solvents in the metal/engineering industry, and soldering flux and solvents in the electronics industry appears to be the most common occupational irritants in the tropics.[2] Poor working habits, poor work condition, and lack of health education of workers are contributing factors for the high incidence of irritant dermatitis. It is not uncommon to see workers using solvents and detergents as skin cleansers at their workplace.

Excessive heat and high humidity play an important role in occupational dermatosis in the tropics. The high heat and humidity provoke workers to discard protective clothing. This exposes workers' skin to irritants and allergens. In addition, the high heat and humidity cause workers to sweat profusely. Sweating predisposes workers to develop skin maceration and dermatitis. Intertrigo is prevalent in the tropics. Furthermore, sweating enhances dissolution of dyes from clothing, and the dissolved dyes can sensitize.[24]

Chromate appears to be the most common occupational allergen in most tropical countries,[9,25-27] and cement is its most common source. Goh and Kwok[28,29] reported that most cements used in tropical Southeast Asia contain hexavalent chromate (3 to 23 μ/g); the concentrations are comparable to those found in Europe. They also reported that the hexavalent chromates in these cements can be removed by adding 0.35% w/w of ferrous sulfate into the cements,[30] as was observed by Fregert[31] in Sweden with Swedish cements. They suggested that addition of ferrous sulfate into cements in tropical Southeast Asian countries, as is done in Sweden and Denmark, can reduce the incidence of cement dermatitis among construction workers.

Atmospheric chromate does not appear to play any role in chromate allergy at construction sites in the tropics. Goh et al.,[32] measured the atmospheric chromate concentrations in industrial zones, in suburban and residential areas, and in a dusty cement factory in Singapore and found that the chromate level was negligible.

In some newly industrialized countries in the tropics, epoxy resins appear to be replacing chromate as the most common occupational allergen in the construction industry.[33] This is because of the decline in new construction projects coupled with an increase in repair and maintenance work on existing buildings in these countries. Epoxy resins is frequently used in repair and maintenance of buildings.

B. NONECZEMATOUS OCCUPATIONAL DERMATOSES

Noneczematous skin disorders are less common causes of occupational dermatoses in the tropical countries than eczematous disorders. Heat and high humidity appear to play a role in the promotion of noneczematous occupational dermatoses in the tropics. Acneiform eruptions and miliaria are such dermatoses.

1. Acne Estivalis

Acne estivalis is a condition described in patients who developed acneiform eruption after spending time in the tropical climate. The exact mechanism is unknown, but heat is believed to play a role.[34] Heat and high humidity have been known to affect the pilosebaceous unit.[35,36] Goh and Ho[37] reported an outbreak of acneiform eruption among workers in a polyvinyl chloride manufacturing factory. The high ambient temperature and humidity of the work environment were believed to be contributory factors, since air-conditioning of the workplace resulted in the resolution of acneiform eruption among the workers. The mechanism was ascribed to the swelling of the keratin lining of the pilosebaceous orifice by heat and high humidity, thereby promoting comedone formation and resulting in acneiform eruptions in their workers.

2. Miliaria

Miliaria is a common occupational skin disorder occurring in individuals working in hot and poorly ventilated workplaces. It is a common disorder among workers operating furnaces and boilers in factories in the tropics. The obstruction probably results from swelling of the keratin lining of the sweat ducts due to heat and high humidity, which leads to obstruction of the sweat ducts and, hence, sweat retention.

3. Skin Infections

Other noneczematous occupational dermatoses commonly seen in the tropics include intertrigo and skin infections. The high heat and humidity in the tropics promote sweating and sweat retention, especially on skin folds, resulting in skin maceration. Secondary bacterial and fungal infections on macerated skin are common. Bacterial intertrigo (tinea cruris and tinea pedis) is common among workers engaged in wet work.

In some tropical countries where farming is the major occupation activity, chromomycosis is not uncommon. The spores of the fungi are able to thrive on the warm, moist soils in the tropics. Similarly, creeping larva migrans is also a common skin infestation among farmers in the tropics.

VI. PHOTOCONTACT DERMATITIS

Photocontact dermatitis is not uncommon in some tropical countries. Photosensitizing chemicals, such as tetrachlorsalicylanilide, tribromasalicylanilide, and bithionol, are still being used as ingredients in some OTC medicaments and medicated soaps.

As in temperate countries, musk ambrette has become the most common photosensitizer in some tropical countries. This is because of the widespread use of perfumed preparations among men. For example, in Singapore, musk ambrette was reported to be the most common photosensitizer.[39] The clinical pattern of photosensitivity to musk ambrette in the tropics appears similar to that reported in European countries. It is more common in males, since the most common source of musk ambrette is perfumed preparations for males. Goh and Kwok[39] analyzed 32 different brands of colognes for men sold in tropical Singapore and found that 14 of them contain various concentrations of musk ambrette; the concentrations ranged from 0.02 to 0.39% w/v.

In tropical Southeast Asia, it is common practice for barbers to apply cologne to their customer's scalp, face, and neck after a haircut. These colognes usually contain musk ambrette. They cause periodic photodermatitis in some patients each time after they have a haircut.[40]

VII. CONCLUSIONS

There appear to be some differences in the pattern of exogenous dermatoses in the tropics compared to other regions of the world. These differences are partly attributed to the difference in climatic conditions. The hot, humid climate in the tropics is a predisposing factor for conditions such as contact dermatitis, miliaria, acneiform eruptions, and skin infections. In addition, cultural and traditional differences such as self-medications and use of herbal preparations contribute to the higher prevalence of irritant contact dermatitis from medicaments in tropical countries. The labor-intensive types of industries (poor work environment coupled with poor work habits and lack of health education of workers in many developing countries in the tropics) further contribute to the high prevalence of occupational irritant contact dermatitis.

REFERENCES

1. **Goh, C. L.,** An epidemiological comparison between hand eczema and non-hand eczema, *Br. J. Dermatol.,* 118, 797, 1988.
2. **Goh, C. L.,** Occupational dermatoses in Singapore, *Contact Dermatitis,* 11, 288, 1984.
3. **Quah, S. R.,** Self-medication in Singapore, *Singapore Med. J.,* 26, 123, 1986.
4. **Lee, T. Y. and Lam, T. H.,** Irritant contact dermatitis due to a Chinese herbal medicine Lu-Shen-Wan, *Contact Dermatitis,* 18, 213, 1988.
5. **Olumide, Y.,** Contact dermatitis in Nigeria. Hand dermatitis in women, *Contact Dermatitis,* 17, 85, 1987.
6. **Goh, C. L. and Chia, S. E.,** Skin irritancy to sodium lauryl sulphate (as measured by skin water vapour loss) by race and sex, *Clin. Exp. Dermatol.,* 13, 16, 1988.

7. **Goh, C. L.,** Contact Sensitivity in Singapore, *Skin Res. (Hifu)* 28 (Suppl. 2), 41, 1986.
8. **Goh, C. L.,** Contact sensitivity to topical medicament in Singapore, *Int. J. Dermatol.,* 28, 25, 1989.
9. **Goh, C. L.,** An epidemiology of contact allergy in Singapore, *Int. J. Dermatol.,* 27, 308, 1988.
10. **Bajaj, A. K. and Gupta, S. C.,** Contact hypersensitivity to topical antibacterial agents, *Int. J. Dermatol.,* 25, 103, 1986.
11. **Lee, T. Y. and Lam, T. H.,** Allergic contact dermatitis to Yunnan Paiyao, *Contact Dermatitis,* 17, 59, 1987.
12. **Goh, C. L.,** Contact sensitivity to proflavine, *Int. J. Dermatol.,* 25, 449, 1986.
13. **Nasution, D., Klokke, A. H., and Nater, J. P.,** A survey of occupational dermatoses in Indonesia, *Dermatosen,* 21, 215, 1973.
14. **Goh, C. L.,** Occupational allergic contact dermatitis from Rengas wood, *Contact Dermatitis,* 18, 300, 1988.
15. **Burkill, I. H.,** A Dictionary of the Economic Products of Malay Peninsula, Kuala Lumpur, Vol. 2, Ministry of Agriculture and Cooperatives, Singapore, 1966, 1458.
16. **Corner, E. J. H.,** *Wayside Trees of Malaya,* Vol. 1, 2nd ed., Government Printing Office, Singapore, 1952, 116.
17. **Gan, S. L., Goh, C. L., Lee, C. S., and Hui, K. H.,** Occupational dermatosis among sanders in the furniture industry, *Contact Dermatitis,* 17, 237, 1987.
18. **Goh, C. L.,** Allergic contact dermatitis from cosmetics. A report from Singapore, *J. Derm (Tokyo),* 14, 248, 1987.
19. **Goh, C. L. and Kozuka, T.,** Pigmented contact dermatitis from "kumkum", *Clin. Exp. Dermatol.,* 11, 603, 1986.
20. **Kozuka, T., Tashiro, M., Sano, S., Fujimoto, K., Kakamura, Y., Hashimoto, S., and Nakaminam, G.,** Brilliant Lake Red R as a cause of pigmented contact dermatitis, *Contact Dermatitis,* 5, 297, 1979.
21. **Kozuka, T., Goh, C. L., Doi, T., and Yoshigawa, K.,** Sudan I in "kumkum", *Ann. Acad. Med. Singapore,* 1988, in press.
22. **Goh, C. L.,** Prevalence of contact allergy by sex, race and age, *Contact Dermatitis,* 14, 237, 1986.
23. **Goh, C. L. and Kwok, S. F.,** The influence of temperature on the concentration homogeneity of patch test materials, *Contact Dermatitis,* 15, 231, 1986.
24. **Olumide, Y., Oleru, G. U., and Enu, C. C.,** Cutaneous implications of excessive heat in the work-place, *Contact Dermatitis,* 9, 360, 1983.
25. **Calnan, C. D.,** Contact dermatitis, *Int. J. Dermatol.,* 14, 264, 1975.
26. **Harahap, M., Nasution, D., and Siregar, A. S.,** Contact allergens in north Sumatra (Indonesia), *Contact Dermatitis,* 5, 271, 1979.
27. **Goh, C. L.,** Chromate sensitivity in Singapore, *Int. J. Dermatol.,* 24, 514, 1985.
28. **Goh, C. L. and Kwok, S. F.,** Chromate allergy: chromate content of Asian cement, *J. Derm. (Tokyo),* 13, 393, 1986.
29. **Johnston, A. J. M. and Calnan, C. D.,** Cement dermatitis. I. Chemical aspect, *Trans. St. John's Hosp. Dermatol. Soc.,* 41, 11, 1958.
30. **Goh, C. L. and Kwok, S. F.,** Prevention of cement dermatitis in construction workers with iron sulphate, *Asia Pacific J. Publ. Health,* 1, 91, 1987.
31. **Fregert, S., Gruvberger, B., and Sandahl, E.,** Reduction of chromate in cement by iron sulphate, *Contact Dermatitis,* 5, 39, 1979.
32. **Goh, C. L., Wong, P. H., Kwok, S. F., and Gan, S. L.,** Chromate allergy: total chromium and hexavalent chromate in the air, *Dermatosen,* 34, 132, 1986.
33. **Goh, C. L.,** Occupational dermatoses — an update, *Ann. Acad. Med. Singapore,* 17, 557, 1988.
34. **Rook, A., Wilkinson, D. S., and Ebling, F. J. G.,** *Textbook of Dermatology,* 3rd ed., Blackwell Scientific, Oxford, 1979, 1728.
35. **Lobitz, W. C. and Dobson, R. L.,** Miliaria, *Arch. Environ. Health,* 11, 460, 1965.
36. **Taylor, J. S.,** The pilosebaceous unit, in *Occupational and Industrial Dermatology,* Maibach, H. I. and Gellin, G. A., Eds.,Year Book Medical Publishers, Chicago, 1982, 125.
37. **Goh, C. L. and Ho, S. F.,** An outbreak of acneiform eruption in a polyvinyl chloride manufacturing factory, *Dermatosen,* 36, 53, 1988.
38. **Goh, C. L. and Tham. S. N.,** Photosensitivity to musk ambrette, *Aust. J. Dermatol.,* 26, 133, 1983.
39. **Goh, C. L. and Kwok, S. F.,** A simple method of qualitative analysis for musk ambrette, musk ketone and musk xylene in cologne, *Contact Dermatitis,* 14, 53, 1986.
40. **Goh, C. L. and Ng, S. K.,** Photodermatitis from musk ambrette in barbers' cologne, *Contact Dermatitis,* 12, 229, 1985.
41. **Fregert, S., Hjorth, N., Magnusson, B., et al.,** Epidemiology of contact dermatitis, *Trans. St. John's Dermatol. Soc.,* 55, 17, 1969.
42. **Rudner, E., Clendenning, W. E., Epstein, E., et al.,** Epidemiology of contact dermatitis in North America, *Arch. Dermatol.,* 108, 537, 1973.

43. **Fujiwara, N., Sasaki, M., Higashi, N., et al.,** Standardization of patch tests in Japan, *Contact Dermatitis,* 2, 205, 1976.
44. **Goh, C. L.,** Contact sensitivity to topical antimicrobials. II. Sensitizing potentials of some topical antimicrobials, *Contact Dermatitis,* 21, 166, 1989.

Chapter 30

PREVENTION OF ALLERGIC CONTACT SENSITIZATION: A HISTORICAL PERSPECTIVE

N. Hjorth and T. Menné

TABLE OF CONTENTS

I. INTRODUCTION

Contact sensitization is a serious public health problem. Preventive measures can reduce the frequency of sensitization. Once established, the sensitivity tends to persist, although 30% of those examined seem to have lost their allergy. Many of them have been included in the publications because of false positive reactions.[1] A person sensitized to one substance frequently becomes sensitized to several. Possibly, this is due to a genetic selection.[2] However, at the present moment we cannot argue that the patient sensitive to one substance should avoid contact with other sensitizers. Premarketing prevention of contact dermatitis probably occurs, but it is striking how many strong sensitizers have been let loose on the public by industry.

II. THE 1950s

Contact dermatitis among the workers employed in the production of epoxy resins almost prevented the marketing of epoxy. They were predictive of the epidemics of epoxy sensitizations to come. The industrial climate of the 1950s could not resist such an extremely efficient substance as epoxy resins.

Baker's eczema from perborate flour improver was abolished in Denmark in the late 1930s and about the same time in Holland. This was well known in Germany, where perborate eczema continued to be common. In the mid-1950s both halves of Germany abolished the perborate; after this, baker's eczema became rare.[3]

III. THE 1960s

The 1960s taught us much about the necessity of preventive measures to reduce the frequency of contact sensitization. In 1960 Kligman developed the maximization test to predict the sensitization properties of individual substances in both guinea pigs and humans. These maximization tests with their modifications have basically improved the ability of preventing sensitization by new chemicals.[4]

In 1961 an extensive wave of photodermatitis was noted in Britain. Thanks to Darrell Wilkinson's curiosity, the cause was traced to tetrachlorsalicylanilide.[5] This epidemic could not have been prevented, since industry at that time might examine a new chemical for sensitizing properties but not for photosensitivity. In 1965 the occurrence of photosensitivity from antiseptic soaps was known, and for that reason tribromosalicylanilide (TBS) should not have been introduced on the European and American markets.

IV. THE 1970s

In Denmark and a few other countries we had another wave of contact dermatitis with thousands of sufferers, this time caused by an optical whitener used in about one half of the heavy detergents on the Danish market.[6] The manufacturers investigated the sensitizing properties before marketing, but the Shelansky-Shelansky method could not discriminate the sensitizing properties, and the whitener was released to millions of consumers. This type of textile dermatitis (with the morphology of a prelymphoma) has not been seen since. This type of epidemic could have been prevented by appropriate sensitization testing, which was being done by the other half of the industries in Denmark.

Another epidemic occurred in Scandinavia, but here the various big industries worked together because the cause was unknown. Sultones are formed in the synthesis of laurylethersulfate if the temperature exceeds 220°C. With inadequate cooling, the temperature can rise to 230°C. Sultones are stronger sensitizers than dinitrochlorobenzene.[7-9] The industry now screens for this problem, as the effective epidemic has come to an end.

V. ACTUAL EPIDEMICS

A. METALS

Chromate remains one of the most frequent contact sensitizers in males. Worldwide, the hexavalent chromate in wet cement is the main source of chromate sensitization. The amount of chromate in cement varies with the production place, as does the frequency of cement dermatitis. Addition of ferrous sulfate to the cement reduces the water-soluble hexavalent chromate compounds to insoluble, nonsensitizing trivalent compounds.[10] Since 1981, in Denmark and later in Sweden and Finland, ferrous sulfate has been added to cement. New cases of cement dermatitis are now rare in Denmark. Epidemiological investigations to validate this observation are in progress.

Nickel is still the top sensitizer in women, as it was 30 years ago. The prevalence of nickel sensitization in the female population ranges from 7 to 15%, dependent on age. Ear piercing in school children is presently the main cause of primary nickel sensitization. Nickel-sensitive persons run an increased risk of developing secondary hand eczema compared to nickel-insensitive persons, probably because of contact with trace amounts of the ubiquitous metal. These trace amounts in the general environment do not lead to primary nickel sensitization, but contribute to the chronicity of nickel dermatitis.

Investigations of commonly used nickel alloys with respect to corrosion, stability, and reactivity in nickel-sensitive individuals indicate that alloys exceeding a release of 1 $\mu g/cm^2$/week in synthetic sweat give a high sensitization risk, and those with a nickel release below 0.5 $\mu g/cm^2$/week result in a low risk.[11] Using a different study design, Emmet et al.[12] established that the lowest provocation threshold in most nickel-sensitive individuals was 1.5 μg. These studies defining concentration limits for sensitization and elicitation risk in nickel allergy provide a basis for prevention of these common conditions which is technically and economically possible.

B. PRESERVATIVES

Ideal preservatives for cosmetic products, topical medicaments, and industrial preservatives still need to be developed. Parabens, which are the most widely employed preservative, rarely sensitize in the concentrations used in cosmetics. The incidence of paraben reactivity has been remarkably constant at a low level for decades.[13] However, for some cosmetic products an additional preservation system is needed to meet the microbiological quality criteria. These additional preservatives, such as formaldehyde and formaldehyde-releasing substances, are miscredited because of an unacceptably high level of contact sensitization and carcinogenecity. Consequently, industry has replaced formaldehyde and formaldehyde-releasing substances with other preservatives, leading to a declining trend in frequency of formaldehyde-sensitive patients.[14] In Europe, the Kathon® CG system,[15] together with parabens, is now the dominating preservation system. Clinical studies[16] have established a concentration-dependent induction and challenge sensitization risk for Kathon® CG.

Consumer-based sensitization risk studies[17] define Kathon® CG as safe with respect to contact sensitization. Even though Kathon® CG has been investigated by all available predictive methods, Kathon® is now one of the leading contact sensitizers in some areas of Europe. There is probably more than one explanation, but the most obvious is marketing of cosmetic products containing Kathon® CG in a concentration up to the European Economic Committe-approved 30-ppm limit.[18] This level is obviously too high, and a down-regulation is planned.

VI. CONCLUSION

Epidemics of allergic contact sensitization cannot be prevented based on animal experiences and limited consumer-based trials. The ultimate risk assessment will still be based on large-scale consumer experience over many years. The preventive strategy of the future is not to ban valuable but potentially sensitizing substances, but to define concentration limits which minimize health risks with respect to contact sensitization. Without limitations and regulations, allergic contact sensitization will continue to be a major health problem, as not all parts of the industry will behave responsibly by their own means.

REFERENCES

1. **Bruynzeel, D. P. and Maibach, H. I.,** Excited skin syndrome (angry back), *Arch. Dermatol.,* 122, 323, 1986.
2. **Menné, T. and Holm, N. V.,** Genetic susceptibility in human allergic contact sensitization, *Semin. Dermatol.,* 5, 301, 1986.
3. **Preyss, J. A.,** Zur Beurteilung des Bäkerekzems, *Berufs-Dermatosen,* 8, 68, 1960.
4. **Andersen, K. E. and Maibach, H. I., Eds.,** *Contact Allergy Predictive Tests in Guinea Pigs,* S. Karger, Basel, 1985.
5. **Wilkinson, D. S.,** Further experiences with halogenated salicylamilide, *Br. J. Dermatol.,* 74, 295, 1962.
6. **Osmundsen, P. E.,** Pigmented contact dermatitis, *Br. J. Dermatol.,* 83, 296, 1970.
7. **Magnusson, B. and Gilje, O.,** Allergic contact dermatitis from a dish-washing liquid containing laurethersulphate, *Acta Derm. Venereol.,* 53, 136, 1973.
8. **Ritz, H. L., Connor, D. S., and Sauter, E. D.,** Contact sensitization of guinea-pigs with unsaturated and hologenated sultones, *Contact Dermatitis,* 1, 349, 1975.
9. **Slyvest, B., Hjorth, N., and Magnusson, B.,** Lauryl ether sulphate dermatitis in Denmark, *Contact Dermatitis,* 1, 359, 1975.
10. **Fregert, S., Gruyberger, B., and Sandahl, E.,** Reduction of chromate in cement by iron sulphate, *Contact Dermatitis,* 5, 39, 1979.
11. **Menné, T., Brandrup, F., Thestrup-Pedersen, K., Veien, N. K., Andersen, J. R., Yding, F., and Valeur, G.,** Patch test reactivity to nickel alloys, *Contact Dermatitis,* 16, 255, 1987.
12. **Emmet, E. A., Risby, T. H., Jiang, L., Ng, S. K., and Feinman, S.,** Allergic contact dermatitis to nickel: bioavailability from consumer products and provocation threshold, *J. Am. Acad. Dermatol.,* 19, 314, 1988.
13. **Menné, T. and Hjorth, N.,** Routing patch testing with paraben esters, *Contact Dermatitis,* 19, 189, 1988.
14. **Edman, B. and Möller, H.,** Trends and forecasts for standard allergens in a 12-year patch test material, *Contact Dermatitis,* 8, 95, 1982.
15. **De Groot, A. C. and Weyland, J. W.,** Kathon CG: a review, *J. Am. Acad. Dermatol.,* 18, 350, 1988.
16. **Chan, P. K., Baldwin, R. G. L., Parsons, R. D., Moss, J. N., Stiratelli, R., Smith, J. M., and Hayes, A. W.,** Kathon biocide: manifestation of delayed contact dermatitis in guinea pigs is dependent on the concentration for induction and challenge, *J. Invest. Dermatol.,* 81, 409, 1983.
17. **Schwartz, S. R., Weiss, S., Stern, E., Morici, I. J., Moss, J. N., Goodman, J. J., and Scarborough, N. L.,** Human safety study of body lotion containing Kathon CG, *Contact Dermatitis,* 16, 203, 1987.
18. **Tostie, A.,** Prevalence and sources of Kathon CG sensitization in Italy, *Contact Dermatitis,* 18, 173, 1988.

Chapter 31

CONTACT DERMATITIS ASSOCIATED WITH EXPOSURE TO OILS AND COOLANTS

James R. Nethercott and D. Linn Holness

TABLE OF CONTENTS

I. INTRODUCTION

Soluble oils used in metal-working operations are a well-known cause of occupational skin disease.[1] Although often attributed to the irritant effect of wet work, contact allergy to components in the soluble oil may account for the development of the cutaneous disease.[2-12] The metal itself may be a sensitizer.[13,14]

Exposure to grease and oil may result in cutaneous disease. The association is such that the eponym "mechanic's hands" has been applied to the presence of chronic hand dermatitis in this trade where workers are exposed to grease, oil, and solvents.

The purpose of this communication is to review the authors' experience in the evaluation of 51 workers with dermatitis believed to be associated with exposure to oil and grease, and, specifically, to document the clinical findings and the patch test results in these individuals.

II. MATERIALS AND METHODS

Between August 1987 and August 1988, 354 patients were patch tested using the method recommended by the International Contact Dermatitis Research Group (ICDRG) for the scoring of the test results,[15] employing the chemicals on the North American Contact Dermatitis Group (NACDG) screening tray supplied by Hermal-Chemie, Reinbeck, West Germany. The chemicals were applied to Finn Chambers® (Epitest Ltd. Oy, Hyryla, Finland) which were then secured to the upper back with Scanpor tape (Norgeplaster, Aksjeselskap, Norway). The patch tests were applied and red on Mondays, Wednesdays, or Fridays. They were read on the first two subsequent days after the tests were applied. The chambers were removed at 48 or 72 h, and the initial scoring of the sites was carried out 30 to 60 min later. This was done to allow adequate time for the physical effects of the pressure of the appliance on the skin to resolve. One observer (J.R.N.) scored all of the patch test responses. Sites scored as 1+, 2+, or 3+ were taken as positive; doubtful responses were included with the negative responses. Irritant responses were scored as negative test results.

In addition to the screening tray, other chemicals were applied on the basis of the clinical and exposure history. A total of 51 individuals had exposure to oils and coolants and were further assessed by additional testing with 32 chemicals which are components in oils and coolants. These chemicals and their concentrations (hereafter referred to as the OC tray), which are listed in Table 1, were supplied by Chemotechnique Diagnostics AB, Malmö, Sweden.

The relevance of each response was assessed if the response was 1+, 2+, or 3+. In these cases the patients were categorized into three groups. The first group consisted of patients where the response could be related to the patient's presenting complaint; these were taken to have "present relevance". The second group consisted of patients who presented with a problem where the positive response could be related to a past dermatitis; these patients had a response that had "past relevance". The third group consisted of patients where no association could be found between the positive response and either the patient's presenting complaint or a past dermatitis; these patients are referred to as having patch-test responses of "unknown relevance".

A standardized data collection instrument was used to record the subject's age, sex, job category, site of cutaneous eruption, diagnosis, and whether the eruption was deemed to be work related.

Individuals were taken to have allergic contact dermatitis if the following criteria were met. They had to have eczematous dermatitis at a site of cutaneous contact with a specific substance or substances which, alone or in combination with ultraviolet (UV) light exposure, were associated with an exacerbation of the dermatitis in circumstances where it was estab-

TABLE 1
Common Oil and Cooling Additives and Their Uses

Abietic acid 10% — a component in tall oil used as a deodorizing agent in cooling fluid
4-Chloro-3-cresol (PCMC) 1% — a fungicide in cooling fluids
4-Chloro-3-xylenol (PCMX) 0.5% — a preservative in cooling fluids
Dichlorophene 0.5% — a bactericide, fungicide, and algicide used in cooling fluids
2-Phenylphenol (Dowicide® 1) 1% — used as a preservative in coolant fluids
Propylene glycol 5% — used in cooling fluids
Triethanolamine 2% — surface active agent added to cutting oils
4-*tert*-Butylbenzoic acid 1% — an antioxidant used in cooling fluids
1,2-Benzisothiazolin-3-one 0.05% — a water-soluble preservative in cooling fluids
Hexahydro-1,3,5-tris (2-hydroxyethyl)triazine (Grotan® BK) 1% — a bactericide used in cooling fluids; also a formaldehyde releaser
Bioban® P 1487 1% — a preservative used in cooling fluids, crude oil, diesel fuel, and heating oil; consists of 70% 4-(2-nitrobutyl)morpholine and 20% 4,4-2-ethyl-2-nitro-trimethylene)dimorpholine in the commercial product
Chloroacetamide (Hibitane®) 0.2% — used as a preservative in cooling fluids
N-Methylol chloroacetamide (Grotan® HD) 0.1% — a preservative used in cooling fluids
1H-Benzotriazol 1% — an anticorrosive agent in cooling fluids, fuels, and antifreeze
Ethylenediamine dihydrochloride 1% — used in soluble cutting fluids and is an inhibitor in antifreeze solutions
Mercaptobenzothiazole 1% — a fungicide and corrosion inhibitor used in soluble cutting oils and antifreeze mixtures
Zinc ethylenebis(dithiocarbamate) 1% — a fungicide for use in cooling fluids
Triclosan (Irgasan® DP 300) 2% — used as an antifungal agent in wet areas
1-Aza-3,7-dioxa-5-ethylbicyclo(3,3,0) octane (Bioban® CS-1246) 1% — used to preserve emulsions and cooling fluids
4,4-dimethyloxazolidine 3,4,4-trimethyloxazoline (Bioban® CS-1135) 1% — used to preserve emulsions and cooling fluids
2-Hydroxymethyl-2-nitro-1,3-propandiol (Tris nitro) 1% — a bactericide and slimicide used in cooling fluids
Thiomersal 0.1% — a preservative used in germicides
Hydrazine sulfate 1% — a pressure stabilizer added to some cutting oils
Trichlorocarbanilide 1% — an antimicrobial agent in soaps and other cleansing materials
Formaldehyde 1% — used as a disinfectant and preservative
Amerchol® L 101 100% — the commercial name for products containing lanolin alcohols obtained from hydrolysis of lanolin; used as emulsifiers and emolients
Dipentene 1% — used as a pressure stabilizer in oils
Sodium omadine 0.1% — used in cooling fluids
2-Bromo-2-nitropropane (Bronopol®) 0.25% — a preservative used in cooling fluids
Coconut diethanolamide 0.5% — a mixture of ethanolamides with coconut acid; a component of cooling fluids
5-Chloro-2-methyl-4-isothiazolon-3-one (Kathon® CG) 0.02% — a preservative used in cooling fluids
1,2-Dibromo-2,4-dysianobutane (Tektamer® 38) 0.1% — used to preserve metal-working fluids

lished that the individual would have had contact with the chemical or chemicals and wherein the dermatitic process partially or totally resolved with avoidance of such exposure. The chemical must have been established to have played a causal role on one of two grounds: either there was a temporal association between exposure and recurrence of the eczematous process at 24 to 72 h in a usage situation wherein other exposed individuals did not experience contact dermatitis, or the substance was shown to produce a dermatitic response at the site of exposure under a closed patch test using the method outlined above, with or without irradiation with UV light, at a concentration established not to cause a dermatitic response in a panel of control subjects.

Irritant contact dermatitis was diagnosed if a specific contactant could not be identified by epicutaneous testing, but wherein history was consistent in that there was an appropriate temporal relationship between exposure or contact to a known irritant or irritants and the expression of the dermatitis. Atopic dermatitis was defined as a chronic eczematous eruption, usually in the flexural distribution, with or without a history of infantile eczema or a personal or family history of hay fever or asthma.

A problem was deemed to be work related if there was a temporal relationship between the workplace exposure and the development or exacerbation of the eruption and possible

TABLE 2
Characteristics of Those Tested with Oil and Cooling (O & C) Tray vs. Others

	O & C tray (N = 51)	Other (N = 303)	p value[a]
% Male	86	49	0.0001
Age	39.9 ± 14.3	38.9 ± 13.4	0.6023
Location of eruption (%)			
General	2	8	0.1714
Hands	92	62	0.0002
Face	12	19	0.1939
Feet	12	12	0.6279
Eyelids	2	12	0.0674
Arms	27	19	0.1570
Legs	4	8	0.3216
Trunk	8	4	0.2797
Past history (%)			
Allergies	23	28	0.2292
Hayfever	18	25	0.3619
Asthma	10	9	0.8419
Eczema	13	19	0.3481
Family history (%)			
Allergies	8	19	0.1276
Hayfever	14	29	0.4498
Asthma	5	16	0.0641
Eczema	3	18	0.0429
Diagnosis (%)			
Allergic contact dermatitis	45	42	0.7031
Irritant contact dermatitis	49	28	0.0029
Atopic dermatitis	12	14	0.7311
Work-related dermatitis	96	55	0.0001

[a] p value calculated by logistic regression analysis or unpaired t-test.

irritant(s) or if a specific allergen(s) could be identified in the workplace setting to account for the cutaneous response. The occupations of the 51 workers exposed to oils and coolants were categorized into five groups — namely, machinists, mechanics (both automotive and machine), workers handling metal parts coated with oils (e.g., assemblers, stockroom workers), welders/fitters, and a miscellaneous group.

Statistical comparison of the differences between proportions was carried out using logistic regression analysis. Comparisons between continuous variables were made by using the unpaired t-test. A probability (p) of <0.05 was taken to be statistically significant.

III. RESULTS

The characteristics of those 51 who were tested with both the screening tray and the OC tray compared to those 303 who were tested with the screening tray but not the OC tray are listed in Table 2. They were of similar age, but significantly more males than females (86 vs. 49%, $p = 0.0001$) tested with the OC tray. Males also presented more commonly with hand dermatitis (92 vs. 62 %, $p = 0.0002$). A similar proportion had a diagnosis of allergic contact dermatitis and atopic dermatitis, but significantly more had a diagnosis of irritant contact dermatitis (49 vs. 28%, $p = 0.0029$). Of the 51 workers exposed to oil and coolants, 19 were machinists, 9 were mechanics, 14 worked handling metal parts, 4 were welders/fitters, and 5 had a variety of other jobs. More of the individuals tested with the OC tray had a work-related problem (96 vs. 55%, $p = 0.0001$).

TABLE 3
Percent Positive Responses to Substances on the NACDG Screening Tray

	O & C Tray[a] (n = 51)	Other (n = 303)
Benzocaine 5%	2	0
Mercaptobenzothiazole 1%	10	1*
Colophony 20%	7	3
Diazolidinyl urea 1% pet	0	0
Diazolidinyl urea 1% aq	5	2
Imidiazolidinyl urea 2% pet	2	1
Imidiazolidinyl urea 2% aq	0	1
p-Phenylenediamine 1%	7	8
Cinnamic aldehyde 1%	2	5
Cinnamic alcohol 5%	0	7
Benzoyl peroxide 1%	0	1
Wool alcohols 30%	2	1
Carba mix 3%	2	2
Neomycin sulfate 20%	2	2
Thiuram mix 1%	2	6
Formaldehyde 2% aq	3	5
Ethylenediamine dihydrochloride 1%	2	3
Epoxy resin 1%	0	7
Quaternium-15 2%	5	5
p-tert-Butylphenol formaldehyde resin 1%	2	1
Mercapto mix 1%	7	1**
Black rubber mix	2	1
Thimerosal 0.1%	5	3
Potassium dichromate 0.25%	5	5
Balsam of Peru 25%	5	3
Diaminophenyl methane 0.5%	2	1
Nickel sulfate 2.5%	12	12

Note: * signifies $p = 0.0033$ by logistic regression, ** signifies $p = 0.0098$ by logistic regression; pet = petrolatum, aq = aqueous.

[a] Oil and cooling tray.

A proportion of each group exhibiting positive responses to the chemicals on the NACDG screening tray is outlined in Table 3. Significantly more of the 51 workers tested with the OC tray had positive responses to mercaptobenzothiazole and mercapto mix. The proportion of the patients with positive responses to the various chemicals on the OC tray are listed in Table 4; 37% of the workers responded to at least one chemical on the OC tray. There were positive responses to 16 of the chemicals, the most common responses being to Bioban® P1487, formaldehyde, and abietic acid.

Of the 51 workers, 18 had a diagnosis of allergic contact dermatitis, 23 had a diagnosis of irritant contact dermatitis, and 7 had both. The distribution of eruption in those with allergic contact dermatitis and irritant contact dermatitis is presented in Table 5. Of those with a diagnosis of allergic contact dermatitis, 32% were related to oil and cooling components, 50% to rubber components, and 27% to metals.

The differences in the presentation between the three main work categories are presented in Table 6. Although it is difficult to make statistical comparisons because of the small sample size, it appears that allergic contact dermatitis is overrepresented in the workers handling metal parts, while irritant contact dermatitis is more common in machinists. The duration of the eruption prior to presentation was on average less than 1 year in mechanics, but greater than 5 years in workers handling metal parts.

TABLE 4
Percent Positive Responses to Chemicals on the Oil and Cooling Tray and Relevance of the Responses (N = 51)

		Relevance		
	% Positive	Present	Past	Unknown
Abietic acid 10%	6	33	33	33
4-Chloro-3-cresol 1%	2	100	0	0
4-Chloro-3-xylenol 0.5%	0			
Dichlorophene 0.5%	0			
2-Phenylphenol 1%	2	0	0	100
Propylene glycol 5% aq	0			
Triethanolamine 2%	0			
4-*tert*-Butylbenzoic acid 1%	0			
1,2-Benzisothiazolin-3-one 0.05% alc	0			
Hexahydro-1,3,5-tris(2-hydroxyethyl)triazine 1% aq	4	50	0	50
Bioban® P 1487 1%	12	83	0	17
Chloroacetamide 0.2%	0			
N-Methylol chloroacetamide 0.1%	2	100	0	0
1H-Benzotriazol 1%	0			
Ethylenediamine dihydrochloride 1%	4	0	0	100
Mercaptobenzothiazole 1%	4	100	0	0
Zinc ethylenebis(dithiocarbamate) 1%	0			
Triclosan (Irgasan® DP 300) 2%	0			
Bioban® CS-1246 1%	4	100	0	0
Bioban® CS-1135 1%	4	50	50	0
Tris nitro 1%	0			
Thiomersal 0.1%	2	0	0	100
Hydrazine sulfate 1%	0			
Trichlorocarbanilide 1%	0			
Formaldehyde 1% aq	8	50	50	0
Amerchol® L 101 100%	4	100	0	0
Dipentene 1%	0			
Sodium omadine 0.1% aq	0			
Bronopol® 0.25%	2	100	0	0
Coconut diethanolamide 0.5%	0			
Kathon®CG 0.02% aq	2	0	0	100
Tektamer® 38 0.1%	2	100	0	0

Note: All substances are in petrolatum except those noted "aq" (which are aqueous) and "alc" (which are in alcohol).

IV. COMMENTS

Men outnumber women in occupations where oil and grease exposure might be implicated in the genesis of cutaneous disease, as the male preponderance of subjects in our patients confirms. The age of onset was the same for those with suspect dermatitis from oils and coolants and all other contactants. Hand dermatitis was more common in the oil-exposed workers (92 vs 62%). There was a trend toward less likelihood of these workers having eyelid dermatitis than in the 303 subjects with other possible sources of contact allergy. Though the oil- and coolant-exposed workers were no more likely to report a personal history of atopic disease, they were less likely to report a family history of asthma or eczema than the overall group of patients we evaluated for possible contact dermatitis. The significance of this observation is unclear.

There was not an overrepresentation of patients with atopic dermatitis in the oil-exposed workers in comparison with the other patients, although in both populations the prevalence

TABLE 5
Comparison of Those Tested with the Oil and Cooling Tray with Irritant and Allergic Contact Dermatitis

	Allergic (n = 18)	Irritant (n = 23)	p value[a]
Location of eruption (%)			
Hands	78	91	0.2386
Dorsa	36	58	0.2119
Palms	36	37	0.9469
Sides of fingers	14	42	0.0994
Face	33	4	0.0351
Feet	17	4	0.2178
Arms	56	35	0.1870
Legs	11	4	0.4250
Trunk	28	4	0.0633
Duration of eruption (months)	34 ± 59	25 ± 29	0.6152
Atopic dermatitis (%)	22	4	0.1159

[a] p value calculated by logistic regression analysis or unpaired t-rest.

TABLE 6
Comparison of Different Job Categories

	Machinist (n = 19)	Mechanic (n = 9)	Metal worker (n = 14)
% Male	95	100	57
Age	39.2 ± 14.6	41.6 ± 15.9	39.4 ± 13.7
Location of eruption (%)			
Hands	95	100	93
Face	5	11	14
Feet	16	11	7
Arms	32	0	21
Legs	0	11	0
Trunk	0	0	21
Duration of eruption (months)	28 ± 27	9 ± 5	61 ± 73
Diagnosis (%)	32	44	64
Allergic contact dermatitis			
% due to oil and cooling	20	25	28
% due to rubber	80	75	25
% due to metals	20	25	50
Irritant contact dermatitis	68	44	29
Atopic dermatitis	11	22	7
Work-related dermatitis	100	100	93
% Positive responses to selected chemicals			
Mercaptobenzothiazole	6	0	10
p-Phenylenediamine	12	0	10
Potassium dichromate	6	0	10
Nickel sulfate	6	11	20

of atopic dermatitis is high compared with the 3% prevalence reported for the general population. Atopic dermatitis is likely an important risk factor for the genesis of contact dermatitis from a variety of causes.[16]

The oil-exposed workers were as likely to have evidence of allergic contact dermatitis as our other patients, in contrast to the widely held view that the skin disease associated with such exposure is more commonly irritant contact dermatitis. On the other hand, irritant contact dermatitis was more commonly diagnosed in these individuals than in our suspect contact dermatitis cases generally, likely because they were working with known irritants, and for some cases both an irritant and allergic basis for their skin disease was believed to be present. Not surprisingly, nearly all of the oil-exposed patients had a work-associated dermatitis, as oil and coolant exposure is not that likely outside the workplace environment. It is notable that irritant contact dermatitis was more common in machinists who work with soluble oils where exposure to water, swarf, and microbial breakdown products is likely to be higher.

A specific contactant was often found to establish an allergic basis for the disorder; 45% of the oil-exposed workers were deemed to have allergic contact dermatitis. However, only the rubber chemicals in the mercaptobenzothiazole group stood out as more frequent causes of such reactions in the group of chemicals used for routine screening with the NACDG tray (Table 3).

In keeping with Alomar et al.[2] we were able to establish specific contactants, not infrequently using an aimed tray which included commonly used oil and coolant additives (Table 4). Table 1 outlines some of the possible sources of exposure to these additives. A few additives accounted for many of the responses on the aimed tray. Most of these positive responses were deemed relevant; a few were unexplainable in terms of the workers' known exposure. It should be noted, though, that there were no positive responses elicited to 16 of the substances on the OC tray in our 51 subjects. It is commonly held that only substances producing a response in 1% of tests should be considered appropriate for screening investigation. It may be that some or all of these 16 substances should be deleted from the tray, but a final answer cannot be determined until a larger sample is tested to ascertain if, in fact, there is a 1% response rate. Furthermore, experience has shown that it may be necessary to test with a specific component in a given oil or coolant even though a specific chemical found in such a product may not be a known contact allergen.[9] In such instances, testing with the specific chemical using appropriate control subjects is indicated.

There was a notable difference between the job categories in terms of positive responses to nickel in that 20% of the metal handlers exhibited a positive response to nickel, in contrast to 11% of the mechanics and 6% of the machinists. As the proportion of females in this group and the overall group was similar and only 12% of the overall group had positive responses to nickel, it suggests that the higher proportion of subjects responding to nickel in the metal handlers may relate to their environmental exposure.

This investigation establishes that patch testing is useful in the evaluation of individuals exposed to cutting oils, grease, or oil products where contact dermatitis is suspected. Many cases were established using the aimed tray; 45% of our subjects proved to have allergic contact dermatitis. Although irritation often plays a significant role in the genesis of skin disease in such exposed individuals, the clinician should not overlook the possibility that a specific contactant may be important, as our findings attest. Epicutaneous testing has proved to be a useful tool in our hands.

REFERENCES

1. **Adams, R. M.,** *Occupational Skin Disease,* Grune & Stratton, New York, 1983.
2. **Alomar, A., Conde-Salazar, L., and Romaguera, C.,** Occupational dermatoses from cutting oils, *Contact Dermatitis,* 12, 129, 1985.
3. **Wrangsjo, K., Martensson, A., Widstrom, L., and Sundberg, K.,** Contact dermatitis from Bioban P 1487, *Contact Dermatitis,* 14, 182, 1986.
4. **Romaguera, C., Grimalt, F., and Lecha, M.,** Photoallergic dermatitis from ethylenediamine, *Contact Dermatitis,* 14, 130, 1986.

5. **Fregert, S.,** Colophony in cutting oil and in soap water used as cutting fluid, *Contact Dermatitis,* 5, 52, 1979.
6. **Brown, R.,** Concomitant sensitisation to additives in a coolant fluid, *Contact Dermatitis,* 5, 340, 1979.
7. **Adams, R. M.,** Allergic contact dermatitis due to *o*-phenylphenol, *Contact Dermatitis,* 7, 332, 1981.
8. **Adams, R. M.,** *p*-Chloro-*m*-xylenol in cutting fluids: two cases of allergic contact dermatitis in machinists, *Contact Dermatitis,* 7, 341, 1981.
9. **Pilger, C., Nethercott, J. R., and Weksberg, F.,** Allergic contact dermatitis due to a biocide containing 5-chloro-2-methyl-4-isothiazolin-3-one, *Contact Dermatitis,* 14, 201, 1986.
10. **Lama, L., Vanni, D., Barone, M., Patrone, P., and Antonelli, C.,** Occupational dermatitis to chloroacetamide, *Contact Dermatitis,* 15, 243, 1986.
11. **English, J. S. C., Foulds, I., White, I. R., and Rycroft, R. J. G.,** Allergic contact sensitization to the glycidyl ester of hexahydrophthalic acid in a cutting oil, *Contact Dermatitis,* 15, 66, 1986.
12. **Keczkes, K. and Brown, P. M.,** Hexahydro, 1,3,5, tris (2-hydroxyethyl) triazine, a new bacteriological agent as a cause of allergic contact dermatitis, *Contact Dermatitis,* 2, 92, 1976.
13. **Einarsson, O., Eriksson, E., Lindstedt, G., and Wahlberg, J. E.,** Dissolution of cobalt from hard metal alloys by cutting fluids, *Contact Dermatitis,* 5, 129, 1979.
14. **Samitz, M. H. and Katz, S. A.,** Skin hazards from nickel and chromium salts in association with cutting oil operations, *Contact Dermatitis,* 1, 158, 1975.
15. **Hannuksela, M.,** Epicutaneous testing, *Allergy,* 34, 5, 1976.
16. **Rydstedt, I.,** Hand eczema in patients with a history of atopic manifestations in childhood, *Acta Derm. Venereol.,* 65, 305, 1985.
17. **Cronin, E.,** *Contact Dermatitis,* Churchill Livingstone, Edinburgh, 1980.

Chapter 32

IRRITANT DERMATITIS: DIAGNOSIS AND TREATMENT

Robert L. Rietschel

TABLE OF CONTENTS

I. INTRODUCTION

The diagnosis of irritant contact dermatitis to strong skin irritants is not difficult because of the rapid onset of burning, stinging, redness, blisters, or other skin changes that alert the patient to the cause. Acute treatment for chemical burns of this type is adequately covered in emergency medical care textbooks. The type of irritant contact dermatitis that requires skill and training for proper diagnosis is the subacute, chronic, or cumulative irritant contact dermatitis that is manifested as an eczematous process without a readily apparent cause.

Features of chemicals that enhance irritancy and features of individuals which increase the proclivity to irritant dermatitis are covered elsewhere in this book (see Chapters 5 and 19). The clinician examining a patient must go through a series of decisions to arrive at a proper diagnosis when confronted with an eczematous dermatitis. The first step is to look at the location of skin inflammation, couple that with morphological nuances, and determine whether a probable diagnosis can be reached. Often features overlap, resulting in several considerations. Table 1 gives a conceptual framework with which to begin distinguishing endogenous from exogenous eczema.

If the physical findings are sufficiently characteristic, an endogenous dermatitis will be diagnosed and either supported or complicated by additional history. Irritant dermatitis may be superimposed on all of these conditions by attempts at therapy. This is most commonly encountered when compresses (wet to dry dressings) are used too long or when a weeping dermatitis becomes overly dry and manifests features of asteotosis. Atopic dermatitis and stasis dermatitis are particularly at risk for this type of medical mismanagement and superimposed irritant dermatitis.

When the physical findings do not summon to mind a specific dermatologic diagnosis, external factors are sought to explain the process. Signs of linear streaks of dermatitis or other geometric patterns with sharp angulation strongly suggest that external factors are responsible for the problem. These factors could be allergic or irritant in nature. The bulk of exogenous cases are eventually diagnosed as irritant, based on several studies. A cohort of 2800 people in the Netherlands was studied in 1973 and 1982 and was found to have a prevalence of hand or forearm dermatitis of 5.8 and 7.1%, respectively.[1] Allergic dermatitis was found in 30% of the patients, and 73% were found to have irritant dermatitis (some had both — thus, 103%). When hospital personnel were surveyed for hand dermatitis, 10 to 53% were found to be affected; of those, 59 to 92% of the cases were irritant, while only 8 to 21% of cases were allergic.[2]

II. DIAGNOSIS

Neither irritant nor allergic contact dermatitis is restricted to a specific body region, and either may be localized or generalized depending on the nature of the exposure. A proper history and thorough knowledge of the substances in the environment capable of producing dermatitis facilitate making an accurate distinction between these two sources of exogenous dermatitis. The patient's occupation, hobbies, newly acquired possessions, medication taken and applied (both prescription and over-the-counter type), and at times the products used by other household members are important to determine the substances to which the individual is exposed. Patch testing should then be performed with the environmentally relevant materials to establish whether allergic contact dermatitis is present and, if it is, the substances causing the condition. The most reliable materials for screening for environmental allergens are the standard screening series of contact allergens recommended by the International and North American Contact Dermatitis Groups.

A positive patch test does not identify the cause of dermatitis actively under evaluation as definitely allergic any more than a negative test establishes that the cause is definitely

TABLE 1
Conceptual Framework for Distinguishing Endogenous from Exogenous Eczema

Areas of skin inflammation	Morphological nuances	Probable diagnosis
Palms and/or soles without dorsal surface or other areas of involvement	Small (2—3 mm) vesicles to bulla with little erythema or pustule formation early in the process	Dyshidrosis or dermatophytid reaction
Only popliteal/antecubital fossae, axilla, neck involved	Lichenified, dry skin rather than vesicles with/without hyper- or hypopigmentation	Atopic dermatitis
Scalp, elbows, knees, palms, soles, possibly gluteal crease, but not the face	Sharp borders to areas at papulosquamous change; no vesicles; pustules may be present	Psoriasis
Dorsal hands, feet, extensor surfaces of arms and/or legs	Vesicles, oozing, round, coin-shaped patches of dermatitis with fairly sharply defined borders and multiple lesions present	Nummular eczema
Scalp, malar, and nasolabial regions of the face, central chest, ears, eyebrows	Papulosquamous rather than vesicular, with some areas not sharply circumscribed	Seborrheic dermatitis
Lower legs	Round to irregular patches with a cracked pavement look and dry skin in between; no vesicles, edema, or pigment disturbance	Asteototic dermatitis
Area surrounding a wound or draining site, with lesser amounts of eczema in scattered patches elsewhere on the body	Purulence overt at site of largest area of dermatitis	Infectious eczematoid dermatitis
Large patch of dermatitis present for a prolonged time, accompanied by more recent, widespread, less well-defined patches of dermatitis	Poor circumscription of most lesions, generally small vesicles rather than bulla or papulosquamous changes	Autoeczematization reaction
Lower legs, especially medial surfaces	Pitting edema, varicosities, hyperpigmentation (sometimes hypopigmentation), possibly ulceration, vesicles, and/or papulosquamous change	Stasis dermatitis

irritant. To support the diagnosis of allergic contact dermatitis, a positive patch test accompanied by a history of exposure to substances containing the putative allergen must have occurred in a time frame and manner sufficient to explain the clinical eruption. Alternatively, the positive patch test could be meaningful to past medical events, but irrelevant to the current one, if there is no recent exposure to the positive test substance, yet a clear history of exposure months or years previous.

When patch tests are negative and the signs and symptoms are not compatible with the diagnoses considered in Table 1, the likelihood of irritant dermatitis increases, but more precise subjective and objective criteria are useful in making a sound diagnosis. Subjective major and minor criteria as well as objective major and minor criteria to help make the diagnosis of irritant contact dermatitis[3] are proposed in Table 2. This pattern of clinical decision-making has been useful with other syndromes (such as lupus erythematosus) and helps formalize the thought processes commonly used by other workers in this field.[4]

The need to mesh morphological findings with patient history to render an accurate diagnosis is clearly apparent with a recent report on soluble oil-induced irritant contact dermatitis.[5] This epidemiologic survey found that 21 of 286 metal workers had the clinical features of dyshidrosis. Only three had a positive patch test (one to nickel and two to

TABLE 2
Criteria for Diagnosis of Irritant Contact Dermatitis

Subjective major criteria
1. The onset of dermatitis within minutes to hours of exposure to a substance
2. Pain, burning, stinging, or discomfort exceeding itching, especially early in the clinical course

Subjective minor criteria
1. Onset of dermatitis within 2 weeks of environmental exposure, often a difficult history to elicit except in special settings where the irritant is a relatively novel rather than ubiquitous substance
2. Many people in the environment similarly affected; based solely on history provided by the patient, validity is suspect unless verified by an examining physician

Objective major criteria
1. Macular erythema, hyperkeratosis, or fissuring predominating over vesicular change; vesicles may be present in irritant reactions, especially on palms, with strong irritants; when vesicles predominate, likelihood of allergy increases
2. Glazed, parched, or scalded appearance of the epidermis
3. Healing process proceeds without plateau upon withdrawal of exposure to substances in question; if not, strongly consider endogenous disease, such as atopic dermatitis or psoriasis
4. Patch testing with known environmentally relevant allergies negative

Objective minor criteria
1. Sharp circumscription of the dermatitis
2. Evidence of gravitational influence, such as a dripping effect
3. Lack of tendency for spread of dermatitis; can be properly evaluated only on sequential examinations
4. Vesicles juxtaposed closely to patches of erythema, erosions, bullae, or other morphological changes which suggest that small differences in concentrations or contact time produce large differences in skin damage

TABLE 3
Malten's Treatment of Irritant Contact Dermatitis

1. Decrease the influence of every deleterious factor.
2. Use real soap and, if water is hard, use a water softener such as Calgon®.
3. Use 9 g sodium chloride per liter for hand washing and rinsing.
4. Do not rub dry, but let skin stay a bit moist.
5. Apply the following ointment to moisten skin: 1% lactic acid, 10% glycerine, 10% water, 10% adeps lanae in petrolatum.

From Malten, K. E., *Contact Dermatitis,* 7, 238, 1981. With permission.

formaldyhyde), and only one had a history of findings suggestive of atopic dermatitis. The history and objective data supported the concept that an irritant dermatitis could produce this morphological pattern.[5] The thought processes used to support the diagnosis were similar to those found in Table 2. The more features that are identified, the stronger the diagnosis becomes.

III. TREATMENT

Few authors have attempted to address therapy for subacute to chronic irritant dermatitis as separate and distinct from other dermatitides. Malten[4] has done so (his treatment options are found in Table 3), and my therapeutic suggestions are found in Table 4.

When the treatment suggestions in Tables 3 and 4 are inadequate and creative therapy is called for, several observations can be considered. Prostaglandins have been recovered from human primary irritant dermatitis, suggesting that they may have a role in the pathophysiology.[6] Furthermore, some hand dermatitis is caused by contact urticaria, which may be a nonimmunologic event.[7] Nonimmunologic contact urticaria can be blocked by acetylsalicylic acid and indomethacin, but not by H_1 antagonists, which would again suggest a

TABLE 4
Rietschel's Treatment of Irritant Contact Dermatitis

1. Decrease contact with any known irritant.
2. Follow each hand washing with application of glycerin-formulated moisturizer (such as Neutrogena® Norwegian Formula Hand Cream; avoid lactic acid and urea).
3. Corticosteroids should be used as a true ointment base (petrolatum) usually only twice daily.
4. Consider using Cetaphil® lotion compounded with corticosteroid as a hand cleanser instead of soap.
5. Discourage the use of gloves and cotton liners except for very brief specific periods of unavoidable exposure to irritants, and then use only polyvinyl chloride gloves without liners.

role for prostoglandin as a mediator.[8] Therefore, it would not be unreasonable to consider antiprostaglandin-directed systemic therapy as an adjunct in the management of irritant contact dermatitis.

REFERENCES

1. **Lantinga, H., Nater, J. P., and Coenraads, P. J.,** Prevalance, incidence and course of eczema on the hands and forearms in a sample of the general population, *Contact Dermatitis,* 10, 135, 1984.
2. **Singgih, S. I. R., Lantinga, H., Nater, J. P., Woest, T. E., and Krauyt-Gasperze, J. A.,** Occupational hand dermatoses in hospital cleaning personnel, *Contact Dermatitis,* 14, 14, 1986.
3. **Rietschel, R. L.,** Diagnosing irritant contact dermatitis, in *Irritant Contact Dermatitis,* Jackson, E. M. and Goldner, R. M., Eds., Marcel Dekker, New York, 1990, 167.
4. **Malten, K. E.,** Thoughts on irritant contact dermatitis, *Contact Dermatitis,* 7, 238, 1981.
5. **DeBoer, E. M., Bruynzeel, D. P., and VanKetel, W. G.,** Dyshidrotic eczema as an occupational dermatitis in metal workers, *Contact Dermatitis,* 19, 184, 1988.
6. **Songergaard, J., Greaves, M. W., and Jorgensen, H. P.,** Recovery of prostaglandins in human primary irritant dermatitis, *Arch. Dermatol.,* 110, 556, 1974.
7. **Lahti, A., Von Krogh, G., and Maibach, H. I.,** Contact urticaria syndrome. An expanding phenomena, in *Dermatologic Immunology and Allergy,* Stone, J., Ed., C. V. Mosby, St. Louis, 1984, 379.
8. **Lahti, A.,** Terfenadine does not inhibit non-immunologic contact urticaria, *Contact Dermatitis,* 16, 220, 1987.

Chapter 33

MINERS' GRANULOMA

P. W. Bowers

TABLE OF CONTENTS

I. INTRODUCTION

The western part of the county of Cornwall, England is probably the most highly mineralized zone in the Old World. Tin mining in this area dates back at least 2000 years, for primitive smelters and tin slag from that time have been discovered. By the middle of the 19th century, Cornwall was producing at least one half of the world's tin; however, as less costly alluvial mining in Indonesia, Malaysia, and Bolivia developed, the expensive deep rock mines of Cornwall closed. There are now only three left: South Crofty, Wheal Jane, and Geevor.

The tin ore called cassiterite is found in veins or lodes formed at high temperatures within the granite of the earth's crust. Compressional forces threw the original layers into great folds so that the lodes may end up almost vertical, rather than horizontal. They are worked to depths of down to nearly 800 m and, at Geevor mine, out under the Atlantic sea bed. Usually the vertical shafts are sunk to one side of the lode, and horizontal tunnels are driven from the shaft to the lode. From these, more tunnels are driven along the ore body by drilling upward into the lode. The drill holes are filled with explosive, and after blasting the rock falls downward, to be transported away by electric locomotives.

It is extremely wet everywhere, mainly from groundwater, but also from water sprays used to damp dust and cool the drills. The water is collected in channels and eventually pumped to the surface. The humidity is extremely high, although with modern ventilation this is improving. The temperature of the water varies from 26 to 44°C. Because of this, the men work stripped to the waist, wearing boots, shorts, and hard hats only.

II. MINERS' GRANULOMA

In 1961 my predecessor, Dr. Mark Hewitt, first saw a case of what he called "tuberculosis verrucosa cutis" on the hand of an Italian-born miner from South Crofty mine. He very reasonably thought that the lesion could not have been acquired in the mine, although the patient was sure it was. Shortly afterward this man's workmate appeared in the clinic, threatening to sue the company, as he was so certain the mine was to blame. Over the subsequent 25 years we have identified a total of 23 cases of this condition — 20 from South Crofty mine, 2 from Wheal Jane mine, and 1 from Geevor.

The clinical features are somewhat variable, but are essentially the same as those seen in "swimming pool" or "fish tank" granulomas. The lesions may be small or multiple (Figure 1), red or violaceous in color, apple-jelly smooth, warty or hyperkeratotic, or scaly. They may discharge pus; they may regress spontaneously or even grow slowly over a period of 26 years in spite of extensive treatment.

The sites of infection were:

Dorsum of hand or wrist	13 cases
Elbow	4 cases
Forearm	2 cases
Thigh or knee	2 cases
Forehead	1 case
Two or more sites	4 cases
Not recorded	3 cases

The duration of infection varied from 3 months to 26 years. The incubation period was more difficult to estimate, as not all lesions could be related to specific inoculating traumatic episodes. However, in one case the patient had an accident in September 1962, cutting the skin of his hand, and back and typical lesions had appeared within 3 months.

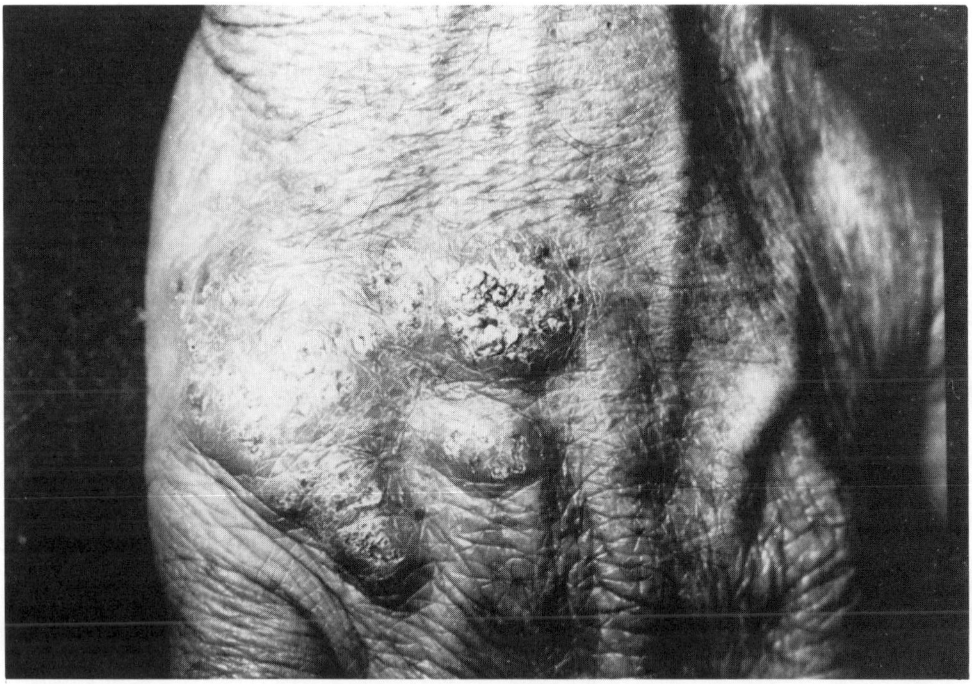

FIGURE 1. Typical *Mycobacterium marinum* granuloma.

III. HISTOLOGY

The histology varied with any or all of the following:

1. Lymphoid aggregates with diffuse inflammation (Figure 2)
2. Pseudo epitheliomatous hyperplasia (Figure 3)
3. Giant cells with organized tubercles or even caseation (Figure 4)
4. Acid-fast bacilli (AFB)

IV. BACTERIOLOGY

The bacteriology proved interesting, with *Mycobacterium marinum* isolated from six patients. In the early days it was not appreciated that *M. marinum* might be the cause, so the lesions were not cultured at the appropriate temperature for *M. marinum*.

M. marinum is a member of the slow-growing group of mycobacteria and forms a yellow pigment (photochromogenosis) upon exposure to light. On primary culture *M. marinum* will grow at 30 to 33°C in 7 to 10 d, but usually not at 37°C. This prediliction for cooler culture is termed ''psychrophilia''. Most of our isolates tended to grow in a higher temperature range than normal, but were differentiated from *M. kansasii* using other criteria.

V. RESPONSE TO TREATMENT

Eight patients are recorded as being treated with streptomycin and isoniazid; seven were cured. One (from whom *M. marinum* was isolated) suffered a severe toxic demyelinating reaction to streptomycin, and the lesions slowly progressed over 26 years until his death from unrelated causes. Three patients resolved spontaneously. Four patients responded rapidly to trimoxazole or minocycline. In eight patients the treatment records have been lost.

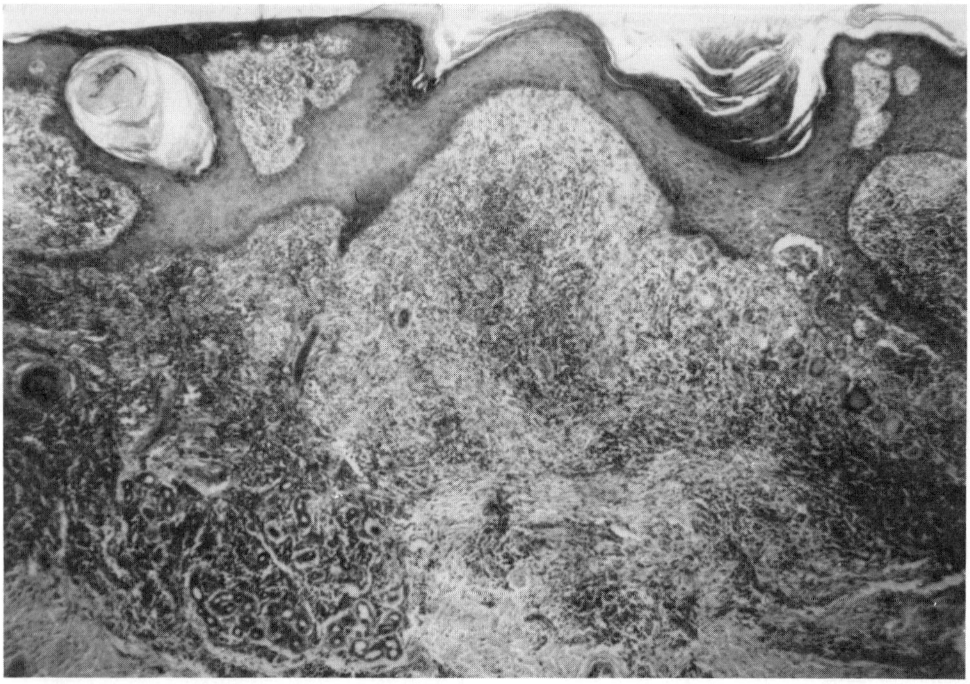

FIGURE 2. Histology of granuloma showing lymphoid aggregates and diffuse inflammation.

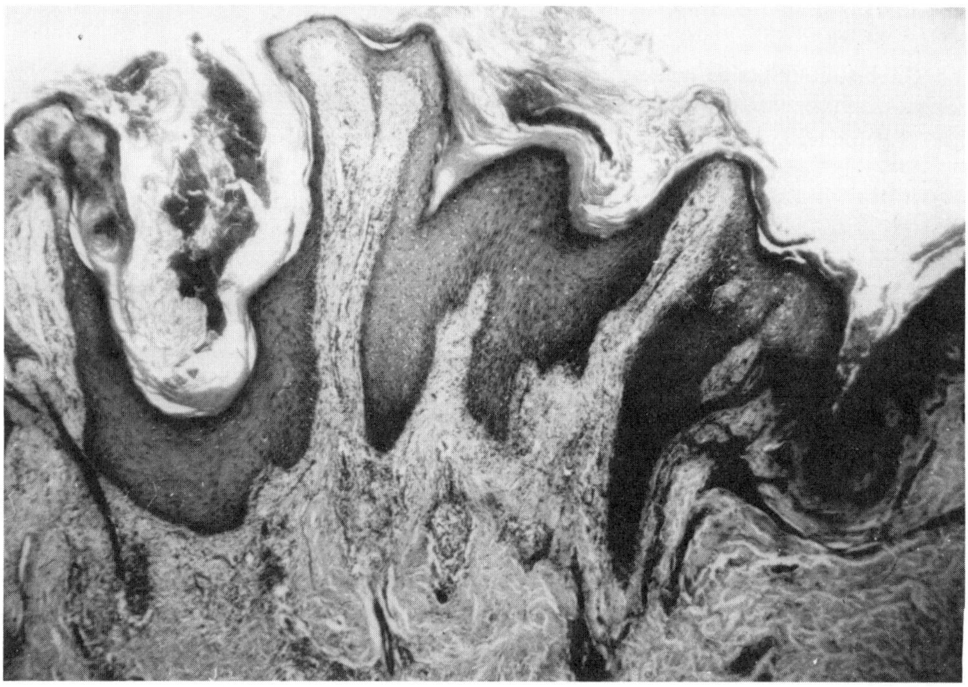

FIGURE 3. Histology of granuloma showing pseudo-epitheliomatous hyperplasia.

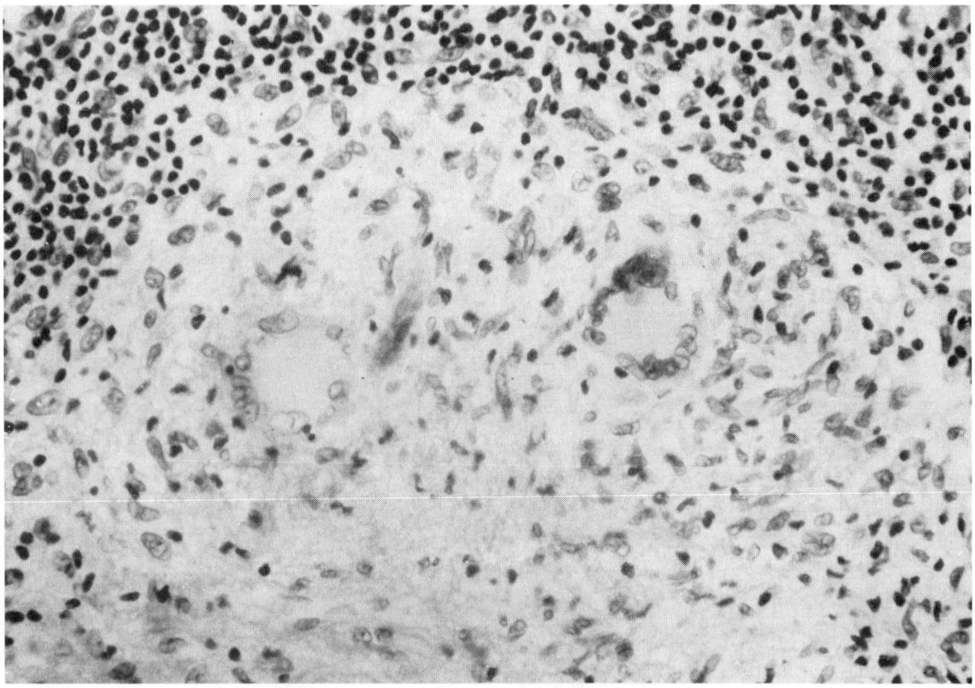

FIGURE 4. Histology of granuloma showing classic giant cell formation.

VI. FURTHER INVESTIGATION

Geevor mine is over 25 miles from South Crofty and Wheal Jane and has no common water source. Indeed, much of the mine extends under the Atlantic Ocean. A patient who worked at Geevor was certain that he had acquired his lesions at a particular spot on the ''Coronation'' lode, following a rock fall 2 years before he presented. The lode is approximately 500 m below the surface and is also below the sea. The ambient temperature is approximately 30°C. Of 12 samples of surface swabs and of water dripping from the rock at this spot, 10 grew *M. marinum* on culture; this was confirmed by the U.K. Public Health Laboratory Service Mycobacterium Reference Unit in Cardiff.

Visits were then made to Wheal Jane. AFB were isolated easily, but these all proved to be of the *M. fortuitum* and *M. avium-intracellulare* group. At South Crofty, initial reports suggested *M. marinum*, but cultures were contaminated and the final isolates were *M. fortuitum, M. avium-intracellulare*, and *M. gordonae*.

VII. DISCUSSION

In 1886 a form of skin tuberculosis was described by Riehl and Paltauf.[1] It usually occurred on the hands and arms and remained localized with a self-limiting course. Similar lesions were reported by Hellerström[2,3] in 1939 and 1951. All six patients had received facial abrasions, and although Koch's bacillus was seen in one case, guinea pig inoculations failed. One of the patients had received his injury in a swimming pool. Most subsequent reports describe infections acquired in association with warm water — through such activities as swimming, fishing, boating, and keeping tropical fish.

In 1954 Linell and Norden[4] described an outbreak affecting 80 people, mainly children, most of whom were keen swimmers. Because it was thought that *M. ulcerans* might have been responsible, the cultures were fortunately incubated at 31°C as well as at 37°C. This lower-temperature bacillus also differed from the classic tuberculous bacillus in producing a yellow pigment on exposure to light. It was given the new name *M. balnei*, from the Latin for ''bath''. This was subsequently changed to *M. marinum*. The largest series has been that of Philpott et al.,[5] who in 1963 presented a series of 290 cases infected from a swimming pool, but they were unable to determine how the pool had become infected. A later report by Swift and Cohen[6] first described a similar ''fish tank'' granuloma in 1962, and since then there have been many others around the world.

Apart from tropical fish keepers, reports of the disease acquired occupationally have been few. In 1970 Klokke[7] described a woman who developed a granuloma on her foot after an injury near a power station cooling outlet. Serological investigation of 108 power workers suggested that 40% had had subclinical infections. Occasional reports of injured fishermen, and even a bite from a captive dolphin,[8] almost all refer to solitary cases.

Other than the series of cases in tin miners discussed in this article, this author can only trace one paper on an outbreak in miners. In 1973 Horacek and Ulicna[9] wrote about 16 coal miners in Czechoslovakia who developed typical lesions. They were able to isolate *M. balnei (marinum)* from one patient and also from 12 samples of mine water which had the relatively high temperature of 25 to 30°C.

This is obviously very similar to the Cornish tin miners' outbreak. The men work under similar hot conditions extracting coal and ore, with many opportunities for minor abrasions. It is perhaps surprising that more granulomas do not occur.

Extensive inquiries have failed to uncover similar reports of these lesions among coal miners in the U.K., although many work in very similar damp, hot surroundings.[10]

In the past, true tuberculosis verrucosa cutis was common in coal miners who frequently got slight abrasions on their hands and inoculated them from their own nose and mouth.[11] Until recently, tuberculosis was also very common in Cornish tin miners, who rarely survived beyond the age of 40 years.

So the question remains, how did *M. marinum* get into these mines? Most mycobacteria are described as ''environmental'' and are commonly found in soil and water; this may explain the presence of *M. avium, M. fortuitum,* and *M. gordonae*. Of these, only *M. gordonae* has been reported to cause cutaneous infections of this type.[12] A recent report[13] of samples taken from four drill holes in the U.S. Department of Energy's Savannah River Laboratory claims that more than 3000 different species of bacteria have been recovered from depths as great as 1000 ft below the surface, many so far unclassified. Fliermans[13] attempted to answer the question of how the bacteria got to these depths. Two possibilities were considered — that they have come from groundwater sources elsewhere and have slowly evolved into new species, or that they were trapped as new sediments were laid down over 70 million years ago. Even at such depths there are traces of organic matter, some of it ancient, which the bacteria can digest. The presence of man in the mines adds to the organic waste, allowing the bacteria to thrive, but whether the miners are also the source of the *M. marinum*, as well as the victims, is unknown. However, the fact that the Cornish isolates are of an unusual psychrophilic pattern perhaps suggests that they have been evolving in the warmth of the wet, hot rock for a long period of time.

ACKNOWLEDGMENTS

I sincerely thank Dr. Mark Hewitt of Truro, who began this investigation; Dr. A. Telfer-Brunton and Dr. P. A. Jenkins of the Public Health Laboratory Service, who performed the microbiological investigations; Dr. D. Gould, who kept encouraging me; and my secretary, Miss R. Bryant, for her patience.

REFERENCES

1. **Riehl, G. and Paltauf, R.,** Tuberculosis Verrucosa Cutis. Einer bisher noch nicht beschriebene Form von Hauttuberculose, *Vierteljahresschr. Dermatol. (Wien),* 13, 1886.
2. **Hellerström, S.,** Contribution a la connaissance de l'infection tuberculeuse de la peau et de la muqueuse, *Acta Derm.,* 20, 276, 1939.
3. **Hellerström, S.,** Collected cases of inoculation lupus vulgaris, *Acta Derm. Venereol.,* 31, 194, 1951.
4. **Linell, F. and Norden, A.,** *Mycobacterium balnei:* a new acid-fast bacillus occurring in swimming pools and capable of producing skin lesions in humans, *Acta Tuberc. Scand. Suppl.,* 33, 1, 1954.
5. **Philpott, J., Philpott, J. A., Woodburne, A. R., Philpott, O. S., Schaefer, W. B., and Mollohan, C. S.,** Swimming pool granuloma: a study of 290 cases, *Arch. Dermatol.,* 88, 158, 1963.
6. **Swift, S. and Cohen, H.,** Granulomas of the skin due to *Mycobacterium balnei* abrasions from a fish tank, *N. Eng. J. Med.,* 267, 1244, 1962.
7. **Klokke, A. H.,** Epidermiological analysis in a patient with skin granuloma caused by *Mycobacterium balnei, Dermatologica,* 140, 303, 1970.
8. **Flowers, D. J.,** Human infection due to *Mycobacterium marinum* after a dolphin bite, *J. Clin. Pathol.,* 23, 475, 1970.
9. **Horacek, J. and Ulicna, L.,** *Mycobacterium balnei* aus kohlengrubenwassern als ursache der Tuberculosis verrucosa cutis, *Hautarzt,* 24, 126, 1973.
10. **Archibald, R.,** Chief Medical Officer, National Coal Board, personal communication, 1988.
11. **Fabry, J.,** On the occurrence of tuberculosis verrucosa cutis in coal miners, quoted in Crocker, H. R., *Diseases of the Skin,* 3rd ed., H. K. Lewis, London, 1905, 741.
12. **Shelley, W. B. and Folkens, A. T.,** *Mycobacterium gordonae* infection of the skin, *Arch. Dermatol.,* 120, 1064, 1984.
13. **Fliermans, C.,** quoted in Anderson, A., "Life below the subsurface", *Times (London),* September 24, 1988.

Chapter 34

CONTACT DERMATOSES IN HEALTH CARE WORKERS

Daniel J. Hogan

TABLE OF CONTENTS

I. INTRODUCTION

The health service sector, one of the largest employers in the U.S., employs over 7 million American workers. Nearly 75% of health care workers are employed in health care facilities other than hospitals. Health care workers are employed by offices, clinics, nursing homes, hospices, home health care, emergency response teams, and research facilities. Health care workers have a wide variety of occupations. More than one half of hospital employees are not involved in direct patient care.[1] Hospitals employ dietary workers, clerical workers, cleaners, carpenters, and other support staff.

More workers develop occupational skin diseases in the health care sector than workers in any other employment category in the U.S. The incidence rate of occupational skin diseases in the health services sector is 29%, higher than the average for all industries combined.[2] Most of the relatively few studies of occupational skin diseases in health care workers are based on hospital workers. There is a need for further studies of occupational skin diseases in health care workers, especially those outside the hospital.

II. OCCUPATIONAL HEALTH PROGRAMS FOR HEALTH CARE WORKERS

The first priority of health care workers is the care of their patients. It is often assumed that health care workers work in a clean environment and enjoy preferential access to the best in health care. The reality in North America is different. Most hospitals do not have effective employee health programs.[1] Health care workers employed outside hospitals have even less access to adequate employee health services. Health care workers, particularly nurses and physicians, may treat their occupational dermatoses rather than consult a dermatologist. Informal "corridor consultations" do a disservice to the health care worker because the dermatologist is not allowed to investigate the dermatosis adequately. There is clearly a need for dermatologists to play an active role in the diagnosis, management, and prevention of occupational skin diseases in their fellow health care workers. This chapter will review occupational dermatoses in health care workers.

III. IRRITANT CONTACT DERMATITIS

Irritant contact dermatitis is the most common occupational skin disease in health care workers.[3-8] Many health care workers must wash their hands frequently. Cleaners and other health care workers may use strong disinfectants. Alcohol solvents are widely used in hospitals.

Cleaners, X-ray technicians, and kitchen workers appear to be the health care workers at highest risk for occupational irritant contact dermatitis.[4-6] Operating room nurses are reported to be at higher risk for hand dermatitis than other nurses.[8] Occupational contact dermatitis in health care workers has a poor to fair medical prognosis, but most workers can continue to work.[7,8]

In a questionnaire survey, Nilsson[3] found that about 40% of hospital nurses, kitchen workers, and cleaners had had hand eczema. This was twice the rate found in office workers. A total of 38% of kitchen workers and cleaners and 24% of nurses had consulted a physician because of hand eczema; 10% had missed work because of hand eczema. Only about 5% of affected workers had had to change jobs because of dermatitis. A history of hand eczema increased the risk of developing hand eczema 13-fold for hospital nurses, kitchen workers, and cleaners. A history of atopic dermatitis increased the risk of developing hand eczema threefold. Those with a history of atopic dermatitis were at an elevated risk for the development of severe hand eczema. The absence of a dishwashing machine and the presence of young children at home also doubled the risk of developing hand eczema. This study supplies quantitative data that the dermatologist can use to advise patients about their suitability for hospital wet work.

A. HEALTH RISKS OF HAND DERMATITIS

The risk of acquiring hepatitis B is increased among patients with chronic hand dermatitis or psoriasis.[9] Health care workers with active dermatitis should refrain from all direct patient care and from handling patient care equipment because of the risk of acquiring the human immunodeficiency virus (HIV) infection in the workplace.[10]

B. ETHYLENE OXIDE

Ethylene oxide may be the greatest chemical hazard in hospitals today. It is an alkylating agent, toxic to microorganisms, that is used in nearly all American Hospitals. By volume it is one the most widely used chemical compounds in the U.S.[11] Medical sterilization is one of the major uses of ethylene oxide. Several hundred different types of medical devices, ranging from surgical drapes to heart valves, are sterilized with ethylene oxide.[12] In hospitals, ethylene oxide gas sterilizers are used to sterilize delicate medical devices that cannot be put in a steam autoclave or soaked in a sterilizing solution. Ethylene oxide sterilizers vary in size from small tabletop models to large industrial models. Ethylene oxide is a very strong cutaneous irritant. It is necessary for articles sterilized with ethylene oxide to be aerated adequately before being used. Slow-airing polyvinyl chloride (PVC) materials retain ethylene oxide the longest. Ethylene oxide in nonwoven fabric is much less likely to produce skin irritation.[12] Fisher[11] reported a hospital worker who suffered burns to his hands after wearing heavy-duty rubber gloves that had been sterilized with ethylene oxide and still contained traces of ethylene oxide. Burns of the skin have been reported in patients from contact with gowns, drapes, and anesthesia masks with excessive residual amounts of ethylene oxide. The clinical appearance may resemble localized toxic epidermal necrolysis.[11] Unfortunately, ethylene oxide is handled carelessly in some hospitals. It may be vented from gas sterilizers directly into the room, where it is breathed at high peak levels by hospital employees. Chronic exposure to ethylene oxide is associated with neurologic dysfunction and an increased risk of leukemia.[13]

C. ANTINEOPLASTIC AGENTS

Many antineoplastic drugs are cutaneous irritants.[14] Mechlorethamine, chlorambucil, *N,N*-bis(2-chloroethyl)-*N*-nitrosourea (BCNU), *N*-(2-chloroethyl)-*N'*-cyclohexyl-*N*-nitrosourea (CCNU), 5-(dimethyltriazene)imidazole-4-carboxamide (DTIC), methotrexate, vindesine, actinomycin D, daunorubicin, bleomycin, mitomycin C, 1-asparginase, hexamethylmelamine, and hydroxyurea are reported to cause irritant contact dermatitis. 5-Fluorouracil (5-FU) and cyclophosphamide were considered to be less irritating to intact skin. Antineoplastic drugs are potential mutagens and teratogens for health care workers. Protective clothing and disposable gloves should be used during preparation and administration of antineoplastics. Chemotherapy solutions should be prepared in specialized work areas by pharmacists with special training.[14]

D. IRRADIATED LATEX GLOVES

Shumes and Darby[15] reported a phlebotomist who developed disabling hand dermatitis which they attributed to residual bacterial endotoxin in latex gloves sterilized by irradiation. They state that gloves sterilized by irradiation may contain significant endotoxin levels because the endotoxin is not radiosensitive.

E. CHILD-RESISTANT PACKAGING

Prescription and nonprescription drugs that are hazardous to children should be dispensed in child-resistant containers. A pharmacist may have to unscrew 100 caps from prescription containers daily before filling the containers and replacing the caps. This cumulative trauma may cause an irritant contact dermatitis.[16] Palmar psoriasis also may be induced by this repetitive trauma.

IV. ALLERGIC CONTACT DERMATITIS

A. NICKEL

Lammintausta and colleagues[17] found that 46% of the hospital employees (predominantly cleaners, nurses, and kitchen employees) they surveyed had present or previous hand dermatitis. The incidence of allergic contact dermatitis was 21%. Nickel was the most common allergen; 60% of hospital employees who were allergic to nickel had present or previous hand dermatitis. Nickel penetrates rubber gloves.[18] Great quantities of stainless steel are found in hospitals. Detergents and rubbing during cleaning may help release nickel from stainless steel. The concentration of nickel in water in wash buckets and towels gradually increases as cleaning progresses. The concentration of nickel reached can exceed the concentration of nickel required to provoke allergic contact dermatitis in individuals allergic to nickel.[19] Detergents and water also reduce the protective barrier of the skin and increase the penetration of nickel. Allergic contact dermatitis to nickel may complicate irritant contact dermatitis, which is common among hospital cleaners. Hospital cleaners with severe or chronic hand dermatitis should be patch tested to nickel and other occupational allergens.

B. CHROMIUM

Rudzki[20] reported two surgeons with occupational contact dermatitis to chromium in chromic catgut.[20] All protective gloves evaluated by Mouridsen and Faber[21] protected individuals allergic to chromate from 0.5% potassium dichromate.[21]

C. FORMALDEHYDE

A half million American hospital workers are routinely exposed to formaldehyde.[22] Exposure may occur in areas such as autopsy rooms, pathology laboratories, and renal dialysis units. Formaldehyde is a strong cutaneous irritant. Allergic contact dermatitis to formaldehyde used to sterilize hemodialysis equipment and thermometers has been reported in nurses.[23,24] Formaldehyde was the most common cause of allergic contact dermatitis among medical personnel at the Muncipal Hospital Schwabing in Munich and among nurses in Warsaw.[20,25] Formaldehyde has been discontinued as a disinfectant in some hospitals and dental offices where allergic contact dermatitis to formaldehyde has become less commonly reported.[17] Most work gloves appear to provide adequate protection against formaldehyde.[21]

D. GLUTARALDEHYDE

Glutaraldehyde is an important cause of allergic contact dermatitis among health care workers. Glutaraldehyde is used in a 2% solution to sterilize surgical and dental instruments, bronchoscopes, and cytoscopes. Cidex is the main source of allergic contact dermatitis to glutaraldehyde. Cidex is a 2% acidic glutaraldehyde which is stable for long periods if stored in a cool place.[26] Sodium bicarbonate is added to Cidex to alkalinize glutaraldehyde. This alkaline Cidex is more active than acidic glutaraldehyde; it loses its antimicrobial activity within 2 to 3 weeks. Glutaraldehyde does not appear to cross-react with formaldehyde.[26] Allergic contact dermatitis to glutaraldehyde has been reported in nurses, radiologists, X-ray technicians, dental assistants, and respiratory technicians.[26-28] Nethercott and colleagues[28] were impressed by the tendency of health care workers with allergic contact dermatitis to glutaraldehyde to develop chronic dermatitis. One half of their patients had to leave their occupations; many still suffered from persistent hand eczema afterward. Several workers in their series were also sensitized to other occupational contactants such as rubber and formaldehyde.

E. ANTIBIOTICS

1. Streptomycin

Allergic contact dermatitis to streptomycin has been reported in nurses performing injections of streptomycin. Some nurses were obliged to stop nursing because of the dermatitis. Allergic contact dermatitis to streptomycin was most likely to occur in wards where streptomycin was not administered routinely and where nurses did not routinely wear gloves and masks when administering streptomycin.[29]

2. Penicillin and its Derivatives

Penicillin is a potent sensitizer, and topical use of this antibiotic is not recommended. Occupational allergic contact dermatitis to penicillin may occur in those who prepare or inject solutions of it.[30] Rudzki[20] reported a nurse with occupational contact dermatitis to penicillin who had a positive patch test reaction to 100,000 μg/l of penicillin in water. Boonk[30] stated that allergic contact sensitivity to penicillin is usually combined with immediate hypersensitivity to it. Health care workers who are allergic to penicillin must avoid handling all preparations containing tis drug or drugs that cross-react with it. Rudzki[20] reported two nurses with occupational contact dermatitis to ampicillin who had positive patch test reactions to 20% ampicillin in water. The same author also reported two cases of occupational contact dermatitis to dicloxacillin with positive patch test reactions to 10% dicloxacillin in water and two cases of occupational contact dermatitis to cephradine with positive patch test reactions to 2.5% cephradine in water.

3. Chloramphenicol

Rudzki[20] reported one case of occupational contact dermatitis to chloramphenicol who had a positive patch test reaction to 5% chloramphenicol in yellow paraffin.

4. Phenothiazines

Chlorpromazine is a phenothiazine and a common cause of allergic contact dermatitis in nurses who administer injections of it and allow the solution to touch their skin.[31,32]

Pharmacists and others who handle chlorpromazine powder or tablets may also become sensitized. Nurses used to crush chlorpromazine tablets between spoons and dissolve the drug in boiling water for oral administration.[31] Severe patch test reactions with id reactions have been reported in nurses patch tested to 2.5% chlorpromazine.[32] Seville[31] emphasized the importance of performing a 96-h reading in patients patch tested to chlorpromazine. Chlorpromazine is also a well known cause of photoallergic contact dermatitis.[33]

Allergic contact dermatitis to a phenothiazine derivative in a laboratory worker was reported by Goodman and Cahn.[34] She was also allergic to chlorpromazine, prochlorperazine, and perphenazine. She became allergic to promethazine during their studies.

5. Ethylenediamine

Ethylenediamine is a well-known sensitizer. The majority of individuals allergic to ethylenediamine were sensitized by ethylenediamine present as a stabilizer in triamcinolone acetonide, nystatin, and neomycin cream (Mycolog®, Kenacomb®, Tri-Adcortryl®, and Assacort® creams).[35] Ethylenediamine is freely soluble in water and is present as a stabilizer in injectable aminophylline. The first case of allergic contact dermatitis to ethylenediamine was reported in a pharmacist preparing aminophylline suppositories.[36] Occupational allergic contact dermatitis to ethylenediamine has also been reported in a nurse who prepared and administered injectable theophylline.[35]

6. Antineoplastic Agents

Bleomycin and *cis*-platinum are reported to be allergenic.[14] Allergic contact dermatitis to 5-FU and nitrogen mustard has been reported.[37] Nitrogen mustard is a well-known vesicant that penetrates rubber and polyethylene gloves.[38] Vinyl gloves afford the skin better protection.

7. Miscellaneous Drugs

Allergic contact dermatitis to maclofenoxate has been reported in nurses administering injections of this medication in France.[39] Rudzki[20] reported one case of occupational contact dermatitis to tubocurarine and two cases of occupational contact dermatitis to propranolol.[20]

V. DENTISTRY-RELATED PROBLEMS

A. MERCURY

It has been estimated that an average of 2.5 lb of mercury is used each year by every dentist.[40] In a recent study a dental student and two biochemists ''patch tested'' dental students to an aqueous solution of mercuric chloride and obtained positive reactions in one third.[41] They stated that these reactions indicated allergy to mercury. A previous study, similar in design, conducted by another dentist found that 2% of prefreshmen and 10.8% of senior dental students had positive reactions to mercuric chloride.[40]

Mercury is the only metal that is a liquid. It evaporates readily at room temperature. Mercury vapor is absorbed through the lungs and skin. Diffuse erythematous eruptions with a predilection for the axillae, popliteal fossae, antecubital fossae, and upper medial thighs have been described in some individuals 1 or 2 d after exposure to mercury vapor.[42] Miliary pustules may develop. The eruption resolves in 2 weeks. Positive patch test reactions to several mercurials were obtained in all patients examined. A broken thermometer has been the source of mercury vapor in most instances. Mercury exposure also may occur from spills of stains in histology laboratories or during the repair of biomedical equipment. A delayed-type allergic reaction is suspected to play a major role in mercury exanthem. Chronic exposure to mercury may produce central nervous system symptoms. Symptomatic exposed workers should have a 24-h urinary excretion of mercury and a blood-mercury level performed.[22]

B. PROCAINE

Allergic contact dermatitis to benzocaine is well known.[37] Allergic contact dermatitis to local and topical anesthetics, especially procaine, has been reported in dentists.[43-46] A survey in 1948 found that the large majority of dentists with occupational dermatitis attributed their condition to contact with local anesthetics.[44] The majority of dentists patch tested had positive reactions to procaine, larocaine, tutocaine, monocaine, butyn, and pontocaine.[45] One dentist had positive patch test reactions limited to pontocaine and apothesine.[44] A large questionnaire survey of dentists found that 20% of dentists who stated they were allergic to procaine were less than 30 years old; 40% of dentists who stated they were not allergic to procaine were under the age of 30.[43] Older dentists would have had a longer period of exposure to procaine. Dentists who used commercial cartridges of procaine were less likely to complain of symptoms from procaine than dentists preparing injections of that anesthetic from procaine powder.[43] More than one half of topical dental medications were reported to contain benzocaine or tetracaine. Benzocaine, procaine, and tetracaine are *p*-aminobenzoic acid (PABA) derivatives. The risk of becoming sensitized to lidocaine (Xylocaine) is much less. Local anesthetics that are esters of PABA should be avoided by health care workers. The use of lidocaine has greatly decreased the problem of allergic contact dermatitis to procaine among dentists.[47]

C. ACRYLIC MONOMER

Acrylic bone cement and dental bonding cement are a mixture of polymethyl methacrylate (acrylic monomer). Acrylic bone cement is used for fixation of orthopedic prostheses to bone in surgical procedures of the hip joint. In the U.S., 95% of dentures are made of acrylic resin by mixing the acrylic liquid monomer methyl methacrylate with polymethyl methacrylate powder. Dentures are usually cured with heat. Self-curing acrylic resin is employed by the dentist to repair or build up a portion of a denture and to build small temporary bridges, crowns, and fillings.[39] Acrylic monomer is a potent sensitizer that penetrates surgical rubber gloves with ease.[48] It is also capable of causing paresthesias that may persist after any dermatitis has subsided.[39] It is essential that surgeons and orthodontists do not touch this monomer.

Benzoyl peroxide and hydroquinone have been used as activators or catalysts for methacrylate. Both have been responsible for allergic contact dermatitis in dentists and dental technicians working with methacrylate.[48,49]

D. DENTAL BONDING

Dental bonding employs an epoxy resin and a methacrylate. The reactants are bisphenol A and glicidyl methacrylate. Both the epoxy and methacrylate are potent sensitizers; the dentist should endeavor to touch neither. The bonding material Contacto does not contain epoxy resin or methyl methacrylate.[39]

E. DENTAL IMPRESSION MATERIALS

Allergic contact dermatitis to dental impression materials has been reported. The sensitizer appears to be the Impregnum catalyst methyl dichlorobenzene sulfonate.[39] Permlastic may be a suitable substitute for those allergic to Impregnum.[50]

F. TEETH SEALING

Allergic contact dermatitis to methyl *p*-toluenesulfonate, the catalyst for Scutan has been reported. Scutan is used for sealing teeth. Texton is an alternate for Scutan that contains methyl methacrylate.[39]

G. ESSENTIAL OILS

Allergic contact dermatitis to probanidid, a derivative of eugenol, has been reported in an anesthetist.[51] Allergic contact dermatitis to eugenol has been reported in dentists and stomatologists.[20,39] Eugenol was the most common cause of allergic contact dermatitis in stomatologists in Warsaw.[20] Clove oils, cinnamon oil, oil of spearmint, peppermint, wintergreen, menthol, balsalm of Peru, and eucalyptus oil are other sensitizers used in dentistry.[39]

VI. SURGICAL SCRUBS

Irritant contact dermatitis is the most common adverse effect from surgical scrubs, but allergic contact dermatitis has been reported from a number of skin cleansers. Chlorhexidine and chlorprovidone are excellent cleansers that rarely sensitize except when used in the treatment of leg ulcers.[52]

A. BENZALKONIUM CHLORIDE

Benzalkonium chloride is a quaternary ammonium cationic detergent that has been used as a preoperative skin disinfectant and as a surgical instrument disinfectant. Solutions of benzalkonium chloride support microbial growth. The Association for Practitioners in Infection Control (APIC) did not approve the use of benzalkonium chloride as a surgical scrub.[53] Benzalkonium chloride has produced allergic contact dermatitis in surgical personnel who handled instruments soaked in this antiseptic.[39] Benzalkonium chloride is present in many over-the-counter products. Health care workers sensitized to benzalkonium chloride need to avoid both occupational and nonoccupational exposures to this detergent.

B. HEXACHLOROPHENE

Allergic reactions to hexachlorophene are rare. Photosensitivity to hexachlorophene has been reported.[39] The APIC does not recommend the use of hexachlorophene as a preoperative scrub. Hexachlorophene should not be used on infants or on pregnant women. This product has minimal activity against gram-negative bacteria.[53]

C. ALCOHOLS

Alcohols are cutaneous irritants. Allergic contact dermatitis to lower aliphatic alcohols and alcohol swabs has been reported rarely.[54-57] Individuals allergic to ethyl alcohol also may be allergic to amyl, butyl, methyl, and isopropyl alcohols.[54,57] Two of these patients had positive patch test reactions to gas chromatographically pure 2-propanol and 2-butanol.[55] Some of those allergic to ethyl alcohol had to avoid ingestion of alcohol.[54]

VII. NURSING UNIFORMS

Allergic contact dermatitis due to a green anthraquinone dye used as a shading component in blue uniforms worn by nursing sisters in England has been reported.[58]

VIII. RUBBER GLOVES

Many hospital workers must wear rubber gloves. The use of rubber gloves has increased among hospital workers since the onset of the acquired immunodeficiency syndrome (AIDS) epidemic. Thiurams are the main causes of occupational allergic contact dermatitis due to rubber gloves among hospital workers (Table 1).[59,60] Occupational allergic contact dermatitis due to PVC gloves has also been reported in hospital workers.[60] Patch testing is required to diagnose allergic contact dermatitis to rubber gloves, as clinical diagnosis is inaccurate. The worker should be patch tested to pieces of the work glove(s) as well as to standard

TABLE 1
Chemicals Responsible for Allergic Contact Dermatitis to Rubber and Plastic Gloves

Tetramethylthiuram monosulfide
Tetramethylthiuram disulfide
Tetraethylthiuram disulfide
Dipentamethylenethiuram disulfide
Zinc diethyldithiocarbamate
Hexamethylenetetramine
N-Phenyl-*N*-isopropyl-*p*-phenylenediamine
Mercaptobenzothiazole
N-Phenyl-*N*-cyclohexyl-*p*-phenylenediamine
4,4-Dihydroxydiphenyl
N-Cyclohexylbenzothiazyl sulfenamide
2-*p*-Methylcyclohexyl 4,6-dimethylphenol
Irgalite orange

chemicals found in rubber gloves. False negative patch test reactions to surgical gloves may occur.

IX. CONTACT URTICARIA

A. RUBBER GLOVES

Occupational contact urticaria to rubber gloves has been reported frequently in health care workers.[61-69] Turjanmaa[67] screened 512 hospital employees with a latex glove scratch chamber test; 29% were proven to have contact urticaria to rubber gloves, and all were nurses and doctors. Atopy, hand eczema, and working in the operating room predisposed to contact urticaria from rubber gloves. Symptoms of contact urticaria from rubber range from pruritus to anaphylaxis.[62,63,69] This contact urticaria is usually mediated via IgE antibodies against natural latex rubber. IgE antibodies to *Hevia brasiliensis* have been detected in the sera of these patients.[63,64,66,69] The manufacturing process used to make the gloves may influence the antigenic capacity of rubber latex.[69] Workers allergic to natural latex rubber gloves may wear gloves made entirely of plastic or artificial rubber (styrene-butadiene polymer, Elastyren®).[66] Occupational contact urticaria due to cornstarch surgical glove powder has also been reported.[61,65] Powder-free surgical gloves are available (Pristine®).[65] A case of occupational contact urticaria to rubber gloves has been attributed to zinc diethyldithiocarbamate,[64] which is an accelerator in the vulcanization of rubber.

Workers suspected of occupational contact urticaria to rubber gloves may be tested by having them wear a wetted finger of the rubber glove for 15 min. There is a small risk of an anaphylactic reaction even if only one finger is exposed to the suspect rubber glove. Use tests are necessary since contact urticaria to rubber gloves may be due to antigens other than rubber latex. A latex glove allergen for use in prick testing may be made by incubating pieces of the rubber glove in saline.[67-69] The radioallergosorbent test (RAST) is less sensitive than prick testing.[69] Contact urticaria to rubber may be life threatening; it is essential that the worker be advised appropriately on allergen substitution, particularly if the patient is to undergo a major surgical procedure or an examination with palpation of a mucosal surface. Rubber contraceptives and balloons have produced anaphylaxis in these patients.

B. DRUGS

1. Phenothiazines

Contact urticaria to chlorpromazine, promethazine, and levomepromazine has been reported.[70] These are all closely related phenothiazines. Immediate allergic photosensitivity to chlorpromazine has been reported;[33] an urticarial reaction was obtained immediately after irradiation of the skin.

2. Alcohols

Contact urticaria to xylene has been reported in a woman with a recurrent facial dermatitis.[71]

3. Antibiotics

Occupational contact urticaria to cephalosporins was reported in a chemist.[72] He had a strong immediate reaction to cephalotin and to coded experimental cephalosporins, but had no reactions to penicillin or ampicillin. Contact urticaria to streptomycin has been reported in a nurse who worked in a tuberculous sanitorium.[73] Inhalation of or contact with penicillin may produce anaphylactic reactions in those with contact sensitivity to this antibiotic.[30] Dermatologists testing patients for contact urticaria must be prepared for the possibility of inducing an anaphylactic reaction.

Nonoccupational contact urticaria to chloramphenicol, penicillin, neomycin, gentamycin, and bacitracin has been reported.[74,75] Several of these patients had the risk factors of atopy or stasis dermatitis.

4. Procaine

Ten of the dentists surveyed by Charter and colleagues[43] complained of urticaria from procaine. Two of them complained of both urticaria and dermatitis.

5. Nitrogen Mustard

Severe contact urticaria to nitrogen mustard has been reported in patients treated with this potent contact sensitizer.[37] Occupational contact urticaria to nitrogen mustard apparently has not been reported yet.

6. Phenylmercuric Propionate

Urticaria, angioedema, and asthma due to phenylmercuric propionate, used as an antibacterial fabric softener for hospital linens, has been reported in a resident physician.[76]

7. Nicotinyl Alcohol

Nicotinyl alcohol is a peripheral vasodilator. Severe contact urticaria to this drug was reported in a nurse who also developed a delayed patch test reaction to this drug.[75]

8. Chlorhexidine

Chlorhexidine may produce allergic dermatitis, contact uticaria, combined contact urticaria, and allergic contact dermatitis and anaphylaxis.[77].

REFERENCES

1. **Emmett, E. A. and Baetz, J. H.,** Health in the heath care industries, *Occup. Med. State of the Art Rev.*, 2, 9, 1987.
2. **Mathias, C. G. T. and Morrison, J. H.,** Occupational skin diseases, United States, *Arch. Dermatol.*, 124, 1519, 1988.
3. **Nilsson, E.,** Individual and environmental risk factors for hand eczema in hospital workers, *Acta Derm. Venereol.*, 66 (Suppl. 128), 1, 1986.
4. **Kavli, G., Angell, E., and Moseng, D.,** Hospital employees and skin problems, *Contact Dermatitis,* 17, 156, 1987.

5. **Hansen, K. S.,** Occupational dermatoses in hospital cleaning women, *Contact Dermatitis,* 9, 343, 1983.
6. **Gawkrodger, D. J., Lloyd, M. H., and Hunter, J. A.,** Occupational skin disease in hospital cleaning and kitchen workers, *Contact Dermatitis,* 15, 132, 1986.
7. **Dahlquist, I. and Fregert, S.,** Occupational dermatoses in hospital personnel, *Berufs-Dermatosen,* 18, 261, 1970.
8. **Lammintausta, K., Kalimo, K., and Aanta, S.,** Course of hand dermatitis in hospital workers, *Contact Dermatitis,* 8, 327, 1982.
9. **Guadagnino, Y., Ayala, F., Chiianni, A., et al.,** Risk of hepatitis B virus infection in patients with eczema or psoriasis of the hand, *Br. Med. J.,* 284, 84, 1982.
10. **Churchill, E.,** AIDS Update, Recommendations for preventing transmission of HIV in health care settings, *Can. Med. Assoc. J.,* 138, 213, 1988.
11. **Fisher, A. A.,** Burns of the hands due to ethylene oxide used to sterilize gloves, *Cutis,* 42, 267, 1988.
12. **Shupack, J. L., Andersen, S. R., and Romano, S. J.,** Human skin reactions to ethylene oxide, *J. Lab. Clin. Med.,* 98, 723, 1981.
13. **Becker, C. E.,** Recognizing the health hazards of ethylene oxide, *West. J. Med.,* 148, 75, 1988.
14. **Rodgers, B.,** Health hazards of personnel handling antineoplastic agents, *Occup. Med. State of the Art Rev.,* 2, 513, 1987.
15. **Shumes, E. and Darby, T.,** Contact dermatitis due to endotoxin in irradiated gloves, *Contact Dermatitis,* 10, 240, 1984.
16. **Fisher, A. A.,** Occupational palmar psoriasis due to safety prescription container caps, *Contact Dermatitis,* 5, 56, 1979.
17. **Lammintausta, K., Kalimo, K., and Havu, V. K.,** Occurrence of contact allergy and hand eczemas in hospital wet work, *Contact Dermatitis,* 8, 84, 1982.
18. **Wall, L. M.,** Nickel penetration through rubber gloves, *Contact Dermatitis,* 6, 461, 1980.
19. **Clemmensen, O. J., Menne, T., Kaaber, K., and Solgaard, P.,** Exposure of nickel and the relevance of nickel sensitivity among hospital cleaners, *Contact Dermatitis,* 7, 14, 1981.
20. **Rudzki, E.,** Occupational dermatitis among health care workers, *Dermatosen,* 27, 112, 1979.
21. **Mouridsen, H. T. and Faber, O.,** Penetration of protective gloves by allergens and irritants, *Trans. St. John's Hosp. Dermatol. Soc.,* 59, 230, 1973.
22. **Patterson, W. B., Craven, D. E., Schwartz, D. A., et al.,** Occupational hazards to hospital personnel, *Ann. Intern. Med.,* 102, 658, 1985.
23. **Sneddon, I. B.,** Dermatitis in an intermittent haemodialysis unit, *Br. Med. J.,* 1, 183, 1968.
24. **Rostenberg, A., Bairstow, B., and Luther, T. W.,** A study of eczematous sensitivity to formaldehyde, *J. Invest. Dermatol.,* 19, 459, 1952.
25. **Agathos, V. M. and Bernecker, H. A.,** Handdermatitis bei medizinischen Personal, *Dermatosen,* 30, 43, 1982.
26. **Fisher, A. A.,** Reactions to glutaraldehyde, *Cutis,* 28, 113, 1981.
27. **Bardazzi, F., Melino, G., Alagna, G., and Veronesi, S.,** Glutaraldehyde dermatitis in nurses, *Contact Dermatitis,* 6, 319, 1986.
28. **Nethercott, J. R., Holness, L., and Page, E.,** Occupational contact dermatitis due to glutaraldehyde in health care workers, *Contact Dermatitis,* 18, 193, 1988.
29. **Wilson, H. T. H.,** Streptomycin dermatitis in nurses, *Br. Med. J.,* 1, 1378, 1958.
30. **Boonk, W. J.,** Dermatologic hazards from hidden contacts with penicillin, *Dermatosen,* 29, 131, 1981.
31. **Seville, R. H.,** Chlorpromazine dermatitis in nurses, *Br. J. Dermatol.,* 68, 332, 1956.
32. **Lewis, G. M. and Sawicky, H. H.,** Contact dermatitis from chlorpromazine, *J. Am. Med. Assoc.,* 157, 909, 1955.
33. **Horio, T.,** Chlorpromazine photoallergy, *Arch. Dermatol.,* 111, 1469, 1975.
34. **Goodman, D. and Cahn, M. M.,** Contact dermatitis to phenothiazine drugs, *J. Invest. Dermatol.,* 33, 27, 1959.
35. **dal Monte, A., De Benedictis, E., and Laffi, G.,** Occupational dermatitis from ethylenediamine hydrochloride, *Contact Dermatitis,* 17, 254, 1987.
36. **Tas, J. and Weissberg, D.,** Allergy to aminophylline, *Acta Allergol.,* 12, 39, 1958.
37. **Cohen, S. V.,** Skin disease in health care workers, *Occup. Med. State of the Art Rev.,* 2, 565, 1987.
38. **Thomsen, K. and Mikkelsen, H. I.,** Protective capacity of gloves for handling nitrogen mustard, *Contact Dermatitis,* 1, 268, 1975.
39. **Fisher, A. A.,** Contact dermatitis in health personnel, in *Contact Dermatitis,* 3rd ed., Lea & Febiger, Philadelphia, 1986, 515.
40. **White, R. R. and Brandt, R. L.,** Development of mercury hypersensitivity among dental students, *J. Am. Dent. Assoc.,* 92, 1204, 1976.
41. **Miller, E. G., Perry, W. L., and Wagner, M. J.,** Prevalence of mercury hypersensitivity in dental students, *J. Prosthet. Dent.,* 58, 235, 1987.
42. **Nakayama, H., Shono, M., and Hada, S.,** Mercury exanthem, *J. Am. Acad. Dermatol.,* 11, 137, 1984.

43. **Charter, W. V., Schlack, C. A., and Ware, R. L.,** A study of novocain (procaine) sensitivity among navy dental officers, *J. Dent. Res.,* 28, 337, 1949.
44. **Strauss, M. J.,** Group sensitivity to local anesthetics, *J. Invest. Dermatol.,* 8, 403, 1947.
45. **Laden, E. L. and Wallace, D. A.,** Contact dermatitis due to procaine: a common occupational disease of dentists, *J. Invest. Dermatol.,* 12, 299, 1949.
46. **Anon.,** Council reports progress in study of occupational dermatitis among dentists, *J. Am. Dent. Assoc.,* 38, 148, 1949.
47. **Samitz, M. H. and Shumes, E.,** Occupational dermatoses in dentists and allied personnel, *Cutis,* 5, 180, 1969.
48. **Pegum, E. S. and Medhurst, F. A.,** Contact dermatitis from penetration of rubber gloves by acrylic monomer, *Br. Med. J.,* 1, 141, 1971.
49. **Calnan, C. D. and Stevenson, C. J.,** Studies in contact dermatitis. XV. Dental materials, *Trans. St. John's Hosp. Dermatol. Soc.,* 48, 9, 1963.
50. **van Ketel, W. G.,** Reactions to dental impression materials, *Contact Dermatitis,* 3, 55, 1977.
51. **Sneddon, I. B. and Caldwell, R. C.,** Contact dermatitis due to propanidid in an anesthesist, *Practitioner,* 211, 321, 1973.
52. **Osmudsen, P. E.,** Contact dermatitis to chlorhexidine, *Contact Dermatitis,* 8, 81, 1982.
53. **Soule, B., Ed.,** *The APIC Curriculum for Infection Control Practice,* Vol. 2, 1983, 853.
54. **Fregert, S., Groth, O., Hjorth, N., et al.,** Alcohol dermatitis, *Acta Derm. Venereol.,* 49, 493, 1969.
55. **Fregert, S., Groth, O., Gruvberger, B., et al.,** Hypersensitivity to secondary alcohols, *Acta Derm. Venereol.,* 51, 271, 1971.
56. **Richardson, D.,** Allergic contact dermatitis to "alcohol" swabs, *Cutis,* 5, 1115, 1969.
57. **van Ketel, W. G. and Tan-Lim, K. N.,** Contact dermatitis from ethanol., *Contact Dermatitis,* 1, 7, 1975.
58. **Wilson, H. T. H. and Cronin E.,** Dermatitis from dyed uniform, *Br. J. Dermatol.,* 85, 67, 1971.
59. **Nater, J. P.,** Uberempfindilichkeit gegen gummi, *Berufs-Dermatosen,* 23, 161, 1975.
60. **Estlander, T., Jolanki, R., and Kanerva, L.,** Dermatitis and urticaria from rubber and plastic gloves, *Contact Dermatitis,* 14, 20, 1986.
61. **Assalve, D., Ciconi, C., Perno, P., and Lisi, P.,** Contact urticaria and anaphylactoid reaction from cornstarch surgical powder, *Contact Dermatitis,* 19, 61, 1981.
62. **Carrillo, T., Cuevas, M., Munoz, T., et al.,** Contact urticaria and rhinitis from latex surgical gloves, *Contact Dermatitis,* 15, 69, 1986.
63. **Wrangsjo, K., Mellstom, G., and Axelsson, G.,** Discomfort from rubber gloves indicating contact urticaria, *Contact Dermatitis,* 15, 79, 1986.
64. **Helander, I. and Makela, A.,** Contact urticaria to zinc diethyldithiocarbamate (ZDC), *Contact Dermatitis,* 9, 327, 1983.
65. **Fisher, A. A.,** Contact urticaria due to cornstarch surgical glove powder, *Cutis,* 38, 307, 1986.
66. **Frosch, P. J., Wahl, R., Bahmer, F. A., and Maasch, H. J.,** Contact urticaria to rubber gloves is IgE mediated, *Contact Dermatitis,* 14, 241, 1986.
67. **Turjanmaa, K.,** Incidence of immediate allergy to latex gloves in hospital personnel, *Contact Dermatitis,* 17, 270, 1987.
68. **Turjanmaa, K., Reunala, T., and Räsänen, L.,** Comparison of diagnostic methods in latex surgical glove contact urticaria, *Contact Dermatitis,* 19, 241, 1988.
69. **Wrangsjo, K., Wahlberg, J. E., and Axelsson, G.,** IgE mediated allergy to natural rubber in 30 patients with contact urticaria, *Contact Dermatitis,* 19, 264, 1988.
70. **Johansson, G.,** Contact urticaria from levomepromazine, *Contact Dermatitis,* 19, 304, 1988.
71. **Altman, A. T.,** Facial dermatitis, *Arch. Dermatol.,* 113, 1460, 1977.
72. **Tuft, L.,** Contact urticaria from cephalosporins, *Arch. Dermatol.,* 111, 1609, 1975.
73. **Rudzki, E., Rebanel, P., and Rogozinski, T.,** Contact urticaria from rat tail, guinea pig, streptomycin and vinyl pyridine, *Contact Dermatitis,* 7, 186, 1981.
74. **Kozakova, M.,** Sub-shock bei der epikutanprobe mit chloramphenicol, *Berufs-Dermatosen,* 25, 134, 1977.
75. **Maucher, O. M.,** Anaphylaktische reaktionen beim epicutantest, *Hautarzt,* 23, 139, 1972.
76. **Matthews, K. P.,** Immediate type hypersensitivity to phenylmercuric compounds, *Am. J. Med.,* 44, 310, 1968.
77. **Fisher, A. A.,** Contact urticaria from chlorhexidine, *Cutis,* 43, 17, 1989.

Chapter 35

CEMENT ECZEMA

Christian Avnstorp

TABLE OF CONTENTS

I. INTRODUCTION

The association between cement eczema and cutaneous chromium hypersensitivity was established in 1950.[1] Chromium contact sensitivity depends on the concentration of water-soluble hexavalent chromate in the cement. More than 2 ppm of water-soluble chromate in the cement is probably needed to induce hypersensitivity.[2] The threshold concentrations for elicitation of hypersensitive reactions approximates 1 ppm.[3]

Most chromium-hypersensitive workers develop chronic hand eczema due to the ubiquitous occurrence of chromium compounds[4] and the primary irritant effect caused by the alkaline, abrasive, and hygroscopic properties of the cement itself.[5] In some workers hand eczema clears due to hardening.[6-8] Some workers have latent hypersensitivity without manifestations of hand eczema.[6]

II. EVOLUTION OF CEMENT

The concrete used today is based on mixtures of Portland cement, sand, stones, and water. The mortar is based on mixtures of Portland cement, sand, water, and often hydrated lime. The exact composition depends on the purpose for which the product is to be used.

More than 2000 years ago the Etruscans used calcined limestone to make concrete and mortar. In ancient Rome a mixture of calcined limestone, vulcanic ash, sand, stone, and water was used in construction.[9] The Latin word "cementum" means crushed stone. In 1824 Joseph Aspdin, a British builder, patented a cement produced by firing a mixture of limestone and clay.[10] He called his product Portland cement because it was similar in appearance to a building stone quarried on the island of Portland off the south coast of England. Aspdin heated his mixture in small, primitive, vertical kilns[10] in which no chromate was formed.[11] Chromate was not formed until the introduction of the modern horizontal kiln.[11]

III. CHROMATE IN CEMENT

A. THE ORIGIN OF CHROMATE

The potential source of chromium in cement is in the trivalent compounds (Cr_2O_3) in the raw materials from which it is produced. The trivalent chromium compounds are oxidized to the hexavalent chromate compounds (CrO_4^{2-}) during heating in the kilns at temperatures about 1400 to 1500°C. The total content of chromium in cement is determined by the presence of chromium compounds in the raw materials, by the kiln lining, and by chromium steel abrasion during the grinding processes.[12] The chromate content of cement varies from country to country according to variations in the amount of chromium in the raw materials.[13]

B. REDUCTION AND SOLUBILITY

Ferrous sulfate reduces the hexavalent chromium (Cr^{6+}) in chromate to trivalent chromium (Cr^{3+}) (Figure 1). Chromate is highly water soluble, and the solubility increases with the alkalinity of the cement.[14,15] In ordinary Danish Portland cement, the total chromium content is 40.5 ppm; of this, 9.6 ppm is water-soluble chromate.[16] Trivalent chromium compounds are precipitated in the alkaline environment as insoluble chromium hydroxide.[13] The addition of 0.35% (w/w) ferrous sulfate to cement reduces a water-soluble chromate content of 20 ppm to undetectable concentrations.[17]

$$Cr^{6+} + 3Fe^{2+} \longrightarrow Cr^{3+} + 3Fe^{3+}$$

FIGURE 1. The main chemical reaction in the reduction of water soluble chromate by the addition of ferrous sulfate.

FIGURE 2. A bricklayer constructs the wall of a building.

IV. DESCRIPTION OF WORK PROCESSES

A. BRICKLAYERS

Skilled workers who lay bricks and stones use various bonding materials, especially cement (Figure 2). Their contact with wet cement is primarily on the hands and forearms. Their primary work sites are indoor or outdoor sites of new construction or renovation; there are often considerable concentrations of dust in the air at the latter sites due to the destruction of old walls. Bricklayers rarely use gloves and often put their hands into hydrochloric acid solutions in water in order to neutralize the alkaline cement on their hands. For some work processes they use products containing epoxy resins.

B. BRICKLAYERS' ASSISTANTS

Semiskilled workers mix the mortar used by the bricklayers. The mixing is done with special tools and is typically performed once or twice a day. Bricklayers' assistants also handle building materials and assist in the actual laying of bricks. Their contact with wet cement is primarily on the hands, forearms, and legs. Their work sites are the same as those of bricklayers. Assistants, however, sometimes use leather gloves; however, like bricklayers, they often put their hands into hydrochloric acid.

C. CONCRETE WORKERS

Unskilled workers are engaged at the building sites. During construction they work largely outdoors where they pour the premixed concrete into molds containing steel rein-

FIGURE 3. Concrete workers laying a floor.

forcements. The concrete is then vibrated manually. These workers often walk on newly poured concrete foundations while wearing rubber boots (Figure 3). Their contact with wet cement is primarily on the hands, forearms, feet, and legs. They often use leather or rubber gloves.

D. PRECAST CONCRETE COMPONENT FACTORIES

Each factory is comprised of two separate areas. In one area, the carpenters, painters, iron winders, and molding smiths, who have no contact with wet cement, carry out their work. In the other area, concrete prepared in the batch plant is transported by cranes to the molding bays, where it is poured into molds and vibrated automatically or manually. The workers then do the leveling (Figure 4) and sometimes the steam curing of the concrete components. Finishing and repairing is done with mixtures of cement and epoxy resins (Figure 5). The workers in the batch plant primarily have contact with wet cement on their hands, but the entire body can be exposed to the cement when spills occur. Most of the work processes are carried out in large workshops, and those workers who have daily contact with wet cement often use leather or rubber gloves.

V. CHROMIUM HYPERSENSITIVITY

Chromium allergy in patients with cement eczema represents a delayed-type contact hypersensitivity reaction.[18] Hexavalent chromate compounds are strong sensitizers,[19] whereas the chromium metal itself is considered nonsensitizing.[20]

A. CONCENTRATION AND EXPOSURE

The concentration of chromate in concrete and mortar is about four to five times lower than in dry cement. Because there can be great variations in the composition of the final product, the concentrations of chromate in the original dry cement are used here as the point of reference.

FIGURE 4. A worker levels a concrete building component.

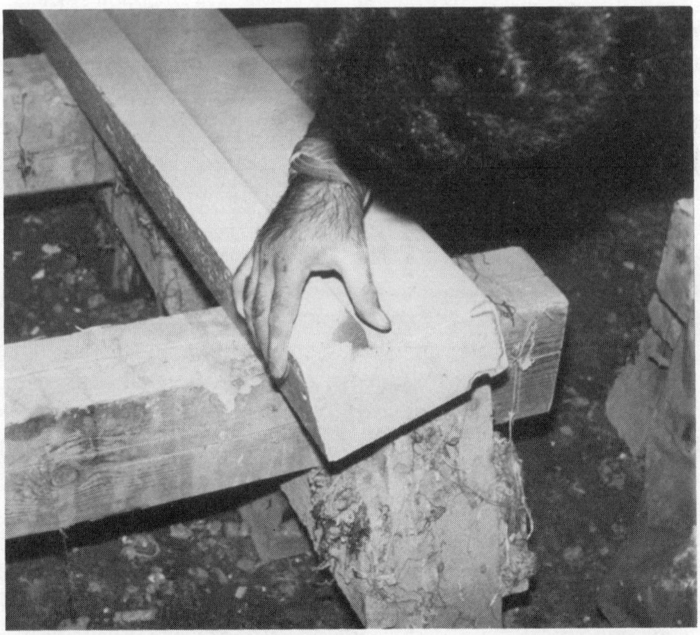

FIGURE 5. A worker repairing a concrete building component.

If the content of water-soluble chromate in dry cement does not exceed 2 ppm, the risk of inducing chromium hypersensitivity following skin contact with the cement in wet form is comparable to the risk seen among workers who have had no contact with wet cement.[2] In a study in San Francisco among workers exposed to cement containing no more than 0.1 ppm water-soluble chromate, few workers became hypersensitive to chromium, although more than one third had evidence of mild to moderate cement eczema.[21] In East Germany, the concentration of water-soluble chromate in cement needed to induce chromium hypersensitivity was estimated to be between 0.4 and 24 ppm.[22]

The risk of developing chromium hypersensitivity increases with the intensity and duration of exposure. The risk of developing chromium allergy is not influenced by age and exposure time to the same degree as by exposure due to certain work processes.[23] Cement burns, which usually occur among those working with cement as a hobby, rarely cause chromium allergy.[24-27]

Mold oils are used in the manufacture of precast concrete components in factories. Oil dermatitis is largely of an irritant nature, but oils for industrial use may contain chromium.[28,29] Such oils may remove the lipids on the surface of the skin, thus facilitating the penetration of chromium compounds.[30]

B. PENETRATION OF THE HAPTEN AND ANTIGEN FORMATION

Hexavalent chromate compounds can penetrate the skin.[31] This means that even a low concentration of chromate in cement can be of importance for the induction of contact sensitivity. A relatively larger proportion of the available chromate will be absorbed as the concentration decreases.[32]

As trivalent chromium compounds are not soluble in a mixture of alkaline cement and water,[13] their ability to induce hypersensitivity is limited. In addition, trivalent chromium salts have a strong affinity for the epithelium, thus reducing their ability to penetrate intact skin.[31] Trivalent chromium compounds are more likely to penetrate damaged skin.[33]

The reduction of Cr^{6+} to Cr^{3+} in the epidermis and the formation of protein-Cr^{3+} complexes are possibly the first steps in the sensitization.[31] A common antigenic determinant is probably involved in the chromium contact sensitivity, rather than independent entities corresponding to each valency state. This determinant is suggested to be formed by chromium in the trivalent form.[34]

The reduction of Cr^{6+} to Cr^{3+} in the skin is caused by the sulfur-containing amino acids.[35] Cr^{3+} then binds to skin proteins[36] and to components of the Langerhans' cells.[37] Chromium-reducing abilities have also been exhibited by lactic acid, hemoglobin, and gamma globulin.[35]

C. CAUSES OF CHRONICITY

In workers already hypersensitive to chromium, the concentration of water-soluble chromate in cement needed to elicit allergic cement eczema is probably lower than the concentration needed to induce sensitivity.[3] Elicitation of allergic cement eczema from even very low concentrations may be further enhanced if the skin is damaged as a result of irritation from wet cement. Concentrations of water-soluble chromate below 2 ppm may cause elicitation, although this concentration should not induce hypersensitivity. Cement in which the chromate content has not been reduced, used as a filler in repair work, may represent a hapten source.[38]

Activated T-lymphocytes can be seen 4 h after a challenge.[39] Recently activated and antigen-specific T-lymphocytes may enter sites of inflammation nonspecifically to produce specific local immunological memory which may last for several months.[40] This may, in part, explain the chronicity of cement eczema, since contact with chromate does not have to be constant and contact with the alkaline cement itself could maintain the inflammation.

Chromium is concentrated at the dermoepidermal junction and in the upper dermis;[41] 2 months after intracutaneous injection of a ^{51}Cr-labeled chromium compound, 3 to 5% was still left.[42] Chromium compounds are ubiquitous in the natural human environment and elicitation may occur from skin contact, not only with hexavalent (Cr^{6+}) compounds, but also with trivalent (Cr^{3+}) compounds.[43,44]

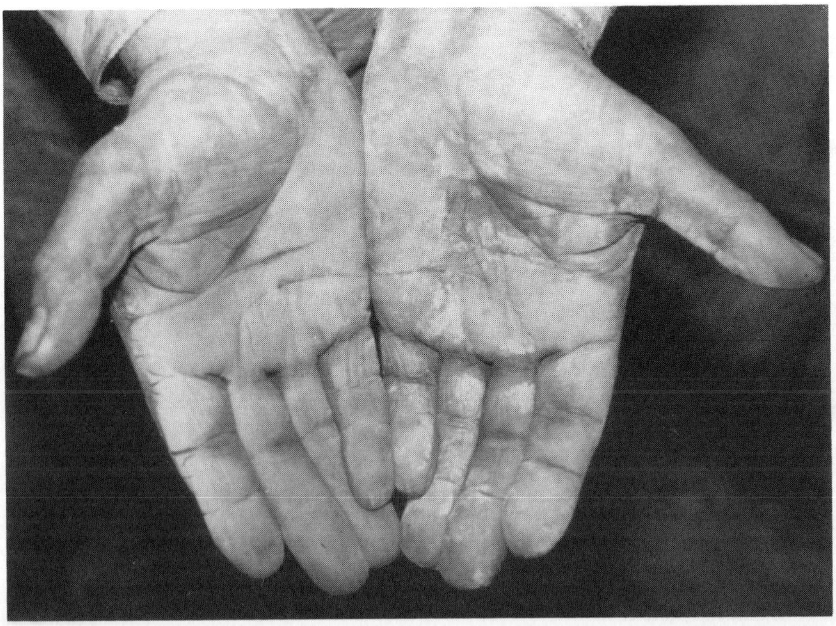

FIGURE 6. The cement-covered hands of a worker who has just finished making a concrete component.

Systemic reactions may occur in patients previously sensitized by percutaneous absorption of chromium.[45] Inorganic chromium compounds are poorly absorbed, and the amount absorbed from an oral dose varies from 1 to 25%[46,47] Chromium-rich foods are brown sugar, animal fats, and butter.[48] Oral challenge with 2.5 mg potassium dichromate produced a dermatitis flare in 21 patients with symmetrical vesicular hand eczema. In these patients the systemic exposure seemed to be more important than skin contact with the hapten, since the dermatitis of 11 of these patients cleared or improved after all had followed a diet restricting the ingestion of chromium-rich foods for 1 month.[49] Systemic contact dermatitis to chromium also may occur after inhalation of chromium in welding fumes.[50,51]

D. CONCOMITANT SENSITIVITIES

The chronicity also may be attributable to coexisting contact allergies.[52,53] Concomitant sensitivities to cobalt and rubber chemicals has been found significantly more often among chromium-sensitive workers than among those not sensitive to chromium.[23] This is in agreement with the findings of Høvding,[6] Fregert and Rorsman,[54] and Fregert.[55]

Cobalt occurs in cement as water-insoluble oxides, but forms complexes with amino acids in eczematous skin, thereby forming haptens.[56] Nickel is a contaminant of cement in the form of insoluble nickel oxide, which (in contrast to the cobalt oxides) is not allergenic.[57] This was confirmed in a recent study where no statistically significant association between nickel allergy and cement eczema was found.[2,23] Sensitization to rubber chemicals is thought to be a secondary sensitization because workers with hand eczema use rubber gloves to protect their hands from the wet cement. A similar correlation has been described among workers from a factory manufacturing precast concrete building components in Singapore.[58]

VI. CLINICAL PATTERNS OF CEMENT ECZEMA

A. IRRITANT CEMENT ECZEMA

Cement eczema may initially develop as a result of daily irritation of the skin by contact with wet cement[23,59] (Figure 6). Susceptibility to cumulative insult dermatitis depends on

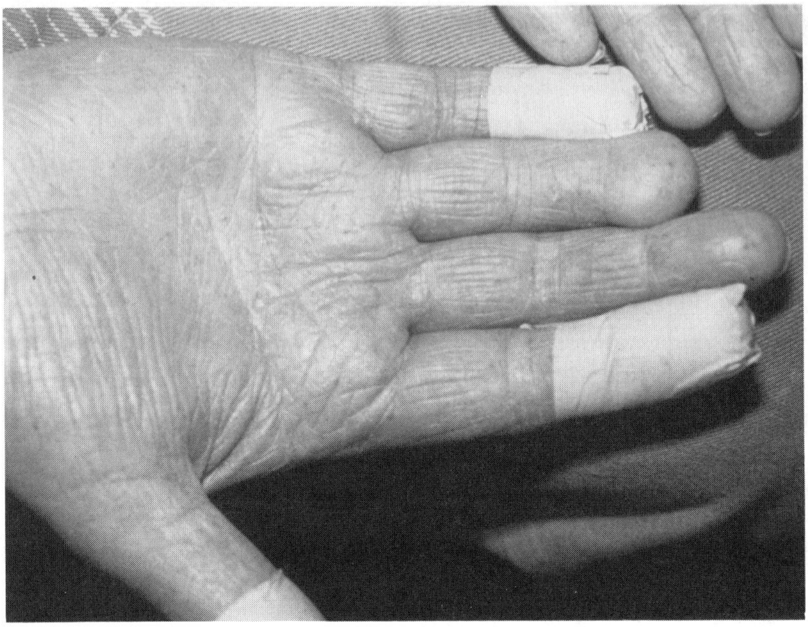

FIGURE 7. A worker covering fissures on the fingers with plastic tape.

the degree of exposure, the skin site exposed, and individual predisposition to atopy. Coexisting or recently healed eczema makes the skin more vulnerable. Chronic irritant dermatitis may, therefore, be due to the sum of various adverse reactions.[60] The influence of daily contact with wet cement has been illustrated by the fact that cement eczema is more common among working bricklayers than among unemployed bricklayers.[23] Among hospital "wet workers", trivial irritants and endogenous factors were found to play a greater role in the aggravation of current hand eczema than contact sensitivity.[61]

Mild, reversible skin damage is commonly seen among workers who have daily contact with wet cement.[23] Minor skin complaints are accepted by these workers as normal. Mechanical traumas, such as the handling of bricks, tools, and machines, causes occupational stigmata in the form of hyperkeratoses with fissures on the palmar aspects of the hands and fingers. Workers often cover such fissures with plastic tape (Figure 7). The skin of those who have daily contact with wet cement becomes irritated by the hygroscopic and alkaline constitution of the cement (pH 12.6) and by washing of the hands with special heavy-duty cleansers. The skin's barrier layer becomes more permeable, and repeated contact with abrasive materials may lead to irritant hand eczema.[62] Roughness and erythema of the skin have been induced experimentally by the application of the irritant sodium lauryl sulfate. In the study in question, roughness could be correlated to increased transepidermal water loss.[63]

B. ALLERGIC CEMENT ECZEMA

Allergic cement eczema is primarily located on the hands and fingers.[6,23,64] Eczema on the doral aspects of the right hand, in particular, is shown to be associated with chromium sensitivity (Figure 8).[23] Allergic cement eczema may spread to the forearms, the feet, and sometimes the face and parts of the trunk. In cases where nummular eczema is also present, systemic contact dermatitis may be suspected.[45]

Heat and high humidity exacerbated allergic cement eczema among workers in Kuwait. This phenomenon was explained by the fact that excessive perspiration could lead to increased

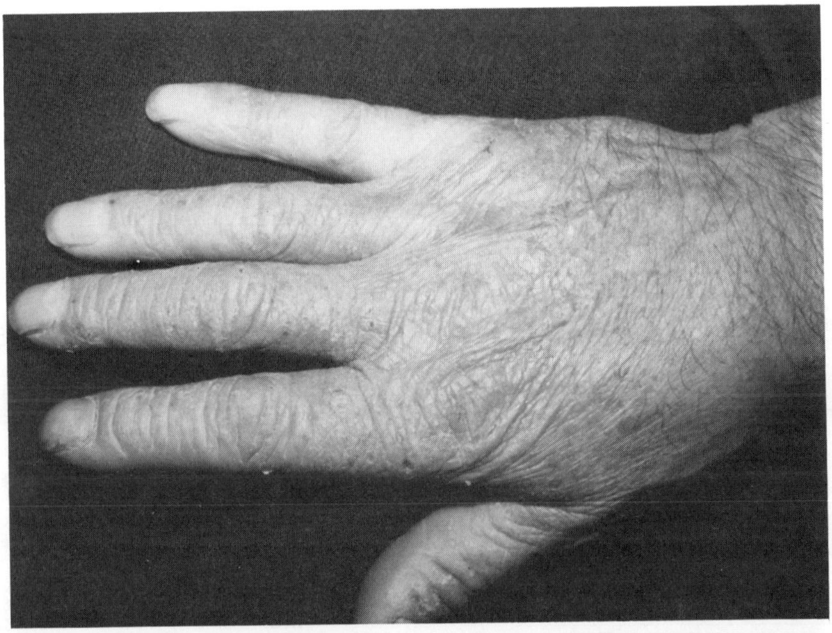

FIGURE 8. Allergic cement eczema.

leaching of chromium through the skin.[65] In Singapore, where the climate is hot and humid year round, the number of heat-related dermatoses was, nevertheless, lower than expected, possibly due to the use of air-conditioning and good factory ventilation.[66]

VII. EPIDEMIOLOGY OF CEMENT ECZEMA

Chromium sensitivity affects men more often than women, thus reflecting the fact that sensitization occurs in jobs more frequently occupied by men (e.g., in the building industry).[5] In some countries, a high incidence of chromium sensitivity among women with hand dermatitis has been found. This association has been explained by the fact that there are small amounts of chromate in detergents and bleaches used in household tasks.[67,68]

The prevalence of chromium sensitivity in men in Finland has varied from 2% in the general population to 6.8% among patients in a hospital dermatology out-patient clinic to 15.5% among workers in the building industry.[69] Most epidemiological data are obtained from dermatological clinics and therefore represent results obtained among highly selected, nonrepresentative materials. Studies carried out at work sites provide more accurate epidemiological information. Such studies have been done previously in Norway,[6] Sweden,[70] the Netherlands,[71] and Australia.[72] These studies gave attention to the high prevalence of irritant dermatitis among workers in the building and construction industries and to the fact that sensitivity to chromium may be of more importance than irritancy in causing cement dermatitis.

A. PREVENTION OF CEMENT ECZEMA

Fregert et al.[17] suggested a method of reducing the amount of water-soluble chromate in cement by adding ferrous sulfate. As of September 1981, ferrous sulfate has been added to all the cement produced for use in Denmark, thus reducing the content of water-soluble chromate in cement to not more than 2 ppm at a cost of about 1% of the total value of the cement.

TABLE 1
Prevalence of Hand Eczema and Chromium Allergy in the Occupational Groups in the 1981 Study among Workers in Denmark with Daily Contact with Wet Cement

	No. of workers examined	Hand eczema No. (%)	No. of workers patch tested	Chromium allergy No. (%)
Bricklayers	525	26 (5.0)	499	23 (4.6)
Bricklayers' assistants	192	13 (6.8)	184	7 (3.8)
Concrete workers	100	3 (3.0)	94	5 (5.3)
Factory workers with daily contact with wet cement	196	23 (11.7)	190	20 (10.5)
Retired bricklayers and bricklayers' assistants	113	4 (3.5)	105	11 (10.5)

TABLE 2
Prevalence of Hand Eczema and Chromium Allergy in 1981 and 1987 among Factory Workers with Daily Contact with Wet Cement

	1981	1987	*p* value[a]
Hand eczema	23 of 196 (11.7)	10 of 229 (4.4)	0.008
Chromium allergy	20 of 190 (10.5)	6 of 227 (2.6)	0.002

Note: Percentages are given in parentheses.

[a] Yates' correction applied to *p* values (χ^2 test).

The prevalence of hand eczema and of chromium allergy in 1981 in a study carried out in Denmark is shown in Table 1.[23] Among workers with daily contact with wet cement at the factories covered by the study, the decrease in the prevalence of these conditions from 1981 to 1987 was statistically significant (Table 2).[2] The decrease was found to correlate to the reduction of the content of water-soluble chromate in Danish cement. The reduction of the chromate content of the cement also was found to be significantly more influential in preventing allergic cement eczema than such traditional methods as the use of gloves. The estimated relative risk of contracting chromium allergy decreased from 8.3 (95% confidence interval 2.6 to 26.9) among workers exposed to cement containing water-soluble chromate in concentrations of about 10 ppm in 1981 to 1.0 among workers exposed exclusively to cement with concentrations of not more than 2 ppm in 1987 (Figure 8).[2]

B. MEDICAL PROGNOSIS

The addition of ferrous sulfate to the cement did not influence the medical status of already chromium-sensitive workers, most of whom had chronic hand eczema. In contrast, few workers with irritant cement eczema developed chronic hand eczema.[52] The prognosis of workers who develop irritant cement eczema in the future may, therefore, be improved if ferrous sulfate is added to the cement.[2]

The persistence of allergic cement eczema as previously described is, in part, associated with daily irritation of the skin from wet cement. If a worker is advised to change occupations or to be retrained, it is essential that the new job should not involve contact with abrasive "wet" work processes or contact with other irritants such as, for example, cutting oils.

Whether or not allergic cement eczema can be improved by a change in occupation is not clearly shown in the literature. Some authors have found the medical prognosis to be

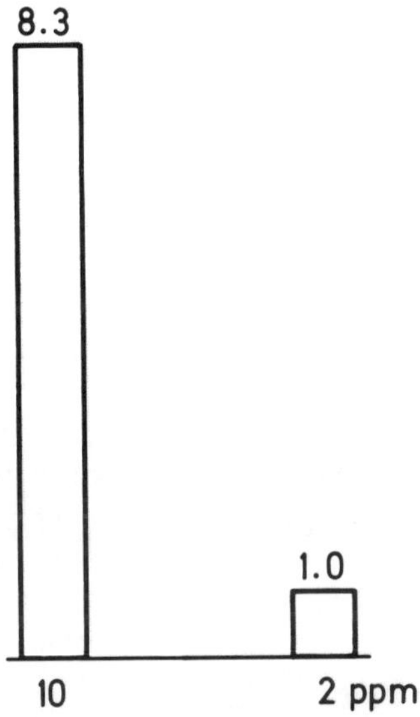

FIGURE 9. Relative risk of chromium allergy after exposure to cement containing 10 ppm or not more than 2 ppm water-soluble chromate.

poor in spite of a change,[55,73] while others have found that a job change had a favorable influence on the course of the disease.[74,75] In one study, the status of workers who had continuous contact with wet cement containing no more than 2 ppm was almost equal to the status of workers who had no contact with wet cement.[52]

If the worker remains in his original occupation, the prognosis also may be favorable. In his cohort study of chromium allergic bricklayers and bricklayers' assistants in Bergen, Norway, Høvding[6] found that 4 of 15 eczemas cleared after 5 years. The cement eczema of 8 of 17 bricklayers in Geneva, Switzerland cleared despite the fact the workers continued to have the same jobs.[7] No reduction was made in the chromate content of the cement used by the workers in either of these studies. No conclusive explanation has been given for this adaptive phenomenon seen in some workers. It may be that some workers develop immunological tolerance,[8] and some may experience hyporeactivity induced by repeated irritation of the skin.[76]

C. SOCIAL PROGNOSIS

The social prognosis of workers with chromium allergy may depend on whether or not they are skilled or unskilled. Bricklayers, for example, are skilled workers who may find it difficult to change occupations. Employed and unemployed bricklayers in Denmark had an equal prevalence of chromium allergy, thus illustrating that chromium allergy itself had no influence on their employment status.[23] Høvding[6] found that healthy bricklayers and bricklayers' assistants changed occupations to a greater degree than did workers with eczema and that workers with eczema did not lose their jobs more often than those without eczema.

The fact that unskilled workers had chromium allergy had no significant influence on whether or not they changed jobs, nor was it a cause of unemployment.[52] Unskilled chromium-sensitive workers are sometimes given the opportunity to change to a job within the factory where they will have no contact with wet cement,[52,71] thus enabling them to remain employed.

The social prognosis of chromium-sensitive workers is influenced by their age, with workers below the age of 40 having the best opportunity to change jobs or be successfully retrained.[52,77] In a German study, most workers experienced a loss of social standing upon entering a retraining program, especially if they were 50 years of age or older.[78] If older workers could retire with little loss of income from the age of 60, it might enable them to maintain their social standing. Social standing is here considered to be the sum of earnings as well as the maintenance of employment status in the peer group.[52]

D. ECONOMIC ASPECTS

In Singapore, allergic cement eczema was the only dermatosis severe enough to result in sick leave among workers in the construction industry, where absence due to occupational dermatoses was generally low.[79] Studies in the Netherlands[80] and the U.K.[64] have shown that the loss of working days in the building and construction industries due to occupational dermatitis have had significant financial consequences for these industries.

In Denmark, fewer workers consulted a dermatologist, and the need for topical steroid treatment decreased in 1987 as compared to 1981. These benefits were thought to result from the reduction in the chromate content of the cement.[2]

REFERENCES

1. **Jäeger, H. and Pelloni, E.,** Tests épicutanés aux bichromates, positifs dans l'eczema au ciment, *Dermatologica,* 100, 207, 1950.
2. **Avnstorp, C.,** Prevalence of cement eczema in Denmark before and since addition of ferrous sulfate to Danish cement, *Acta Derm. Venereol.,* 69, 151, 1989.
3. **Pirilä, V.,** On the role of chrome and other trace elements in cementeczema, *Acta Derm. Venereol.,* 34, 136, 1954.
4. **Nriagu, J. O.,** Production and uses of chromium, in *Chromium in the Natural and Human Environments,* Nriagu, J. O. and Nieboer, E., Eds., John Wiley & Sons, New York, 1988, 81.
5. **Burrows, D.,** Adverse chromate reactions on the skin, in *Chromium: Metabolism and Toxicity,* Burrows, D., Ed., CRC Press, Boca Raton, FL, 1983, 137.
6. **Høvding, G.,** Cement Eczema and Chromium Allergy, an Epidemiologic Investigation, Doctoral thesis, University of Bergen, Bergen, Norway, 1970.
7. **Hunziker, N. and Musso, E.,** A propos léczéma au ciment, *Dermatologica,* 121, 204, 1960.
8. **Peck, S., Gant, J. Q., and Schwartz, L.,** "Hardening" in industrial allergic dermatitis, *Ind. Med.,* 14, 214, 1945.
9. Portland cement, Betonteknik, Cementfabrikkernes Tekniske Oplysningskontor, Ålborg, 1980.
10. **Bogue, R. H., Portland cement,** in *Encyclopedia of Chemical Technology,* Vol. 3, Finle, G. J., Ed., 1949, 411.
11. **Fregert, S.,** Chromium valencies and cement dermatitis, *Br. J. Dermatol.,* 105 (Suppl. 21), 7, 1981.
12. **Skomorowski, E.,** Chromium in Portland cement, *Cement Wapno-Gips,* 1, 1, 1985.
13. **Fregert, S. and Gruvberger, B.,** Chemical properties of cement, *Dermatosen,* 20, 238, 1972.
14. **Brun, R. M.,** Contribution a l'étude des chromates du ciment. Nouvelle technique pour le test épicutané au ciment, *Dermatologica,* 129, 79, 1964.
15. **Fregert, S. and Gruvberger, B.,** Correlation between alkali sulphate and water soluble chromate in cement, *Acta Derm. Venereol.,* 53, 225, 1973.
16. **Lund, M. H.,** Chrom i cement, *Nord. Betong,* 6, 28, 1977.
17. **Fregert, S., Gruvberger, B., and Sandahl, E.,** Reduction of chromate in cement by iron sulfate, *Contact Dermatitis,* 5, 39, 1979.
18. **Baer, R. L.,** The mechanism of allergic contact dermatitis, in *Contact Dermatitis,* 3rd ed., Fisher, A. A., Ed., Lea & Febiger, Philadelphia, 1986, 1.
19. **Magnusson, B. and Kligman, A. M.,** The identification of contact allergens by animal assay. The guinea pig maximization test, *J. Invest. Dermatol.,* 52, 268, 1969.
20. **Cronin, E.,** *Contact Dermatitis,* Churchill Livingstone, Edinbourgh, 1980, 288.
21. **Perone, V. B., Moffitt, A. E., Possick, P. A., Key, M. M., Danzinger, S. J., and Gellin, G. A.,** The chromium, cobalt and nickel contents of American cement and their relationship to cement dermatitis, *Am. Ind. Hyg. Assoc. J.,* 35, 301, 1974.

22. **Reifenstein, H., Lück, H., Pätzold, M., and Harms, U.,** Zur häufigkeit des zementeksems bei der verarbeitung chromatarmer zemente, *Z. Gesamte Hyg. Ihre Grenzgeb.,* 32, 559, 1986.
23. **Avnstorp, C.,** Cementeksem Blandt Danske Arbejdere Beskæftiget pa Byggepladser og i Industrien, Ph.D. thesis, University of Copenhagen, Copenhagen, Denmark, 1983 (English summary included).
24. **Rostenberg, A.,** Primary irritant dermatitis versus allergic eczematous contact dermatitis, in *Dermatoses Due to Environmental and Physical Factors,* Rees, R. B., Ed., Charles C Thomas, Springfield, IL., 1962, 15.
25. **Vickers, H. R. and Edwards, D. H.,** Cement burns, *Contact Dermatitis,* 2, 73, 1976.
26. **Hannuksela, M., Suhonen, R., and Karvonen, J.,** Caustic ulcers caused by cement, *Br. J. Dermatol.,* 95, 547, 1976.
27. **Flowers, M. W.,** Burn hazard with cement, *Br. Med. J.,* 1, 1250, 1978.
28. **Anderson, F. E.,** Cement and oil dermatitis. The part played by chrome sensitivity, *Br. J. Dermatol.,* 72, 108, 1960.
29. **Wahlberg, J. E., Lindsted, G., and Einarsson, Ö,** Chromium, cobalt and nickel in Swedish cement, detergents, mould and cutting oils, *Dermatosen,* 25, 220, 1977.
30. **Fisher, A. A.,** Allergic dermatitis of the hands due to industrial oils and fluids, *Cutis,* 23, 131, 1979.
31. **Mali, J. W. H., Van Kooten, W. J., Van Neer, F. C. J.,** Some aspects of the behavior of chromium compounds in the skin, *J. Invest. Dermatol.,* 41, 111, 1963.
32. **Mali, J. W. H., Van Kooten, W. J., and Van Neer, F. C. J., and Spruit, D.,** Quantitative aspects of chromium sensitization, *Acta Derm. Venereol.,* 44, 44, 1964.
33. **Flesch, P.,** The role of epidermal components in cutaneous sensitization to some chemical agents, in *Proc. Congr. Hungarian Dermatological Society,* 1965, 113.
34. **Siegenthaler, U., Laine, A., and Polak, L.,** Studies on contact sensitivity to chromium in the guinea pig. The role of valence in the formation of the antigenic determinant, *J. Invest. Dermatol.,* 80, 44, 1983.
35. **Samitz, M. H. and Katz, S.,** A study of the chemical reactions between chromium and skin, *J. Invest. Dermatol.,* 43, 35, 1964.
36. **Samitz, M. H., Katz, S., Scheiner, D. M., and Gross, P. R.,** Chromium — protein interactions, *Acta Derm. Venereol.,* 49, 142, 1969.
37. **Shelley, W. B. and Juhlin, L.,** Langerhans cells form a reticuloepithelial trap for external contact antigens, *Nature (London),* 261, 46, 1976.
38. **Lück, H. and Jentsch, G.,** Chromium dermatitis caused by epoxy resin, *Contact Dermatitis,* 19, 154, 1988.
39. **Avnstorp, C., Ralfkjae, E., Jørgensen, J., and Wantzin, G. L.,** Sequential immunophenotypic study of lymphoid infiltrate in allergic and irritant reactions, *Contact Dermatitis,* 16, 239, 1987.
40. **Scheper, R. J., von Blomberg, M., Boerrigter, G. H., Bruynzeel, D., van Dinther, A., and Vos, A.,** Induction of immunological memory in the skin. Role of local T cell retention, *Clin. Exp. Immunol.,* 51, 141, 1983.
41. **Lidén, S. and Lundberg, E.,** Penetration of chromium in intact human skin *in vivo, J. Invest. Dermatol.,* 72, 42, 1979.
42. **Pedersen, N. B., Bertilsson, G., Fregert, S., Lidén, K., and Rorsman, H.,** Disappearance of chromium injected intracutaneously, *Int. Arch. Allergy Appl. Immunol.,* 36, 82, 1969.
43. **Polak, L.,** Immunology of chromium, in *Chromium: Metabolism and Toxicity,* Burrows, D., Ed., CRC Press, Boca Raton, FL, 1983, 51.
44. **Zelger, J.,** Zur klinik und pathogenese des chromat-ekzems, *Arch. Klin. Exp. Dermatol.,* 218, 499, 1964.
45. **Menné, T. and Maibach, H. I.,** Systemic contact allergy reactions, *Semin. Dermatol.,* 6, 108, 1987.
46. **Schroeder, H. A., Balassa, J. J., and Tipton, I. H.,** Abnormal trace metals in man — chromium, *J. Chronic Dis.,* 15, 941, 1961.
47. **Hansky, J. and Connell, A. M.,** Measurement of gastrointestinal transit using radioactive chromium, *Gut,* 3, 187, 1962.
48. **Love, G.,** Chromium — biological and analytical considerations, in *Chromium: Metabolism and Toxicity,* Burrows, D., Ed., CRC Press, Boca Raton, FL, 1983, 1.
49. **Veien, N. K., Hattel, T., Justesen, O., and Nørholm, A.,** Oral challenge with metal salts. I. Vesicular patch-test-negative hand eczema, *Contact Dermatitis,* 9, 402, 1983.
50. **Fregert, S. and Övrum, P.,** Chromate in welding fumes with special reference to contact dermatitis, *Acta Derm. Venereol.,* 43, 119, 1963.
51. **Shelley, W. B.,** Chromium in welding fumes as cause of eczematous hand eruption, *J. Am. Med. Assoc.,* 189, 772, 1964.
52. **Avnstorp, C.,** Follow-up of workers from the prefabricated concrete industry after the addition of ferrous sulfate to Danish cement, *Contact Dermatitis,* 20, 365, 1989.
53. **Förström, L., Pirilä, V., and Huju, P.,** Rehabilitation of Workers with Cement Eczema due to Hypersensitivity to Bichromate, Rep. 69, Institute of Occupational Health, Helsinki, Finland, 1969.

54. **Fregert, S. and Rorsman, H.,** Allergy to chromium, nickel and cobalt, *Acta Derm. Venereol.,* 46, 114, 1966.
55. **Fregert, S.,** Occupational dermatitis in a 10-year material, *Contact Dermatitis,* 1, 96, 1975.
56. **Fregert, S. and Gruvberger, B.,** Solubility of cobalt in cement, *Contact Dermatitis,* 4, 14, 1978.
57. **Wahlberg, J. E., Lindsted, G., Einarsson, Ö.,** Chromium, cobalt and nickel in Swedish cement, detergents, mould and cutting oils, *Dermatosen,* 25, 220, 1977.
58. **Goh, C. L. and Gan, S. L.,** Rubber allergy among construction workers in a prefabrication construction factory, *Clin. Exp. Dermatol.,* 12, 332, 1987.
59. **Adams, R. M.,** *Occupational Skin Disease,* Grune & Stratton, New York, 1983, 397.
60. **Wilkinson, J. D. and Rycroft, R. J. G.,** Contact Dermatitis, in *Textbook of Dermatology,* 4th ed., Rook, A., Wilkinson, D. S., Ebling, F. J. G., Champion, R. H., and Burton, J. L., Eds., Blackwell Scientific, Oxford, 1986, 435.
61. **Nilsson, E.,** Contact sensitivity and urticaria in "wet" work, *Contact Dermatitis,* 13, 321, 1985.
62. **Orris, L. and Tesser, M.,** Dermatoses due to water, soaps, detergents and solvents, in *Occupational and Industrial Dermatology,* Maibach, H. I. and Gellin, G. A., Eds., Year Book Medical Publishers, Chicago, 1982, 23.
63. **Van Neste, D., Mahmoud, G., and Masmoudi, M.,** Experimental induction of rough dermatitic skin in humans, *Contact Dermatitis,* 16, 27, 1987.
64. **Burrows, D. and Calnan, C. D.,** Cement dermatitis. II. Clinical aspects, *Trans. St. John's Hosp. Dermatol. Soc.,* 51, 27, 1965.
65. **Kanan, M. W.,** Cement dermatitis and atmospheric parameters in Kuwait, *Br. J. Dermatol.,* 86, 155, 1972.
66. **Goh, C. L. and Soh, S. D.,** Occupational dermatoses in Singapore, *Contact Dermatitis,* 11, 288, 1984.
67. **Feuerman, E. J.,** Housewives' eczema and the role of chromates, *Acta Derm. Venereol.,* 49, 288, 1969.
68. **Garcia-Perez, J., Martin-Pascual, A., and Sanches-Misiego, A.,** Chrome content in bleaches and detergents. Its relationship to hand dermatitis in women, *Acta Derm. Venereol.,* 53, 353, 1973.
69. **Peltonen, L. and Fräki, J.,** Prevalence of dichromate sensitivity, *Contact Dermatitis,* 9, 190, 1983.
70. **Wahlberg, J. E.,** Yrkeshudsjukdomar hos Byggnadsarbetare, Byggnadsindustrins Forskningsrapporter och Uppsatser, Stockholm, Sweden, 1968.
71. **Coenraads, P. J.,** Prevalence of eczema and other dermatoses of the hands and forearms in construction workers in the Netherlands, in Prevalence of Hand Eczema. Association with Occupational Exposure, Especially in Construction Workers, Thesis, University of Groningen, Groningen, The Netherlands, 1983, 86.
72. **Varigos, G. A. and Dunt, D. R.,** Occupational dermatitis. An epidemiological study in the rubber and cement industries, *Contact Dermatitis,* 7, 105, 1981.
73. **Dooms-Goossens, A., Ceuterich, A., Van Maele, N., and Degreff, H.,** Follow-up study of patients with contact dermatitis caused by chromates, nickel and cobalt, *Dermatologica,* 160, 249, 1980.
74. **Geiser, J. D. and Girard, A.,** Remarques sur les cas déczéma au ciment observés á la clinique de dermato-vénéreologie de Lausanne de 1947 á 1961, *Dermatologica,* 131, 93, 1965.
75. **Peter, K.,** Über das Schicksal der Patienten mit Gewerbeeksem, *Dermatologica,* 136, 236, 1968.
76. **Lammintausta, K., Maibach, H. I., and Wilson, D.,** Human cutaneous irritation: induced hyporeactivity, *Contact Dermatitis,* 17, 193, 1987.
77. **Hjorth, N. and Avnstorp, C.,** Rehabilitation in hand eczema, *Dermatosen,* 34, 74, 1986.
78. **Carrie, C. and Kühl, M.,** Rehabilitation von chronisch Hautkranken, *Dermatosen,* 16, 351, 1968.
79. **Goh, C. L.,** Sickness absence due to occupational dermatosis in a prefabrication construction factory, *Contact Dermatitis,* 15, 28, 1986.
80. **Coenraads, P. J.,** Sickness and absence from work due to skin diseases in the construction industry: review of literature, in Prevalence of Hand Eczema. Association with Occupational Exposure, Especially in Construction Workers, Thesis, University of Groningen, Groningen, The Netherlands, 1983, 70.

Chapter 36

SELECTION OF CHEMICAL PROTECTIVE CLOTHING

Stephen P. Berardinelli and Thomas R. Bender

TABLE OF CONTENTS

I. INTRODUCTION

Many chemical substances used or produced in workplaces can cause severe damage to human skin if contact is allowed to occur. Skin disorders, including injuries such as chemical burns and diseases such as various types of dermatitis, are very common, very costly, and often severely disabling disorders. Additionally, chemicals may permeate the skin and contribute to disease processes in other organ systems through percutaneous absorption. Thus, the skin is not only a target organ for injury and disease, but also provides a route of entry similar to respiration and ingestion.

Preferably, worker exposures to harmful chemical substances are either eliminated, as through substitution of nontoxic substances, or reduced to safe levels by design, as through the implementation of engineering or administrative controls. However, many view chemical protective clothing (CPC) as the most economical means of protecting workers from skin-damaging chemicals; therefore, CPC is widely used. If CPC materials fail, the worker may be at immediate risk of experiencing harmful exposures; consequently, CPC has been called the "last line of defense".

II. BACKGROUND

CPC is constructed from plastic and elastomeric materials. Selection of CPC is based on the performance of these materials as a barrier. Selection requires evaluation of three interactive parameters: chemical resistance of garment materials, physical properties of garment materials, and human factors.

III. CHEMICAL RESISTANCE

CPC chemical resistance testing evaluates the interaction between challenge chemicals and the garment material. There are three possible interactions: (1) chemical degradation, a breakdown of the garment's physical structure due to garment-chemical incompatibility; (2) chemical penetration, the bulk chemical flow through garment imperfections or through discontinuities such as seams and closures; and (3) chemical permeation, the molecular flow of chemicals through garment material.

The American Society for Testing and Materials (ASTM) F23 Committee has developed and validated consensus standard methods of measuring the chemical permeation of protective clothing (F739)[1] and the chemical penetration of protective clothing (F903).[1,2]

Degradation, the deterioration of one or more physical properties of CPC, can cause a glove to crack, to tear easily, or, in extreme cases, to dissolve in liquid. For example, surgical latex gloves are prone to chemical degradation; they generally soften and shrivel when exposed to organic chemicals. Such failures can result in large amounts of liquid chemical being transported through the glove to the skin. Usually, chemical degradation is caused by the interaction of chemicals with the CPC material. Chemical degradation data are used by the industrial hygienist to eliminate those gloves that are degraded by a particular liquid or mixture.[3] Those gloves that are not degraded are then evaluated for chemical penetration and chemical permeation.

Tests for chemical penetration rely on visual observation to ascertain whether liquid flows through imperfections (e.g., pinholes) in the garment material, or through seams and closures. The total test time is 15 min — 5 min at atmospheric pressure, then 10 min at 1 $lb_f/in.^2$ (gauge pressure).[2] The result of a chemical penetration test is either pass or fail. In general, limited penetration data are available from CPC manufacturers and CPC selection guides.

Permeation is the movement of a chemical through intact CPC material. This is a three-step process: (1) sorption of the liquid onto the outside of the glove, (2) diffusion of the liquid through the glove, and (3) desorption of the liquid from the inside of the glove. In this way, small amounts of liquid are transported to the skin. Two pieces of information are obtained by permeation testing: breakthrough time and steady-state permeation rate. The breakthrough time is the duration of time required for the liquid to be transported through CPC and be seen by the detector. The steady-state permeation rate is an equilibrium condition wherein the CPC is saturated with liquid so that there is a constant flow of liquid through the CPC. Both parameters are useful in selecting chemical protective clothing. The breakthrough time indicates the potential usable time; theoretically, if the CPC is removed before breakthrough, minimal exposure occurs. The steady-state permeation rate can be used to estimate the dermal exposure if the CPC is worn beyond the breakthrough time.[4]

The use of breakthrough time and steady-state permeation rate in glove selection is best described by an illustration. A worker may have dermal exposure for 20 min to a highly toxic chemical. Glove A and glove B have a breakthrough time of 60 min. Glove A has a very high steady-state permeation rate, while the permeation of glove B is much lower against the chemical. An industrial hygienist would select glove B because of its lower steady-state permeation rate (lower exposure).

Various brands of the same generic elastomer may have different permeation properties for the same liquid. This means that selection of chemical protective gloves must be based on specific permeation data.[5] Furthermore, the chemical permeation properties of chemical mixtures must be determined by testing, not inferred from the permeation characteristics of the individual mixture components.[6] Most research on chemical permeation and chemical penetration has been limited to pure liquids, yet in the workplace most liquids are mixtures. Industrial hygienists with a knowledge of chemistry may, in many cases, conduct chemical resistance testing themselves.[7,8]

IV. PHYSICAL PROPERTIES

Physical properties of CPC are important to barrier performance. Key physical properties are resistance to flexing, tearing, abrasion, cut, and puncture. Required physical properties must be considered in the selection process. The ASTM F23 Committee on Protective Clothing promulgates consensus standards. There are active F23 subcommittees on human factors, chemical resistance, and physical properties.[1]

Chemical protective clothing is made from elastic polymers, or elastomers, the general physical properties of which are listed in Table 1. Some ergonomic evaluations, such as tactility and grip, involve physical properties that are governed by glove thickness. This is especially true of dexterity and tactility. Surface texture is another important property; grip is enhanced by a rough surface. The physical requirements of the task must be balanced against the chemical resistance requirements and human factors based on the task analysis.

V. HUMAN FACTORS

There is an active ASTM F23 subcommittee on human factors. Human factors involve the interaction between worker and CPC. CPC, while protecting the worker, must not unduly restrict worker performance. One human factor of concern is the physiological stress to workers induced by the wearing of CPC. Other human factors are concerned with garment design: appropriate fit, acceptable manual dexterity, tactility, and grip. Ill fit, poor dexterity, poor tactility, and poor grip can mean that work takes longer to perform. Also, the risk of traumatic injury is increased.

CPC ensembles (level A, B or C) reduce work performance. Heat stress is a prime concern while wearing CPC outdoors or in hot indoor environments. An important stressor

TABLE 1
Physical Characteristics of Elastomers[a]

Material (designation in matrices)	Flexibility	Resistance						Relative cost
		Abrasion	Cut	Heat	Ozone	Puncture	Tear	
Butyl rubber (butyl)	G	F	G	E	E	G	G	High
Chlorinated polyethylene (CPE)	G	E	G	G	E	G	G	Low
Natural rubber	E	E	E	F	P	E	E	Medium
Neoprene	G	E	E	G	E	G	G	Medium
Nitrile rubber (nitrile)	E	E	E	G	F	E	G	Medium
Nitrile rubber/polyvinyl chloride (nitrile/PVC)	G	G	G	F	E	G	G	Medium
Polyethylene	G	F	F	F	F	P	F	Low
Polyurethane	E	E	G	G	G	G	G	High
Polyvinyl alcohol (PVA)	P	F	F	G	E	F	G	Very high
Polyvinyl chloride (PVC)	F	G	P	P	E	G	G	Low
Styrene-butadiene rubber (SBR)	G	E	G	G	F	F	F	Low
Viton	G	G	G	G	E	G	G	Very high

Note: E = excellent, G = good, F = fair, P = poor.

[a] Ratings are subject to variation depending on formulation, thickness, and whether the material is supported by fabric.

when outdoors is solar radiant heat. There are two ways to reduce such stress: (1) wear comfortable CPC, i.e., CPC which allows some evaporative cooling; and (2) wear a cooling garment under the CPC. More comfortable CPC is lightweight, usually woven fabric, which "breathes" or allows evaporative cooling (but is less chemically resistant). Traditional elastomeric CPC does not allow evaporative cooling, so heat stress may occur. References 9 to 15 comprise an excellent primer in this specialized area.

There are some tests, such as anthropometric sizing and fit testing and a manual dexterity test battery, that scientifically determine the effect which CPC may have on work performance.[12,13] However, in practice, candidate CPC is given to the workers. Workers then perform their usual assignments, noting how the CPC affects their job performance. An industrial hygienist oversees this operation. This is a simple but effective way to determine the effect of CPC on job performance and the worker acceptability of CPC. CPC must be worn if it is to perform.

VI. LEVELS OF PROTECTION

In the workplace, the presence of skin-irritating chemicals may be accompanied by the presence of respiratory irritants. Thus, protection against both types of irritants is usually required. The following describes how both types of protection are implemented.

First, an industrial hygienist conducts an on-site survey or inspection to establish chemical agents and potential worker exposure to these agents. Analysis of the toxicology of the chemical agents, the level and duration of exposure coupled with the probability of respiratory and dermal contact (contact time, body part exposed), permits selection of protective ensembles. There are four ensemble protective levels defined by the U.S. Environmental Protection Agency (EPA).[16]

Level A affords the worker the highest level of respiratory and dermal protection. The worker is in a controlled microenvironment, protected from the environment by a fully encapsulating chemically resistant suit and a life support system (positive pressure supplied-air respirator). Level A must be worn when there could be an exposure to high concentrations of pressurized toxic liquid, toxic vapor, or gas which can damage skin or which permeates skin.

Level B requires a high level of respiratory protection, as in level A, but lesser dermal protection. Chemical protective clothing, such as one- or two-piece splash suits or overalls and jacket, are typical of level B.

Level C requires the use of an air-purifying respirator to filter the ambient atmosphere. This means that the oxygen concentration is not less than 19.5% (by volume), and airborne contaminant concentrations are relatively low. The level C chemical protective clothing ensemble is the same as level B.

Level D is the lowest level of dermal and respiratory protection; no respiratory protection is worn and dermal protection is minimal. Using gloves to wash household dishes or to do gardening demonstrates level D protection. These levels are discussed in more detail in Reference 16.

Up to this point, the worker's potential risk has been evaluated and a level of protection has been selected. However, the composition of the CPC material has *not* been selected. Chemical resistance and physical properties of materials should be evaluated for proper selection of CPC. These parameters are evaluated during the on-site survey. Each chemical agent interacts with a given chemical protective garment in a unique way, such that generalities have many exceptions.

VII. SELECTION PROCESS

In practical terms, the physical and chemical properties of CPC are derived from tables, charts, and general references and are used to select candidate gloves. Chemical resistance data that are specific to the brand of CPC are usually available from the manufacturer. References 16 and 17 list CPC manufacturers and distributors. The former also includes tables that can be used to select chemical protective clothing based on chemical degradation and chemical permeation. Also, it includes a list of resource persons who have knowledge of chemical protective clothing. A new pocket guide which has color-coded chemical resistance tables is also available.[18]

Finally, the candidate CPC must be evaluated under actual use conditions against the specific chemicals or mixtures used. Candidate CPC should be used by workers, who in turn report on its performance to the industrial hygienist. In this way, the appropriate human factors and physical properties required for adequate job performance can be evaluated. As stated earlier, chemical resistance is frequently evaluated by on-site testing.

VIII. SELECTION EXAMPLE

As an example, a young gas company employee is found to have allergic contact dermatitis caused by diglycidyl ether of bisphenol A (monomers and oligomers), a common binder in epoxy resin cement used to join segments of the plastic gas lines he installs. He wears disposable polyethylene gloves during this work. Skin patch testing to epoxy resin is positive regardless of whether polyethylene or polyvinyl chloride glove materials are worn. He does not react to the glove materials themselves. No respiratory protection is needed because exposure is limited to dermal exposure of the hands.

Survey analysis shows that moderate dexterity and tactility is needed to apply the epoxy resin and join the pipes. The extent of contact with the resin is considerable (several minutes many times a day). No other physical factors are significant considerations. The exposure analysis shows a mixture of epoxy resin cement with diglycidyl ether bisphenol A binder.

Permeation data reveal that polyethylene/ethylene vinyl alcohol laminate is a suitable glove material, with a breakthrough time for epoxy mixture of more than 8 h.[18] All other gloves tested have a breakthrough time of less than 10 min.

Polyethylene/ethylene vinyl alcohol gloves are selected since they have appropriate physical properties and the best chemical resistance. The gloves are discarded every 4 h — at lunch and at the end of the work shift, long before breakthrough could occur.

IX. CONCLUSION

Selection of CPC requires the services of a knowledgeable safety and health professional. Evaluation of the workplace dictates the level of protection (e.g., level A). Evaluation of chemical resistance of garment materials and physical properties of garment materials dictates the elastomeric or plastic composition of that level of protection (garment). Evaluation of human factors dictates how well and/or how long the users can perform their job assignment while wearing CPC.

DISCLAIMER

Mention of brand names does not constitute endorsement by the National Institute for Occupational Safety and Health, the Centers for Disease Control, the U.S. Public Health Service, or the Department of Health and Human Services.

REFERENCES

1. **Norman, H., III,** Chairman, American Society for Testing and Materials Committee F-23 (Protective Clothing), Haskell Laboratory, E. I. DuPont deNemours, Newark, DE.
2. **Mansdorf, S. Z. and Berardinelli, S.,** Chemical protective clothing standard test method development. I. Penetration test method, *Am. Ind. Hyg. Assoc. J.*, 49(1), 21, 1988.
3. **Coletta, G. C., Mansdorf, S. Z., and Berardinelli, S.,** Chemical protective clothing standard test. Chemical protective clothing standard test method development. II. Degradation test method, *Am. Ind. Hyg. Assoc. J.*, 49(1), 26, 1988.
4. **Berardinelli, S. P.,** Prevention of occupational skin disease through use of chemical protective gloves, *Occup. Derm. Dermatol. Clin.*, 6(1), 115, 1988.
5. **Mickelsen, R. L. and Hall, R. C.,** A breakthrough time comparison of nitrile and neoprene glove materials produced by different glove manufacturers, *Am. Ind. Hyg. Assoc. J.*, 48(11), 941, 1987.
6. **Mickelsen, R. L., Roder, M. M., and Berardinelli, S.,** Permeation of chemical protective clothing by three binary solvent mixtures, *Am. Ind. Hyg. Assoc. J.*, 47(4), 236, 1986.
7. **Berardinelli, S., Ruscyck, R. A., and Mickelsen, R. L.,** A portable chemical protective clothing test method: applications at a chemical plant, *Am. Ind. Hyg. Assoc. J.*, 48(9), 804, 1987.
8. **Berardinelli, S. and Roder, M.,** Chemical protective clothing field evaluation methods, in *Performance of Protective Clothing,* ASTM STP 900, Barker, R. and Coletta, G., Eds., American Society for Testing and Materials, Philadelphia, 1986, 250.
9. **Tayyari, F.,** Evaluation of a prototype microclimate cooling system, *Am. Ind. Hyg. Assoc. J.*, 50(4), 229, 1989.
10. **Parell, J. M. and Rosenthal, F.,** Heat strain and heat stress for workers wearing protective suits at a hazardous waste site, *Am. Ind. Hyg. Assoc. J.*, 48(5), 458, 1987.
11. **White, M. K. and Hodous, T. K.,** Physiological responses to wearing of fire fighter's turnout gear with neoprene and Gore-Tex barrier liners, *Am. Ind. Hyg. Assoc. J.*, 49(10), 523, 1988.
12. **Mansdorf, S. Z., Sager, R., and Nielsen, A. P.,** *Human Factors in Performance of Protective Clothing,* 2nd Symp., STP 989, American Society for Testing and Materials, Philadelphia, 1988.
13. **Barker, R. and Coletta, G.,** *Heat Stress Fit Testing and Other Performance Requirements for Protective Clothing in Performance Protective Clothing,* STP 900, American Society for Testing and Materials, Philadelphia, 1986.

14. **Ronk, R.,** Personal Protective Equipment for Hazardous Materials Incidents: A Selection Guide, Doc. No. PB85-222230, National Technical Information Service, Springfield, VA, 1984.
15. **Plog, B. A.,** Recognition of hazards temperature extremes, in *Fundamentals of Industrial Hygiene,* 3rd ed., Part III, National Safety Council, Chicago, IL, 1988.
16. **Schwope, A. D.,** Guidelines for the Selection of Chemical Protection Clothing, 3rd ed., U.S. Environmental Protection Agency, Washington, D.C., 1988.
17. *Best's Safety Directory,* Vol. 1 and 2, Best AM Co., Oldwick, NJ, 1989.
18. **Forsberg, K. and Mansdorf, S. Z.,** *Quick Selection Guide for Chemical Protective Clothing,* Van Nostrand Reinhold, New York, 1989.

Chapter 37

DECONTAMINATION OF CHEMICAL PROTECTIVE CLOTHING

Stephen P. Berardinelli and Thomas R. Bender

TABLE OF CONTENTS

I. INTRODUCTION

Chemical protective clothing (CPC) is worn to prevent harmful chemical agents from contacting the skin. Normally, CPC comes in contact with such chemical agents, however, and becomes contaminated. Surface contamination can lead to internal (interstitial) contamination by way of permeation. Usually, in order for contaminated CPC garments to be removed (doffed) without a harmful amount of the contaminant contacting the wearer's skin, the amount of contaminant on the surface of the CPC must be removed or reduced. In order for contaminated CPC to be reused, both surface and internal contamination must be removed or reduced. The process of removing or reducing the amount of contaminants from CPC garments is known as decontamination.

II. BACKGROUND

In order to understand decontamination of CPC, one must understand the two modes of contamination: (1) Surface or outer contamination and (2) internal or interstitial contamination. The doffing procedure requires surface decontamination.[1] Reuse of CPC requires internal decontamination. Time is important because chemicals will permeate from the surface to the interstices of the CPC material. Surface contaminants are removed more easily than internal contaminants. Also viscous liquids and semisolid sludges (fudges) are the most difficult contaminant to remove, followed by easy-pouring liquids, solids, and then gases.[2]

III. DECONTAMINATION AGENTS

A decontamination agent must meet certain criteria:

1. Contaminants are soluble in it.
2. It causes no permanent physical damage to the CPC.
3. It has low toxicity.
4. It is cost effective.
5. It has low flammability.
6. It is stable.

Aqueous solvents, organic solvents, and even air (aeration) may be effective decontaminating agents. Aqueous solvents are soaps or detergents dissolved in water. Typically, aqueous decontaminating agents are not reused. Organic solvents such as Stoddard solvent and perchloroethylene, which meet the six criteria above, have been used. Some organic solvents are captured (redistilled) and reused. Rapidly moving air, usually heated, may be a decontaminating agent. However, a contaminant must have a moderate to high vapor pressure to dissolve in air (volatilize).[2]

IV. DECONTAMINATION PROCEDURES

A. SURFACE DECONTAMINATION

Surface decontamination removes gross surface contaminants so that the garment may be removed. Surface decontamination is much like home laundering. Soaps and detergents (aqueous solvents) cleanse garment surfaces by dispersion (emulsification) of low-viscosity liquids or powders.[1] Multiple aqueous solvent washes and rinses are more effective than a single wash and rinse.[3,4] Sticky viscous liquid or semisolids may never be fully removed from surfaces by this method.[5] Surface decontamination using organic solvents (i.e., dry cleaning) demonstrates mixed results. The organic solvents tend to permeate CPC materials. Also, organic solvents may cause chemical degradation.[2] Therefore, surface decontamination with organic solvents should not be practiced.

Surface decontamination using warm or hot air is effective for liquids with moderate to high vapor pressure; however, sticky liquids and semisolids, which do not vaporize readily due to low vapor pressure, require other decontamination methods.

B. INTERNAL DECONTAMINATION

Internal or interstitial contamination occurs when the chemical has permeated into the CPC. The degree of internal contamination is determined by the solubility of the chemical agent in the garment material. Matrix release, or persistent permeation, has been investigated; this is the worst case of internal contamination.[6] Controlled release technology has demonstrated that organic chemical reservoirs can be created in elastomers or plastics (materials used in protective clothing) which will slowly release chemicals for extended periods of time. That is, permeation rates will be low and permeation will continue for days to months.[2] Aqueous solvents are ineffective decontaminating agents against internal contamination.[7,8]

Organic solvents must permeate the CPC material to dissolve, then remove, other permeants. As a result, washing with organic solvents is usually counterproductive.[4,8] These organic solvents may not only cause chemical degradation, but also internally contaminate the CPC themselves, requiring their removal before the affected garment can be safely reused.[6]

Hot air drying (thermal decontamination) to remove internal permeants has been shown to be effective by several studies;[6,7-9] however, all of these studies used only a piece, or swatch, of material, rather than the entire garment. Hot air drying is suitable only for contaminants with moderate or high vapor pressure. Furthermore, variables such as temperature have not been evaluated. The higher the temperature, the larger the vapor pressure — hence, the more effective the decontamination. However, elevated temperatures can adversely affect one or more physical properties of the CPC.[9] In order to reuse CPC, effective internal decontamination must be demonstrated.

C. ESTABLISHING WORK ZONES

It is difficult to determine what level of contamination of the work environment is permissible, even in specific situations, and there are no standards for guidance. However, the "work zones" concept contains or limits contamination.[10] Three zones or levels of contamination are established for work areas. These zones are an inner or most contaminated zone, sometimes called the "hot zone", where harmful contact to chemicals may occur; the middle zone, or contamination reduction zone, which is a transition area where personnel enter into the contamination zone or exit from the contamination zone (CPC is donned and doffed in this zone); and the outermost zone or support zone, which should be a clean or noncontaminated area. Work zones may be thought of as three concentric circles. Use of work zones will restrict contamination to specific work areas and will avoid cross-contamination. This concept should be adopted routinely.

V. CONCLUSION

The most prudent recommendation is single use, then disposal, of CPC. The optimal decontamination method is the use of aqueous agents to clean the surface, followed by hot air drying for internal decontamination.

Many sticky liquid or semisolid chemicals that permeate cannot be removed satisfactorily from the CPC interior. In these cases, it is best to dispose of the CPC rather than attempt cleaning.

DISCLAIMER

Mention of brand names does not constitute endorsement by the National Institute for Occupational Safety and Health, the Centers for Disease Control, the Public Health Service, or the U.S. Department of Health and Human Services.

REFERENCES

1. National Institute for Occupational Safety and Health, Occupational Safety and Health Administration, U.S. Coast Guard, and U.S. Environmental Protection Agency. Occupational Safety and Health Guidance Manual for Hazardous Waste Site Activities, DHHS(NIOSH) No. 85-115, National Institute for Occupational Safety and Health, Cincinnati, OH, 1985.
2. Chemical Protective Clothing Workshop Manuel, S. Z. Mansdorf & Associates, Inc., Cuyahoga Falls, OH, 1989.
3. **Hearchall, J. C., Laughlin, J., and Gold, R.,** Effect of laundering procedures and functional finishes on removal of insecticides from three chemical classes, in *Performance of Protective Clothing,* Barker, R. C. and Coletta, G. C., Eds., ASTM STP 900, American Society for Testing and Materials, Philadelphia, 1986, 162.
4. **Schlatter, C. N.,** Effects of water rinsing on subsequent permeation of rubber chemical protective gloves, in *Performance of Protective Clothing: Second Symposium,* Mansdorf, S. Z., Sager, R., and Nielsen, A. P., Eds., ASTM 989, American Society for Testing and Materials, Philadelphia, 1988, 376.
5. **Stampfer, J. F., McLeod, M. J., Betts, M. R., Martinez, A. M., and Berardinelli, S.,** Permeation of polychlorinated biphenyls and solution of these substances through protective clothing materials, *Am. Ind. Hyg. Assoc. J.*, 45, 634, 1984.
6. **Garland, C. E. and Torrence, A. M.,** Protective clothing materials: chemical contamination and decontamination concerns and possible solutions, *Am. Ind. Hyg. Assoc. J.*, 45, 368, 1984.
7. **Berardinelli, S. and Hall, R.,** Decontamination of chemical protective clothing exhibiting matrix release, in *Performance of Protective Clothing: Second Symposium,* Mansdorf, S. Z., Sager, R., and Nielsen, A. P., Eds., ASTM 989, American Society for Testing and Materials, Philadelphia, 1988, 359.
8. **Vahdat, N. and Delaney, R.,** Decontamination of chemical protective clothing, *Am. Ind. Hyg. Assoc. J.*, 50, 152, 1989.
9. **Perkins, J. L., Johnson, J. S., Swaregen, P. M., Sackett, C. P., and Weaver, S. C.,** Residual spilled solvents in butyl protective clothing and usefulness of decontamination procedures, *Appl. Ind. Hyg.*, 2, 179, 1987.
10. **Schwope, A. D., Costas, P. P., Jackson, J. O., Stull, J. O., and Weitzman, D. J.,** Guidelines for the Selection of Chemical Protection Clothing, 3rd ed., U.S. Environmental Protection Agency, Washington, D.C., 1987.

Chapter 38

FLUORESCENT LIGHT AND SKIN CANCER RISK

Dennis F. Gross and Darrell S. Rigel

TABLE OF CONTENTS

I. INTRODUCTION

The use of fluorescent light in the workplace has increased steadily since the late 1940s, when it replaced less efficient tungsten lamps. Whether long-term occupational exposure to fluorescent light increases the risk of malignant melanoma remains controversial. Several clinical studies support this conclusion,[1-3] although the majority of series reported in the literature do not find an association.[4-8] Still, other biological and clinical observations suggest a carcinogenic potential of fluorescent light in doses consonant with occupational exposure.

Fluorescent bulbs emit visible light (wavelength of 400 to 700 nm) and low levels of electromagnetic radiation in the ultraviolet (UV) range (wavelength of 280 to 320 nm). Solar radiation which contains ultraviolet-B (UV-B) is a major risk factor for melanoma as well as for basal cell and squamous cell carcinomas.[9-13] Although the UV-B emission by fluorescent bulbs is minuscule relative to the sun, the cumulative yearly exposure to certain UV-B wavelengths is greater for individuals working under fluorescent lights than for those spending equal time in the sun.[14,15] Low-wavelength UV-B (290 to 295 nm) is emitted selectively by some fluorescent tubes, whereas such rays emitted from the sun are generally refracted by the ozone layer away from the surface of the earth.

Laboratory experiments have demonstrated that procaryotic and eucaryotic cell cultures develop genetic mutations in a dose-related fashion when exposed to fluorescent light.[16-21] Similar studies found a malignant transformation in mouse embryo cell cultures induced by fluorescent lights.[22] Clinical investigations have shown fluorescent lights to cause cutaneous burns similar to sunburns among female factory workers[23,24] and to possibly increase the risk of benign skin cancers.[25] The following chapter will attempt to evaluate the results of epidemiologic and basic science research in assessing whether an occupational exposure to fluorescent light can significantly increase the risk of malignant melanoma.

II. FLUORESCENT LIGHT AS A POTENTIAL CARCINOGEN: RELEVANT PHYSICS AND BIOLOGY

The association between malignant melanoma and fluorescent light exposure is supported by the demonstration of UV-B emission from fluorescent bulbs.[14,15] UV-B exposure has been implicated in the pathogenesis of melanoma and nonmelanoma skin cancers.[9-13] As shown in Figure 1, the nonionizing radiation emitted by the sun includes (in order of increasing wavelengths) UV-B, ultraviolet-A (UV-A), visible light, and infrared rays. Wavelengths shorter than 290 to 295 nm do not reach the earth's surface because they are absorbed by ozone in the stratosphere.[26] However, some fluorescent bulbs emit these short UV-B wavelengths and may result in more exposure to these rays than one would receive with equal hours of sun exposure.[14] This phenomenon might partially explain the carcinogenic potential of fluorescent light reported by several investigators.

A fluorescent bulb is a glass-enclosed chamber containing mercury vapor at low pressure and electric coils at each end (Figure 2). The inner surface of the glass tube is coated with phosphor. When the fluorescent bulb is discharged, the predominant electromagnetic energy produced is UV radiation and very little visible light. The dominant wavelength produced is 254 nm (UV-C), with additional peaks in the UV-A and UV-B spectra.[2] As the UV radiation exits the tube it passes through the phosphor coat, which converts it to wavelengths within the visible light spectra (400 to 700 nm) (Figure 2). However, this transformation is not 100% efficient, and UV-B wavelengths may be emitted. Styrene diffusers which fully cover the fluorescent bulbs markedly reduce the final index of UV-B emission without lowering the irradiance of visible light.[14] Diffusers of the open lattice type or those made of acrylic, however, remove much less UV-B.

Maxwell and Elwood[14] calculated the average daily dose of UV-B according to wave-

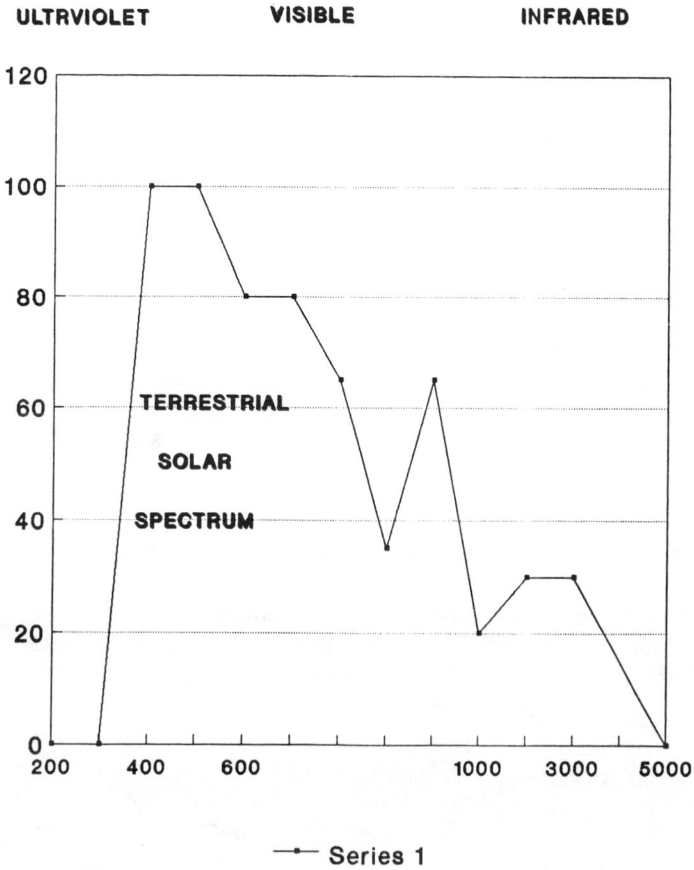

FIGURE 1. Relative intensity and wavelength of solar radiation.

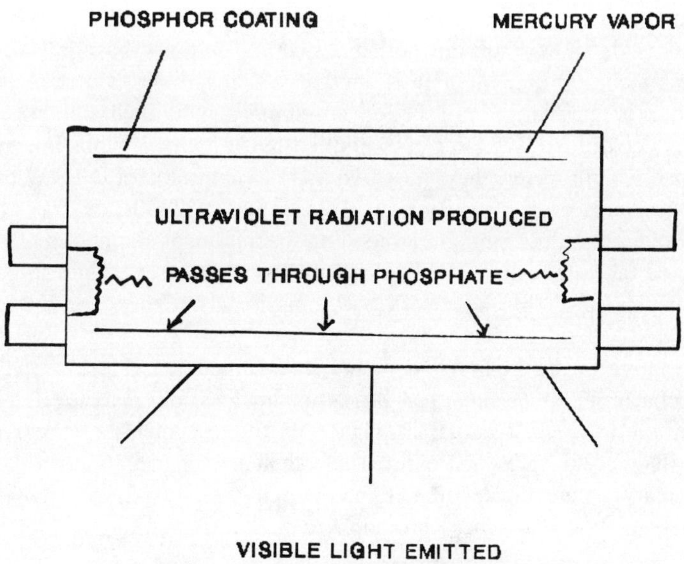

FIGURE 2. Structure of fluorescent light bulb.

TABLE 1
Comparison Average Daily Ultraviolet Exposure to Specific Wavelengths Produced by Sunlight (34° Latitude) and Nine Commercially Available Fluorescent Lights

Wavelength (nm)	Solar exposure[a]	Fluorescent bulb[a] 1	2	3	4	5	6	7	8	9
290	0.4	4	8	8	4	4	4	12	6	8
295	92	*	106	112	*	*	*	*	*	*
300	471	*	*	*	*	*	*	*	*	*

Note: Undiffused fluorescent light sources were used. (*) Denotes emission by fluorescent bulb is less than solar exposure.

[a] Units of measure $= \frac{\mu J \times mm}{cm^2}$

length for nine different commercially available fluorescent lightbulbs. This calculated yearly exposure assumes a constant exposure for an 8-h work day, 5 days a week for 50 weeks. For comparison, Maxwell calculated the average daily UV-B dose based on sun exposure at 34° latitude (Sydney, Australia) from sunrise to sunset, every day of the year. As shown in Table 1, the ozone layer is responsible for a solar exposure of 0 for UV-B wavelengths 280 and 285 nm and for a minimal exposure of 0.4 for wavelength 290 nm. Because the ozone shield becomes increasingly permeable to wavelengths greater than 290 nm, the solar exposure to these rays increases exponentially. Maxwell demonstrated that, compared to solar exposure, all fluorescent bulbs tested emitted 10 to 30 times more UV-B at wavelength 290 nm. Some tubes exceeded solar exposure for wavelength 295 nm as well, although this difference never exceeded one logarithm. In addition, these short-wavelength 290- to 295-nm UV-B rays possess more energy than do the longer rays and, therefore, may be more deleterious to melanocytes in the epidermis. The potential to cause DNA damage is also greater with short-wavelength UV-B because cellular DNA of whole epidermis has a maximal absorption around 260 nm.[25] This threat to basal melanocytes is somewhat countered by the aromatic amino acids of the stratum corneum, which partially absorb UV radiation of wavelengths less than 300 nm. Melanin is another UV-absorbing epidermal chromophore, but this protection is markedly diminished in fair-skinned individuals.

Rigel et al.[5] determined that the UV-A emitted by fluorescent bulbs is essentially negligible. Using a UV light meter, they measured a UV-A emission of 0.7 $\mu W/cm^2$ from two 35-W fluorescent bulbs at 1.5 m. By contrast, a value of 2300 $\mu W/cm^2$ of UV-A was measured from indirect indoor sunlight at noon on the autumnal equinox in New York. The authors calculated the total yearly UV-A exposure from fluorescent light to equal only 40 min of direct sun exposure. This degree of UV-A emission is unlikely to produce an increased risk of melanoma.

Webb and Malina[17] were the first to demonstrate that moderate fluorescent light exposure is mutagenic to bacteria. *Escherichia coli* had a maximal 18-fold increase in mutations that was directly proportional to the total irradiance of fluorescent light. These results were attained using fluorescent bulbs that emitted less than 1% of their light in the UV range. Webb's preliminary data indicated that light did not directly mutate DNA, but instead photodynamically activated another chromophore (e.g., riboflavin) that intercalated into DNA.

Experiments have demonstrated that fluorescent light is mutagenic to cultured mammalian cells and can lead to malignant transformation. Kennedy et al.[22] showed that mouse embryo cell cultures exposed to fluorescent light had a significant increase in DNA pyrimidine dimer formation and resulted in malignant transformation. This effect was dose dependent, and a maximal tenfold increase in transformation frequency occurred with 5 h of fluorescent light exposure. Gantt et al.[16] exposed mouse cell cultures to two 15-W fluorescent bulbs for 20 h, which resulted in significant chromatid breakage and approximately 50 chromosomal cross-links per cell. Areas of dense cell growth showed fewer DNA aberrations than smaller cell clusters, indicating a density-dependent mechanism for either the penetration of fluorescent light or the absorption of a photoactivated mutagen.

Ritter and Williams[19] showed that fluorescent light caused pyrimidine dimers, a known carcinogenic mutation, in fibroblasts of patients with xeroderma pigmentosum. Similar mutations were produced in normal human fibroblast cultures, but these were repaired within 24 h of discontinuing fluorescent light exposure. The rate of pyrimidine dimer formation increased linearly with increased fluorescent light exposure. Other work by Parshad and Sanford[27,28] demonstrated a significantly increased proliferation of *in vitro* human fibroblasts exposed to fluorescent lights. The increased mitotic rate was reversible and peaked at 160% above unexposed controls. When cultures were subjected to one 4-h exposure, the increased proliferative rate reverted to baseline within 1 week in the absence of light. However, intermittent exposure for 1 month required more than 1 week to revert to a lower mitotic rate. The authors concluded that fluorescent light exposure had a cumulative effect which simulated hyperplasia *in vitro*.

III. EPIDEMIOLOGY

The first major clinical study to suggest an increased risk of melanoma due to occupational fluorescent light exposure was reported by Beral et al. in 1982.[1] This retrospective study predominantly utilized women, 18 to 54 years old, as its original goal was to investigate whether oral contraception increases the risk factor of melanoma. The 274 women with malignant melanoma and the 549 age- and sex-matched controls all lived in New South Wales, Australia. Each patient was interviewed using a standardized questionnaire that included questions about their occupation and their exposure to fluorescent light. Other variables known to affect the risk of melanoma were also reviewed, including skin type, hair color, number of nevi, and sunlight exposure.

An occupational history ascertained by Beral determined whether work was conducted mainly indoors or outdoors for any job lasting at least one year. Earlier work revealed higher melanoma rates in Australia and Britain among professional and office workers[29,30,39] and lower rates for outdoor workers. If fluorescent lights were used in the workplace, the subject was asked whether the lights were on the majority of the day. Duration of exposure was calculated as the number of years fluorescent lights were turned on most of the time. Statistical analysis compared the relative risk of melanoma among indoor workers working under fluorescent lights vs. indoor workers who were never exposed to fluorescent lights.

The results of Beral's investigation revealed a significant dose-dependent increased risk of melanoma with occupational exposure to fluorescent lighting (Table 2). The relative risk (RR) significantly increased with greater cumulative exposure as follows: 1 to 9 years exposure RR = 1.9, 10 to 19 years RR = 2.5, and ≥20 years exposure RR = 2.6 (all intervals $p < 0.05$). Any exposure to fluorescent light doubled (RR = 2.1) the risk of melanoma among women working indoors. In addition, fluorescent lighting in an office (RR = 2.6) was more deleterious than exposure in other indoor work settings such as a warehouse or factory (RR = 1.8). The greatest overall risk (RR = 4.3) was found for women working in an office with fluorescent light exposure for 20 or more years. Finally, women working indoors with fluorescent lights had a greater risk for melanoma than did women who had at some time worked outdoors (RR = 1.8 vs. RR = 0.9, respectively).

When the anatomic sites of melanoma were examined for the exposed and control groups,

TABLE 2
Relationship of Risk of Melanoma with Occupational Fluorescent Light Exposure

Years of exposure	Melanoma patients[a] (*n*)	Controls[b] (*n*)	Relative risk[c] (95% confidence intervals)
0	25	94	1.0
1—9	134	269	*1.9 (1.2—3.0)
10—19	80	121	*2.5 (1.5—4.2)
⩾20	14	30	*2.6 (1.2—5.9)

Note: * = statistically significant (p <0.05).

[a] Melanoma patients were among 253 females aged 18—54 who had always worked indoors and were exposed to fluorescent light.
[b] Controls were age-matched females who always worked indoors but were never exposed to fluorescent light.
[c] Relative risk is the risk of melanoma among exposed individuals compared to those never exposed to fluorescent lights.

From Beral, V., Shaw, M., Evans, S., et al., *Lancet,* 2, 29, 1982. With permission.

a striking relative increase in malignant melanomas was found on the trunk in exposed patients compared to nonexposed indoor workers (24% vs. 4%, respectively; p <0.05). The rates of melanoma on other anatomic areas were not significantly different for the two groups.

Further analyses were performed to determine whether the increased risk of melanoma attributed to fluorescent light might have been caused by other confounding variables. The only group in whom fluorescent light did not remain significant was for women with red hair in childhood. Other variables modified the relative risk attributable to fluorescent light, but did not change its statistical significance. For example, fluorescent light exposure had a greater impact on indoor workers whose childhood sun exposure was rated as slight vs. great (RR = 7.3 vs. RR = 1.8, respectively). Similarly, individuals with a birthplace other than Australia or those with subjectively more nevi than average had a two- or threefold greater relative risk of melanoma due to fluorescent light than did their respective counterparts. Interestingly enough, however, no increased risk of melanoma was found to be due to fluorescent light or television in the home. A limited study with men, ages 18 to 56, also revealed a significantly increased risk of melanoma due to an occupational exposure to fluorescent light, but only with exposure ⩾10 years (RR = 4.4; p <0.05).

The above results of Beral suggested an increased risk of melanoma with occupational fluorescent light exposure. While other investigators have also found an association,[2,3] the results by Beral are the most suggestive, and for this reason her work is the most widely criticized. In their review of the subject, Maxwell and Elwood[14] allude to the dose relationship observed by Beral as the strength of the study. A causation of melanoma by fluorescent light is supported by the clear-cut trend shown in Table 2. However, because only a moderate relative risk was found (RR = 2.5 for greater than 10 years of exposure), these results may be spuriously due to methodological error. For example, another confounding risk factor inadequately analyzed in the study may be responsible for the attained results. Similarly, an underlying bias in either patient selection or data collection could have given falsely elevated results.

Sunlight exposure is the most often cited confounding variable or bias that may have lead to Beral's result.[4-6] Critics suggest that Beral's finding an increased rate of melanomas on the trunk was not consistent with the anatomic sites of fluorescent light exposure. These sites would be more causally related to sun exposure than exposure to fluorescent light. Beral explains, however, that while the trunk is covered at work the skin there is less habituated to solar radiation than are the face and extremities. This renders the trunk more susceptible to the carcinogenic wavelengths of fluorescent light which theoretically might have permeated lightweight clothing worn at work.[31] Consistent with this explanation is Beral's finding a reduced (but still significant) relative risk of melanoma among women exposed to fluorescent light who were also heavily exposed to sunlight.

Other investigators disagree with these interpretations. Sun exposure is difficult to quantitate and probably plays a major role in causing melanoma, especially in Australia.[2,6,7] Furthermore, acute rather than chronic exposure to sunlight poses the most risk.[32] Thus, an alternative interpretation of Beral's data is that indoor office workers are more likely to receive intense recreational sun exposure during weekends, and this accounts for their increased risk of melanoma. This also would explain why an increased number of melanomas were found on the trunk, as that location is most vulnerable to such acute exposures to sunlight. A deleterious role played by sun exposure is supported in the Beral study, which found a greater relative risk of melanoma among those exposed to fluorescent light who also had a greater than average number of nevi.[1] In addition, indoor office workers also may be of generally higher socioeconomic status, which also increases the risk of melanoma.[33,34] These confounding variables (sun exposure and socioeconomic class) may have led to spurious results in the Beral study. In defense of the Beral study, however, these confounding variables must have been significant only in the group exposed to fluorescent light, as the controls were indoor office workers who were not exposed to fluorescent light. In addition, Beral and Evan[35] showed that no significant difference existed between cases and controls with regard to either social class or education.

The published study by Beral generated further investigation worldwide. Dubin[2] and Rigel[4,5] were the next to publish their findings. Despite their using the same predominantly middle class New York population, these two investigators attained different results. Dubin surveyed 136 patients with primary malignant melanoma and 282 skin clinic controls without melanoma. Fluorescent light exposure at work and home was quantified as the average number of hours per day for three time periods: up to 5 years, 5 to 10 years, and 10 to 20 years. Other potentially confounding variables, similar to those selected by Beral, were included in the statistical analyses. Dubin found a significantly increased relative risk (RR $= 1.87$; $p < 0.01$) among individuals exposed to over 35 h of fluorescent light during the 5 years preceding the diagnosis of melanoma. Exposure beyond 5 years did not increase the risk. Thus, a significant association was demonstrated, although a dose-dependent relationship was not found.

Rigel et al.[5] found no significant difference in fluorescent light exposure between 114 patients with stage I melanoma and 228 age-matched controls. A comprehensive 20-year history of fluorescent light exposure was available, and melanoma patients were found to average 4.9 h of daily exposure compared to 5.4 h for controls. No increased risk of melanoma was found for indoor office workers exposed to fluorescent lights. However, a significant relative risk of 2.5 ($p = 0.005$) was found for indoor workers who had primarily outdoor recreational habits. These observations once again suggest that acute bursts of exposure to the sun, not fluorescent lighting, increases the risk of melanoma among indoor office workers.

Similar conclusions were reached by Sorahan and Grimley,[6] who compiled data using postal questionnaires sent to residents of Birmingham, England. A total of 58 patients with melanoma (excluding these with lentigo maligna melanoma) and 333 controls showed no significant difference in fluorescent light exposure. Indoor office workers exposed to fluorescent lights for 1 to 9 years had a relative risk of 1.2 which rose to 1.7, but then fell to 0.5 for 10 to 19 years and 20+ years exposure, respectively. The results were the same regardless of whether control groups were age-matched hospital patients or a population randomly selected from the general electoral register. The data did suggest that acute intermittent exposure to fluorescent light might be slightly more deleterious than continuous uninterrupted exposure, a phenomenon clearly established for sun exposure.

The next two studies to appear chronologically in the literature are credited with attempting to control the quantity of fluorescent light exposure better. English et al.[3] and Elwood et al.[7] assessed the risk of melanoma associated with exposure to fluorescent lights with and without diffusers, a term referring to the plastic cover over some fixtures which dissipates visible light and reduces emission of UV rays.[36] English also attempted to measure the distance between light fixtures and the exposed individual. Neither group of investigators was able to control for the number of fluorescent bulbs that contributed to exposure.

English interviewed either by telephone (96%) or in person (4%) 337 patients with newly diagnosed melanoma in Perth, Western Australia. This number of melanoma patients constituted 68% of all new melanoma patients age 18 years or older diagnosed in a consecutive 22-month period between 1980 and 1981. Each patient was assigned a control matched for age, sex, and location. During the 15-min interviews, subjects were questioned on their exposure to fluorescent light at home and work, including the average daily exposure and whether diffusers were used. The approximate distance between the subject and the fixture was estimated by determining whether the fluorescent source was a desk lamp or a ceiling fixture in a room, small office, large office, or large workspace such as a warehouse. Only 21% of the subjects recalled the number of tubes in the fixture, and this data was insufficient for statistical analyses. Occupational exposure was analyzed separately, and unexposed indoor workers served as controls. Risk according to histologic type and anatomic location of melanoma was also studied. All analyses were conducted controlling for other confounding risk factors of malignant melanoma.

English,[3] using the above method, found a significant dose-dependent increased risk of histologically unclassified melanoma ($p = 0.02$; $n = 136$), although no increased risk for all melanomas was found with fluorescent light exposure. Occupational exposure, quantified both cumulatively and as a yearly rate, gave a relative risk of <1 for exposure under 10 years, and the largest relative risk of 1.2 ($p = 0.22$) was attained for 10 to 19 years of exposure. In addition, no increased risk was found when cumulative exposure at work and at home were considered together ($p = 0.27$). The presence of diffusers was known for 91% of occupational exposures, and 43% of them answered affirmatively. Analyses specifically addressing exposure to fluorescent lights without diffusers showed no relationship to melanoma ($p = 0.36$). The distance of the fluorescent light fixtures from the subject did not produce a significantly increased risk of melanoma even at close-up distances encountered in small offices. However, the risk of histologically unclassified melanoma ($n = 136$) at any anatomic site was significantly increased and proportional to cumulative fluorescent light exposure: RR = 3.0, 3.9, and 5.1 for intervals of cumulative exposure × 100 h of <75, 75 to 224, and ⩾225, respectively ($p = 0.02$). This finding may be due to chance alone, the authors explain, because this trend was not observed for patients who likely received the most intense exposure, namely, the subgroup of subjects working in small offices with fluorescent lights lacking diffusers. This subgroup of highly exposed subjects did show a dose-dependent trend for developing lentigo maligna melanoma ($n = 126$) on the head, neck, and arms after adjusting for sun exposure variables (RR = 2.7 for cumulative fluorescent light exposure ⩾17,500 h; $p = 0.08$). No trends were observed for superficial spreading or nodular melanoma.

Elwood et al.[7] conducted a study in Nottingham, England and found melanoma patients had significantly more occupational exposure to fluorescent light, but this did not account for a greater relative risk. This study used 83 melanoma patients with precisely matched controls from the hospital data base. The majority of the subjects were of the same social class, and data from all participants was accumulated by postal questionnaires as well as by direct personal interviewers. Occupational exposure to fluorescent light was quantified for total undiffused and diffused light. Although exposure to undiffused, compared to diffused, fluorescent light produced a stronger trend toward greater risk, neither trend was statistically significant. The lifetime cumulative occupational exposure to undiffused fluorescent light was significantly greater among cases than controls (15,447 h vs. 9,451 h, respectively; $p = 0.048$). A statistically significant difference for diffused light exposure did not exist between the two groups.

The Elwood study, like all epidemiological studies, may have been hampered by methodological bias. Although no significant risk of melanoma was attributable to fluorescent light exposure, the degree of exposure reported by the same patient differed depending on whether he or she was responding to a postal questionnaire or to a personal interviewer. Only data attained by personal interview led to a significant difference in exposure of undiffused light between cases and controls. Thus, a bias of overreporting exposure may occur during face-to-face interviews and, as some have suggested,[37] may have led to false conclusions in the original Beral study, which utilized personal interviews. Dubin et al.,[2] in their retrospective study on melanoma risk factors, showed that interviews may bias results significantly. When Dubin interviewed patients, he found that exposure to fluorescent lights for 6+ h a day more than doubled the risk of melanoma in less than 5 years ($p < 0.05$). However, when Dubin compiled data from the same patients using a written questionnaire, no significant association was found. Of course, the possibility exists that underreporting of fluorescent light exposure might occur with the postal questionnaire format. This issue cannot be resolved yet and is further complicated by the fact that both methodological formats have yielded both positive and negative results as shown in Table 3.

Finally, the most recent epidemiological study to date was conducted by Osterlind et al.[8] in Denmark. This study examined 474 patients and their 926 population controls. Patients with lentigo maligna melanoma were excluded. Almost all subjects (96%) were personally interviewed at home by a trained interviewer, unaware of the subject's status, who used a standardized questionnaire. A total of 67% of the subjects (cases and controls) were exposed to fluorescent light at work. Osterlind found no significant increased risk of melanoma associated with age at first exposure, duration of exposure, or type of occupational setting. The study did not, however, control for exposure to diffused or undiffused fluorescent light. In addition, exposure differences in patients with different histologic types of melanomas were not examined, leaving unconfirmed the positive association for histologically unclassified melanomas that had been reported by English et al.[3]

IV. DISCUSSION

Whether an occupational exposure to fluorescent light significantly increases the risk of melanoma continues to remain unclear. Beral was the first to conduct a retrospective study on the subject and her finding — a dose-dependent relationship between risk and exposure — has been confirmed by only one other study in the literature. However, the majority of studies have failed to show a significant occupational risk. The results and findings of the studies presented are summarized in Table 3.

Methodological problems with confounding variables and bias, which make any epidemiologic study difficult, are particularly relevant to the studies discussed in this chapter. Interview bias may explain why, as in both the Elwood et al.[7] and Dubin et al.[2] studies, different risks of melanoma are found when occupational exposure to fluorescent light is assessed by personal interview or postal questionnaire. The method by which exposure data are compiled directly influences the conclusions drawn; no single method was common to all the studies, nor was one method shown to be superior. Other variables known to affect

TABLE 3
Summary of Published Reports on the Possible Increased Risk of Melanoma from an Occupational Exposure to Fluorescent Light

Patient profile			Methods		Results		
Geographic location	Patients (*n*)	Age range (years)	Interview method	Controlled for plastic diffuser covers	Statistically significant increased risk found	Comments	Ref.
Australia	274 females	18—54	Personal interview	No	Yes	Increase risk was dose-dependent; maximum relative risk (RR) = 2.6 for ⩾ years exposure; melanomas predilected the trunk in the exposed group.	1
	27 males	18—56	Unknown	No	Yes	Significant RR = 4.4 was found only for exposure ⩾10 years.	
New York City	136	18+	Personal and postal	No	Yes	Significant RR = 1.87 only for the 5 years exposure preceding diagnosis of melanoma; positive results above attained only when patients personally interviewed; dose-dependent relationship was not found.	2
New York City	114	18+	Telephone	No	No		5
England	58	20—70	Postal questionnaire	No	No	Lentigo maligna melanoma cases were excluded.	6
Australia	337	18+	Telephone 96%, personally 4%	Yes	See comments	No overall increased risk of melanoma; dose-dependent increased risk of histologically unclassified melanoma when analyzed separately; trend found for lentigo maligna melanoma; diffusers did not affect results.	3
England	83	18+	Personal and postal	Yes	No	Cases overall had significantly more exposure to fluorescent light and specifically to undiffused light.	7
Denmark	474	18+	Personal interview	No	No	Lentigo maligna melanoma cases were excluded	8

the risk of melanoma, such as sun exposure and socioeconomic status, may confound the risk of melanoma found attributable to fluorescent light, even in the best controlled study. Several authors question whether Beral's methods adequately deal with these confounding variables. Finally, it remains unclear whether more positive results would have been attained if all studies controlled for nondiffused fluorescent lights, as those studies that did control for this variable report mixed results.

Other basic research studies may strengthen the plausibility of an association between fluorescent light exposure and melanoma. Maxwell and Elwood[14,15] have shown that, compared to sun exposure of equal duration, commercially available fluorescent bulbs may expose an individual to 10 to 30 times more UV-B of certain monochromatic wavelengths (e.g., 290 nm). UV-A emission by fluorescent lights is negligible.[5] The UV-B emitted by fluorescent fixtures may be greatly reduced by encasing fixtures with styrene diffusers and, to a lesser extent, with acrylic diffusers.[14,36] These plastic covers should be used as they do not reduce illumination and may prevent the deleterious effects of fluorescent lights found experimentally. *In vitro* studies demonstrate that monochromatic wavelengths of UV-B, such as those emitted by nondiffused fluorescent light, may cause a dose-dependent malignant transformation of mouse embryo cells.[22] Numerous experiments demonstrate that fluorescent light causes DNA mutations in bacterial and mammalian cell cultures;[16-21,38] these mutations include pyrimidine dimer formation, chromosomal breaks, and cross-linkages. Fluorescent light has also been shown to increase the rate of mitosis in cultured cells, simulating hyperplasia *in vitro*.[27,28]

V. SUMMARY

Whether an occupational exposure to fluorescent light increases the risk of melanoma remains controversial. If an increased risk does exist, the literature generally suggests it to be a two- to fourfold increase. This risk may be largely avoidable if styrene diffusers are placed over fluorescent fixtures.

REFERENCES

1. **Beral, V., Shaw, M., Evans, S., et al.,** Malignant melanoma and exposure to fluorescent lighting at work, *Lancet,* 2, 29, 1982.
2. **Dubin, N., Mosesosn, M., and Pasternack, B.,** Epidemiology of malignant melanoma: pigmentary traits, ultraviolet radiation, and the identification of high risk populations, *Recent Results Cancer Res.,* 102, 56, 1986.
3. **English, D., Rouse, I., and Xu, Z.,** Cutaneous malignant melanoma and fluorescent light, *J. Natl. Cancer Inst.,* 71, 1191, 1985.
4. **Rigel, D., Friedman, R., and Levenstein, M.,** Malignant melanoma and exposure to fluorescent lighting at work, *Lancet,* 1, 704, 1983.
5. **Rigel, D., Friedman, R., Levenstein, M., et al.,** Relationship of fluorescent lights to malignant melanoma: another view, *J. Dermatol. Surg. Oncol.,* 9, 836, 1985.
6. **Sorahan, A. T. and Grimley, R.,** The aetological significance of sunlight and fluorescent lighting in malignant melanoma: a case-control study, *Br. J. Cancer,* 52, 765, 1985.
7. **Elwood, J., Wiliamson, C., and Stapleton, P.,** Malignant melanoma in relation to moles, pigmentation and exposure to fluorescent and other lighting sources, *Br. J. Cancer,* 53, 65, 1986.
8. **Osterlind, A., Tucker, M., Stone, B., et al.,** The Danish case-control study of cutaneous malignant melanoma. II. Importance of UV-light exposure, *Int. J. Cancer,* 42, 319, 1988.
9. **Lee, J.,** Melanoma and exposure to sunlight, *Epidemiol. Rev.,* 4, 110, 1982.
10. **Beral, V. and Robinson, N.,** The relationship of malignant melanoma, basal and squamous skin cancers to indoor and outdoor work, *Br. J. Cancer,* 44, 886, 1981.
11. **Blum, M.,** Sunlight as a causal factor in cancer of the skin of man, *J. Natl. Cancer Inst.,* 9, 247, 1948.

12. **Fitzpatrick, T., et al.,** Prevention of solar degeneration and sun-induced carcinoma of the skin, in *Melanoma and Skin Cancer,* McCarthy, W. H., Ed., International Union Against Cancer, Australian Cancer Society, Sidney, 1972.
13. **Urbach, F., et al.,** Ultraviolet radiation and skin cancer in man, in *Advances in Biology of the Skin,* Vol. 7, *Carcinogenesis,* Montagna, W. and Dobson, R., Eds., Pergamon Press, New York, 1966.
14. **Maxwell, K. and Elwood, J.,** Current issues in melanoma research: fluorescent light, *Recent Results Cancer Res.,* 102, 127, 1986.
15. **Maxwell, K. and Elwood, J.,** UV radiation from fluorescent lights, *Lancet,* 2, 579, 1983.
16. **Gantt, R., Parshad, R., Ewig, R., et al.,** Fluorescent light-induced DNA crosslinkage and chromatid breaks in mouse cell cultures, *Proc. Natl. Acad. Sci. U.S.A.,* 75, 3809, 1978.
17. **Webb, R. and Malina, M.,** Mutagenesis in *Escheria coli* by visible light, *Science,* 156, 1104, 1967.
18. **Bradley, M. and Sharkey, N.,** Mutagenicity and toxicity of visible fluorescent light to cultured mammalian cells, *Nature (London),* 266, 724, 1977.
19. **Ritter, M. and Williams, J.,** Endonuclease-sensitive sites induced by fluorescent lights, *Radiat. Res.,* 70, 668, 1977.
20. **Bradley, M., Erickson, L., and Kohn, K.,** Non-enzymatic DNA strand breaks induced in mammalian cells by fluorescent light, *Biochem. Biophys. Acta,* 520, 11, 1978.
21. **Jostes, R., Dewey, W., and Hopwood, L.,** Mutagenesis by fluorescent light in mammalian cell cultures, *Mutat. Res.,* 42, 139, 1977.
22. **Kennedy, A., Ritter, M., and Little, J.,** Fluorescent light induces malignant transformation in mouse embryo cell cultures, *Science,* 207, 1209, 1980.
23. **Bresler, R.,** Cutaneous burns due to fluorescent light, *J. Am. Med. Assoc.,* 140, 1334, 1949.
24. **James, A.,** Sensitivity of the skin to fluorescent light, *Arch. Dermatol.,* 44, 250, 1941.
25. **Diethelm, R.,** Hautkarzinom dutch Fluoreszenzlampen?, *Schweiz. Med. Wochenshr.,* 100, 1159, 1970.
26. **Kochevar, I.,** Photobiology: basic science, *Dermatol. Clin.,* 4, 171, 1986.
27. **Parshad, R. and Sanford, K.,** Intermittent exposure to fluorescent light extends lifespan of human diploid fibroblasts in culture, *Nature (London),* 268, 736, 1979.
28. **Parshad, R. and Sanford, K.,** Proliferative response of human diploid fibroblasts to intermittent light exposure, *J. Cell. Physiol.,* 92, 481, 1977.
29. **Beral, V. and Robinson, N.,** The relationship of malignant melanoma, basal and squamous skin cancers to indoor and outdoor work, *Br. J. Cancer,* 44, 886, 1981.
30. **Lee, J. and Strickland, D.,** Malignant melanoma: social status and outdoor work, *Br. J. Cancer,* 41, 757, 1980.
31. **Robertson, D.,** Solar ultraviolet radiation in relation to sunburn and skin cancer, *Med. J. Aust.,* 2, 1123, 1968.
32. **Elwood, J., Gallagher, R., Hill, G., et al.,** Cutaneous melanoma in relation to intermittent and constant sun exposure. The Western Canada Melanoma Study, *Int. J. Cancer,* 35, 427, 1985.
33. **Cooke, K., Skegg, D., and Fraser, J.,** Socio-economic status, indoor and outdoor work and malignant melanoma, *Int. J. Cancer,* 34, 57, 1984.
34. **Stern, R.,** Malignant melanoma and exposure to fluorescent lighting work, *Lancet,* 2, 1227, 1982.
35. **Beral, V. and Evan, S.,** Malignant melanoma and exposure to fluorescent lighting at work, *Lancet,* 2, 1227, 1982.
36. **Levin, R., Clark, G., Spears, G., et al.,** Ultraviolet radiation-consideration in interior lighting design, *J. Illum. Eng. Soc.,* 6, 80, 1977.
37. **Brown, A.,** Melanoma and fluorescent light, Letter to the Editor, *Lancet,* 1, 1398, 1982.
38. **Hart, R., Setlow, R., and Woodhead, A.,** Evidence that pyrimidine dimers in DNA can give rise to tumors, *Proc. Natl. Acad. Sci. U.S.A.,* 74, 5574, 1977.
39. **Holman, C., Mulroney, C., and Armstrong, B.,** Epidemiology of preinvasive and invasive malignant melanoma in Western Australia, *Int. J. Cancer,* 25, 317, 1980.

INDEX

A

B

D

H

I

M

N

O

P

Q

R

T

U

G.T 19 No to 75 29/12/05

Box Nos:- 422 to 436

16 Books is each Box

Total Books - 240

Total Boxes :- 15

---

27 | 0 2 1

05 - 1

T - 252